Carotenoids

Carotenoids

Volume 4: Natural Functions

Edited by   G. Britton
                           S. Liaaen-Jensen
                           H. Pfander

Birkhäuser Verlag
Basel · Boston · Berlin

Editors:

Dr. George Britton
University of Liverpool
School of Biological Sciences
Biosciences Building
Crown Street
Liverpool L69 7ZB
UK

Professor Dr. Dr. h.c. Synnøve Liaaen-Jensen
Organic Chemistry Laboratories
Department of Chemistry
Norwegian University of Science and Technology (NTNU)
7491 Trondheim
Norway

Prof. Dr. Hanspeter Pfander
CaroteNature GmbH
Chief Operating Officer
Muristrasse 8e
3006 Bern
Switzerland

Library of Congress Control Number: 2008932322

Bibliographic information published by Die Deutsche Bibliothek
Die Deutsche Bibliothek lists this publication in the Deutsche Nationalbibliografie;
detailed bibliographic data is available in the internet at http://dnb.ddb.de

ISBN 3-7643-7498-3 Birkhäuser Verlag, Basel – Boston - Berlin

© 2008 Birkhäuser Verlag, P.O. Box 133, CH-4010 Basel, Switzerland
Part of Springer Science+Business Media
Printed on acid-free paper produced from chlorine-free pulp. TFC
Printed in Germany

Cover design: Markus Etterich, Basel

ISBN 978-3-7643-7498-3   e-ISBN 978-3-7643-7499-0

9 8 7 6 5 4 3 2   www.birkhauser.ch

Contents

Chapter 1: Special Molecules, Special Properties

George Britton, Synnøve Liaaen-Jensen and Hanspeter Pfander

Chapter 2: Structure and Chirality

Synnøve Liaaen-Jensen

Chapter 3: *E/Z* Isomers and Isomerization

Synnøve Liaaen-Jensen and Bjart Frode Lutnæs

Chapter 4: Three-dimensional Structures of Carotenoids by X-ray Crystallography

Madeleine Helliwell

Chapter 5: Aggregation and Interface Behaviour of Carotenoids

Sonja Köhn, Henrike Kolbe, Michael Korger, Christian Köpsel, Bernhard Mayer, Helmut Auweter, Erik Lüddecke, Hans Bettermann and Hans-Dieter Martin

Chapter 6: Carotenoid-Protein Interactions

George Britton and John R. Helliwell

Chapter 7: Carotenoid Radicals and Radical Ions

Ali El-Agamey and David J McGarvey

Chapter 8: Structure and Properties of Carotenoid Cations

Synnøve Liaaen-Jensen and Bjart Frode Lutnæs

Chapter 9: Excited Electronic States, Photochemistry and Photophysics of Carotenoids

Harry A. Frank and Ronald L. Christensen

Chapter 10: Functions of Intact Carotenoids

George Britton

Chapter 11: Signal Functions of Carotenoid Colouration
Jonathan D. Blount and Kevin J. McGraw

Chapter 12: Carotenoids in Aquaculture: Fish and Crustaceans

Bjørn Bjerkeng

Chapter 13: Xanthophylls in Poultry Feeding
Dietmar E. Breithaupt

Chapter 14: Carotenoids in Photosynthesis
Alison Telfer, Andrew Pascal and Andrew Gall

Chapter 15: Functions of Carotenoid Metabolites and Breakdown Products

George Britton

Chapter 16: Cleavage of β-Carotene to Retinal

Adrian Wyss and Johannes von Lintig

Chapter 17: Enzymic Pathways for Formation of Carotenoid Cleavage Products

Peter Fleischmann and Holger Zorn

List of Contributors

Helmut Auweter
BASF AG
ZKM/D – J542S
67056 Luwigshafen
Germany
(helmut.auweter@basf.com)

Hans Bettermann
Institute for Physical Chemistry
University of Düsseldorf
Universitätstrasse 1
40225 Düsseldorf
Germany
(betterma@uni-duesseldorf.de)

Bjørn Bjerkeng
Nofima Akvaforsk-Fiskeriforskning AS
6600 Sunndalsøra
Norway
(bjorn.bjerkeng@nofima.no)

Jonathan D. Blount
Centre for Ecology and Conservation
School of Biosciences
Unversity of Exeter
Cornwall Campus
Penryn TR10 9EZ
UK
(j.d.blount@ex.ac.uk)

Dietmar E. Breithaupt
Rich. Hengstenberg GMBH & Co KG
Mettingerstr. 109
73728 Esslingen
Germany
(dietmar.breithaupt@hengstenberg.de)

George Britton
University of Liverpool
School of Biological Sciences
Crown Street
Liverpool L69 7ZB
UK
(g.britton@liv.ac.uk)

Ronald L. Christensen
Department of Chemistry
Bowdoin College
Brunswick
ME 04011-8466
USA
(rchriste@bowdoin.edu)

Ali El-Agamey
Chemistry Department
Faculty of Science
Mansoure University
New Damietta
Damietta
Egypt
(a_el_agamey@yahoo.co.uk

Peter Fleischmann
Institute of Food Chemistry
TU Braunschweig
Schleinitzstr. 20
38106 Braunschweig
Germany
(p.fleischmann@tu-bs.de)

Harry A. Frank
Department of Chemistry
University of Connecticut
55 North Eagleville Road
Storrs
CT 06269-3060
USA
(harry.frank@uconn.edu)

Andrew Gall
Commissariat à l'Énergie Atomique (CEA)
Institute de Biologie et Technologies de Saclay (iBiTecS)
Service de Bioeénergétique, Biologie Structural et Mécanismes (SB2SM)
F-91191 Gif sur Yvette
France

John R. Helliwell
University of Manchester
School of Chemistry
Oxford Road
Manchester M13 9PL
UK
(john.helliwell@man.ac.uk)

Madeleine Helliwell
University of Manchester
School of Chemistry
Oxford Road
Manchester M13 9PL
UK
(madeleine.helliwell@man.ac.uk)

Sonja Köhn
Institute of Organic Chemistry and Macromolecular Chemistry
University of Düsseldorf
Universitätstrasse 1
40225 Düsseldorf
Germany

Henrike Kolbe
Instituite of Organic Chemistry and Macromolecular Chemistry
University of Düsseldorf
Universitätstrasse 1
40225 Düsseldorf
Germany

Christian Köpsel
BASF AG
MEM/FO – A110
67056 Ludwigshafen
Germany
(christian.koepsel@basf.com)

Michael Korger
Instituite of Organic Chemistry and Macromolecular Chemistry
University of Düsseldorf
Universitätstrasse 1
40225 Düsseldorf
Germany

Synnøve Liaaen-Jensen
Organic Chemistry Laboratories
Department of Chemistry
Norwegian University of Science and Technology (NTNU)
7491 Trondheim
Norway
(slje@chem.ntnu.no)

Erik Lüddecke
BASF AG
MEP/HC – O820
67056 Ludwigshafen
Germany
(erik.lueddecke@basf.com)

Bjart Frode Lutnæs
BFL
Borregaard Ind. Ltd.
Borregaard Corporate R&D
P. O. Box 162
1701 Sarpsborg
Norway
(bjart.frode.lutnes@borregaard.com)

Hans-Dieter Martin
Instituite of Organic Chemistry and Macromolecular Chemistry
University of Düsseldorf
Universitätstrasse 1
40225 Düsseldorf
Germany
(martin@uni-duesseldorf.de)

Bernhard Mayer
Instituite of Organic Chemistry and Macromolecular Chemistry
University of Düsseldorf
Universitätstrasse 1
40225 Düsseldorf
Germany
(mayer@uni-duesseldorf.de)

David J. McGarvey
School of Physical and Geographical Sciences
Lennard-Jones Laboratories
Keele University
Keele
Staffordshire ST5 5BG
UK
(d.j.mcgarvey@chem.keele.ac.uk)

Kevin J. McGraw
School of Life Sciences
Arizona State University
Tempe
AZ 85287-4501
USA
(kevin.mcgraw@asu.edu)

Andrew Pascal
Commissariat à l'Énergie Atomique (CEA)
Institute de Biologie et Technologies de Saclay (iBiTecS)
Service de Bioeénergétique, Biologie Structural et Mécanismes (SB2SM)
91191 Gif sur Yvette
France
(andrew.pascal@cea.fr)

Hanspeter Pfander
CaroteNature GmbH
Muristrasse 8e
3006 Bern
Switzerland
(hanspeter.pfander@carotenature.com)

Alison Telfer
Division of Molecular Biosciences
Imperial College London
South Kensington Campus
London SW7 2AZ
UK
(a.telfer@imperial.ac.uk)

Johannes von Lintig
Case Western Case University
Department of Pharmacology
10900 Euclid Avenue
Cleveland
Ohio 44106
USA
(Johannes.vonLintig@case.edu)

Adriam Wyss
DSM Nutritional Products
R & D-Human Nutrition and Health
Bldg 205/219B
P. O. Box 2676
4002 Basel
Switzerland
(Adrian.Wyss@dsm.com)

Holger Zorn
Technical Biochemistry Workgroup
Fachbereich Bio- und Chemieingenieurwesen
Universität Dortmund
Emil-Figge-Str. 68
44227 Dortmund
Germany
(h.zorn@bci.uni-dortmund.de)

Preface

Twenty years after the idea of this *Carotenoids* book series was first discussed, we are finally reaching the end of the project. *Carotenoids Volumes 4* and *5* now move us into the great field of biology, and cover the functions of carotenoids and the actions of carotenoids in nutrition and health. In the classic 1971 Isler book *Carotenoids*, carotenoid functions were covered in just one chapter. Now, thanks to technical developments and multidisciplinary approaches that make it possible to study functional processes in great detail, this subject is the most rapidly expanding area of carotenoid research, and occupies two full volumes. These can be used as stand-alone books, but they are really planned to be used together and to complete the coverage of the carotenoid field begun in earlier volumes, as the final part of a coordinated series.

The general philosophy and strategy of the series, to have expert authors review, analyse and present information and give guidance on practical strategies and procedures is maintained in *Volumes 4* and *5*, though the subject matter does not lend itself to the kind of detailed Worked Examples that were featured in earlier volumes. It has also been the aim that these publications should be useful for both experienced carotenoid researchers and newcomers to the field.

The material presented in the earlier volumes is all relevant to studies of biological functions and actions. Biological studies must be supported by a rigorous analytical base and carotenoids must be identified unequivocally. It is a common view that carotenoids are difficult to work with; this may be daunting to newcomers to the field, especially if they do not have a strong background in chemistry and analysis. There are difficulties; carotenoids are less stable than most natural products but, as we emphasize in *Chapter 1*, ways to overcome the difficulties and to handle these challenging compounds are well established and are described and discussed in *Volume 1A*, which, together with *Volume 1B*, gives a comprehensive treatment of the isolation, analysis and spectroscopic characterization of carotenoids that is an essential foundation for all carotenoid work. This is complemented by the *Carotenoids Handbook* (2004), which was produced in association with this series and provides key analytical data for each of the 750 or so known naturally-occurring carotenoids.

Volume 2 describes methods for the chemical synthesis of carotenoids that are needed as analytical standards and on a larger scale for biological trials. Functions and actions are inextricably linked with biosynthesis and metabolism, covered in *Volume 3*.

There are many other major publications in the carotenoid field which are still extremely valuable sources of information. The history of key publications up to around 1994 was outlined in the preface to the series, in *Volume 1A*. Since then there have been other progress reports, notably the published proceedings of the International Carotenoid Symposia in 1996, 1999, 2002 and 2005. References to specialized monographs and reviews on particular topics can be found in the following Chapters.

Volume 4 and its companion, *Volume 5*, which deals with human nutrition and health, are the last volumes in the *Carotenoids* series, and in many ways point the way to the future of carotenoid research. If the insight that these books provide stimulates chemists, physicists and biologists to understand and talk to each other and thus serves as a catalyst for interdisciplinary studies that will bring great advances and rewards in the future, then the editors will feel that their time and effort has been well spent.

Editors' Acknowledgements

We repeat our comment from the earlier Volumes. Although we are privileged to be the editors of these books, their production and publication would not be possible without the efforts of many other people.

The dedicated work of the authors, their attention to requests and questions and their gracious acceptance of the drastic editing that was sometimes needed to avoid duplication and to meet the stringent limitations of space, is gratefully acknowledged. The job of the editors is made so much easier when authors provide carefully prepared manuscripts in good time.

We thank Klara Nagy of the Department of Biochemistry and Medical Chemistry, University of Pécs, Hungary, for her continuing help during the preparation of manuscripts, and Detlef Klüber and the editorial staff at Birkhäuser, especially Kerstin Tüchert who was responsible for the 'hands-on' work in the final push to get the book into publication.

Discussions with carotenoid colleagues during the planning of these Volumes were very useful, and we especially appreciate much valuable advice from Norman Krinsky (Tufts University, Boston).

Finally, we again express our gratitude to DSM and BASF for the financial support without which this project would not have been possible.

Authors' Acknowledgements

Chapter 5

The authors acknowledge helpful discussions with Drs. J. Benade and S. Beutner, and financial support from Deutsche Forschungsgemeinschaft (SFB 666, Photostability, Photoprotection and Photoreactivity), and from the Collaborative Research Center 663 of the German Research Foundation (SFB/DFG 663).

Chapter 9

The authors thank Mary Grace Galinato, Dariusz Niedzwiedski, Zeus Pendon, Robert Birge, Richard Cogdell, George Gibson and Tomáš Polivka for help, discussions and collaborations. RLC acknowledges support from the Bowdoin College Porter Fellowship Program, the NSF-ROA program (MCB-0314380 to HAF) and the Petroleum Research Fund administered by the American Chemical Society. HAF acknowledges support from the National Institutes of Health (GM-30353), the National Science Foundation (MCB-0314380) and the University of Connecticut Research Foundation.

Chapter 15

The author acknowledges the authors of *Chapter 5* for some material that was originally submitted in *Chapter 5* but incorporated in abbreviated form in *Chapter 15*.

Editors' Notes on the Use of this Book

The *Carotenoids* books are planned to be used together with the *Carotenoids Handbook*, which lists and gives data for the 750 or so known natural carotenoids. Whenever a known natural carotenoid is mentioned, its number in the *Handbook* is given in bold print. Other compounds, including purely synthetic carotenoids that do not appear in the *Handbook*, are numbered separately in italics in sequence as they appear in the text for each chapter, and their formulae are shown. The numbers are not given at every mention of a particular compound but may be repeated for clarity or to aid comparison.

The *Carotenoids* books form a coordinated series, so there is substantial cross-referencing between *Volumes 4* and *5*, and to earlier *Volumes* in the series. *Chapters* in earlier *Volumes* are not included in reference lists unless they are referred to several times in a particular chapter.

Carotenoid nomenclature

The IUPAC semi-systematic names for all known naturally occurring carotenoids are given in the *Carotenoids Handbook*. Trivial names for many carotenoids are, however, well-established and convenient, and are generally used in biological publications, including β-carotene rather than β,β-carotene. These common, trivial names are used throughout these volumes.

The *E/Z* and *trans/cis* deniminations for describing the stereochemistry about a double bond are not always equivalent. The terms *cis* and *trans* are also used in most cases to designate geometrical isomers, especially of carotenoids in general. The *E/Z* system is reserved mainly for naming geometrical isomers of a particular compound.

Naming of organisms

The correct classification and naming of living organisms is essential. The editors have not checked all these but have relied on the expertise of the authors to ensure that classification schemes and names in current usage are applied accurately, and for correlation between new and old names.

Abbreviations

The abbreviations listed are ones that occur in more than one place in the book. Abbreviations defined at their only place of mention are not listed.

Indexing

For many purposes the *List of Contents* is sufficient to guide the reader to a particular topic. The subject *Index* at the end of the book complements this and lists key topics that occur, perhaps in different contexts, in different places in the book. No author index, index of compounds or index of organisms is given.

Abbreviations

ABA	Abscisic acid
ADC	Apparent digestibility coefficient
AFM	Atomic Force Microscopy
AM 1	Austin model 1
APCI	Atmospheric pressure chemical ionization
Bchl	Bacteriochlorophyll
BHT	Butylated hydroxytoluene (2,6-di-*t*-butyl-p-cresol)
Bphaeo	Bacteriophaeophytin
CD	Circular dichroism
Chl	Chlorophyll
DLS	Dynamic light scattering
DMPC	Dimyristoylphosphatidylcholine
DMSO	Dimethyl sulphoxide
DPA	Diphenylamine
DPPC	Dipalmitoylphosphatidylcholine
EM	Electron microscopy
ENDOR	Electron nuclear double resonance
EPR	Electron paramagnetic resonance
FCP	Fucoxanthin-chlorophyll protein
GC-MS	Gas chromatography-mass spectrometry
HDL	High density lipoprotein
HOMO	Highest occupied molecular orbital
HPLC	High performance liquid chromatography
IR	Infrared
IUPAC	International Union of Pure and Applied Chemistry
kDa	kiloDalton
LB	Langmuir-Blodgett
LDL	Low density lipoprotein
LH	Light harvesting
LOX	Lipoxygenase
LUMO	Lowest occupied molecular orbital
MALDI-TOF	Matrix-assisted laser desorption ionization - time of flight
MS	Mass spectrometry
NIR	Near infrared
NMR	Nuclear magnetic resonance
NPQ	Non-photochemical quenching
ORD	Optical rotatory dispersion
PCP	Peridinin-chlorophyll protein
PCR	Polymerase chain reaction
PDAD	Photodiode array detector
PPAR	Peroxysome proliferator activated response
pRBP	Plasma retinol-binding protein
PSI	Photosystem I
PSII	Photosystem II
RA	Retinoic acid
RAR	Retinoic acid receptor
RC	Reaction centre
RPE	Retinal pigment epithelium
RXR	Retinoid receptor
SDS	Sodium dodecyl sulphate
SENS	Sensitizer
SLS	Static light scattering

STM	Scanning tunnelling microscopy
TEM	Transmission electron microscopy
TLC	Thin-layer chromatography
UV	Ultraviolet
UV/Vis	Ultraviolet/visible
VADD	Vitamin A deficiency disorder
WAXS	Wide-angle X-ray scattering
XBP	Xanthophyll-binding protein

Carotenoids
Volume 4: Natural Functions
© 2008 Birkhäuser Verlag Basel

Chapter 1

Special Molecules, Special Properties

George Britton, Synnøve Liaaen-Jensen and Hanspeter Pfander

A. Introduction

This is a book about functions of carotenoids. The reader may therefore be surprised to find that the first half of the book is chemistry. This should not be a surprise, however. The target for researchers now is not simply to discover and describe functions and actions, but to understand their mechanisms. This requires understanding of the underlying fundamental principles and appreciation of the application of the advanced techniques now used to elucidate details of structure and of processes that may occur on a very short timescale.

Carotenoids Volumes 4 and *5* take the carotenoid story into the realm of biology, dealing with the functions of carotenoids and the actions of carotenoids in nutrition and health. They can be used as stand-alone books, but they are really planned to complete the coverage of the carotenoid field begun in earlier volumes, as the final part of a coordinated series.

The material presented in the earlier volumes is all relevant to studies of biological functions and actions. The common view that carotenoids are difficult to work with may be daunting to newcomers to the field, especially if they do not have a strong background in chemistry and analysis. There are difficulties; carotenoids are less stable than most natural products, but ways to overcome the difficulties and to handle these challenging compounds are well established. These are described and discussed in *Volume 1A*, which, together with *Volume 1B*, gives a comprehensive treatment of the isolation, analysis and spectroscopic characterization of carotenoids that is an essential foundation for all carotenoid work.

Biological studies must be supported by a rigorous analytical base. Carotenoids must be identified unequivocally and the various analytical procedures described in *Volumes 1A* and

1B must be understood and applied correctly, whether they are being used for quantitative analysis, identification or in complex studies of carotenoids *in situ*.

Volume 2 describes methods for the chemical synthesis of carotenoids on a laboratory and a commercial scale. Synthetic compounds are needed as analytical standards and on a larger scale for biological trials. Functions and actions are inextricably linked with biosynthesis and metabolism, as covered in *Volume 3*.

Biologists recognize the value of Field Guides for identifying species they encounter. The *Carotenoids Handbook* is the 'Field Guide' for the identification of carotenoids.

B. Structure, Properties and Function

The carotenoids have special properties that no other group of substances possesses. The functions and actions of carotenoids depend on these special properties. These properties are determined by the structural features of the molecules. Understanding the intricacies of the relationships between structure, properties and function is therefore essential for understanding the importance of carotenoids in a biological context, as well as for devising and optimizing commercial applications.

The *Carotenoids Handbook* lists more than 700 carotenoids that have been isolated from natural sources. The structures of about 500 of these have been fully elucidated. So why is it necessary to address structure again? In *Volume 4*, the structural features that are identified as being most important for determining the biological roles of carotenoids are emphasized. The overall molecular geometry (size, three-dimensional shape, presence of functional groups) is vital for ensuring that the carotenoid fits into cellular, sub-cellular and molecular structures in the correct location and orientation to allow it to function efficiently. Then the conjugated double-bond system determines the light-absorption properties and chemical reactivity that form the basis of most functions. Specific interactions with other molecules in the immediate vicinity strongly influence the properties of a carotenoid and are also crucial to functioning.

1. Three-dimensional shape

Carotenoids are not simply flat two-dimensional structures, like formulae drawn on a page. They have a precise three-dimensional shape which is an important determinant of their functions. Several different stereochemical factors contribute to the shape of the molecule and must be considered when describing and defining the three-dimensional structure.

a) Configuration: geometrical isomers

Any carotenoid can exist in a number of geometrical (*cis/trans* or *E/Z*) isomeric forms. There is currently much interest in *cis* isomers and whether their different shape, solubility and stability compared with the linear all-*trans* isomer give rise to different biological properties. The topic of geometrical isomers and isomerization is treated in detail in *Chapter 3*.

b) Absolute configuration: chirality

Most of the known carotenoids have structures that contain at least one chiral centre or axis, and can therefore exist as different optical isomers, including enantiomers. With carotenoids as with other compounds, biological actions may be specific for one enantiomer.

Structural aspects of chirality are treated in *Chapter 2*.

c) Conformation

In principle, rotation is possible about any C-C single bond, so a carotenoid can, theoretically, adopt an infinite number of shapes or conformations. In practice, however, the carotenoid will exist in a particular preferred, low-energy conformation. The application of X-ray crystallography methods to define the linear extended conformation of the rigid polyene chain, ring shape, and the preferred angle of twisting about the C(6)-C(7) single bond in carotenoids with cyclic end groups is described in *Chapter 4*.

2. The conjugated double-bond system

The extended delocalized π-electron system that characterizes the central part of the structure is the key to many important properties of carotenoids.

a) Light absorption and photochemical properties

In carotenoids, the π-electrons of the conjugated double-bond system are highly delocalized, so the energy required to bring about the transition to the comparatively low-energy excited state is relatively small and corresponds to light in the visible region in the wavelength range 400-500 nm. This gives rise to the yellow, orange and red colours generally associated with carotenoids. The relationship between electronic structure and light-absorption properties is developed more fully in *Chapter 9* and in *Volume 1B, Chapter 1*. The relationship between the polyene chromophore and absorption spectrum, used in the analysis of carotenoids, is discussed in *Volume 1B, Chapter 2*.

The energy levels of carotenoid singlet and triplet states are ideally positioned for the carotenoids to participate in energy-transfer processes. This singlet-singlet and triplet-triplet energy transfer is the basis for both the accessory light-harvesting role and the photoprotective roles of carotenoids, respectively. The fundamental photochemistry and photophysics of carotenoids and their roles in energy transfer and related processes are explained in *Chapter 9*. The important functioning of carotenoids in light-harvesting and photoprotection in photosynthesis are described in detail in *Chapter 14*. Photoprotection in other systems is outlined in *Chapter 10*. Examples of photoprotection by carotenoids in the eye and in skin can be found in *Volume 5, Chapters 15* and *16*.

b) Reactivity

The susceptibility of the electron-rich polyene chain to attack by electrophilic reagents and oxidizing free radicals is the basis for the behaviour of carotenoids as antioxidants or pro-oxidants (see *Chapter 10*). The chemical principles of this are treated in *Chapter 7*, and the biological significance for health in *Volume 5, Chapter 12*.

This susceptibility to oxidation also has important practical implications. Pure, even crystalline, carotenoids may be broken down rapidly if stored in the presence of traces of oxygen. Even if there are no obvious signs of bleaching, stored samples may contain strongly pro-oxidant carotenoid peroxides. Standards should be monitored for purity.

This instability can have serious consequences for large-scale trials of carotenoids for biological activity. Extreme care should be taken to ensure that samples used in such investigations are free from peroxides and other degradation products, otherwise misleading results may be obtained.

c) Carotenoid radicals

Carotenoid radicals and radical ions are stabilized by delocalization of the unpaired electron along the polyene chain and have distinctive properties that are relevant to the functioning of the carotenoids, for example, in photosynthesis and as antioxidants/pro-oxidants. Under-standing the structure, properties and reactivity of these radicals, as presented in *Chapter 7* is therefore essential. The remarkable light-absorption properties of radical cations are treated in *Chapter 8*.

3. Molecular interactions

Following decades of extensive investigation by chemists, much is known about the properties of carotenoids in simple organic solution, but this only tells part of the story about carotenoid properties *in vivo*. *In situ* in living organisms, and in formulations for commercial applications, carotenoids are part of a much more complex system, and are frequently found in organized and ordered sub-cellular structures. The physical and chemical properties of a carotenoid are inevitably influenced by interactions with other molecules in its immediate vicinity, such as proteins and lipids. Complementary to this, the carotenoid may influence these neighbouring molecules and the structure and properties of the matrix in which it is located. Understanding these interactions and their effects is thus a major part of under-standing how carotenoids function and act *in vivo*, and is addressed in *Chapters 5* and *6*.

a) Aggregation

Being highly hydrophobic, carotenoids show a strong tendency to aggregate and crystallize, especially in aqueous media. Aggregation changes the properties of carotenoids, *e.g.* light absorption and chemical reactivity, as well as their effective size and ease of solubilization. *In vivo*, therefore, it may be such aggregates that are the biologically active or influential factors rather than the simple carotenoid monomers. The aggregation of carotenoid molecules is described in detail in *Chapter 5*.

b) Carotenoids in membranes

Carotenoids are highly hydrophobic compounds which will therefore tend to be associated with lipid (oil, fat) or in hydrophobic structures such as membranes. Hydrophobic molecules are often located in natural membranes, and constitute an integral part of the complex membrane structure. Predictably, carotenoids are associated with the lipid core of the membrane bilayer, but the concentration that can be achieved, the orientation of a particular carotenoid in the membrane, and its effect on membrane properties, depend on structural features such as the size and shape of the carotenoid and the presence of functional groups. The topic of carotenoids in membranes is treated in *Chapter 10*, and the aggregation of carotenoids in membranes is addressed in *Chapter 5*.

c) Carotenoid-protein interactions

Interactions between carotenoids and proteins occur in all kinds of living organisms. Carotenoids are usually much more stable *in vivo* than they are after isolation. The interactions can alter the chemical or physical properties of the carotenoid. A striking example of this is provided by the blue carotenoproteins found in many invertebrate animals. These carotenoproteins and other examples of carotenoid-protein interactions are discussed in detail in *Chapter 6*, including associations with blood lipoproteins for carotenoid transport. The structures of the photosynthetic pigment-protein complexes that maintain correct positioning of carotenoids with respect to other moleclues, and allow efficient energy transfer between carotenoid and chlorophyll, are described in *Chapter 14*.

C. Functions of Carotenoids

With the fundamental principles of structure and properties in mind, the latter part of *Volume 4* then describes the main functions ascribed to carotenoids. *Chapter 10* gives an overview of the main known and proposed functions of intact carotenoids, including the contribution that carotenoids make to natural colouration along with other pigments and structural colours, effects on membrane structure and properties, and various 'photofunctions' such as photoprotection. *Chapter 11* takes a novel approach, assessing the significance of colouration by carotenoids from the perspective of behavioural science and ecology. Commercial applications of carotenoids as colourants in aquaculture and poultry production are then evaluated in *Chapters 12* and *13*.

The roles of carotenoids in photosynthesis are discussed in detail in *Chapter 14*. This advanced treatment shows what can be achieved by the application of a battery of powerful spectroscopic and other physical techniques.

D. Metabolites and Breakdown Products

Any discussion of the biological significance of carotenoids must not overlook the role of carotenoids as precursors of metabolites that have important biological roles and of breakdown products that could be the actual biologically active molecules in functions and actions attributed to the carotenoids themselves. The roles of vitamin A and its derivatives in vision, cellular regulation and hormone action have been known for a long time. There is now increasing interest in the role of carotenoid fragments (norisoprenoids) as highly effective components of natural perfumes and flower fragrances and of flavour/aroma of tea, wine *etc*. Functions of carotenoid metabolites and fragments are surveyed in *Chapter 15*, and the importance of vitamin A and retinoids in human health is treated in *Volume 5, Chapters 8* and *9*. The enzymic formation of retinoids and other norisoprenoids is described in *Chapters 16* and *17*, respectively. The extensive treatment of carotenoids and human health in *Volume 5* at times alludes to the possibility that some of the biological effects seen could be attributable to breakdown products. This possibility is evaluated in *Volume 5, Chapter 18*.

E. Conclusions

Volume 4 and its companion, *Volume 5*, which deals with human nutrition and health, are the last volumes in the *Carotenoids* series, and in many ways point the way to the future of carotenoid research. They highlight how biologists are not only discovering new phenomena, but are looking at functions in a new light, and striving to elucidate details of the underlying mechanisms that explain their observations. They also show how chemistry is moving in new directions relevant to studies *in vivo* of biological actions and how new techniques are being developed to study these increasingly complex and sophisticated systems. If the insight that these books provide stimulates chemists, physicists and biologists to understand and talk to each other and thus provides a catalyst for interdisciplinary studies that will bring great advances and rewards in the future, then the editors will feel that their time and effort has been well spent.

Carotenoids
Volume 4: Natural Functions
© 2008 Birkhäuser Verlag Basel

Chapter 2

Structure and Chirality

Synnøve Liaaen-Jensen

A. Introduction

Naturally occurring carotenoids display considerable structural diversity, including different carbon skeletons, various types of oxygen functions and variable degree of unsaturation (double-bond equivalents) due to cyclization, double, triple or allenic bonds. In the *Carotenoids Handbook* the structures and properties of the known naturally occurring carotenoids (published up to 2003) are presented in two groups, (i) those considered to be proved unequivocally (Main List, some 500 compounds) and (ii) those requiring additional supporting evidence or proof of natural occurrence (Supplementary List, some 225 compounds).

B. Three-dimensional Carotenoid Structures

A feature often overlooked but crucial to the localization and functioning of carotenoids in biological systems is their 3-dimensional shape, especially chirality. Even achiral carotenoids are 3-dimensional molecules. Around a quarter of the known carotenoids are achiral, the majority chiral. Achiral carotenoids are encountered particularly among C_{40} carotenes and their apo-carotenoids, C_{30} diapocarotenoids and acyclic xanthophylls (oxygenated carotenoids) from phototrophic and some other bacteria.

Achiral carotenoids are identical with their own mirror image, in contrast to the chiral carotenoids, which are different from their mirror image and contain chiral centres or a substituents, *e.g.* H, OH, CH and CH$_3$. The structure of a chiral carotenoid is not fully established until the stereochemistry at each chiral centre is determined. Frequently, chirality is assumed by analogy with other carotenoids that occur in the same source, or with structurally related carotenoids. The terms 'constitution' or 'planar structure' are used until the chirality is settled. Chirality (handedness) or absolute configuration, defined by the *R/S* convention [1], is the basis for optical isomerism. The *R/S* convention is treated in *Volume 1, Chapter 2* but, for convenience, a brief summary of the application of the *R/S* system to the carotenoids, exemplified by the zeaxanthin (**119**) and astaxanthin (**406**) end groups (a) and (b), is given in Scheme 1.

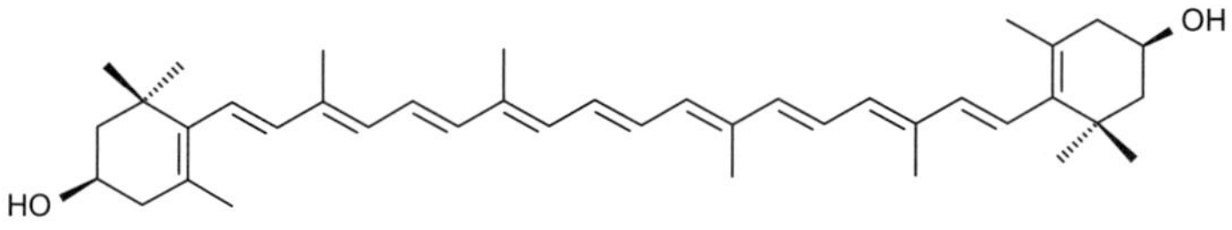

(**119**) zeaxanthin

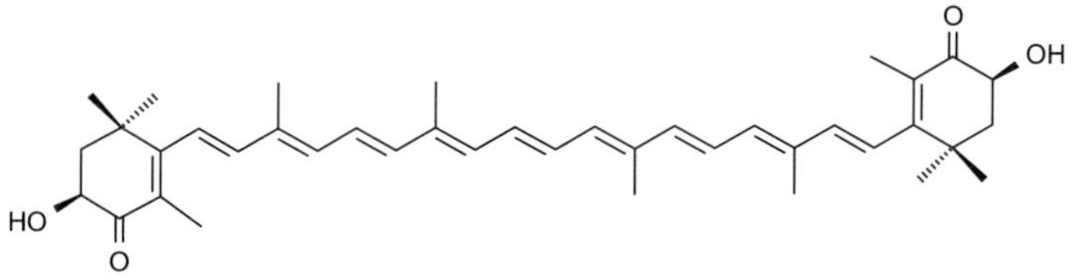

(**406**) astaxanthin

(a) HO > C(2) > C(4) ⟶ (3*R*)   (b) HO > C(4) > C(2) ⟶ (3*S*)

Scheme 1

Looking towards the chiral carbon atom C(3) with the substituent of lowest priority (H) pointing back, the direction traced by the other three substituents, going from the highest to the lowest priority, is clockwise and leads to (3*R*) designation for (a) and anticlockwise, leading to (3*S*) designation for (b), in spite of the fact that the hydroxy groups are pointing out from the plane of the paper in both cases. This shows that the *R/S* designation must be worked out for each chiral centre and compound and cannot be assumed by analogy.

Enantiomers have the opposite chirality at each chiral centre. *Meso* forms are internally compensated due to symmetry, *e.g.* (3*R*,3'*S*, *meso*)-zeaxanthin (**120**). The *R/S* convention is also applicable to substituted allenes, as neoxanthin (**234**).

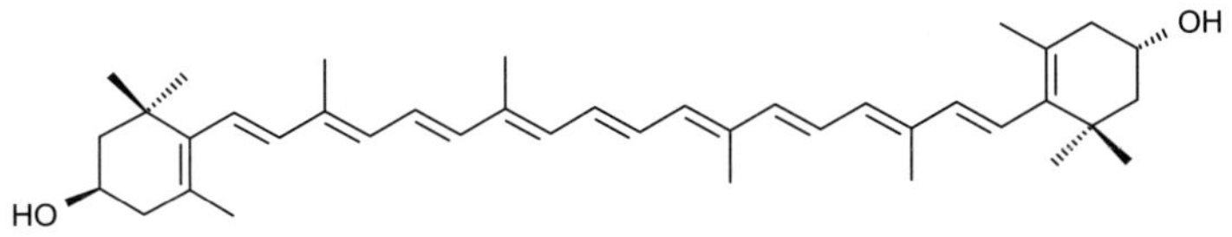

(**120**) *meso*-zeaxanthin

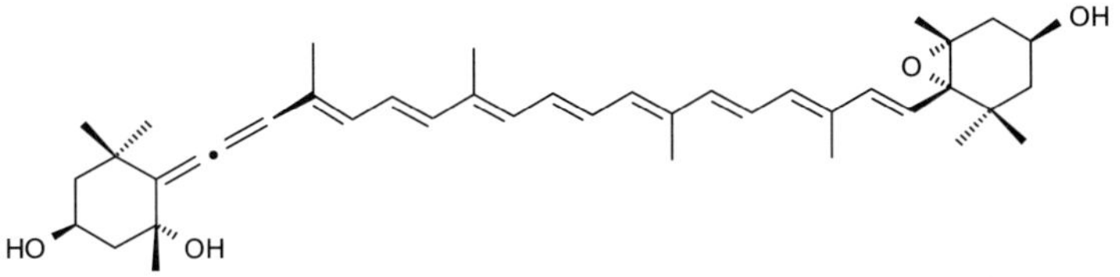

(**234**) neoxanthin

Geometrical isomerism is referred to by the *cis/trans* or more recent *E/Z* system, as described in *Chapter 3* and *Volume 1A, Chapter3*. The terms 'configuration' and 'conformation' are frequently misused. Changing the configuration of a carotenoid necessitates the breaking and reforming of bonds, *e.g.* going from an *R* to an *S* configuration. No bonds are broken in conformational changes, which occur by twisting or bending the molecule, *e.g.* going from an s-*cis* to an s-*trans* conformation of a single bond as illustrated in Scheme 2.

6-s-*cis*   6-s-*trans*

Scheme 2

C. Methods for Structure Determination

Relative configuration may be determined by circular dichroism (CD) or nuclear magnetic resonance (NMR). Chirality (absolute configuration) can be determined by a combination of these two techniques or directly by X-Ray analysis (*Chapter 4*). The application of CD in the carotenoid field is discussed in detail in *Volume 1B, Chapter 3*. There have been no significant methodological advances in this for unbound carotenoids since then. NMR spectroscopy, discussed in detail in *Volume 1B, Chapter 6*, serves to determine relative configuration, and also *E/Z* configuration of double bonds. In recent years, there has been significant progress in developing new 2-dimensional (2D) NMR techniques [2], in particular COSY, ROESY, HSQC and HMBC, which are now used routinely. Their application is outlined briefly in *Chapter 3*.

Recent examples where modern NMR techniques have been used to establish complicated carotenoid structures, including relative stereochemistry, can be found in the *Carotenoids Handbook, e.g.* the C_{37} norcarotenoids peridinin (**558**) and pyrrhoxanthin (**556**), the lactoside P457 (**280.1**), isolated as the 13-*cis* (13*E*) isomer, the 3,6:3'6'-diepoxide cycloviolaxanthin (**259.3**) and the allenic murillaxanthin (**421.1**) as well as carotenoids with peculiar and complex carbon skeletal substituents, such as pittosporumxanthin A (**259.1**), botryoxanthin A (**283.1**) and the braunixanthins (**294.1**).

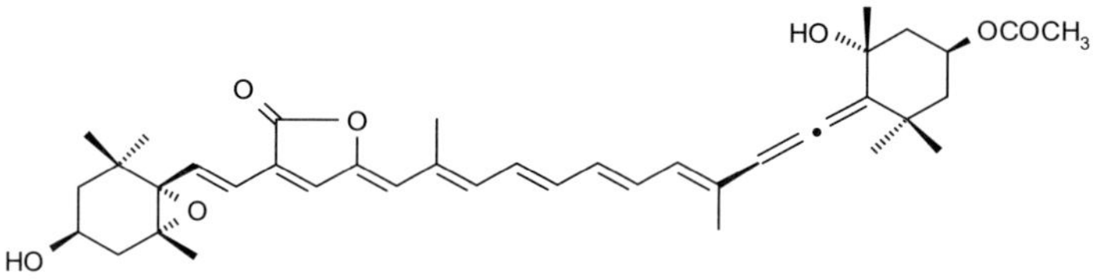

(**558**) peridinin

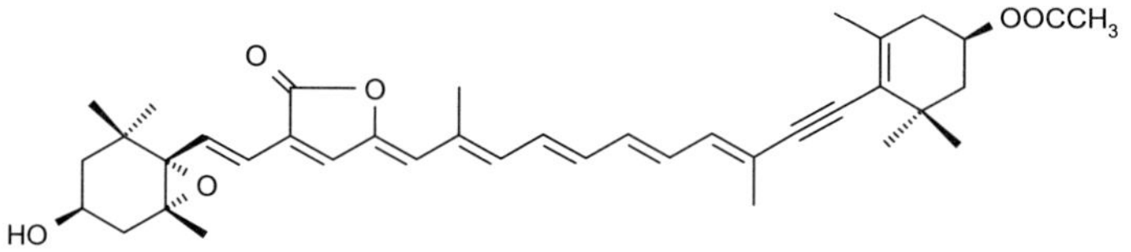

(**556**) pyrrhoxanthin

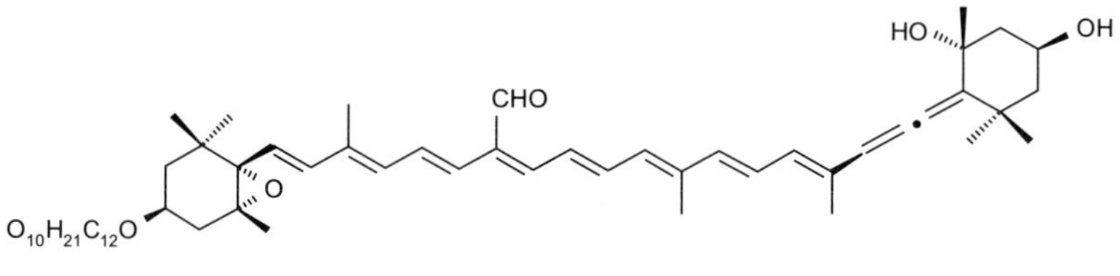

(**280.1**) 'P457'

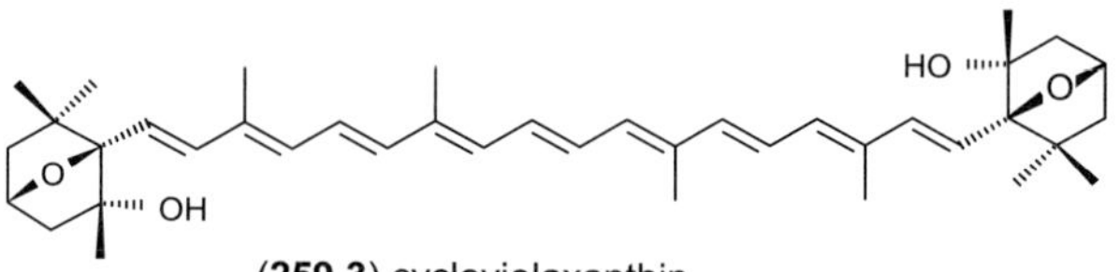

(**259.3**) cycloviolaxanthin

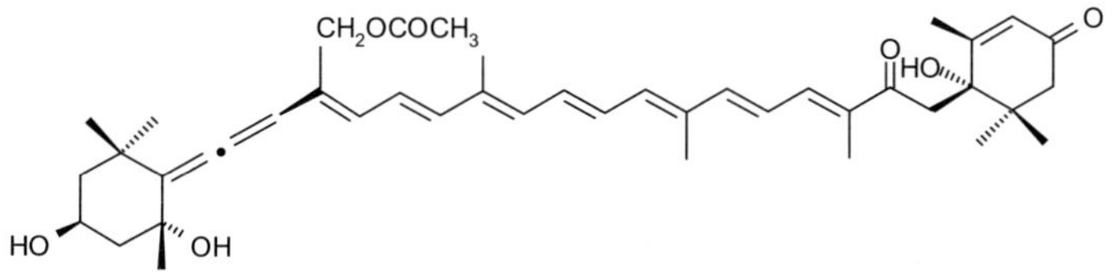

(**421.1**) murillaxanthin

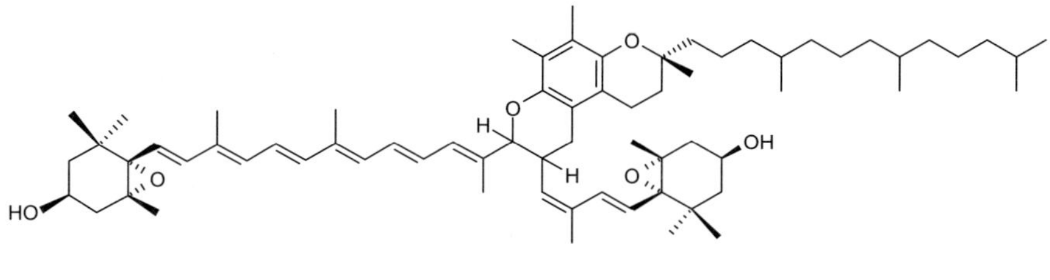

(**259.1**) pittosporumxanthin A

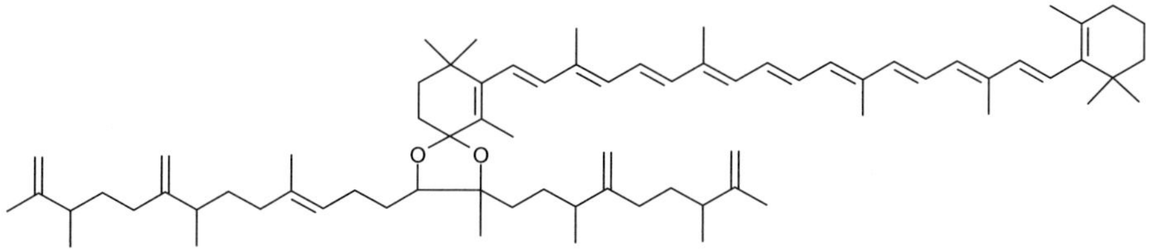

(**283.1**) botryoxanthin A

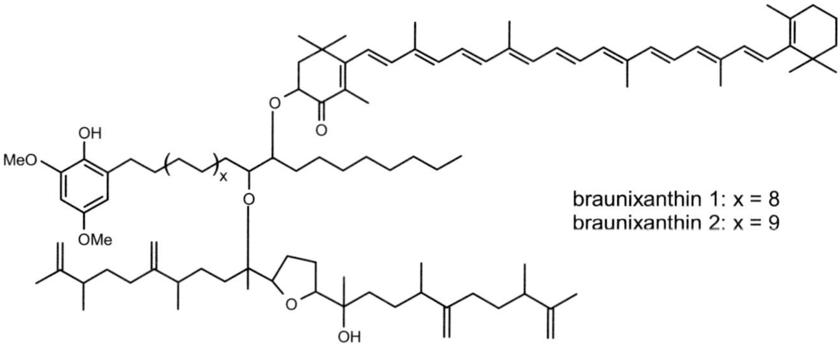

braunixanthin 1: x = 8
braunixanthin 2: x = 9

(**294.1**) braunixanthins

There are more recent examples of the application of 2D-NMR for assignment of structure and relative stereochemistry. These include additional examples of carotenoids with 2-hydroxy-β rings, namely 2-hydroxytorularhodin (*1*, the suggested *2R* configuration is not compatible with relevant CD models) from a red yeast [3] and (*2R*)-2-hydroxyastaxanthin (**417.1**) produced *via* (*3S,3'S*)-astaxanthin by combinatorial biosynthesis [4].

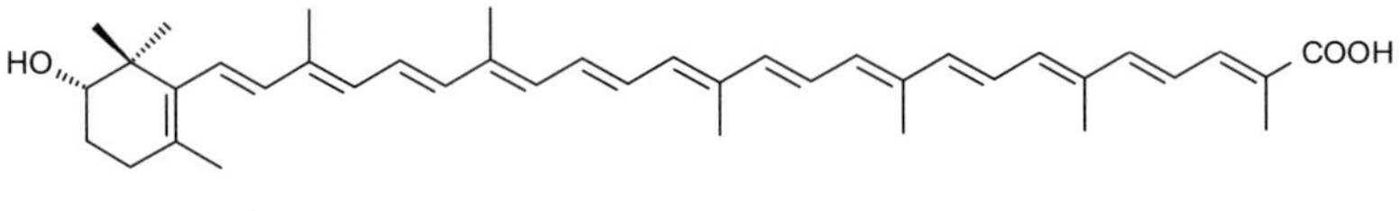

(*1*) 2-hydroxytorulene

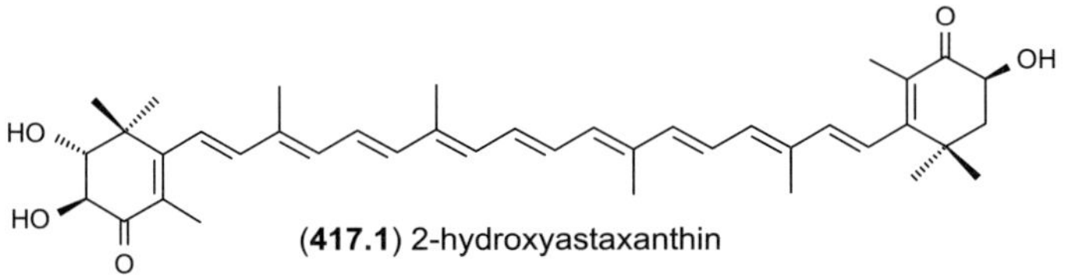

(**417.1**) 2-hydroxyastaxanthin

A new acetylenic carotenoid claimed to show cytotoxic activity towards cancer cells was unequivocally assigned the structure 3,5,3',18'-tetrahydroxy-6,7,7',8'-tetradehydro-5,6,7,8-tetrahydro-β,β-caroten-8'-one (*2*) with *3R,5S,3'R* configuration, by advanced NMR data in combination with the modified Mosher method [5,6]. The IUPAC name given in the publications is based on wrong end-group designations and is incorrect. 5',6'-Deepoxy-7',8'-didehydro-neoxanthin (*3*) is another documented new acetylenic/allenic carotenoid [7]. A new C_{37} skeletal peridinin derivative is hydratoperidinin (*4*), reported from clams, together with several other new carotenoids, including corbiculataxanthin acetate (*5*) with unexpected configuration at C(3') [7].

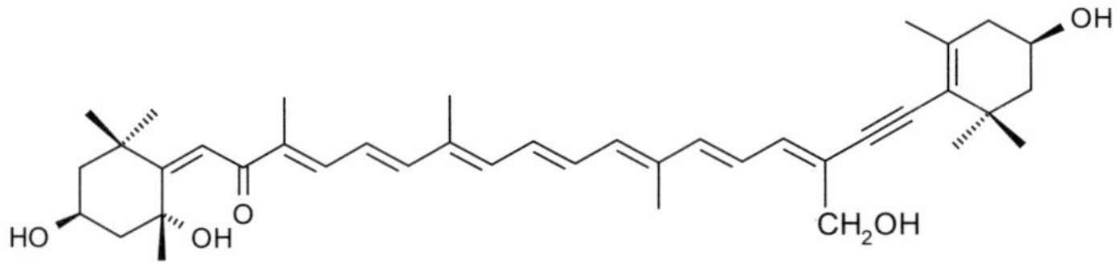

(*2*)

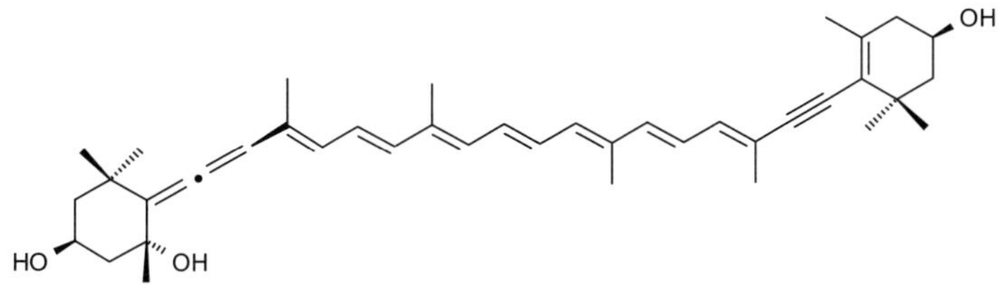

(3) 5',6'-deepoxy-7',8'-didehydroneoxanthin

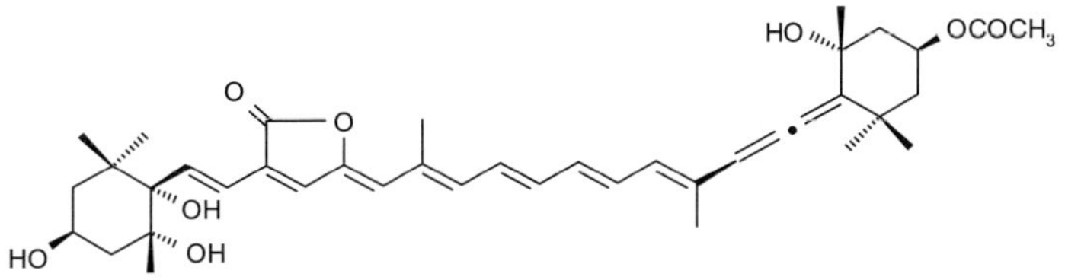

(4) hydratoperidinin

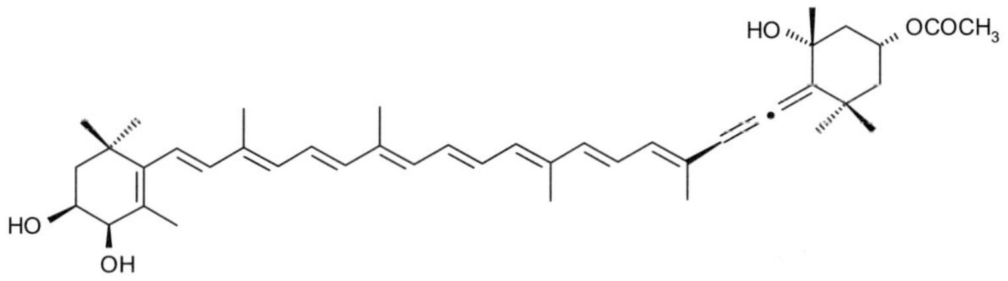

(5) corbiculataxanthin acetate

Finally a warning is given about the need for modern NMR methods in structural assignments. Two examples with UV/Vis absorption spectra suggestive of carotenoid structures have been shown to be non-isoprenoid polyenes. Isobutyl xanthomonadin I (6) is a dibrominated aryl-polyene pigment of bacterial origin [8], and granadiene (7), a more sophisticated polyene deri-vative with special solubility properties, was isolated recently from other bacterial sources [9].

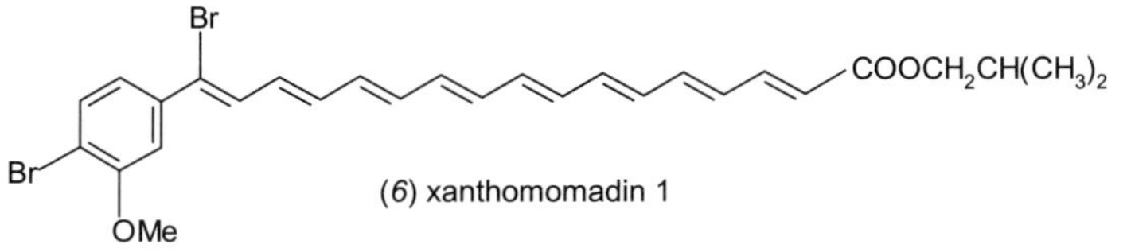

(6) xanthomomadin 1

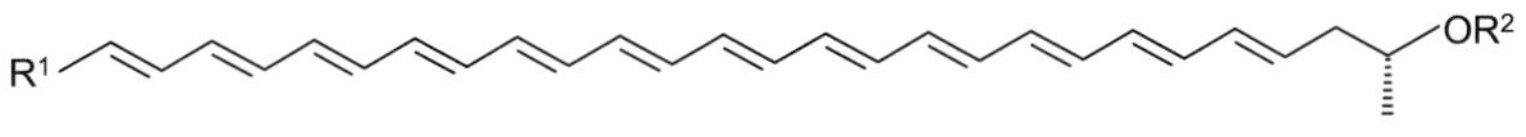

(7) granadiene; $R^1 = H_3N^+(CH_2)_3CH(COO^-)NHOC$ R^2 = glycosyl

References

[1]   R. S. Cahn, *J. Chem. Educ.*, **41**, 116 (1964).

[2]   H. Friebolin, *Basic One- and Two-Dimensional NMR Spectroscopy, 3rd Edition*, Wiley-VCH, Weinheim (1998).

[3]   R. W. S. Weber, A. Madhour, H. Anke, A. Mucci and P. Davoli, *Helv. Chim. Acta*, **88**, 2960 (2005).

[4]   Y. Nishida, K. Adachi, H. Kasai, Y. Shiguri, K. Shindo, A. Sawabe, S. Kommushi, W. Miki and N. Misawa, *Appl. Environ. Microbiol.*, **71**, 4286 (2005).

[5]   E. W. Rogers and T. F. Molinsky, *J. Nat. Prod.*, **68**, 450 (2005).

[6]   I. Ohtani, T. Kusumi, Y. Kashman and H. Kakisawa, *J. Am. Chem. Soc.*, **113**, 4092 (2005).

[7]   T. Maoka, M. Takashi, F. Yasuhiro, H. Keiji and A. Naoshiga, *J. Nat. Prod.*, **68**, 1341 (2005).

[8]   A. G. Andrewes, C. L. Jenkins, M. P. Starr, J. Shepherd and H. Hope, *Tetrahedron Letts.*, 4023 (1976).

[9]   M. Rosa-Fraile, J. Rodriguez-Granger, A. Haidoue-Benamin, J. Manuel-Cuerva and A. Sampedro, *Appl. Environ. Microbiol.*, **72**, 6367 (2006).

Carotenoids
Volume 4: Natural Functions
© 2008 Birkhäuser Verlag Basel

Chapter 3

E/Z Isomers and Isomerization

Synnøve Liaaen-Jensen and Bjart Frode Lutnæs

A. Introduction

The natural occurrence of several carotenoid *cis* isomers and their biological significance were not anticipated in 1962, when the classical monograph on *cis-trans* isomeric carotenoids [1] was published. More recent research has demonstrated that various *cis* isomers occur naturally in bacteria, plants, algae and invertebrate animals, and are present in human blood and tissues. The participation of *cis* isomers in the biosynthethic route to coloured carotenoids is well established (*Volume 3, Chapter 2*). Important biological functions of (15*Z*)-carotenoids in photosynthesis have been revealed [2]. In relation to health aspects of carotenoids, the bioavailability of *cis* isomers may be higher than that of the all-*trans* isomer [3], and accumulated evidence suggests that *cis/trans* isomerization may occur in biological tissues, particularly of lycopene (**31**) in human serum [4] (*Volume 5, Chapter 7*).

With the increasing appreciation of the biological importance of *cis* isomers, efficient methods for their analysis and identification, optimized working practices to minimize unwanted isomerization, and understanding of the mechanisms of isomerization are essential.

B. *E/Z* Isomers

1. Geometrical isomerism: Definition and nomenclature

Geometrical isomerism deals with the relative position of substituents around a planar carbon-carbon double bond. According to the IUPAC nomenclature for carotenoids [5], the *cis-trans* convention may still be used to denote geometrical isomerism in the polyene chain, but the more precise *E/Z* designation, based on the sequence rules, may also be used. In this Chapter, the *E/Z* system is preferred, but *cis/trans* designations are maintained when referring to some published work, when *E* and *trans* do not correspond (see below), and when the geometrical shape of the polyene chain is of major concern. The two alternative nomenclature systems are illustrated in Fig. 1.

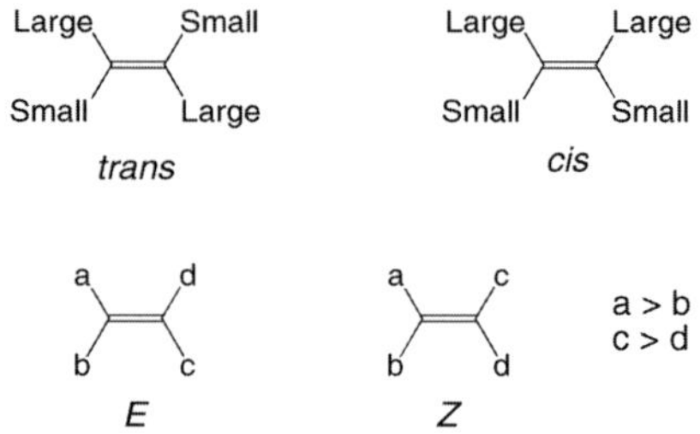

Fig. 1. Illustration of the *E/Z* and *trans/cis* nomenclature systems. Substituent **a** takes precedence over **b**, and **c** over **d**, according to the Cahn – Ingold – Prelog rules.

For carotenoids, *E* (entgegen) in most cases corresponds to *trans*, and *Z* (zusammen) to *cis*, except when there are oxygen substituents in the polyene chain, *e.g.* peridinin (**558**), (Fig. 2a), and for exocyclic double bonds, *e.g.* rhodoxanthin (**424**), (Fig. 3). Peridinin (**558**) also illustrates the stereochemistry of a substituted allene with two cumulative double bonds, a chiral structural element which is designated by the *R/S* convention, (Fig. 2b).

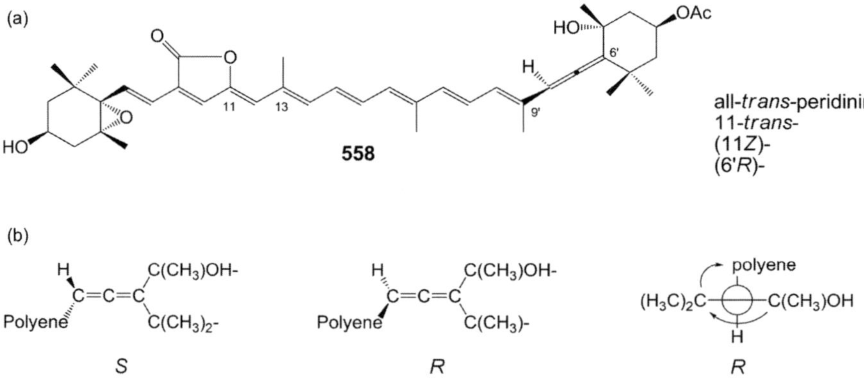

Fig. 2. Illustration of (a) *E/Z* versus *trans/cis* nomenclature for an oxygen substituted polyene, and (b) the *R/S* nomenclature for allenic double bonds.

Geometrical isomers may have more than one *cis* double bond (di-*cis*, tri-*cis*, poly-*cis*). The so-called sterically hindered *cis* double bonds [6,7] are energetically less favoured because there is a steric conflict between methyl and hydrogen in C(CH$_3$)-CH=CH-CH in contrast to CH-CH=CH-CH. This is illustrated for the tetra-*Z* isomer prolycopene (**31.1**) in Fig. 4, which also shows central [(15*Z*)-**31**)] and terminal [(5*Z*)-**31**] *Z* double bonds.

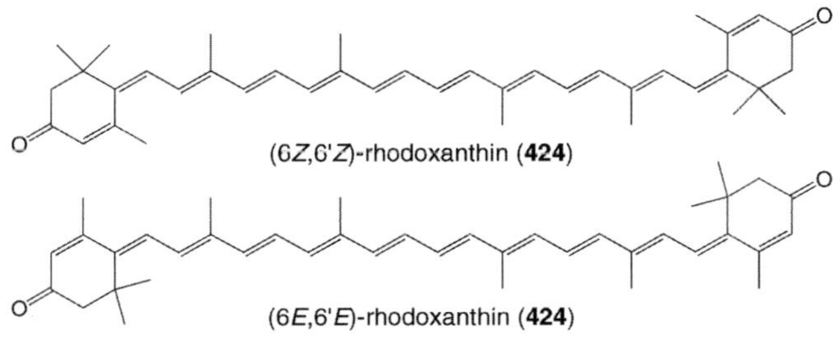

(6*Z*,6'*Z*)-rhodoxanthin (**424**)

(6*E*,6'*E*)-rhodoxanthin (**424**)

Fig. 3. Illustration of *E/Z* nomenclature for carotenoids with exocyclic double bonds.

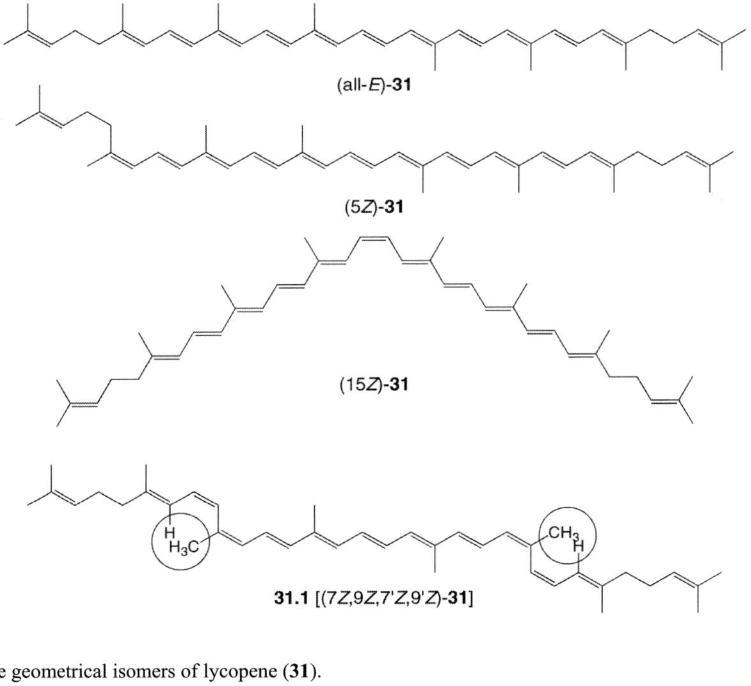

(all-*E*)-**31**

(5*Z*)-**31**

(15*Z*)-**31**

31.1 [(7*Z*,9*Z*,7'*Z*,9'*Z*)-**31**]

Fig. 4. Some geometrical isomers of lycopene (**31**).

The term 'stereochemical set' is used for all *E/Z* isomers of a carotenoid [1]. The number of *E/Z* isomers, *N*, may be calculated from the expressions below [7], where *n* is the number of double bonds:

$N = 2^n$	for unsymmetrical systems
$N = 2^{(n-1)/2} \bullet (2^{(n-1)/2} + 1)$	for symmetrical systems, *n* odd
$N = 2^{(n/2)-1} \bullet (2^{n/2} + 1)$	for symmetrical systems, *n* even

For lycopene (**31**), with a symmetrical undecaene chromophore (n = 11), there are 1056 possible geometrical isomers. However, the number of *Z* isomers that can be isolated is always significantly lower and depends on the relative stability of the isomers, and the separation techniques used. In contrast to most carotenoids with different absolute stereochemistry (chirality), geometrical isomers may be interconverted in organic solvents. This process is called isomerization or stereomutation and may also be effected for allenic carotenoids under particular conditions [8].

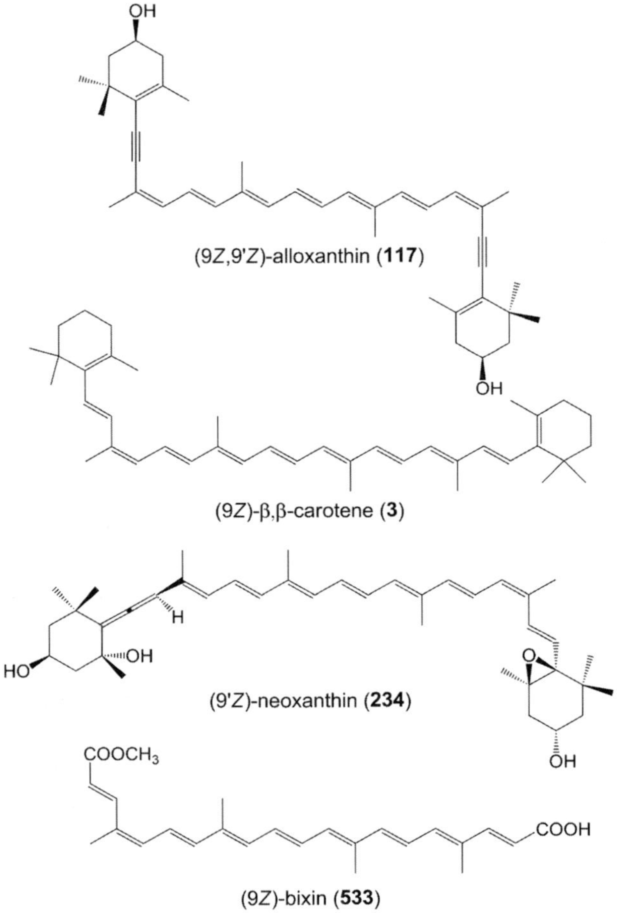
OH
OH
(9Z,9'Z)-alloxanthin (**117**)
(9Z)-β,β-carotene (**3**)
H
HO
OH
O
OH
(9'Z)-neoxanthin (**234**)
COOCH3
COOH
(9Z)-bixin (**533**)

Fig. 5. Some naturally occurring carotenoid *Z* isomers.

2. Structural examples of common *Z* isomers

Selected members of the lycopene (**31**) stereochemical set, shown in Fig. 4, illustrate the influence of *Z* double bonds on the molecular shape. Depicted are terminal (5*Z*) and central (15*Z*) mono-*Z* isomers, and the naturally occurring prolycopene (**31.1**), which is a tetra-*Z* isomer, (7*Z*,9*Z*,7'*Z*,9'*Z*)-**31**, with two sterically hindered *cis* double bonds [Δ(7),Δ(7')] [9].

Examples of other naturally occurring *Z* isomers, illustrated in Fig 5, are bixin (**533**) from *Bixa orellana*, (9*Z*)-β,β-carotene (**3**) from *Dunaliella bardawil*, (9'*Z*)-neoxanthin (**234**) from spinach, and (9*Z*,9'*Z*)-alloxanthin (**117**) which is an artefact readily formed from the all-*E* isomer. Two exocyclic isomers of rhodoxanthin (**424**) are shown in Fig. 3.

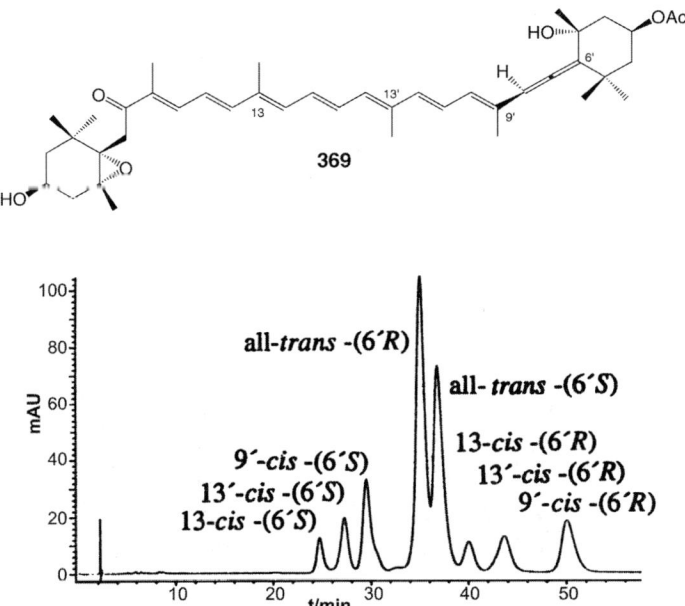

Fig. 6. HPLC separation of eight geometrical (*E/Z*, *trans/cis*) and allenic (*R/S*) isomers of fucoxanthin (**369**) on a Spherisorb S5W column with hexane/isopropyl acetate/isopropanol/*N*-ethyldiisopropylamine, 83.9:14:2:0.1, as the mobile phase. Flow rate 1.5 ml/min. Detection wavelength 490 nm [11,19].

3. Physical properties

a) Chromatography

Separation of *E/Z* isomers by HPLC is superior to that achieved by other chromatographic systems. (See *Volume 1A, Chapter 6*, where examples reported before 1994 are given for the separation of *E/Z* isomers of various carotenes and xanthophylls, and *Volume 5, Chapter 2*).

Excellent separation of *E/Z* isomeric carotenes has subsequently been achieved on three coupled Nucleosil columns [10], and silica columns have functioned well for various xanthophylls [11-13]. Routine analyses are now usually performed with reversed phase and/or nitrile phases [14]. The application of two systems may be required for complete resolution of complex mixtures. Particularly useful for the separation of *E/Z* isomers are C_{30} columns [15-17], and experiments with C_{34} stationary phases are promising [18]. Simultaneous recording of UV/Vis spectra by means of a photodiode array detector (PDAD) is essential for tentative identification of the isomers. A recommended procedure for semi-preparative HPLC separations of *E/Z* isomers of fucoxanthin (**369**) with multiple injection, collection and detection has been described [19]. An example, the separation of *E/Z* and allenic isomers of fucoxanthin (**369**) is shown in Fig. 6 [11].

b) Solubility

The non-linear *cis* isomers generally have greater solubility in organic solvents then the all-*trans* isomer. Higher solubility of *cis* isomers than of the all-*trans* isomer in lipophilic solution may facilitate transport of *cis* isomers within cells and between tissues.

4. Spectroscopy

a) UV/Vis spectroscopy

An early comprehensive treatment of *cis-trans* isomeric carotenoids includes UV/Vis spectra of several *cis* carotenoids [1]. The important diagnostic features have been summarized, with illustrations, in *Volume 1B, Chapter 2*. The main points are given below.

An all-*trans* carotenoid absorbs at longer wavelength, exhibits higher absorption coefficient and more pronounced spectral fine structure. Individual mono-*cis*-isomers display λ_{max} hypsochromically shifted by around 4 nm for the main absorption band (II). Upon *cis*-isomerization, a so-called *cis*-peak occurs at 142 nm lower wavelength than the absorption maximum (III) at longest wavelength (in hexane). A double *cis*-peak (peaks denominated A, B) is characteristic of aliphatic and monocyclic chromophores. The relative size of the *cis*-peak is greater when the *cis* double bond is located near the centre of the molecule and is greatest for carotenoids with a central *cis* double bond. In the absence of ^1H NMR data, the relative size of the *cis*-peak is commonly used for tentative structural assignment of *cis* carotenoids by HPLC/PDAD.

Carotenoids such as prolycopene (**31.1**, Fig. 4) with sterically hindered *cis* double bonds have greatly reduced λ_{max} and spectral fine-structure (%III/II) [20], (*Volume 1B, Chapter 2*).

The fact that carotenoid *cis* isomers generally have lower absorption coefficients than the all-*trans* isomer frequently results in underestimation of the amount of *cis* isomers in a *cis/trans* mixture, if the absorption coefficient of the all-*trans* carotenoid is used. Only in rare cases have the coefficients for individual *cis* isomers been determined. The following data are reported [21] for lycopene (**31**) (2% CH_2Cl_2 in hexane), as nm (ε x 10^{-3}): (all-*E*) 470 (187); (5*Z*) 470 (184); (5*Z*,5'*Z*) 470 (182); (9*Z*,9'*Z*) 459 (168); (7*Z*) 469 (154); (7*Z*,7'*Z*) 466 (128); (7*Z*,9*Z*) 444 (115); (7*Z*,9*Z*,7'*Z*,9'*Z*) (= **31.1**) 437 (105); (15*Z*) 468 (110). For zeaxanthin (**119**), corresponding data are compiled in *Volume 2, Chapter 1, Part III*. Cited are nm (ε x 10^{-3}): (all-*E*) 452 (133) hexane; (7*Z*) 459 (115) chloroform; (9*Z*) 458 (95) chloroform; (13*Z*) 458 (86) chloroform; (15*Z*) 449 (77) hexane, showing a systematic reduction in absorption coefficient in the series (all-*E*) > (9*Z*) > (13*Z*) > (15*Z*).

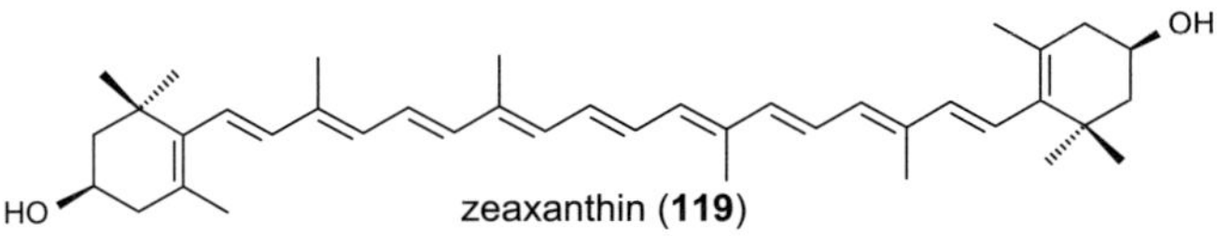

zeaxanthin (**119**)

b) NMR spectroscopy

^{1}H NMR data are essential for unequivocal structure determination of geometrical isomers; ^{13}C NMR provides little diagnostic evidence. The extensive treatment of NMR in *Volume 1B, Chapter 6* includes geometrical isomers. Protons in close proximity to a *cis* double bond are particularly affected. A useful table has been published [22] giving chemical shift differences ($|\delta_{cis} - \delta_{trans}|$ >0.05 ppm) of olefinic protons of *cis* isomers (9-*cis*, 11-*cis*, 13-*cis*, 9,13-di-*cis* and 11,13-di-*cis* isomers) of C_{15} model compounds. Some approximate average values, given in Fig. 7, show the general trend that, in *cis* isomers, protons on the inside of the loop are deshielded (positive $\delta_{cis} - \delta_{trans}$) and those on the outside shielded, relative to the *trans* isomer.

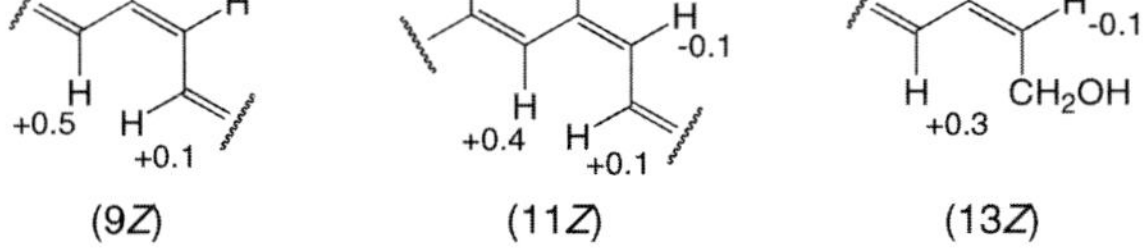

(9*Z*)   (11*Z*)   (13*Z*)

Fig. 7. Approximate ($\delta_{cis} - \delta_{trans}$) chemical shift differences for (9*Z*), (11*Z*), and (13*Z*) carotenoids (adapted from [22]).

Individual members of several isomeric sets have been investigated by ^{1}H NMR and structures assigned unequivocally, *e.g.* lycopene (**31**) [21], fucoxanthin (**369**) [12], peridinin (**558**) [13], and neoxanthin (**234**) [23]. The recent application of coupled on-line HPLC-NMR represents an important development [17].

c) IR and resonance Raman spectroscopy

Reference is made to an early treatment [1] of IR spectroscopy, a technique which is currently not employed for studies of *cis* isomers. Resonance Raman spectroscopy provides information on *cis/trans* double bonds, including sterically hindered *cis* double bonds [9]. The application of these techniques is outlined in *Volume 1B, Chapters 4 and 5*.

d) Mass spectrometry

The expulsion of toluene and xylene, causing the characteristic M-92 and M-106 ions in electron impact MS (*Volume 1B, Chapter 7*) requires a *trans-cis* isomerization of the polyene chain. Mass spectrometry therefore does not provide information on the location of *cis* double bonds. Combined HPLC-MS is, however, an important tool to confirm that a peak corresponds to a component of the isomeric set for a particular carotenoid [17].

e) Circular dichroism

The effect of *cis* double bonds on the CD spectra of carotenoids is treated in *Volume 1B, Chapter 3*. For carotenoids exhibiting so-called conservative CD spectra, a reversal of the sign of the Cotton effect results from isomerization of one double bond. Mixtures of *cis-trans* isomers of carotenoids with conservative CD spectra consequently show reduced Cotton effects. For di-*cis* isomers, the effect is reversed again.

5. Total synthesis

The *Z* isomers of carotenoids may be obtained by isomerization, or by total synthesis (*Volume 2, Chapter 3, Part V*). Methods are available for construction of sterically unhindered disubstituted (15*Z*), sterically hindered disubstituted (7*Z* and 11*Z*) and trisubstituted (5*Z*, 9*Z* and 13*Z*) double bonds.

Note that for acetylenic (7,8-didehydro) carotenoids, the 9*Z* configuration is thermodynamically favoured. Stereoselective formation of the stable 9*Z* isomer is therefore readily achieved by synthesis, *e.g.* (9*Z*,9′*Z*)-alloxanthin (**117**), (Fig. 5).

C. *E/Z* Isomerization

The process of *E/Z* isomerization, *i.e.* the interconversion of *E/Z* isomers, occurs readily, and may be achieved by thermal methods, photochemical methods in the presence or absence of catalysts, acids, contact with active surfaces and enzymes [1,7]. This process is also referred to as *cis-trans* isomerization.

1. Thermodynamic and kinetic aspects

An isomerization reaction can take place under either kinetic or thermodynamic control. In a kinetically controlled isomerization, the composition of the stereoisomeric set formed is determined by the relative rate of formation for each isomer. In an isomerization reaction where thermodynamic equilibrium is reached, the composition of the stereoisomeric set is decided by the relative stability of each isomer.

a) Thermodynamic equilibrium

Controlled photochemical isomerization in the presence of catalytic amounts of iodine has long been the method of choice [1,7,11-13,24]. More recently, diphenyl diselenide has been investigated as an alternative catalyst [14,23,25]. Since an equilibrium mixture of identical composition is obtained with either of these catalysts, it is assumed that thermodynamic equilibrium is reached [14].

At thermodynamic equilibrium, the all-*trans* isomer usually dominates over mono-*cis* isomers, and with lesser amounts of di-*cis* isomers. The 15*Z* isomer is formed only in a minor amount. Structural modifications in the polyene chain may influence the preferred bond configurations. In-chain oxidized methyl groups and acetylenic bonds favour *cis*-configuration of the adjacent double bond, whereas *cis* double bonds adjacent to keto groups appear not to be favoured [12] (see also *Volume 2, Chapter 3, Parts IV* and *V*). At thermodynamic equilibrium, no sterically hindered *cis*-isomers are present. Prolycopene (**31.1**) and the naturally occurring (7*Z*)-isomer of renieratene (**26**) ('renieracistene') [26] are not part of the isomeric sets at equilibrium.

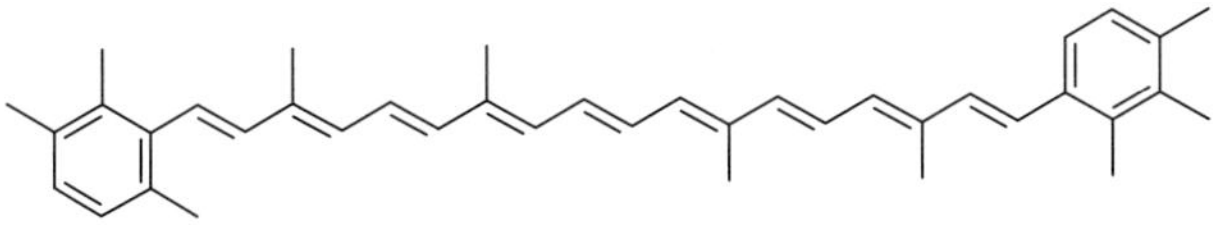

renieratene (**26**)

Other isomerization conditions, such as heat, or light with no catalyst, result in slower reactions, and appear to be kinetically controlled. Triplet-state photoisomerization is catalysed by chlorophyll or other sensitizer (*Chapter 9*).

b) Reversibility test

Sterically hindered *cis* isomers are irreversibly isomerized. Also, the thermodynamically stable 9*Z* isomers of acetylenic carotenoids cannot be converted into the all-*E* isomer. Apart from these known exceptions, iodine-catalysed isomerization of any geometrical isomer will result in the same qualitative and quantitative composition of the isomerization mixture. In combination with HPLC, this is a useful test on the micro scale to demonstrate, in the absence of NMR data, that the compound in question is indeed a *Z* isomer belonging to that particular stereoisomeric set.

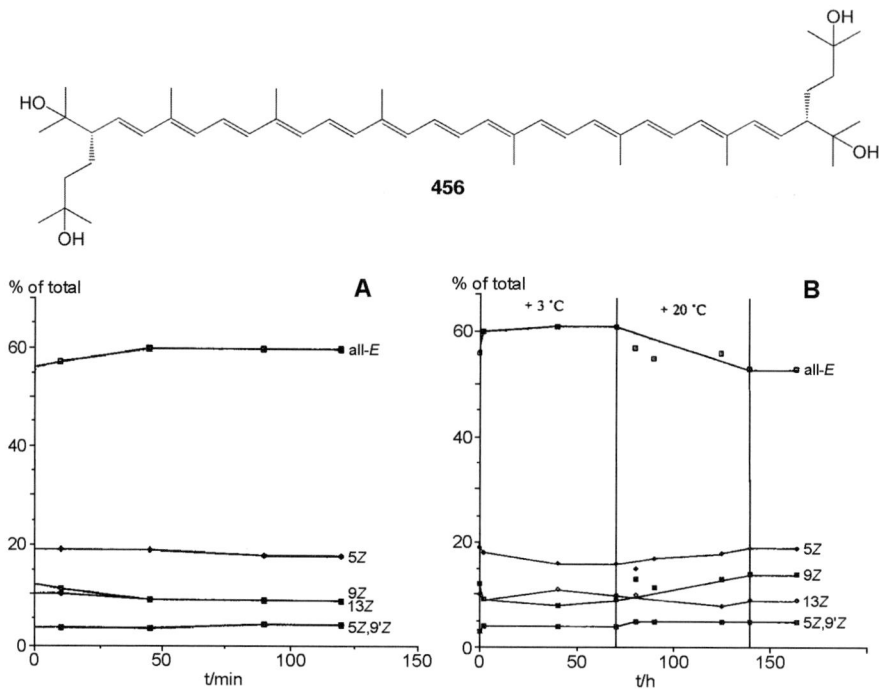

Fig. 8. A) Extracted mixture of bacterioruberin (**456**) isomers (8 min), further isomerization in darkness at 3 °C, and extrapolation to zero time. B) Further isomerization in darkness at 3 °C for 70 h, then at 20 °C [27].

c) Kinetically controlled isomerization

After Zechmeister's classical work [1], kinetic studies on *E/Z* isomerization of carotenoids promoted by heat have continued [27,28]. A kinetic study of *E/Z* isomerism of the individual all-*E*, 5*Z*, 9*Z*, 13*Z*, and 15*Z* isomers of the aliphatic tridecaene bacterioruberin (**456**) was performed in ethanol solution in the dark, and in the absence of iodine (Fig. 8). Rate constants, equilibrium constants and ΔG° values were calculated [29]. The rate of isomerization from the all-*E* isomer to individual mono-*Z* isomers was 13*Z* > 9*Z* > 15*Z*, suggesting lowest activation energy for isomerization of the Δ(13) double bond. The comparative thermodynamic stability of the individual stereoisomers was, judged by the composition of the iodine-catalysed equilibrium mixture, all-*E* > 5*Z* > 9*Z* > 13*Z* [27]. The calculated changes in free energy ($\Delta G°_{298K}$) of 2.27 kJ/mol for all-*E* to 9*Z*, 2.4-2.75 kJ/mol for all-*E* to 13*Z*, and 7.25 kJ/mol for all-*E* to 15*Z* [29] support the relative thermodynamic stability estimated from the equilibrium data [27].

It may thus be predicted that, for kinetically controlled isomerizations, where equilibrium is not yet reached, the 13*Z* isomers are predominant relative to 9*Z* isomers. Thus thermal isomerization of synthetic (all-*E*)-*syn,syn*-violaxanthin gave mainly the 13*Z* isomer, and synthetic (all-*E*)-*syn-anti*-violaxanthin mainly the 13*Z* and 13'*Z* isomers [28]. (all-*E*)-5,6-Diepikarpoxanthin (**201.1**) provided the 13*Z*, the 13'*Z*, and also the labile 15*Z* isomer upon thermal isomerization [30].

A comprehensive study of thermal *E/Z* isomerization, involving all-*E*, 9*Z*, 13*Z* and 15*Z* isomers of zeaxanthin (**119**), and all-*E*, 9*Z*, 13*Z*, 15*Z*, 9'*Z* and 13'*Z* isomers of violaxanthin (**259**), capsanthin (**335**), capsorubin (**413**), and lutein epoxide (**232**), was conducted at 333-368 K. The Arrhenius and activation parameters were derived from the specific rate constants, calculated on the basis of four kinetic models. The effects of different end groups and solvents were addressed [31].

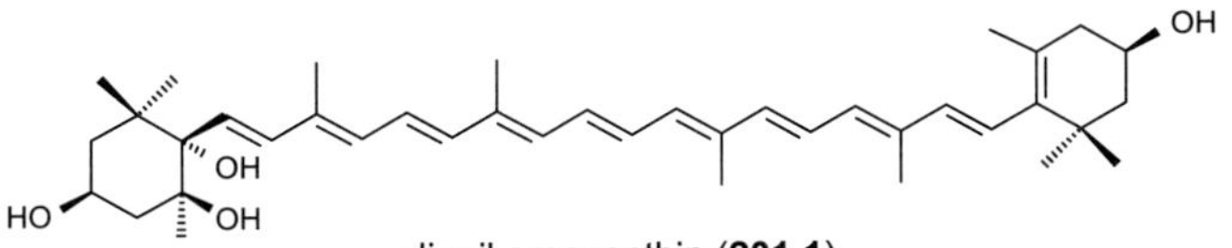

diepikarpoxanthin (**201.1**)

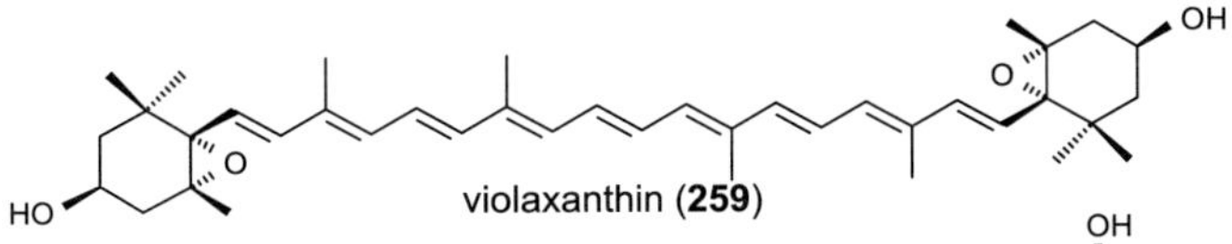

violaxanthin (**259**)

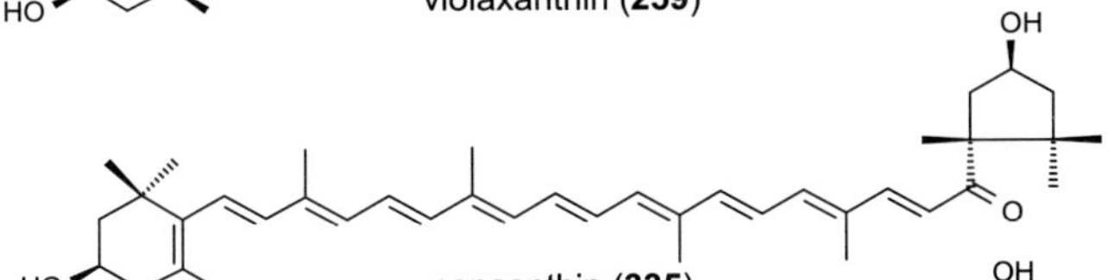

capsanthin (**335**)

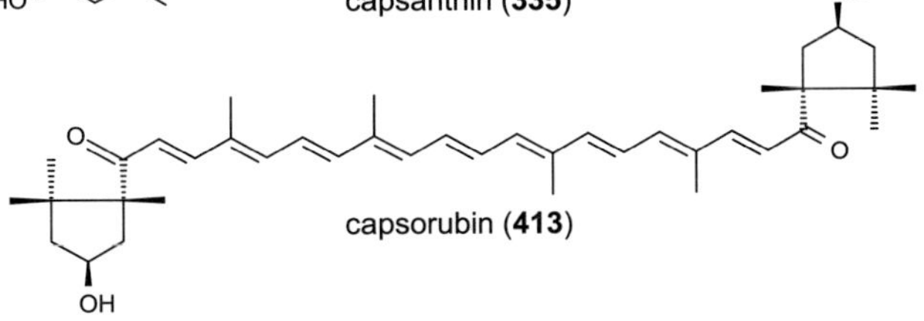

capsorubin (**413**)

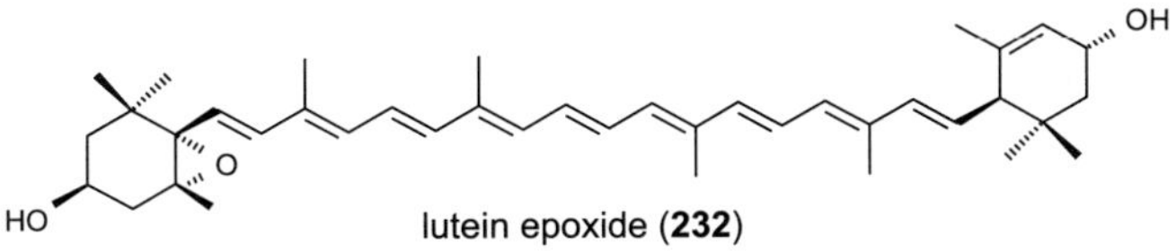

lutein epoxide (**232**)

2. Photochemical isomerization catalysed by iodine

a) Conditions for *E/Z* isomerization

The isomerization is carried out under nitrogen in hexane, benzene [1] or dichloromethane [12] solution in diffuse daylight (window, no sunshine) using catalytic amounts of iodine. Recommended concentrations are: approx. 0.1 mg carotenoid/ml solvent (0.2 mM) and iodine (1-2 wt% of the carotenoid) [1]. For larger-scale isomerizations, carotenoid (21 mg) in benzene (500 ml) with iodine (0.3 mg) and sunshine exposure have been used [12].

The course of the isomerization is conveniently monitored by UV/Vis spectroscopy. Photo-stationary equilibrium is usually reached within 15-60 min [1] on an analytical scale, and up to 2 hours [12] on the semi-preparative scale. The resulting *E/Z* mixture is analysed by HPLC/PDAD. Tentative identification of *cis* isomers is based on λ_{max} shifts and relative *cis*-peak intensity, and preferably by co-chromatography, though few standards of authentic *cis* isomers are available commercially. For structure determination of individual *cis* isomers, fractions separated by semi-preparative HPLC must be collected at 0°C in darkness for subsequent ^{1}H NMR analysis [12], unless LC-NMR equipment is available. All *cis* isomers must be stored cold in darkness prior to NMR and CD analysis.

As an example, the established *cis/trans* composition in the equilibrium mixture of fucoxanthin (**369**) [12] is given in Table 1. The semi-preparative separation was carried out on an Ultrasphere TH-cyano column [12,19]. The detection wavelength was 490 nm [16] and no correction was made for the unknown absorption coefficients of the individual *Z* isomers. The isomers were identified by ^{1}H NMR data, supplemented by CD spectra.

Table 1. Composition of the iodine-catalysed stereoisomerization mixture of (3*S*,5*R*,6*S*,3'*S*,5'*R*,6'*R*)-fucoxanthin (**369**) and UV/Vis absorption characteristics of the individual geometrical isomers in the HPLC eluent (hexane/isopropyl acetate/propan-1-ol/*N*-ethyldiisopropylamine, 83.9:14:2:0.1) [12].

Geometrical isomer	% of total	VIS λ_{max}	%III/II	%A_B/A_{II}
		nm		
15-*cis*	1	325, (420), 439, (458)		55
13,9'-di-*cis*	4	329, (415), 435, (459)		40
13,13'-di-*cis*	3	329, (418), 435, (457)		26
9',13'-di-*cis*	6	329, (420), 437, (461)		34
all-*trans*	42	330, (427), 445, 471	6	7
13-*cis*	11	329, (420), 437, (463)		45
13'-*cis*	14	329, (422), 441, (466)		52
9'-*cis*	20	327, (424), 443, 469	2	12

b) Conditions for simultaneous *E/Z* and allenic *R/S* isomerization

An allenic bond may also be isomerized photochemically in the presence of iodine, but this requires higher iodine concentration and stronger light exposure [11-13].

A recommended procedure based on studies of fucoxanthin (**369**) and peridinin (**558**), involves the use of 2% (w/w) iodine relative to the carotenoid, carotenoid concentration 40 µg/ml benzene, and irradiation in sunlight at room temperature for 3 hours under nitrogen [11,12]. The separation by HPLC of the equilibrium mixture of fucoxanthin (**369**) is shown in Fig. 6. For peridinin (**558**), the ratio of the sum of geometrical isomers of the allenic 6'*R* series *versus* the 6'*S* series was determined as 55:45 at equilibrium, reflecting slightly higher thermodynamic stability of the naturally occurring 6'*R* isomer [13]. A similar result was given by fucoxanthin (**369**), compatible with AM1 calculations [8].

Other epoxidic, allenic carotenoids such as neoxanthin (**234**) quickly turn blue under the conditions specified, but may undergo allenic isomerization with diphenyl diselenide as catalyst [32] (see Section **C**.3.a).

c) Mechanistic aspects

i) Photoisomerization in the absence of catalyst. The photochemistry of carotenoids is treated in *Chapter 9*. Knowledge of the electronic structure of carotenoids, essential for the understanding of the mechanism of geometrical isomerization, has been considered in *Volume 1B, Chapter 1*. The isomerization is most effective when light corresponding to the main absorption band of the carotenoid is used [1]. Isomerization in an excited state with low isomerization barrier is assumed, but a clear picture remains to be established.

It has been suggested that planar carotenoid cations or radical cations are intermediates in the isomerization process (*Chapter 8*). AM1 calculations have shown that the energy barrier for the radical cation (*ca.* 24 kcal/mol) and the dication (*ca.* 3 kcal/mol) is much lower than for the neutral carotenoid (*ca.* 55 kcal/mol) [33].

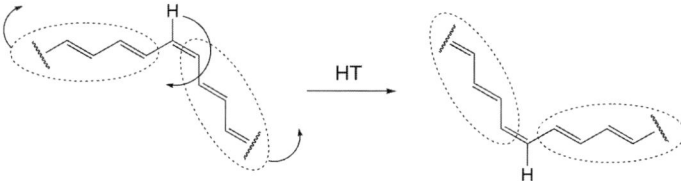

Fig. 9. The Hula-twist (HT) mechanism for *E/Z* isomerization [34].

The Hula-twist (HT) process is a volume-conserving reaction mechanism serving to rationalize rapid isomerization of the sterically hindered 11*Z* double bond in solid solutions or other constrained conditions (Fig. 9). It involves sweeping translocation of a single C-H unit with the net result of simultaneous conformational and configurational changes of two adjacent bonds. The possible involvement in the photochemistry of carotenoids has been suggested [34].

ii) E/Z and allenic R/S photoisomerization in the presence of iodine. In general, the electronic configuration upon photoisomerization promoted by a triplet sensitizer is better understood than stereoisomerization by direct irradiation. However, no accepted mechanism for the iodine-catalysed stereoisomerization of polyenes has appeared.

The main factors that influence iodine-catalysed stereoisomerization are the concentrations, the ratio between carotenoid and iodine, intensity and wavelength of light, exposure time, and the solvent. Light absorbed by iodine (450-600 nm) is more effective than light absorbed by the carotenoids (350-500 nm) [35]. It has been suggested that for a carotene ($C_{40}H_{56}$) a charged complex, with $C_{40}H_{56}\cdots I^+$, λ_{max} *ca.* 1000 nm, predominant in the electronic ground state, and $C_{40}H_{56}{}^+\cdots I$, predominant in the excited state, with the latter form as the active species, is involved in iodine-catalysed stereoisomerization [36,37]. The structure of this complex has recently been studied in more detail (see *Chapter 8*). Decomposition of this complex resulted in extensive *E/Z* isomerization [16].

Kinetic studies on the iodine-catalysed stereoisomerization of several carotenoids indicated that the reaction occurs through a series of first-order, reversible reaction steps [29]. Mechanisms of the *E/Z* isomerization *via* electron transfer and energy transfer have been discussed (*Volume 1B, Chapter 1*). However, the photochemical events during the *E/Z* iso-merization of carotenoids in the presence of catalytic amounts of iodine remain unknown. A mechanism has been suggested for iodine-promoted *R/S* photoisomerization of allenic carotenoids (Fig. 10).

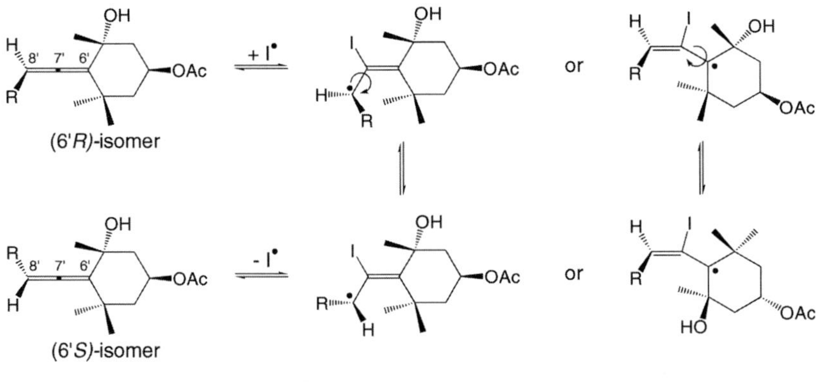

Fig. 10. Proposed mechanism for *R/S* photoisomerization of allenic carotenoids in the presence of high concentrations of iodine [38].

Attack of an iodine radical I° on C(7) was the key step. AM1 calculations favour *R/S* isomerization *via* the iodine radical and not *via* radical cation or dications, and explain the experimental observation that the ease of allenic isomerization is peridinin (**558**) > fucoxanthin (**369**) > neoxanthin (**234**) [38].

3. Photochemical isomerization catalysed by diphenyl diselenide

a) Conditions

It has been demonstrated that diphenyl diselenide (Ph$_2$Se$_2$) is a good alternative to iodine as a promoter of photochemical stereoisomerization. Geometrical *E/Z* as well as allenic *R/S* isomerization may be achieved under selected and reproducible conditions [14,25,32]. Suitable conditions for *E/Z* isomerization of zeaxanthin (**119**), the epoxidic violaxanthin (**259**), canthaxanthin (**380**), and the allenic fucoxanthin (**369**) were reported [25]. Since the presence of base decreased the isomerization rate in the absence of catalyst, base may serve to decrease undesirable *E/Z* stereoisomerization of base-stable carotenoids [25]. Moreover, in contrast to iodine, Ph$_2$Se$_2$ tolerates the presence of Hünig's base upon stereoisomerization of acid-sensitive carotenoids.

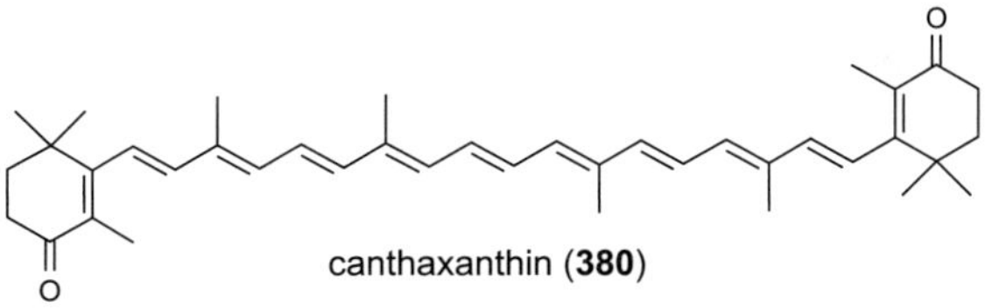

canthaxanthin (**380**)

Recommended procedure [25]: Carotenoid (40 µg/ml) in benzene (10 ml) and Ph$_2$Se$_2$ (2%, w/w) is irradiated with light of medium intensity (1600 µE m^{-2}s^{-1} or outdoor sunlight for 13 hours) and the reaction is monitored by HPLC [14,25]. For acid-sensitive carotenoids such as the epoxides violaxanthin (**259**) and neoxanthin (**234**), 1% (v/v) *N*-ethyldiisopropylamine (Hünig's base) should be added. For *E/Z* and simultaneous allenic *R/S* isomerization, a higher amount of Ph$_2$Se$_2$ (100%, w/w) and high light intensity (*ca.* 3100 µE m^{-2}s^{-1}, or outdoor sunlight for 13 hours) should be used [14,25].

b) Thermodynamic equilibrium

Kinetic and thermodynamic aspects of the Ph$_2$Se$_2$-mediated photoisomerization of the allenic carotenoids fucoxanthin (**369**) and peridinin (**558**) have been discussed [14]. The effects of light quality and intensity, and of Ph$_2$Se$_2$ concentration, were studied in kinetic experiments. Stereoisomerization promoted by UVA light, corresponding to the absorption region of

Ph$_2$Se$_2$, was particularly effective. Stationary conditions, with 6'R:6'S ratio *ca.* 12:88, were achieved for peridinin (**558**) within 2 h with 40 mol% Ph$_2$Se$_2$. The quantitative compositions of geometrical E/Z isomers in the R and S series were identical, consistent with a thermo-dynamic equilibrium. Accumulated evidence [11-14,25] supports the conclusion that the activation energy for geometrical E/Z isomerization is lower than that for allenic R/S isomerization. The high yield of allenic 6'R isomers is rationalized mechanistically below.

c) Mechanistic aspects

The increased reaction rate under UVA radiation and the response towards a radical quencher were taken as support for a radical mechanism for allenic R/S isomerization, initiated by photolysis of Ph$_2$Se$_2$. The observation that the 6'S allene is the major isomer present after diselenide-promoted isomerization suggests a thermodynamic preference for one conformer of the radical intermediate [14] (Fig. 11).

For E/Z stereoisomerization promoted by Ph$_2$Se$_2$, a radical mechanism *via* attack at an sp^2 carbon with possible formation of a tertiary radical has been suggested [14]. Mild conditions that result only in E/Z isomerization have been defined [14,32].

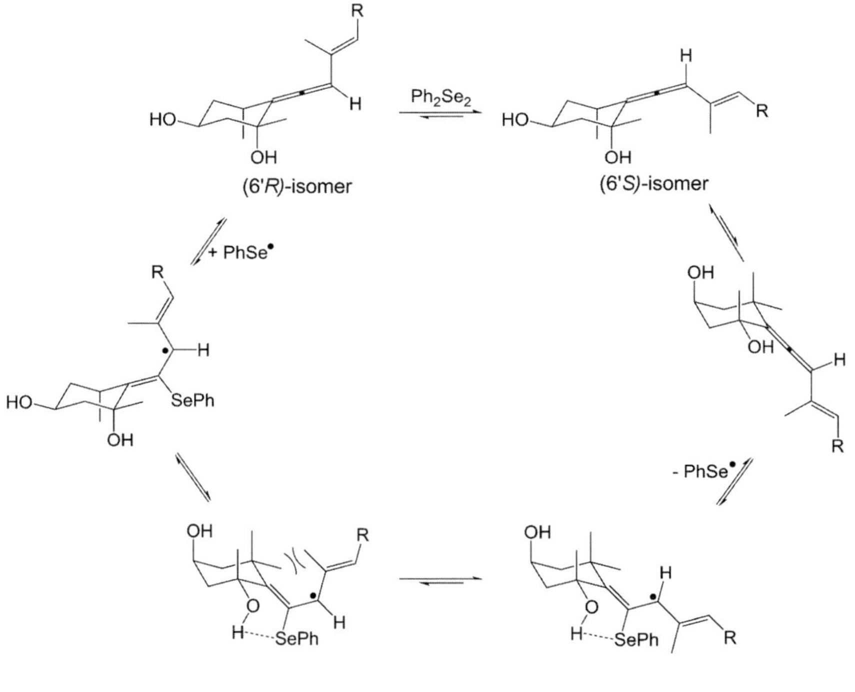

Fig. 11. Proposed mechanism for R/S photoisomerization of allenic carotenoids in the presence of high concentrations of diphenyldiselenide. For further details, see [14].

4. Avoiding unwanted *E/Z* isomerization

Thermal *E/Z* isomerization occurs even at room temperature. To avoid undesirable *E/Z* isomerization, manipulations and storage should be carried out at the lowest practical temperature. Even diffuse daylight is sufficient to cause slow *E/Z* isomerization. The extent of isomerization may be reduced by working in dim light/darkness. Black cloth is used for covering columns, TLC plates and evaporators. Acids and active surfaces can cause *E/Z* isomerization.

5. Isolation artefacts

a) Precautions for avoiding *E/Z* isomerization

The most common artefacts encountered in the carotenoid field are *cis* isomers [39]. General precautions have been treated in *Volume 1A, Chapter 5*, and Section **C**.4. Matrix solid-phase dispersion (MSPD) may serve as a mild, rapid, complete and reproducible extraction technique [17,40]. The biological material, with BHT as stabilizer, is mixed with a MSPD C_{30} sorbent material, and blended to a free-flowing powder. It is then packed into a solid phase extraction (SPE) column. After conditioning and elution of impurities, the carotenoids are eluted with acetone, and analysed by HPLC.

b) Proof of natural occurrence of *cis* isomers

In general, all-*trans* isomers are the most common in Nature. Rapid isolation with all necessary precautions is required in order to claim the natural occurrence of *cis* isomers.

In a recommended procedure for isolating the sterically labile bacterioruberin (**456**) from a *Halobacterium* sp. [27], the carotenoids were extracted at 3°C in darkness. HPLC analysis was performed within 10 min and at intervals afterwards, and with extrapolation to zero time (Fig. 8A). The low extent of further isomerization at 3°C in darkness for 70 h and increased isomerization at 20°C is demonstrated in Fig. 8B.

D. Biological Implications of Carotenoid *Z* Isomers

1. Biosynthesis

The biosynthesis of carotenoids is treated in *Volume 3, Chapter 2*. The pathway proceeds *via* (15*Z*)-phytoene (**44**) in plants and some bacteria. Subsequent *E/Z* isomerization may occur at the phytofluene (**42**) or ζ-carotene (**38**) level. A biosynthetic scheme for the biosynthesis of the sterically hindered prolycopene (7*Z*,9*Z*,7′*Z*,9′*Z*)-lycopene (**31.1**) has been proposed

[*Volume 3, Chapter 2*]. It has been suggested that prolycopene is an intermediate in the biosynthesis of dicyclic carotenes. This would involve further isomerizations.

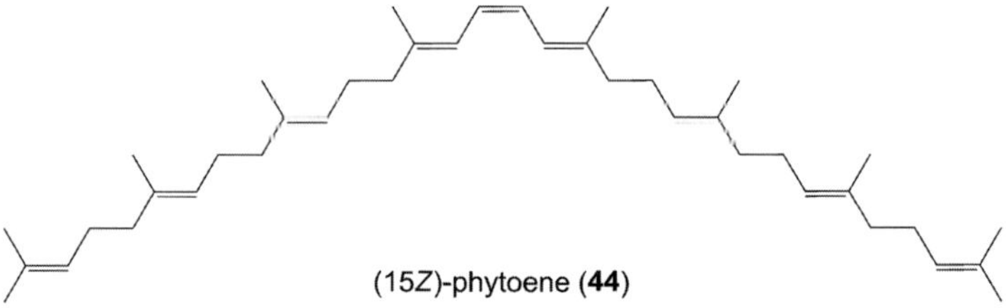

(15Z)-phytoene (**44**)

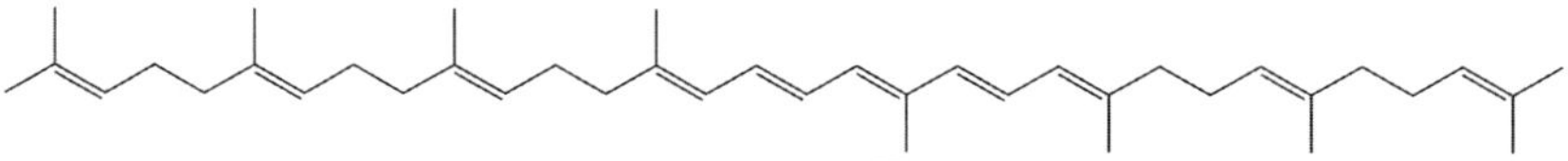

phytofluene (**42**)

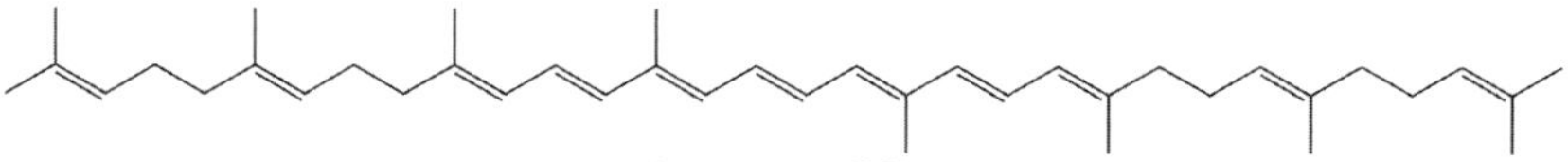

ζ-carotene (**38**)

Some microalgae and bacteria living in extreme light conditions are known to accumulate *Z* carotenoids. A proposal for the formation of (9*Z*)-β,β-carotene (**3**) in *Dunaliella bardawil* by a separate desaturation sequence has not been substantiated. The direct biosynthesis of *Z* isomers of bacterioruberin (**456**) in halophilic bacteria [27] is indicated by the data illustrated in Fig. 8.

2. Photosynthesis

Thermodynamically unstable 15*Z* isomers are functional in the photosynthetic reaction centres of green plants and phototrophic bacteria [3] (see *Chapter 14*).

3. Transport, accumulation, and *E/Z* isomerization in biological tissues

a) Human serum

Lycopene (**31**) occurs in tomatoes as the all-*E* isomer together with the previously overlooked 5*Z* isomer [41]. Processing of the tomatoes results in *E/Z* isomerization and higher bioavailability. Some ten geometrical isomers of lycopene (**31**) have been detected in human serum, including the 5*Z*, 9*Z*, 13*Z* and 15*Z* isomers [4]. The *Z* isomers accounted for around 50% of total lycopene (**31**). An even higher proportion (around 80%) has been found in prostate tissue [42]. An *E/Z* isomerization of lycopene (**31**) *in vivo* has been inferred from experiments involving diets supplemented with tomatoes devoid of these *Z* isomers [4].

For β,β-carotene (**3**) there is conflicting evidence concerning the absorption of the 9*Z* isomer, detected in serum. The isomerization from 9*Z* to all-*E in vivo* has been demonstrated in studies where [^{13}C]-labelled β,β-carotene (**3**) was administered orally to humans [43]. The kinetically favoured (13*Z*)-β,β-carotene (**3**) was detected in serum of patients suffering from erythropoetic porphyria, after injection with synthetic **3** [43].

Supplementation of humans with astaxanthin (**404-406**) (all-*E*:9*Z*:13*Z* ratio 7:9:17) resulted in an isomer ratio of 49:13:37 in serum, ascribed to selective absorption [44]

b) Salmonid fishes

The fish species investigated (*Salmo salar* and *Salvelinus alpinus*) selectively accumulated (all-*E*)-astaxanthin (**404-406**) in plasma and muscle and (13*Z*)-astaxanthin in the liver [45], when the diet was supplemented with an *E/Z* mixture (all-*E*:9*Z*:13*Z* ratio 75:3:22), in contrast to the results with human serum [44].

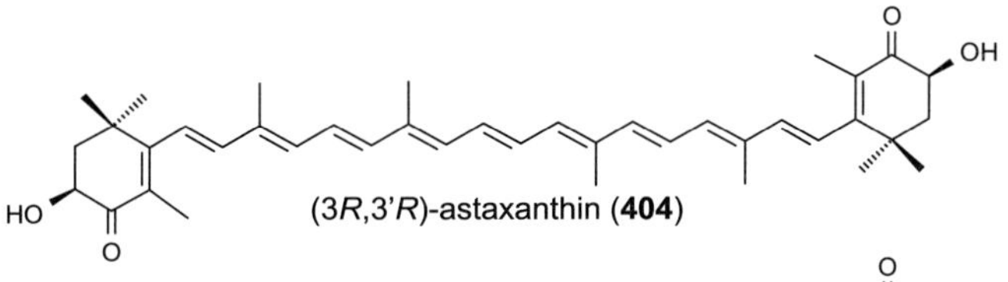

(3*R*,3'*R*)-astaxanthin (**404**)

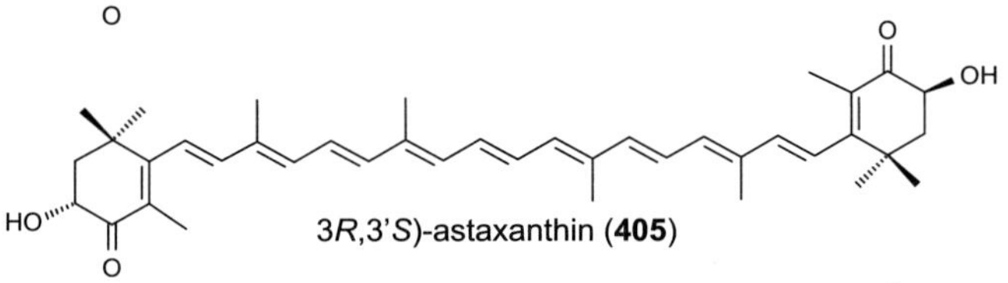

3*R*,3'*S*)-astaxanthin (**405**)

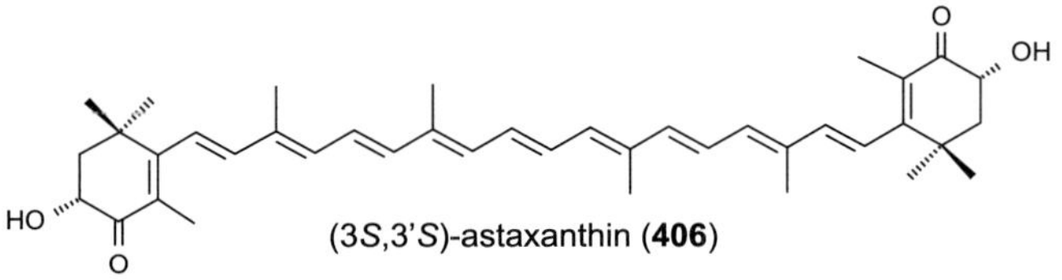

(3*S*,3'*S*)-astaxanthin (**406**)

4. Metabolic conversions

(all-*E*)-β,β-Carotene (**3**) is the best provitamin A and provides vitamin A (retinol) upon metabolic central or excentric cleavage (*Chapter 16* and *Volume 3, Chapter 6*). (9*Z*)-β,β-Carotene (**3**) is a less efficient provitamin A; around 25% of the efficiency of the all-*E* isomer has been reported for rats [46].

The sterically hindered (11Z)-retinal is involved in the visual process [47]. The biologically active (9Z)-retinoic acid can be formed *in vivo* from (9Z)-β,β-carotene (**3**) [48] (*Volume 5, Chapter 8*).

(9'Z)-Neoxanthin (**234**) is a biological precursor of the plant growth regulator abscisic acid, which retains the Z configuration (*Chapter 17* and *Volume 3, Chapter 4*).

5. Antioxidant properties

The antioxidant properties of carotenoids have received much attention (see *Volume 5, Chapter 12*). Recently, the effect of the *cis* configuration has been addressed. Astaxanthin (**404-406**) has strong antioxidant properties [49]. Mixtures of astaxanthin *E/Z* isomers appeared to be a better antioxidant *in vitro* than (all-*E*)-astaxanthin in model experiments with cod liver oil and fatty acid ethyl esters enriched in eicosapentaenoic acid and docosahexanoic acid [50].

6. Formation during food and feed processing

It is generally accepted that *E/Z* isomerization takes place during processing of food and feed where exposure to light, heat treatment, cooking, cell breakage, *etc.* are involved.

7. Conclusion

The view held until around 1970, that carotenoids generally occurred naturally as the all-*trans* isomer is now modified, and carotenoid *cis* isomers are given increasing importance in a biological context. Effects on functional, nutritional and health aspects deserve further attention, now that geometrical *E/Z* isomers of carotenoids are readily separated and characterized by chromatographic and spectroscopic methods.

References

[1]   L. Zechmeister, *Cis-trans Isomeric Carotenoids, Vitamins A and Arylpolyenes*, Springer Verlag, Wien (1962).
[2]   D. E. Holloway, M. Yang, G. Paganga, C. A. Rice-Evans and P. M. Bramley, *Free Radic. Res.*, **32**, 93 (2000).
[3]   Y. Koyama and Y. Mubai, in *Biomolecular Spectroscopy, Part B (Advances in Spectroscopy*, Vol. *21)* (ed. R. J. H. Clark and R. E. Hester), p. 49, Wiley, Chichester (1993).
[4]   S. J. Schwartz, in *Pigments in Food – A Challenge to Life Science* (ed. R. Carle, A. Schieber and F. S. Stintzing), p. 114, Shaker Verlag, Aachen (2006).
[5]   IUPAC Commision on the Nomenclature of Organic Compounds and IUB Commission on Biochemical Nomenclature, *Pure Appl. Chem.*, **41**, 405 (1975).

[6]   L. Pauling, *Fortschr. Chem. Org. Naturst.*, **3**, 203 (1939).

[7]   B. C. L. Weedon, in *Carotenoids*, (ed. O. Isler), p. 29, Birkhäuser, Basel (1971).

[8]   S. Liaaen-Jensen, *Pure Appl. Chem.*, **69**, 2027 (1997).

[9]   Y. Hu, H. Hashimoto, G. Moine, U. Hengartner and Y. Koyama, *J. Chem. Soc. Perkin Trans. 2*, **12**, 2699 (1997).

[10] J. Schierle, W. Bretzel, I. Buhler, N. Faccin, D. Hess, K. Steiner and W. Schüep, *Food Chem.*, **59**, 459 (1997).

[11] J. A. Haugan and S. Liaaen-Jensen, *Tetrahedron Lett.*, **35**, 2245 (1994).

[12] J. A. Haugan, G. Englert, E. Glinz and S. Liaaen-Jensen, *Acta Chem. Scand.*, **46**, 389 (1992).

[13] J. A. Haugan, G. Englert, T. Aakermann, E. Glinz and S. Liaaen-Jensen, *Acta Chem. Scand.*, **48**, 769 (1994).

[14] T. Refvem, A. Strand, B. Kjeldstad, J. A. Haugan and S. Liaaen-Jensen, *Acta Chem. Scand.*, **53**, 114 (1999).

[15] L. C. Sander, K. E. Sharpless, N. E. Craft and S. A. Wise, *Anal. Chem.*, **66**, 1667 (1994).

[16] B. F. Lutnaes, J. Krane and S. Liaaen-Jensen, *Org. Biomol. Chem.*, **2**, 2821 (2004).

[17] T. Glaser, A. Lienau, D. Zeeb, M. Krucker, M. Dachtler and K. Albert, *Chromatographia*, **57**, S/19 (2003).

[18] C. M. Bell, L. C. Sander and S. A. Wise, *J. Chromatogr. A*, **757**, 29 (1997).

[19] J. A. Haugan, T. Aakermann and S. Liaaen-Jensen, *Meth. Enzymol.*, **213**, 231 (1992).

[20] B. Ke, F. Imsgard, H. Kjoesen and S. Liaaen Jensen, *Biochim. Biophys. Acta*, **210**, 139 (1970).

[21] U. Hengartner, K. Bernhard, K. Meyer, G. Englert and E. Glinz, *Helv. Chim. Acta*, **75**, 1848 (1992).

[22] W. Vetter, G. Englert, N. Rigassi and U. Schwieter, in *Carotenoids* (ed. O. Isler), p. 189, Birkhäuser, Basel (1971).

[23] A. Strand, K. Kvernberg, A. M. Karlsen and S. Liaaen-Jensen, *Biochem. Syst. Ecol.*, **28**, 443 (2000).

[24] T.-R. Lindal and S. Liaaen-Jensen, *Acta Chem. Scand.*, **51**, 1128 (1997).

[25] A. Strand and S. Liaaen-Jensen, *Acta Chem. Scand.*, **52**, 1263 (1998).

[26] S. Hertzberg, G. Englert, P. Bergguist and S. Liaaen-Jensen, *Bull. Soc. Chim. Belg.*, **95**, 801 (1986).

[27] M. Rønnekleiv and S. Liaaen-Jensen, *Acta Chem. Scand.*, **46**, 1092 (1992).

[28] P. Molnár, J. Deli, F. Zsila, A. Steck, H. Pfander and G. Tóth, *Helv. Chim. Acta*, **87**, 11 (2004).

[29] R. Riesen, *Dr. phil. thesis.*, University of Bern, Bern (1991).

[30] P. Molnár, J. Deli, G. Tóth, A. Haeberli and H. Pfander, *Helv. Chim. Acta*, **85**, 1327 (2002).

[31] P. Molnár, T. Kortvelyesi, Z. Matus and J. Szabolcs, *J. Chem. Res.*, **4**, 120 (1997).

[32] A. Strand and S. Liaaen-Jensen, *J. Chem. Soc. Perkin Trans. 1*, **4**, 595 (2000).

[33] G. Gao, C. C. Wei, A. S. Jeevarajan and L. D. Kispert, *J. Phys. Chem.*, **100**, 5362 (1996).

[34] R. S. H. Liu, *Pure Appl. Chem.*, **74**, 1391 (2002).

[35] F. P. Zscheile, R. H. Harper and H. A. Nash, *Arch. Biochem.*, **5**, 211 (1944).

[36] J. H. Lupinski, *J. Phys. Chem.*, **67**, 2725 (1963).

[37] N. T. Ioffe, A. A. Engovatov and V. G. Mairanovskii, *Zh. Obshch. Khim.*, **46**, 1638 (1976).

[38] Z. He, G. Gao, E. S. Hand, L. D. Kispert, A. Strand and S. Liaaen-Jensen, *J. Phys. Chem. A*, **106**, 2520 (2002).

[39] S. Liaaen-Jensen, in *Carotenoids: Chemistry and Biology* (ed. N. I. Krinsky, M. M. Mathews-Roth and R. F. Taylor), p. 149, Plenum Press, New York (1989).

[40] M. Dachtler, T. Glaser, K. Kohler and K. Albert, *Anal. Chem.*, **73**, 667 (2001).

[41] D. S. McLaren, *Sight Life Newsletter*, **2**, 17 (2000).

[42] S. K. Clinton, C. Emenhiser, S. J. Schwartz, D. G. Bostwick, A. W. Williams, B. J. Moore and J. W. Erdman, *Cancer Epidemiol. Biomarkers Prev.*, **5**, 823 (1996).

[43] J. von Laar, W. Stahl, K. Bolsen, G. Goerz and H. Sies, *J. Photochem. Photobiol. B*, **33**, 157 (1996).

[44] M. Østerlie, B. Bjerkeng and S. Liaaen-Jensen, *J. Nutr. Biochem.*, **11**, 482 (2000).

[45] B. Bjerkeng and G. M. Berge, *Comp. Biochem. Physiol.*, **127B**, 423 (2000).

[46] A. R. Kemmerer and G. S. Fraps, *J. Biol. Chem.*, **161**, 305 (1945).

[47] K. Nakanishi, *Pure Appl. Chem.*, **63**, 161 (1991).

[48] R. S. Parker, *FASEB J.*, **10**, 542 (1996).

[49] H.-D. Martin, C. Jager, C. Ruck, M. Schmidt, R. Walsh and J. Paust, *J. Prakt. Chem.*, **341**, 302 (1999).

[50] M. Østerlie, B. Bjerkeng and S. Liaaen-Jensen, *Abstr. 13th International Symposium on Carotenoids, Honolulu*, p. 61 (2002).

Carotenoids
Volume 4: Natural Functions
© 2008 Birkhäuser Verlag Basel

Chapter 4

Three-dimensional Structures of Carotenoids by X-ray Crystallography

Madeleine Helliwell

A. Survey of Previously Reported Carotenoid Crystal Structures

The number of crystal structures of carotenoid molecules and carotenoid derivatives deposited in the Cambridge Crystallographic Data Centre [1] is still relatively small, but has increased compared with the previous survey [2]. The list is summarized in Table 1.

Table 1. Crystal structures of carotenoids in the Cambridge Crystallographic Data Centre (CCDC).

CCDC code	Carotenoid Name Formula	Ref
ASOTIJ	(all-*E*)-13-Demethyl-14-methyl-8'-apo-β-caroten-8'-al $C_{30} H_{40} O_1$	[3]
ASOTIL	(11*Z*)-13-Demethyl-14-methyl-8'-apo-β-caroten-8'-al $C_{30} H_{40} O_1$	[3]
ASOTUX	(*erythro*)-7,8-Didehydro-5,6-dihydro-11-apo-β-carotene-5,6,11-triol $C_{15} H_{24} O_3$	[3]
AZIJAU	(3*R*,3'*R*,6'*R*)-Lutein (**133**) 3,3'-di[(1*R*)-menthylcarbonate] $C_{62} H_{92} O_6$	[4]
AZIJEY	(3*R*,3'*R*)-Zeaxanthin (**119**) 3,3'-di[(1*R*)-menthylcarbonate] $C_{62} H_{92} O_6$	[4]
BAHYEP BAHYEP01	(all-*E*)-Astaxanthin (**404-406**) $C_{40} H_{52} O_4$	[5,6]

CANTHX10 CANTHX11 CANTHX12	Canthaxanthin (**380**) $C_{40} H_{52} O_2$	[6,7]
CAPSBB10	Capsanthin (**335**) di-*p*-bromobenzoate $C_{54} H_{62} Br_2 O_5$	[8]
CARTEN CARTEN01 CARTEN02	β-Carotene (**3**) $C_{40} H_{56}$	[3,9,10]
DCANTX10	15,15'-Didehydrocanthaxanthin $C_{40} H_{50} O_2$	[11]
DCAROT	15,15'-Didehydro-β-carotene $C_{40} H_{54}$	[12]
FCOXTN	Fucoxanthin (**369**) $C_{42} H_{58} O_6$	[13]
GIFSIX	8'-Apo-β-caroten-8'-al (**482**) [α-form, 6-s-*cis*] $C_{30} H_{40} O_1$	[14]
GIFSOD	8'-Apo-β-caroten-8'-al (**482**) [β-form, 6-s-*trans*] $C_{30} H_{40} O_1$	[14]
HCAROT	7,7'-Dihydro-β-carotene (**49**) $C_{40} H_{58}$	[15]
KUJGAX	N,N'',N'''':N',N''',N'''''-bis(2,2',2''-Nitrilotriethyl)-tris-(all-*E*)-8,8'-diapo-ψ,ψ-carotene-8,8'-diimine)dichloromethane solvate $C_{72} H_{96} N_8, C_1 H_2 Cl_2$	[16]
QAZHOO	1,8-bis-(1-Adamantyl)-1,3,5,7-octatetrayne: 2-butanone: (all-*E*)-8'-apo-β-caroten-8'-al (**482**) clathrate $C_{28} H_{30}, C_4 H_8 O_1, 0.01(C_{30} H_{40} O_1)$	[17]
TUKTEY	Bixin (**533**) $C_{25} H_{30} O_4$	[18]
TUKTIC ZZZSGW ZZZSGW01	Methyl bixin (**535**) $C_{26} H_{32} O_4$	[18-20]
VOLYEA	Jiocarotenoside A1 monohydrate $C_{21} H_{34} O_9, H_2 O_1$	[21]
XEZJAO	Astaxanthin (**404-406**) chloroform solvate $C_{40} H_{52} O_4, 2(C_1 H_1 Cl_3)$	[6]
XEZJES	Astaxanthin (**404-406**) pyridine solvate $C_{40} H_{52} O_4, 2(C_5 H_5 N_1)$	[6]
XEZJIW	(3*R*,3'*S*,*meso*)-Zeaxanthin (**120**) trihydrate $C_{40} H_{56} O_2, 3(H_2 O_1)$	[6]
YOYXOZ	Dimethylcrocetin (**539**) $C_{22}H_{28}O_4$	[22]
ZZZMBY	β-Carotenone (**562**) $C_{40} H_{56} O_4$	[23]

The review of crystallographic studies of carotenoids in *Volume 1B* [2] describes carotenoid crystal structures determined up to about 1992. There are some general features, that are also seen in carotenoid structures determined later. (i) The all-*trans* carotenoids have a pronounced S shape, which is thought to arise from the steric crowding of the methyl groups along the polyene chain, which can itself be slightly bent. (ii) The bond lengths alternate; the difference between the single-bond and double-bond lengths tends to decrease towards the centre of the polyene chain, because of the greater conjugation towards the centre, and to be more pronounced when the polyene chain is more nearly planar. (iii) In most cases, the ring-chain conformation of β end groups has been found to be 6-s-*cis*, attributed to steric crowding between the ring methyl substituents and the hydrogen atoms bonded to C(7) and C(8). Calculations on the all-*E* carotenoids β-carotene (**3**), zeaxanthin (**119**), astaxanthin (**404-406**) and canthaxanthin (**380**) predict the most stable structures to have the 6-s-*cis* conformation twisted out of the plane of the polyene chain by between 40° and 50°, given by the torsion angle about C(6)-C(7) [5]. For violaxanthin (**259**), which has two C(5)-C(6) epoxide groups, and tunaxanthin (**150**), which has two ε end groups, the predicted torsion angles about C(6)-C(7) are -84.8° and -106.7° respectively. The crystal structures of the latter two carotenoids have not been determined, but predicted ring-chain conformations are generally borne out by experiment, with a few exceptions. Of those for which crystal structures have been determined to reasonable precision, exceptions are the β-form of 8'-apo-β-caroten-8'-al (**482**) (GIFSOD) [14], which has a (6-s-*trans*) conformation and C(5)-C(6)-C(7)-C(8) torsion angle of -158.4(8)°, zeaxanthin (**119**) 3,3'-di-[(1*R*)-menthylcarbonate] (AZIJEY) [4], which has the 6-s-*trans* conformation at one end and 6-s-*cis* at the other, with C(5)-C(6)-C(7)-C(8) torsion angles of 144.5(6)° and 48.5(8)° respectively, lutein (**133**) 3,3'-di-[(1*R*)-menthylcarbonate] (AZIJAU) where the respective torsion angles for the β and ε end groups are -68.8(5)° and 118.0(4)° [4], and (3*R*,3'*S*, *meso*)-zeaxanthin (**120**) with a torsion angle of -74.9(3)° [6]. These deviations from the expected values in the solid state are probably due to crystal packing forces. It seems likely from spectroscopic evidence that, in solution, the more stable 6-s-*cis* conformation will be preferred, but this has not been determined experimentally [4,6].

β-carotene (**3**)

(3*R*,3'*R*)-zeaxanthin (**119**)

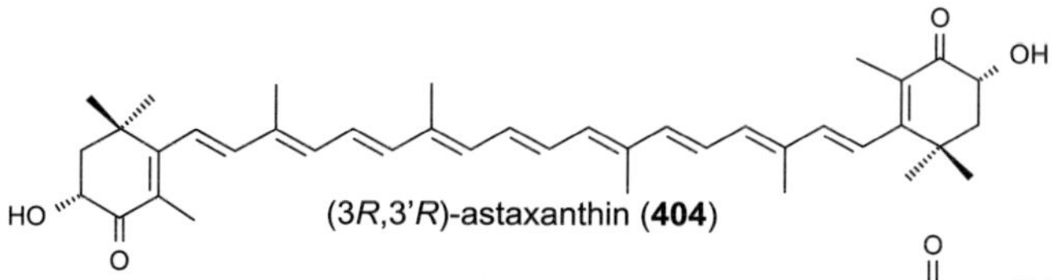

(3R,3'R)-astaxanthin (**404**)

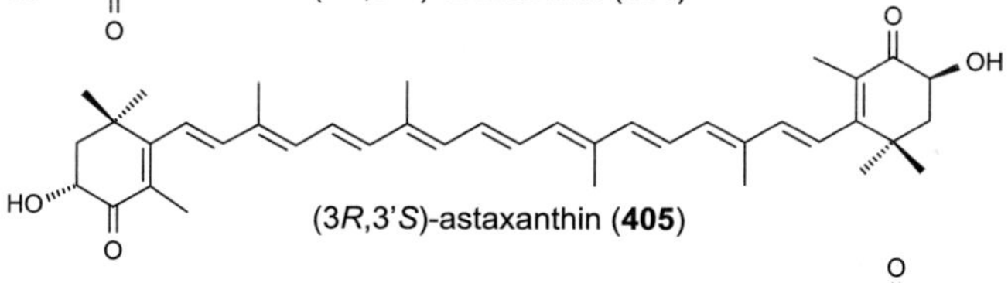

(3R,3'S)-astaxanthin (**405**)

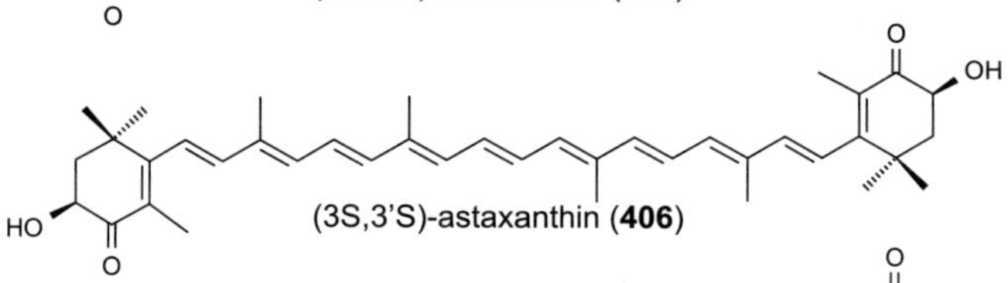

(3S,3'S)-astaxanthin (**406**)

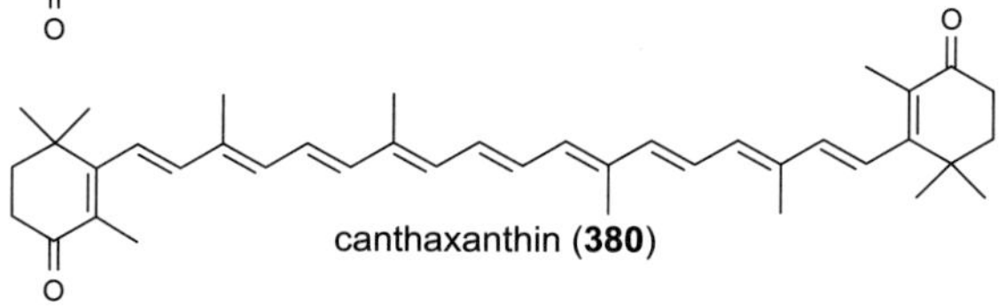

canthaxanthin (**380**)

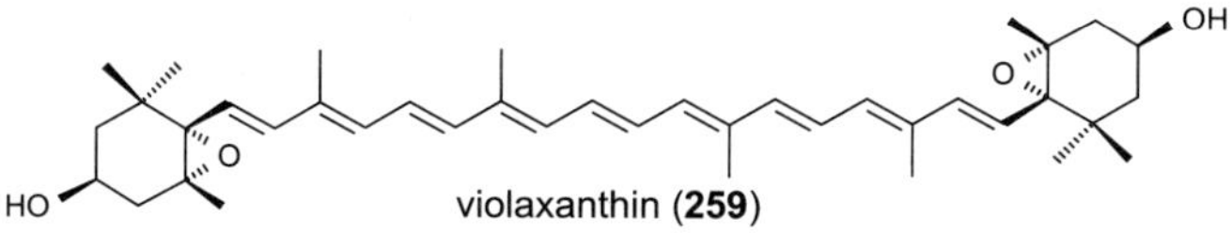

violaxanthin (**259**)

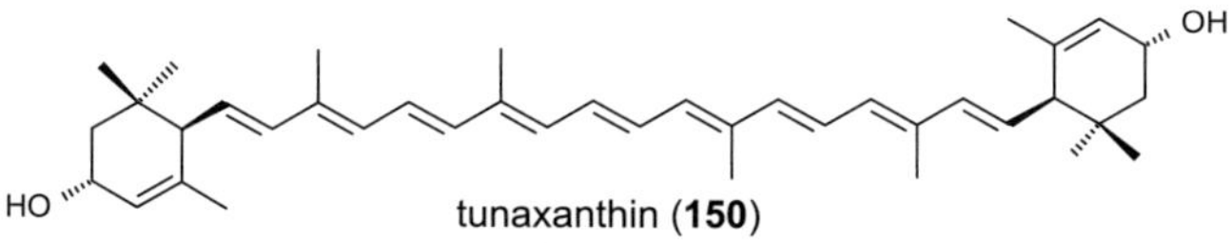

tunaxanthin (**150**)

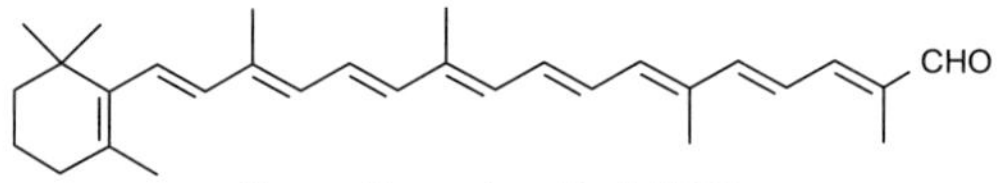

8'-apo-β-caroten-8'-al (**482**)

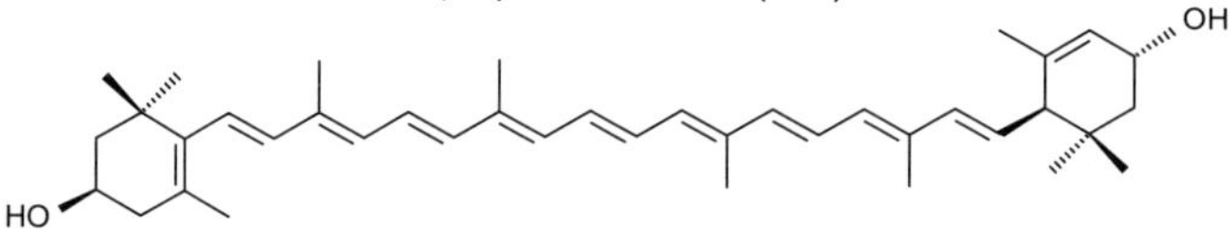

(3R,3'R,6'R)-lutein (**133**)

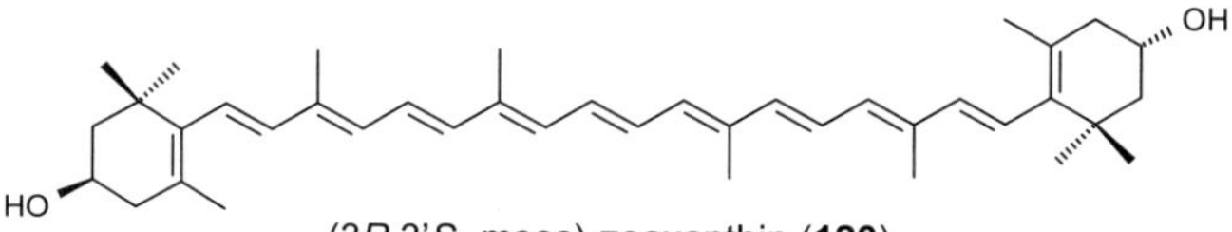

(3R,3'S, *meso*)-zeaxanthin (**120**)

B. New Experimental Methods

1. Methods for crystallization of free carotenoids

Carotenoids tend to be rather difficult to crystallize and, when they do, they often produce weakly diffracting crystals which may be too small for standard laboratory X-ray equipment [2]. In some of the reports of carotenoid crystal structures, the crystallization methods have not been described, but methods that have been reported include slow evaporation techniques, for example from solvents such as benzene for 15,15'-didehydrocanthaxanthin (*1*) [11] and dimethylformamide/benzene for astaxanthin (**404-406**) [5] and canthaxanthin (**380**) [7].

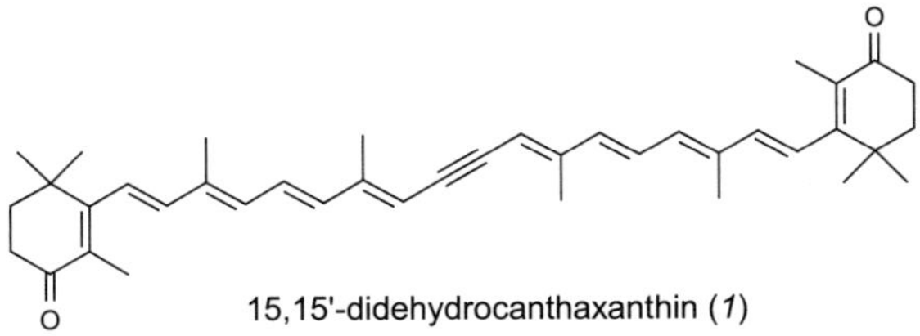

15,15'-didehydrocanthaxanthin (*1*)

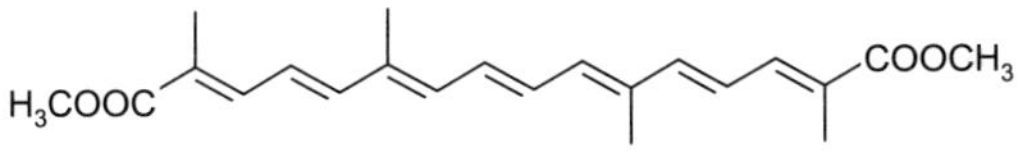

dimethylcrocetin (**539**)

In the confirmation of the structures of the (3*R*,3'*R*) isomers of (6'*R*)-lutein (**133**) and zeaxanthin (**119**), it was necessary to crystallize the (-)-(1*R*)-menthylcarbonate ester derivatives in order to obtain crystals large enough for structure determination by X-ray crystallography; these derivatives were recrystallized from dichloromethane/methylcyclo-hexane and hot tetrahydrofuran/acetonitrile combinations, respectively [4]. Dimethylcrocetin (**539**) was crystallized from dichloromethane and diethyl ether at 4°C [22].

A programmme of crystallization of carotenoids begun recently in the author's laboratory has had particular success with vapour diffusion techniques from solvent combinations such as dichloromethane/hexane, chloroform/hexane, pyridine/water, and pyridine/hexane, often in conjunction with slowing the crystallization by using 'pin hole' punctured covers or reduced temperatures in the refrigerator.

2. Data collection

Generally, the crystal structures have been obtained by use of standard laboratory X-ray diffractometers, with sealed tube Mo Kα or Cu Kα sources. In the author's laboratory, a

CCD diffractometer with a sealed tube Mo Kα X-ray source has been used, with cryocooling to 100 K. The use of cryocooling is particularly important since carotenoid molecules tend to show relatively high atomic displacement parameters (adps) [2], causing the diffraction pattern to be weak, particularly at high resolution. Also, mixtures of isomers which may not be seen clearly at room temperature can be resolved into separate disordered components at 100-150 K as is shown when the room temperature structures of β-carotene (3), CARTEN [9] and CARTEN02 [3], are compared with the 130 K structure, CARTEN01 [10], which showed disordered sites of the C(2) and C(3) positions in a 1:1 ratio. Likewise, since synthetic astaxanthin is a mixture of the (3R,3'R), (3R,3'S), and (3S,3'S) isomers (404, 405, 406, respectively) the crystals also contain a mixture of isomers, resulting in disorder of several atoms of the end groups, as was clearly seen at 100 K in the structures XEZJAO and XEZJES [6]. This disorder was not seen in the room temperature determination BAHYEP [5], although it is implied by an anomalous C(2)-C(3) bond length. In the case of particularly weakly diffracting samples, high intensity synchrotron radiation techniques may be necessary. Strong data have been collected for the crystal structures of 7,8-didehydroastaxanthin (402), 7,8,7',8'-tetradehydroastaxanthin (400) and capsorubin (413) [24].

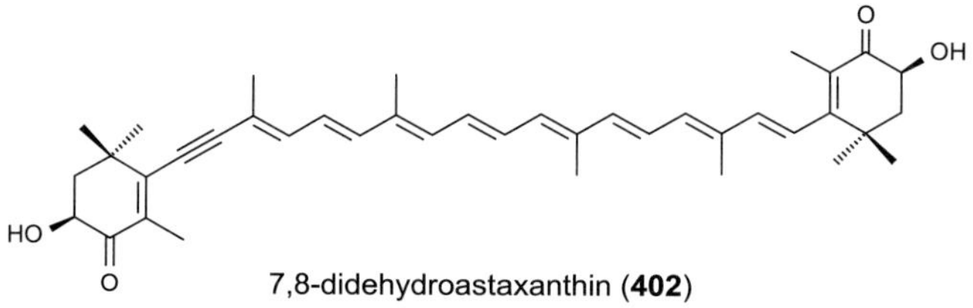

7,8-didehydroastaxanthin (402)

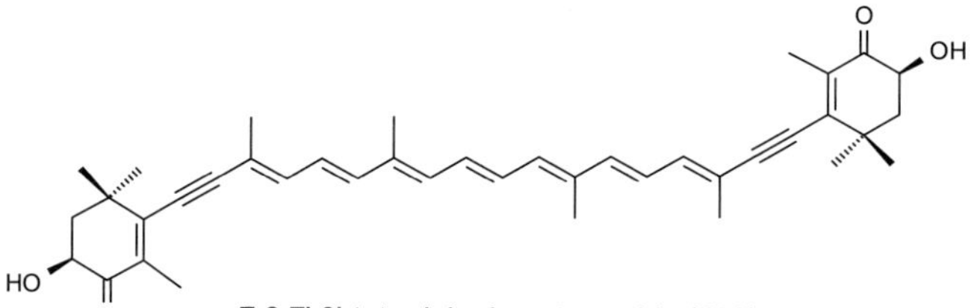

7,8,7',8'-tetradehydroastaxanthin (400)

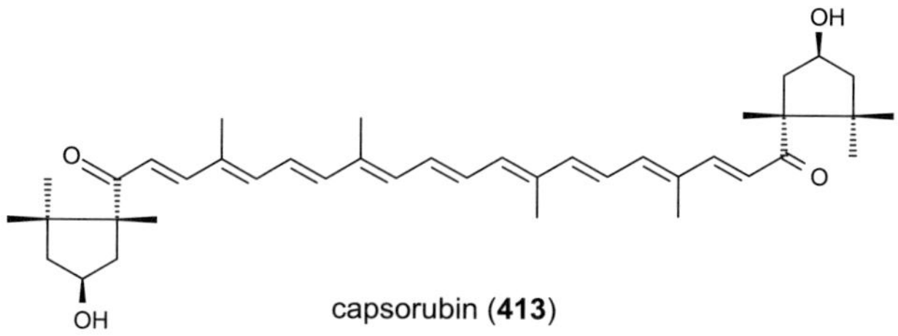

capsorubin (413)

C. New Determinations of Crystal Structures

1. Astaxanthin and related xanthophylls

The crystal structures were determined of three crystal forms of astaxanthin (**404-406**), namely a chloroform solvate, a pyridine solvate and an unsolvated crystal form, and of the related molecules, canthaxanthin (**380**) and (3R,3'S, *meso*)-zeaxanthin (**120**) [6] (Fig. 1). In all these cases, the molecules were found to adopt the 6-*s cis* conformation. The main difference between the conformations arose from the torsion angle between the polyene chain and the end rings which varied by a small amount (from -43° to -53°) between all the structures, and the previously determined structure of β-carotene (**3**) [3,9,10], except for (3R,3'S)-zeaxanthin (**120**), where this angle was quite different at -74.9(3)° [6]. Although the crystals were all distinctly red in colour, the solid state UV/Vis spectrum of (3R,3'S)-zeaxanthin was shifted slightly to shorter wavelength compared with that of β-carotene, which has exactly the same length of polyene chain as zeaxanthin. In solution, the UV/Vis spectra of (3R,3'S)-zeaxanthin and β-carotene are virtually identical, suggesting that, in solution, they have very similar conformations.

Fig. 1 (a)

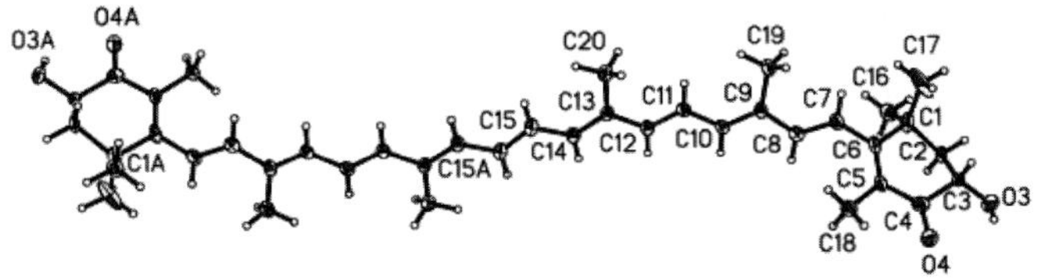

Fig. 1 (b)

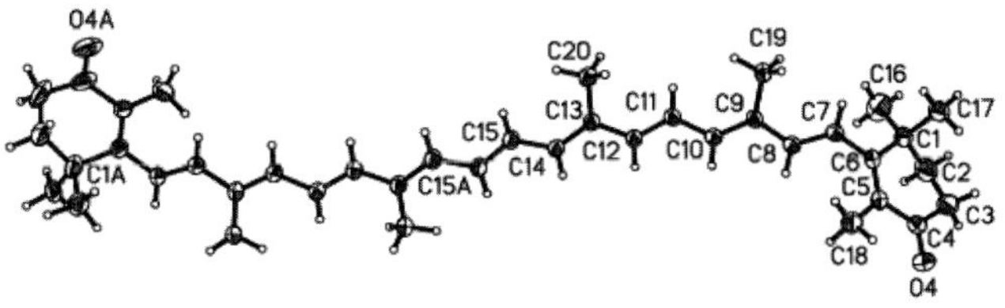

Fig. 1 (c)

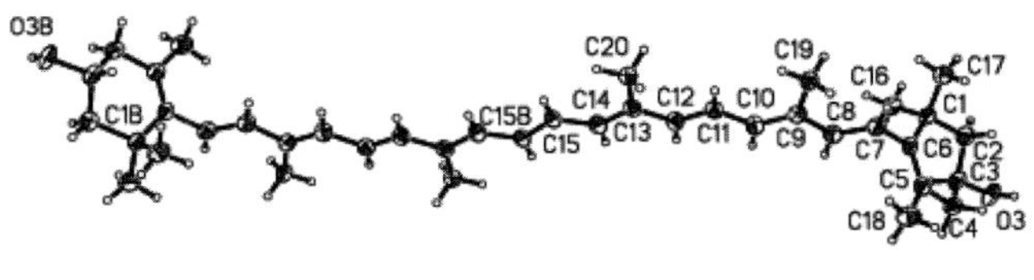

Fig 1 (d)

Fig. 1. Crystal structures of (a) the chloroform solvate of astaxanthin (**404-406**), using 50% probability ellipsoids. The chloroform solvate molecule and the disordered atoms have been omitted for clarity. (b) canthaxanthin (**380**) at 100 K, using 50% probability ellipsoids; the disordered atoms have been omitted for clarity. (c) (3*R*,3'*S*, *meso*)-zeaxanthin (**120**) using 50% probability ellipsoids; the disordered water solvent atoms have been omitted. (d) (3*R*,3'*S*, *meso*)-zeaxanthin (**120**) viewed approximately down the plane of the polyene chain, showing that the end rings are twisted out of the plane of the polyene chain [the C(5)-C(6)-C(7)-C(8) torsion angle is -74.9(3)°].

The chloroform and pyridine solvates of astaxanthin (**404-406**) both form chains of molecules by pairwise end-to-end hydrogen bonding of the hydroxyl and keto oxygen atoms with hydrogen bonding distances of 2.790(3) Å and 2.829(6) Å respectively (Figs. 2a, 2b). In addition, in the chloroform solvate, there is a particularly strong interaction of the chloroform hydrogen atom with the hydroxyl oxygen atom and, in the pyridine solvate, there are weak C-H hydrogen-bond interactions of the pyridine H atoms with the hydroxyl or keto oxygen atoms of the astaxanthin molecules.

Fig. 2 (a)

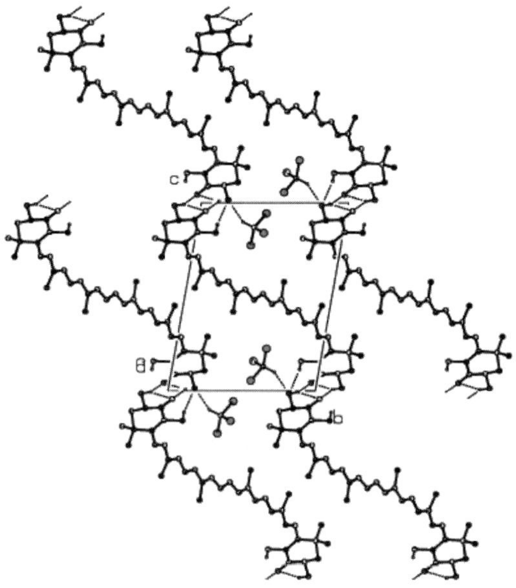

Fig. 2 (b)

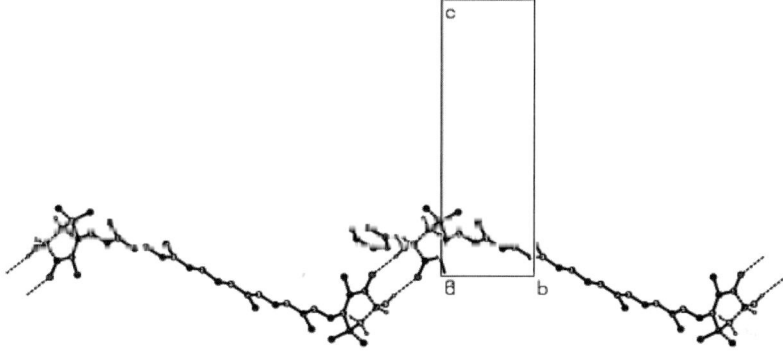

Fig. 2(c)

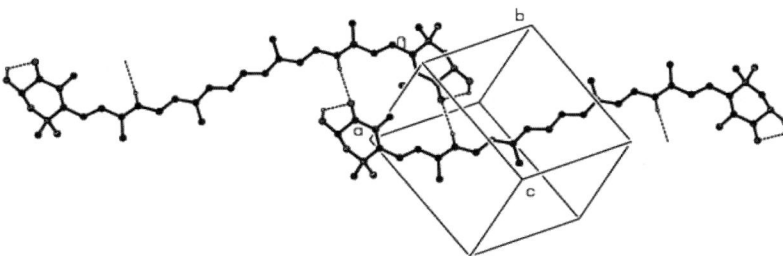

Fig. 2 (d)

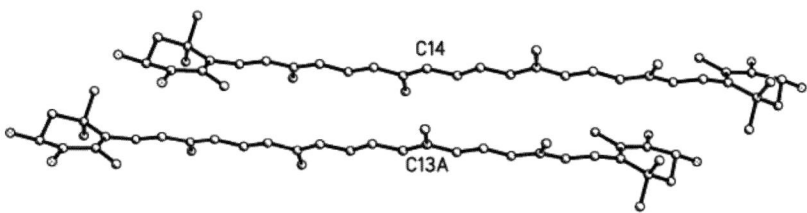

Fig. 2. Packing arrangements of the carotenoid crystal structures; H atoms not involved in H-bonding have been omitted for clarity. From [6] with the permission of IUCr Journals. (a) Packing of the chloroform solvate of astaxanthin (**404-406**) viewed down a. (b) Packing arrangement of the pyridine solvate of astaxanthin viewed down a. (c) Packing arrangement of the unsolvated crystal form of astaxanthin. (d) Plot showing the π-π stacking interactions of the chloroform solvate of astaxanthin.

For the unsolvated astaxanthin structure, the hydroxyl and keto oxygens form an intramolecular hydrogen bond of length 2.656(4) Å and the molecules are linked into chains by much weaker C-H hydrogen bonds (Fig. 2c) similar to those found in the structure of canthaxanthin (**380**) [6]. In addition, for each astaxanthin structure there are further π-π stacking interactions bringing the molecules into close proximity (between 3.61 and 3.79 Å), with the molecules one above the other (Fig. 2d).

The diacetates of (6-s-*cis*)- and (6-s-*trans*)-astaxanthin (**404-406**) have been prepared and used to investigate whether the s-*cis* or s-*trans* conformation has any effect on the colour of the crystals [24]. Recrystallization yielded crystals of the (6-s-*cis*) and (6-s-*trans*) isomers, so that the crystal structures of each could be determined and compared with those of free astaxanthin as well as the astaxanthin molecules bound in the carotenprotein β-crustacyanin. Structural plots of the s-*cis* and s-*trans* ester are shown in Figs. 3 and 4.

Fig. 3 (a)

Fig. 3 (b)

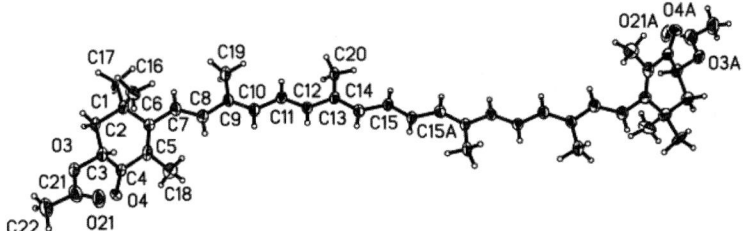

Fig. 3. (a) ORTEP plot of the crystal structure of the diacetate of 6-s-*cis*-astaxanthin (**404-406**); only the highest occupancy disordered component is shown and solvent molecules are omitted for clarity; (b) plot viewed down the plane of the polyene chain.

Fig. 4 (a)

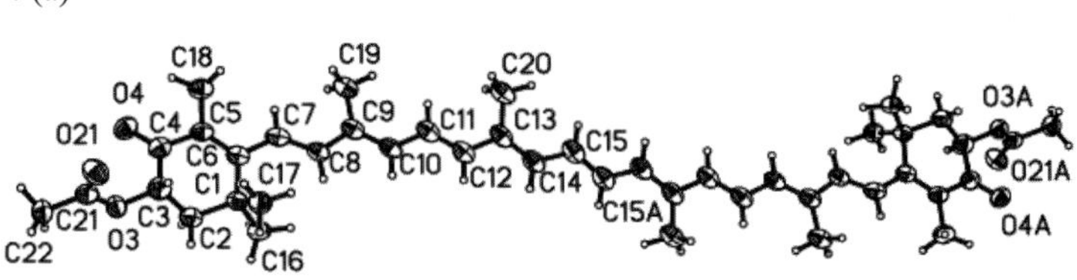

Fig. 4 (b)

Fig. 4. (a) ORTEP plot of the crystal structure of the diacetate of (6-s-*trans*)-astaxanthin (**404-406**); only the highest occupancy disordered component is shown and solvent molecules are omitted for clarity; (b) plot viewed down the plane of the polyene chain.

In general, the geometric parameters of the (6-s-*cis*)- and (6-s-*trans*)-astaxanthin esters agree closely with each other, except for their C(5)-C(6)-C(7)-C(8) torsion angles, which are greatly different. For the 6-s-*cis* isomer, the angle is -49.0(5)°, similar to the angles found for free astaxanthin (**404-406**), canthaxanthin (**380**) and β-carotene (**3**) [6] (Fig. 3c). For the 6-s-*trans* ester, this angle is 178.3(3)°, indicating that the end rings are approximately coplanar with the polyene chain (Fig. 4b), a conformation predicted to be the least favourable of the possible conformations [5]. Interestingly, however, the conformation of astaxanthin in β-crustacyanin (*Chapter 6*) is also 6-s-*trans*, and the C(5)-C(6)-C(7)-C(8) torsion angles are 172.9(2)° and 177.1(2)° for the two ends of one of the two bound astaxanthin molecules, both being similar [25]; the main difference between the protein-bound astaxanthin molecules and the (6-s-*trans*)-astaxanthin ester is the distinct bowing of the polyene chain in the former (Fig. 5).

Fig. 5 (a)

Fig. 5 (b)

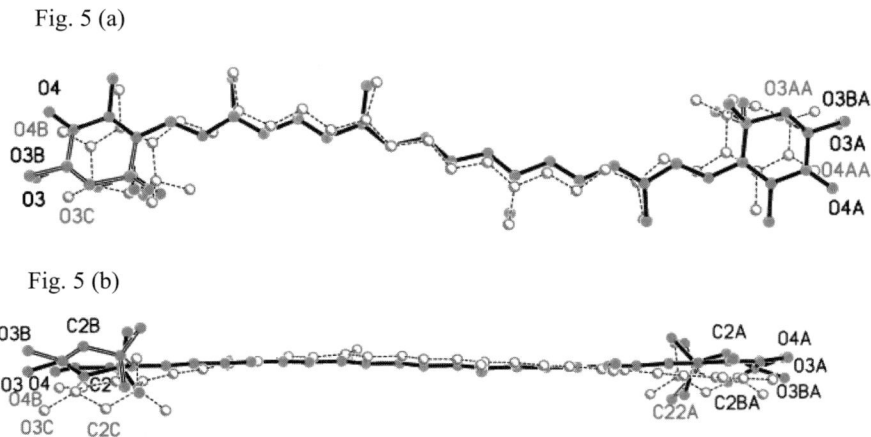

Fig. 5. Comparison of the crystal structure of the diacetate of (6-s-*trans*)-astaxanthin (**404-406**) (in red, bold, omitting the ester groups for clarity) with that of astaxanthin bound in the protein β-crustacyanin (in blue, dotted): (a) looking down on the polyene chain; (b) viewed down the plane of the polyene chain showing the bowing of the protein-bound astaxanthin polyene chain.

The crystal structures of 7,8-didehydroastaxanthin (**402**) (Fig. 6) and 7,8,7',8'-tetradehydro-astaxanthin (**400**) (Fig. 7) also have the β-ring end groups almost coplanar with the polyene chain; the dihedral angles between the polyene chain and the end groups are 4.8(1)° and 14.4(1)° for 7,8-didehydroastaxanthin (**402**) and 17.1(1)° and 6.3(1)° for 7,8,7',8'-tetra-dehydroastaxanthin (**400**). The conformation of the end groups is 6-s-*cis* for **402** (Fig. 6a) and 6-s-*trans* for **400** [24] (Fig. 7a).

Fig. 6 (a)

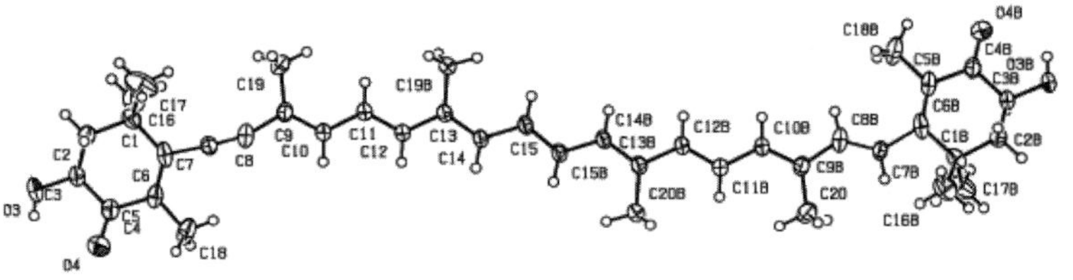

Fig. 6 (b)

Fig. 6. Plot of the crystal structure of 7,8-didehydroastaxanthin (**402**) showing (a) the 6-s-*cis* conformation of the molecule and (b) the end rings coplanar with the polyene chains. Disordered atoms have been omitted for clarity.

Fig. 7 (a)

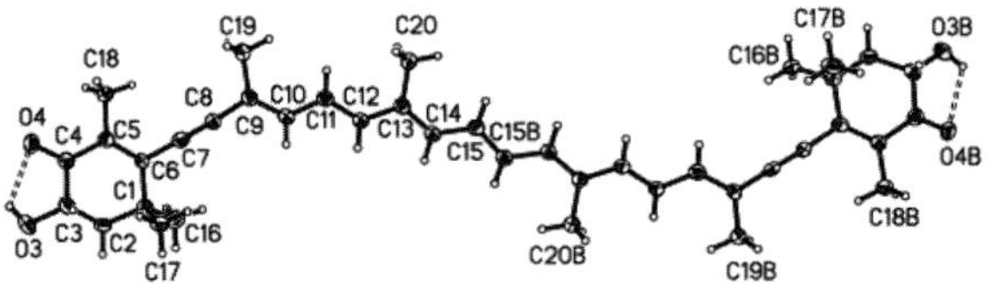

Fig. 7 (b)

Fig. 7. Plot of the crystal structure of 7,8,7',8'-tetradehydroastaxanthin (400) showing (a) the 6-s-trans conformation of the molecule and (b) the end rings almost coplanar with the polyene chains.

The UV/Vis spectra in the solid state of the crystals of each astaxanthin (**404-406**) diester isomer and of 7,8-didehydroastaxanthin (**402**) were obtained. Surprisingly, all absorb at around the same wavelength, 500 nm, and the crystals are clearly red in colour. The change of the molecule from the 6-s-*cis* conformation to the planar 6-s-*trans* conformation of the astaxanthin ester, and the planar 6-s-*cis* conformation of 7,8-didehydroastaxanthin lead to an extension of conjugation into the rings, and therefore would be expected to lead to a bathochromic shift, but it appears that this effect is smaller than expected [25]. The acetylenic analogues of astaxanthin do absorb at slightly longer wavelength in solution.

Overall, differences in ring chain conformation of the free carotenoids, or the various packing arrangements, including the close approach of neighbouring molecules in various configurations, seem to have very little effect on the colour of the crystals, leading to the conclusion that such effects do not contribute much to the bathochromic shift seen with the crustacyanin proteins (*Chapter 6*).

2. Carotenes

The recent systematic programme of crystallizing carotenoids (Section **C**) has led to a number of new carotene structures [26,27]. The crystal structure of (all-*E*)-β-carotene (**3**) has been determined a number of times [3,9,10] and now the crystal structures of the 13*Z* and 15*Z* isomers have been determined (Fig. 8) [26].

In the structure of (13*Z*)-β-carotene (Fig. 8a), the asymmetric unit contains the whole molecule; the methyl groups are all on the same side of the polyene chain and the ring-chain conformation is 6-s-*cis*, bent out of the plane of the polyene chain by 47.6(4)° and 52.2(4)°, respectively, for the two ends. (15*Z*)-β-Carotene (Fig. 8b) is symmetrical, with a 2-fold axis relating the two halves of the molecule. The methyl groups are again arranged on the same side of the polyene chain, the conformation is 6-s-*cis,* and the C(5)-C(6)-C(7)-C(8) torsion angle is -41.4(2)°. In both crystal structures, the cyclohexene end rings are in a half-chair conformation.

Fig. 8 (a)

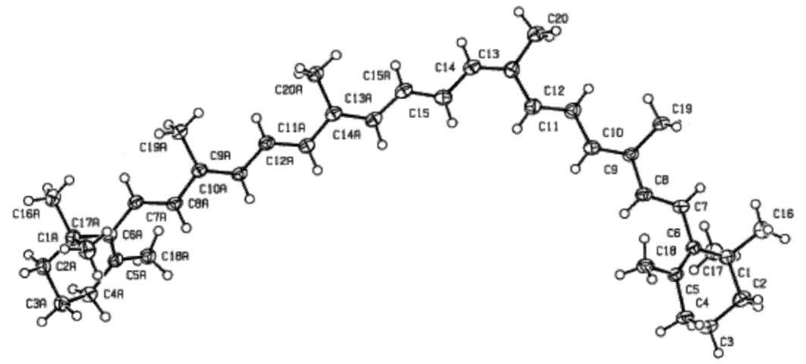

Fig. 8 (b)

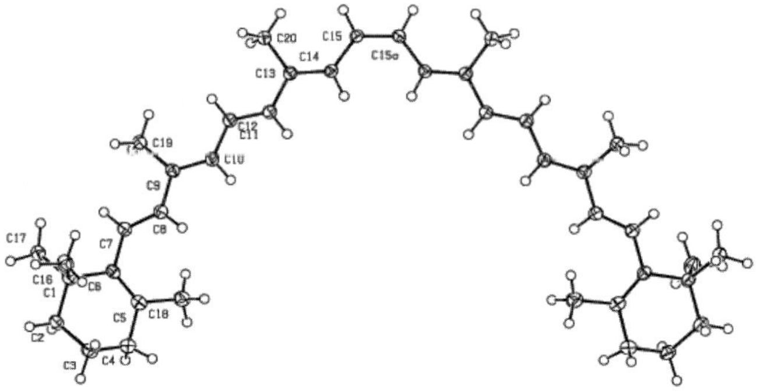

Fig. 8. Plots of the crystal structures of (a) (13*Z*)-β-carotene and (b) (15*Z*)-β-carotene.

The crystal structures of two polymorphs of 20-nor-β-carotene (*2*) have been determined in the monoclinic space group P-2₁ and the triclinic space group P1̄ [27]. In the P-2₁ polymorph (Fig. 9), the asymmetric unit consists of the whole molecule, and there is disorder because the molecules can pack either way round, *i.e.* so that the 20′-methyl group is in the left or the right half of the molecule. The conformation is 6-s-*cis* and the two rings are bent out of the plane of the polyene chain by -53.2(8)° and 47.3(8)° respectively. In the P1̄ polymorph, the asymmetric unit again consists of the whole molecule and the overall backbone is fairly similar to that of the P-2₁ polymorph and to that of β-carotene. Disorder of the 20′-methyl group is again observed, but to a much lower degree, and additional disorder of the C(2) and C(3) atoms of one of the β rings is also seen. The conformation is 6-s-*cis* and the C(5)-C(6)-C(7)-C(8) and C(5A)-C(6A)-C(7A)-C(8A) torsion angles are -43.6(3)° and 56.1(3)°, respectively. The packing arrangements of the two polymorphs are quite different, because of the differing space group symmetries [27].

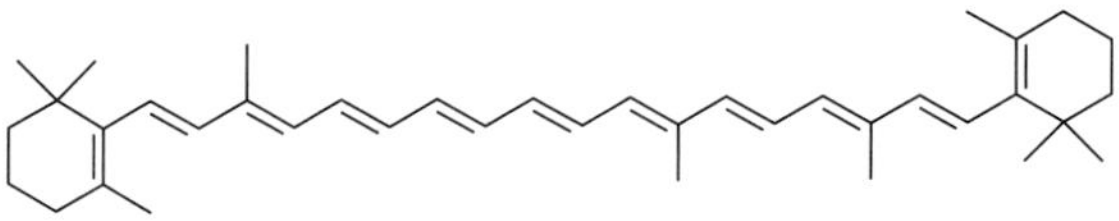

20-nor-β-carotene (*2*)

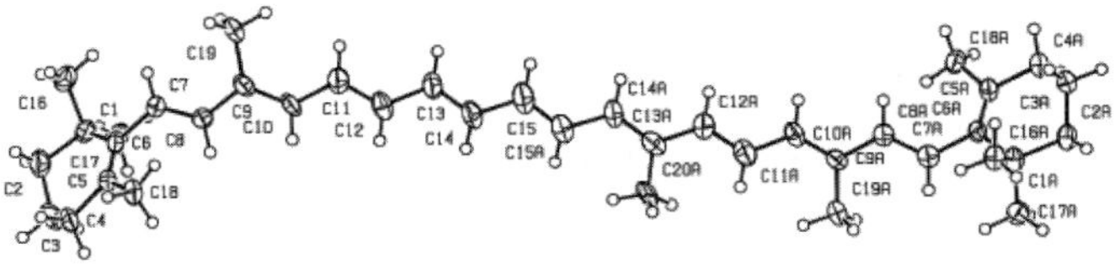

Fig. 9. Plot of the P2$_1$ polymorph crystal structure of 20-nor-β-carotene (2); disordered atoms have been omitted for clarity.

D. Conclusions

The coverage of X-ray crystallographic studies in *Volume 1* [2] drew attention to the fact that, because of the difficulty of obtaining carotenoid crystals of suitable size and quality, fewer X-ray structures had been determined for carotenoids than for most other groups of substances. It was predicted, however, that the improvement of crystallization techniques and the advent of synchrotron radiation sources would make possible structural studies that were previously not feasible. The advances reported in this Chapter validate this prediction. Major advances are also being made in the crystallization and structure determination of carotenoid-protein complexes. Structural studies of the lobster carotenoprotein β-crustacyanin are described in *Chapter 6*, and X-ray structures of photosynthetic pigment-protein complexes in *Chapter 14*. The technology is now in place to allow continuing advances in this area.

References

[1]   F. H. Allen, *Acta Crystallogr.*, **B58**, 380 (2002).
[2]   F. Mo, in *Carotenoids Volume 1B:Spectroscopy*, (ed. G. Britton, S. Liaaen-Jensen and H. Pfander) p. 321, Birkhäuser, Basel (1995).
[3]   M. B. Hursthouse, S. C. Nathani, G. P. Moss and M. Hadorn, *Personal Communication*.
[4]   A. Linden, B. Burgi and C. H. Eugster, *Helv. Chim. Acta*, **87**, 1254 (2004).
[5]   H. Hashimoto, T. Yoda, T. Kobayashi and A. J. Young, *J. Mol. Struct.*, **604**,125 (2002).
[6]   G. Bartalucci, J. Coppin, S. Fisher, G. Hall, J. R. Helliwell, M. Helliwell and S. Liaaen-Jensen, *Acta Cryst.*, **B63**, 328 (2007).
[7]   J. C. J. Bart and C. H. MacGillavry, *Acta Cryst.*, **B24**, 1587 (1968).
[8]   I. Ueda and W. Nowacki, *Z. Kristallogr. Kristallgeom. Kristallphys. Kristallchem.*, **140**, 190 (1974).
[9]   C. Sterling, *Acta Cryst.*, **17**, 1224 (1964).
[10]   M. O. Senge, H. Hope and K. M. Smith, *Z. Naturforsch.*, **47**, 474 (1992).
[11]   J. C. J. Bart and C. H. MacGillavry, *Acta Cryst.*, **B24**, 1569 (1968).
[12]   W. G. Sly, *Acta Cryst.*, **17**, 511 (1964).
[13]   G .P. Moss, *Pure Appl. Chem.*, **51**, 507 (1979).
[14]   G. Drikos, H. Dietrich and H. Ruppel, *Eur. Biophys. J.*, **16**, 193 (1988).
[15]   C. Sterling, *Acta Cryst.*, **17**, 500 (1964).
[16]   J.-M. Lehn, J.-P. Vigneron, I. Bkouche-Waksman, J. Guilhem and C. Pascard, *Helv. Chim. Acta*, **75**, 1069 (1992).
[17]   T. Muller, J. Hulliger, W. Seichter, E. Weber, T. Weber and M. Wubbenhorst, *Chem.-Eur. J.*, **6**, 54 (2000).

[18]   D. R. Kelly, A. A. Edwards, J. A. Parkinson, G. Olovsson, J. Trotter, S. Jones, K. M. A. Malik, M. B. Hursthouse and D. E. Hibbs, *J. Chem. Res.*, **446**, 2640 (1996).

[19]   J. Hengstenberg and R. Kuhn, *Z. Kristallogr. Kristallgeom. Kristallphys. Kristallchem.*, **76**, 174 (1931).

[20]   H. Waldmann and E. Brandenberger, *Z. Kristallogr. Kristallgeom. Kristallphys. Kristallchem.*, **82**, 77 (1932).

[21]   H. Sasaki, H. Nishimura, T. Morota, T. Katsuhara, M. Chin and H. Mitsuhashi, *Phytochemistry*, **30**, 1639 (1991).

[22]   P. A. Tarantilis, M. Polissiou, D. Mentzafos, A. Terzis and M. Manfait, *J. Chem. Cryst.*, **24**, 739 (1994).

[23]   R. F. Hunter, T. R. Lomer, V. Vand and N. E. Williams, *J .Chem. Soc.*, 710 (1948).

[24]   G. Bartalucci, S. Fisher, M. Helliwell, J. R. Helliwell, S. Liaaen-Jensen, J. E. Warren and J. Wilkinson, unpublished results.

[25]   M. Cianci, P. J. Rizkallah, A. Olczak, J. Raftery, N. E. Chayen, P. F. Zagalsky and J. R. Helliwell, *Proc. Natl. Acad. Sci. USA.*, **99**, 9795 (2002).

[26]   G. Bartalucci, C. Delroy, S. Fisher, M. Helliwell and S. Liaaen-Jensen, *Acta Cryst.*, **C64**, o128 (2008).

[27]   M. Helliwell, S. Liaaen-Jensen and J. Wilkinson, *Acta Cryst.*, **C64**, o252 (2008).

Carotenoids
Volume 4: Natural Functions
© 2008 Birkhäuser Verlag Basel

Chapter 5

Aggregation and Interface Behaviour of Carotenoids

Sonja Köhn, Henrike Kolbe, Michael Korger, Christian Köpsel, Bernhard Mayer, Helmut Auweter, Erik Lüddecke, Hans Bettermann and Hans-Dieter Martin

A. Introduction

1. Molecular aggregates

Molecular aggregates attract considerable attention, as they bridge the gap between the physics of single molecules and structurally ordered crystals. Molecular self-assembly in biological systems is highly specific and fundamentally important for correct functioning in living organisms.

Molecular aggregation of cyanine dyes has been examined extensively since 1936, because cyanine or polymethine dyes in general are among the best known self-associating dyes and have many important and outstanding applications [1-5]. These early studies showed that the aggregation of the cyanine dyes in aqueous solutions often results in the appearance of a very intense narrow light-absorption band, red-shifted (to longer wavelength, bathochromic) with respect to the monomer band. This absorption band was named the J-band, and the system a J-aggregate, after **J**elly, a pioneer in the research. Another form, the H-aggregate, is usually characterized by a broader, blue-shifted (to shorter wavelength, **H**ypsochromic) absorption band. Dimers, the smallest members, often display both a strong blue-shifted band and a band of variable intensity, red-shifted relative to the absorption band of the monomer.

The formation of aggregates is forced by weak and reversible bonding by H-bridges, dipole forces, van der Waals interactions and hydrophobic effects of hydrophobic molecules, their polar substitutents and the surrounding solvent, *i.e.* water (Fig. 1). The hydrophobic effect is primarily entropy driven and attributed to the effect of the solute imparting additional structure to the surrounding shell water, thereby reducing the entropy relative to the bulk solvent. Additionally, aggregated molecules have reduced surface of contact with the aqueous solvent compared to the monomeric species [6-8].

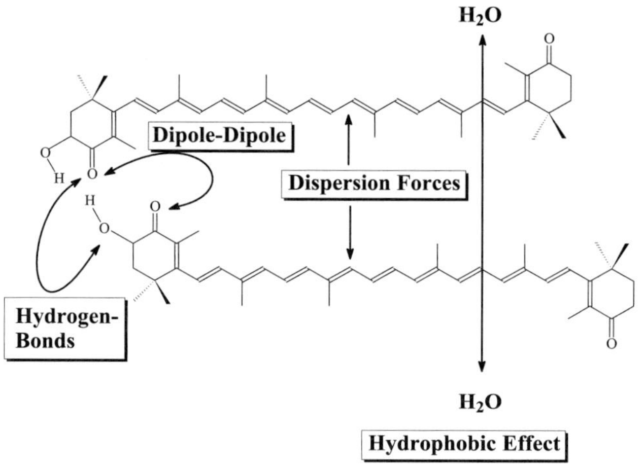

Fig.1. Important interactions responsible for many of the optical properties of carotenoid aggregates.

2. Aggregates of carotenoids

It is well known that carotenoids form aggregates when dissolved in hydrated polar solvents and that this aggregation is characterized by dramatic changes in their absorption spectra. Two types of carotenoid aggregates can be distinguished according to their absorption spectra. The first type, associated with a large blue shift of the absorption spectrum and loss of vibrational structure of the S_2 excited state, is suggested to take the form of H-aggregates, in which the molecules are stacked with the conjugated chains oriented more or less parallel to each other and closely packed. The blue shift of the absorption spectrum is explained in terms of excitonic interaction between the closely packed carotenoid chromophores [1-5]. The second aggregation type, characterized by a red shift of the absorption spectrum while the resolution of vibrational bands is preserved, is attributed to J-type aggregation, in which there is a more head-to-tail organization of conjugated chains, forming a loose association of carotenoid molecules.

Although individual chiral carotenoid molecules usually do not exhibit pronounced rotational strengths, many carotenoid aggregates have been found to be chiral. This phenomenon has been attributed to formation of large carotenoid assemblies that have a helical structure [9]. Aggregation of carotenoids also occurs in various natural and artificial systems. Typically, carotenoids tend to aggregate when present in lipid bilayers, in which long-range organization of carotenoid molecules is believed to influence physical and dynamic properties of lipid membranes and protect the bilayers against lipid peroxidation. The structure of the monomer and the kind and arrangement of neighbouring molecules mainly determine whether the formation of H- or J-aggregates is preferred. In lipid bilayers, some carotenoids usually form H-type aggregates whilst others form both H- and J-aggregates. In carotenoproteins, special interactions are responsible for the aggregation. In artificial systems, H-aggregates are often formed when carotenoids are deposited on surfaces or arranged at interfaces. Since assemblies consisting of carotenoid molecules attached to conducting or semiconducting materials hold promise as photoactive species in dye-sensitized solar cells or as molecular wires, understanding the effects of aggregation on these materials is an important factor in controlling the efficiency of such devices [10]. Carotenoid hydrosol particles, consisting of nanosized chromophore aggregates and crystallites, have considerable commercial and nutritional importance as ingredients for 'nutraceuticals' and 'functional food' [11].

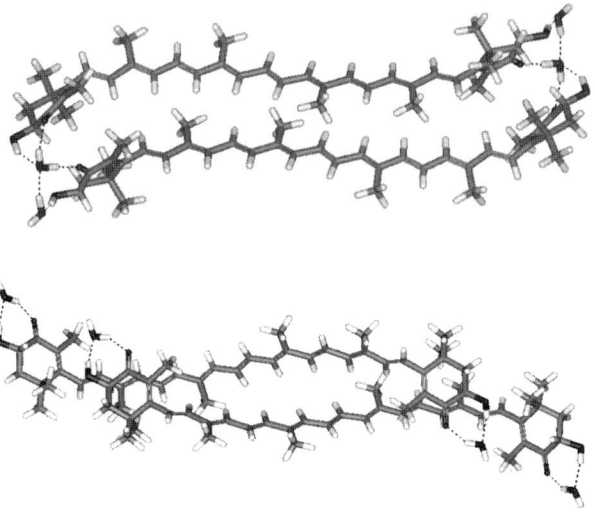

Fig. 2. H-aggregate (upper) and J-aggregate (lower) of astaxanthin (**404-406**), each with four water molecules. (See also Section **C**.3).

The properties of a carotenoid may change considerably from those of the monomer when it is enclosed in an assembled structure. This is important for the functioning of carotenoids in, for example, light absorption, light reflection, light emission, energy collection, energy transfer, energy dissipation, electron transfer and redox properties, protection against light and against damaging species, and mechanical stabilization of structures.

B. Aggregation in Solution

1. Optical and related properties

a) UV/Vis spectra

The UV/Vis absorption maximum of astaxanthin (**404-406**) in aqueous media depends on the refractive index of the medium, on the acidity and on the presence of salts [12,13]. Similar findings have been described [14] for β-carotene (**3**), lutein (**133**), zeaxanthin (**119**), violaxanthin (**259**), neoxanthin (**234**) and lycopene (**31**).

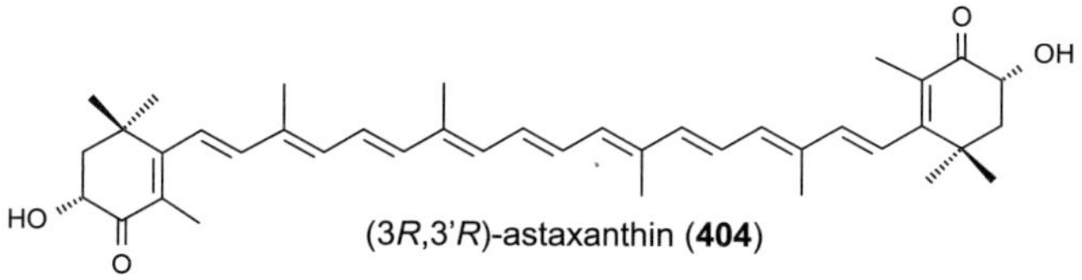

(3R,3'R)-astaxanthin (**404**)

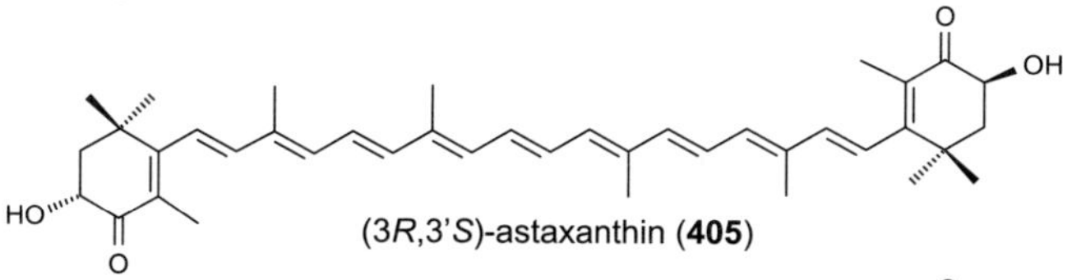

(3R,3'S)-astaxanthin (**405**)

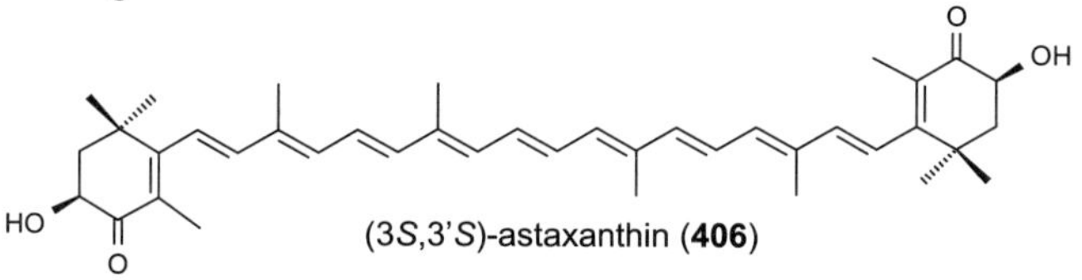

(3S,3'S)-astaxanthin (**406**)

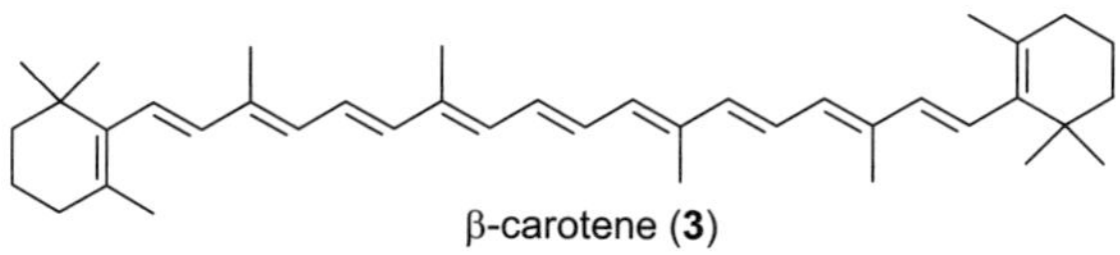

β-carotene (**3**)

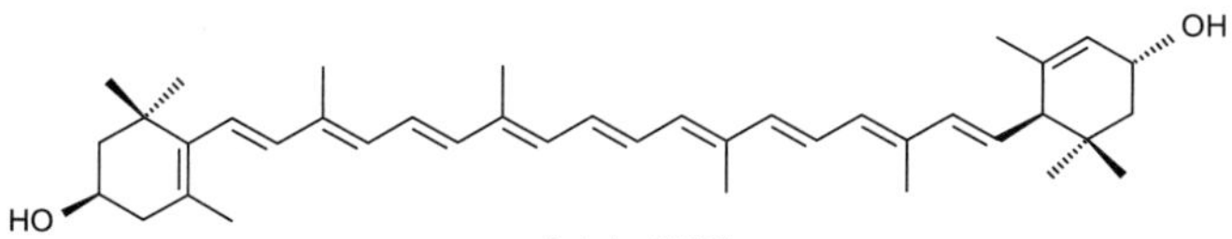

lutein (**133**)

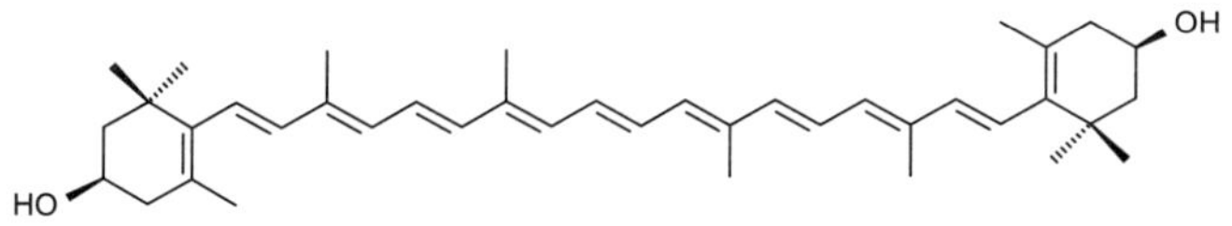

(3R,3'R)-zeaxanthin (**119**)

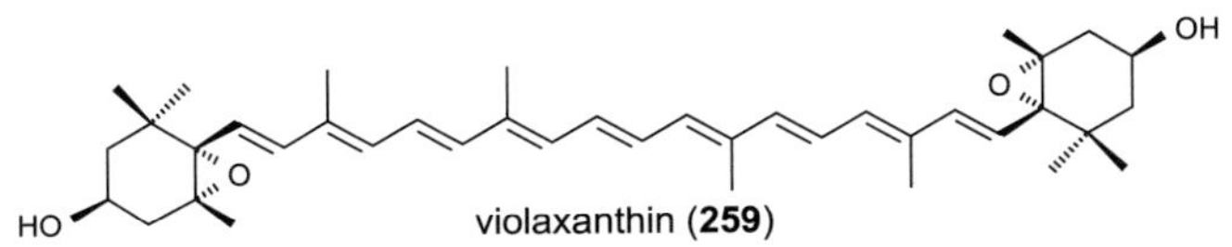

violaxanthin (**259**)

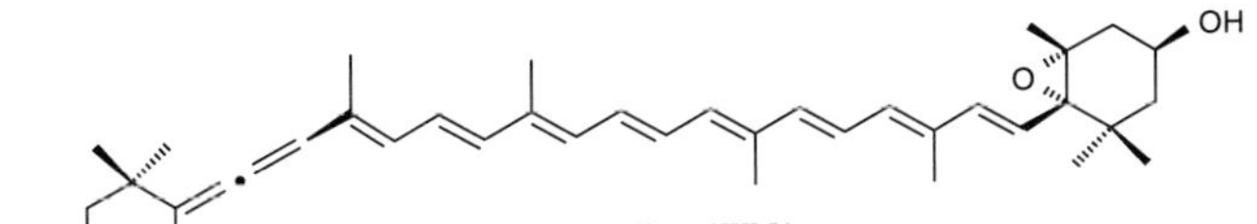

neoxanthin (**234**)

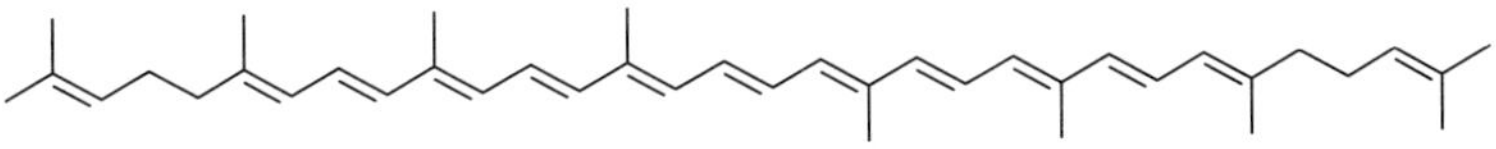

lycopene (**31**)

The effect of added salts with hard and soft anions on aqueous ethanolic dispersions of astax-anthin was investigated [15]. The observed hypsochromic shift to give a band around 400 nm (yellow colour), and the ratio of the absorbance at 407 and 475 nm, are more pronounced with soft anions, *e.g.* perchlorate. A notable exception, however, is sodium hydroxide; this gives the highest ratio with lower concentrations. The efficiency of the anions correlates with the Hofmeister series:

$$ClO_4^- > SCN^- \sim I^- > ClO_3^- > Br^- \sim NO_3^- > Cl^- > HCOO^- \sim H_2PO_4^- \sim CH_3COO^- \sim SO_4^{2-} \sim S_2O_3^{2-} \sim citrate^{3-}$$

A mixture of water with a water-miscible solvent such as ethanol or acetone is necessary to bring about the formation of aggregates, and there is a relationship between the ethanol/water ratio that gives half-maximum aggregation, and the λ_{max} of the longest wavelength band formed upon aggregation [16]. For a series of xanthophylls, the tendency for aggregation cor-relates with λ_{max} of the aggregate, declining in the order zeaxanthin (**119**) > antheraxanthin (**231**), lutein (**133**) > violaxanthin (**259**), lactucaxanthin (**150**) > neoxanthin (**234**).

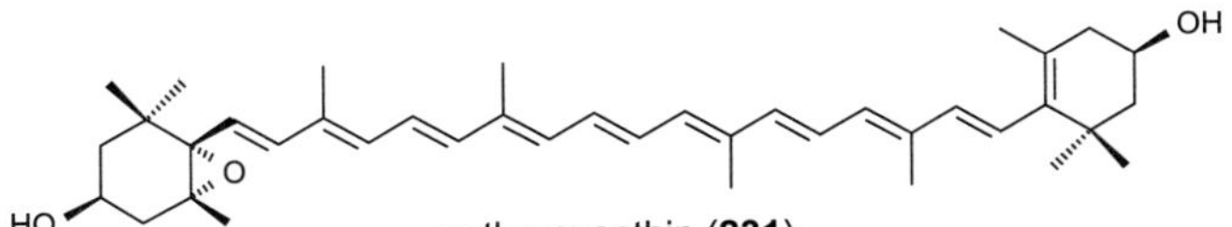

antheraxanthin (**231**)

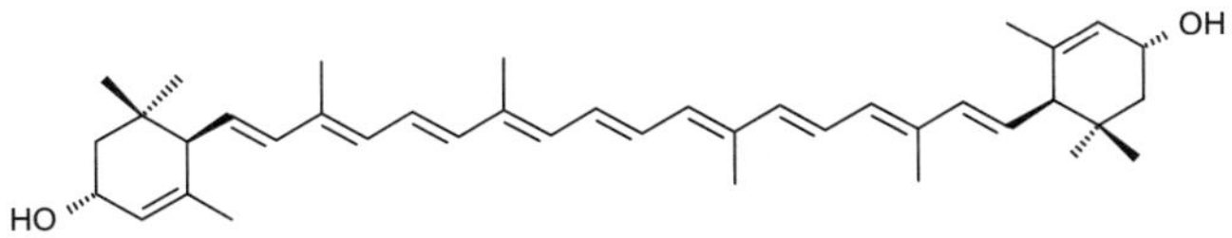

lactucaxanthin (**150**)

It is also noteworthy that, during the operation of the xanthophyll cycle in plants (*Chapter 14*), an increase in energy dissipation has been correlated with a change in λ_{max} from 520 nm to 540 nm. Aggregation of the xanthophyll responsible, zeaxanthin (**119**), gives rise to a similar absorption change [16].

b) Geometry of aggregate formation

The following discussion will use selected carotenoids to elaborate and develop a model for the origin of supramolecular exciton chirality, for the importance of intramolecular and intermolecular hydrogen bonding, and for the subtle steering effects for forming left-handed or right-handed supramolecular helices in solution.

Astaxanthin is a particularly interesting example because structurally different aggregates of different stabilities may be formed easily by changing either the solvent shell or the temperature [17-19].

In 80% water/20% acetone, (3R,3'R)-astaxanthin (**404**) forms H-aggregates, which rearrange to J-aggregates when the acetone concentration is increased (Fig. 3 left).

In 77% water/23% acetone, (3R,3'S)-astaxanthin (**405**) has a slightly but significantly higher tendency to form H-aggregates of at least two kinds than does (3R,3'R)-astaxanthin (**404**) (Fig. 3 right).

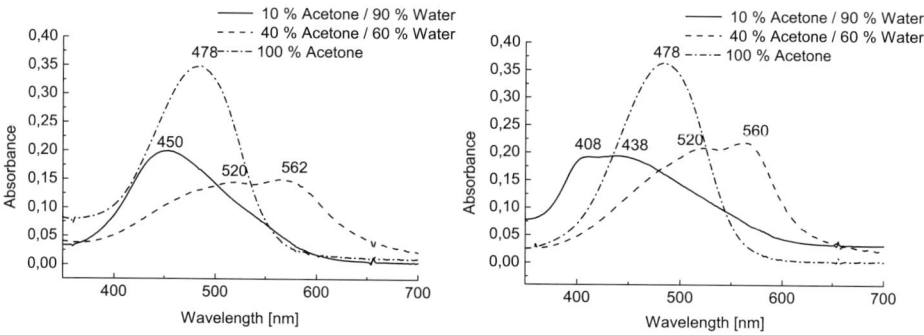

Fig. 3. UV/Vis spectra of (3R,3'R)-astaxanthin (**404**) (left), and (3R,3'S)-astaxanthin (**405**) (right) in acetone/water mixtures that give rise to both H-aggregates and J-aggregates [18].

This behaviour clearly demonstrates not only that the presence of hydroxy groups is conducive to aggregate formation but also that the orientation of these hydroxy groups is decisive in determining the particular aggregate type. Studies with the pepper carotenoids capsorubin (**413**) and capsanthin (**335**) gave similar results (see below).

The temperature dependence of the formation of aggregate mixtures is explained well by assuming different free energies of these H- and J-aggregates and the monomers [17].

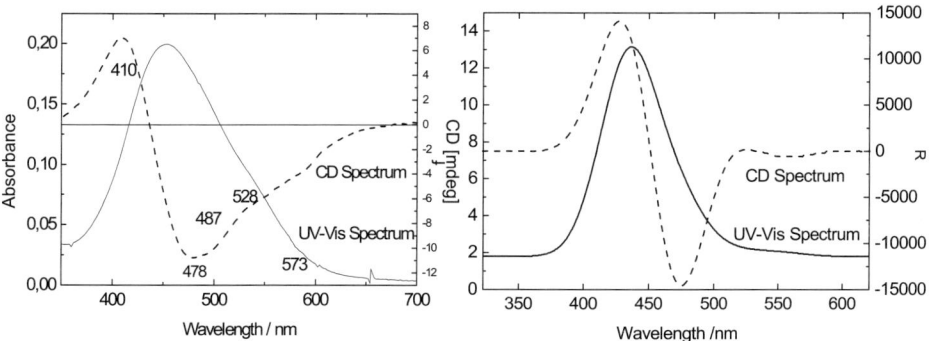

capsorubin (**413**)

capsanthin (**335**)

CD spectroscopy is widely used to investigate the chirality of carotenoid molecules in solution (*Volume 1B, Chapter* 3). It is also powerful and indispensable for investigating the structural relationships between monomers in aggregates. The absorption and CD spectra of the (3R,3'R)-astaxanthin (**404**) H-aggregate are shown in Fig. 4 [18-20].

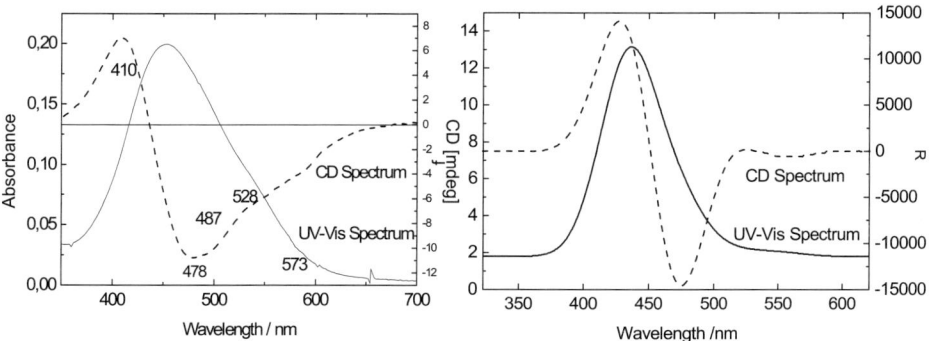

Fig. 4. CD and absorption spectrum of (3R,3'R)-astaxanthin (**404**) H-aggregate: (left) observed (in 10% acetone/90% water; 1x10⁻⁵M) and (right) calculated (for the dodecamer) [18-20].

Molecular modelling calculations (force field, molecular dynamics) of a dodecameric structure yield information on the intermolecular distances and angles. Inclusion of a considerable water shell (2000 water molecules) does not change the structure significantly [18].

This structure of the H-aggregate can be used as input geometry for calculating transition energies, oscillator strengths and the molar ellipticity [21]. Comparison of the spectra in Fig. 4 (left) with the graphs in Fig. 4 (right) shows that theoretically-derived spectra match the experimental spectra and that the structure of H- and other aggregates may be quite satisfactorily described by the structures obtained from the molecular modelling treatment.

J-Aggregates are noteworthy for sometimes considerable bathochromic shifts. Even in these cases, the above-mentioned simulation procedure results in wavelengths and rotatory strengths that are in good accord with the experimental results, as is shown (Fig. 5) for the J-aggregate mixture of $(3R,3'R)$-astaxanthin (**404**) in 30% acetone/70% water [18-20].

The presence of a mixture of two different J-aggregates is proved by fluorescence measurements. With 514,5 nm excitation both absorption bands are involved. If both transitions belonged to the same species, fluorescence would take place from the lowest level, *i.e.* longer wavelength than 560 nm. Since fluorescence emission begins at 543 nm, however, the absorption bands at 520 and 560 nm must be attributed to two different J-aggregates.

The calculated (Fig. 6) and experimental UV/Vis and CD spectra (Fig. 5) agree well.

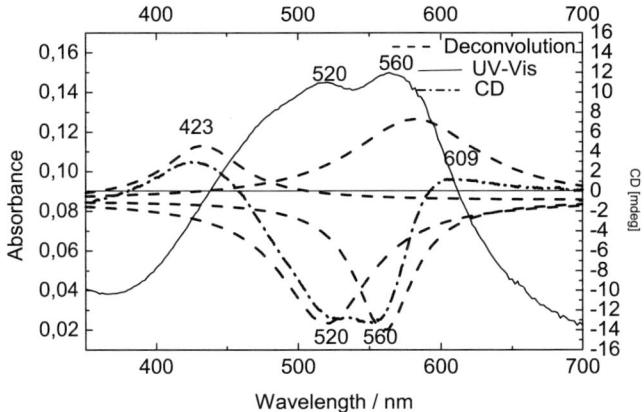

Fig. 5. UV/Vis and CD spectra of $(3R,3'R)$-astaxanthin (**404**) in 30% acetone/70% water (with Gaussian deconvolution).

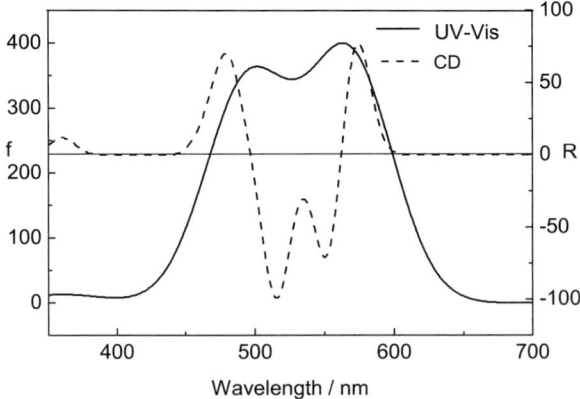

Fig. 6. Calculated CD and UV/Vis spectra for two J-aggregates, J_1 and J_2 (ratio 6:4), of $(3R,3'R)$-astaxanthin (**404**); these calculated spectra compare well with the experimental data in Fig. 5 [18-20].

c) Geometry of H- and J-aggregates

It has been proposed that J-aggregates and H-aggregates on surfaces and in solution may exist as assemblies that can be interpreted in a simplified manner as a brickwork, a ladder, or a staircase arrangement [22-24] (Fig. 7).

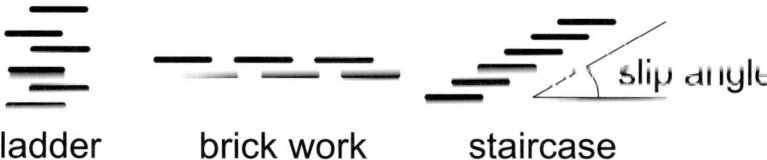

ladder   brick work   staircase

Fig. 7. Various aggregate structures that may be adopted by linear molecules on surfaces or in solution.

The dependence of the aggregate character, J or H, on the slip angle is well founded and allows a quick means of assessment [25,26]. The slip angle defines the slope of the axis through the centres of the monomers: large molecular slippage (<32°) yields a bathochromic shift (J-type aggregate), whereas a small slippage (>32°) results in a hypsochromic shift of the absorption (H-type aggregation) [22-24].

d) Chirality of aggregates

Polyenes with C_{2h} symmetry have a strong, electric-dipole allowed $S_0 \rightarrow S_2$ excitation in the UV, Vis or NIR region. The transition moment μ is polarized along the (all-*trans*) polyene chain axis. The orientation of μ depends on the structures of the aggregates and may be as in the borderline cases shown in Fig. 8 for two molecules, (I) in parallel or anti-parallel, (II) head-to-tail or (III) arbitrary directions.

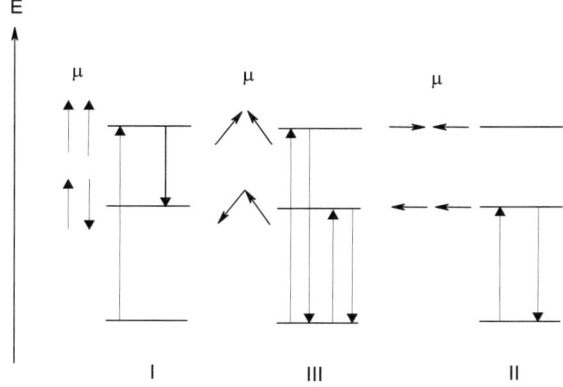

Fig. 8. Borderline cases for the arrangement of the electric transition moment μ: (I) stacked, (II) head to tail, (III) arbitrary. The interaction in the point-dipole approximation depends on distance, angles and transition moments.

Allowed transitions occur with all three aggregate types but with different energies (absorption wavelengths), whenever there is a non-vanishing transition dipole moment. The situation becomes more complicated when more than two monomers are assembled, e.g. as shown in Fig. 9. An important fact is that the corresponding CD effects of these transitions obey different selection rules and, irrespective of allowed or forbidden electric dipole transitions, strong CD couplets or peak manifolds can be seen. The prerequisite, however, is the chiral nature of the aggregate, not necessarily of the monomer.

The steering effect of hydroxy groups discussed above for astaxanthin is particularly conspicuous and important with capsorubin (**413**), capsanthin (**335**) and related compounds.

(3R,3'S,5'R,6'S)-Capsanthol [(6'S)-capsanthol (*1*)] is exceptional. It contains at both ends OH groups that are capable of hydrogen-bonding interactions, yet it still produces a J-type aggregate (left-handed), thereby defying the prediction of a preference toward the H-type. Apparently, the structure of the aggregate is such that the molecules cast some shadow on other molecules. This is compatible with a model containing parallel layers of molecules in which chirality is mainly induced by a twist of neighbouring layers [27-31] (Fig. 9).

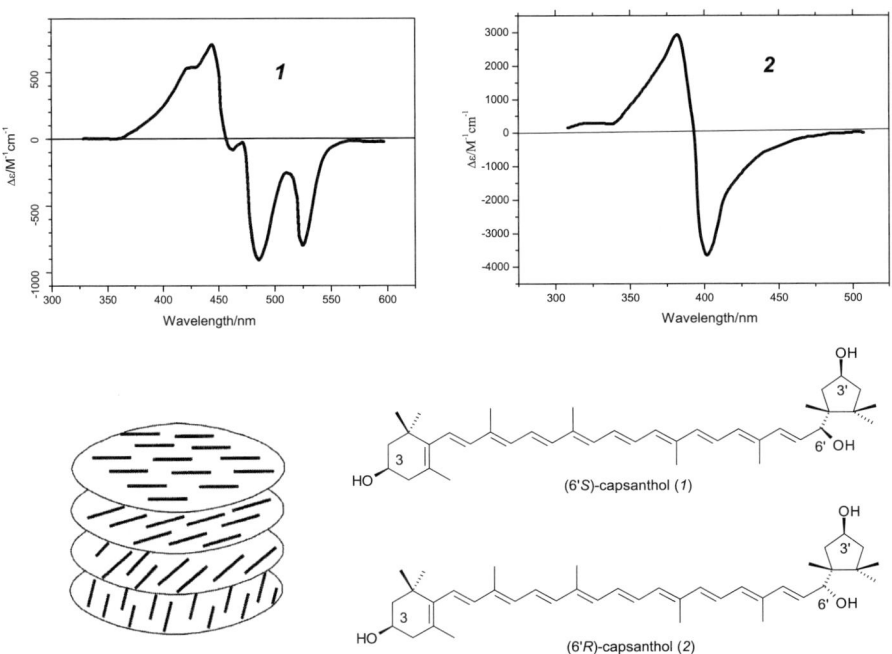

Fig. 9. Top. CD spectra of aggregates of (6'S)-capsanthol (*1*) and (6'R)-capsanthol (*2*). Bottom left. A suggested model of the J-type aggregate of (6'S)-capsanthol (*1*), containing parallel layers of molecules in which chirality is mainly induced by a twist of neighbouring layers [27-31].

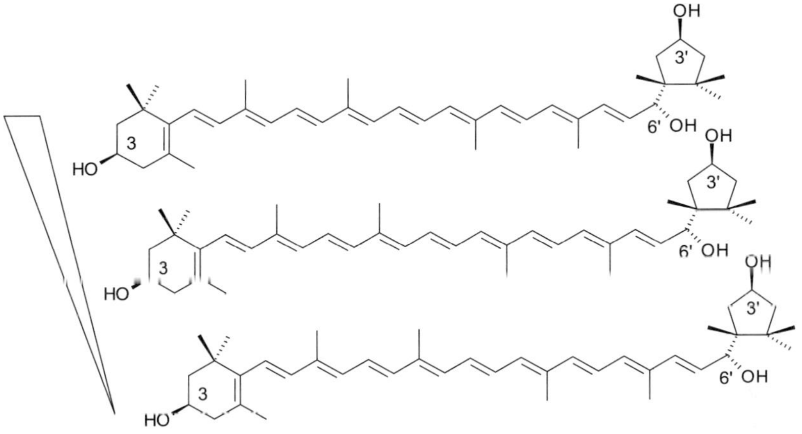

Fig. 10. H-Aggregate formation of (6′R)-capsanthol (2); the 6′-3′ hydrogen bonding leads to helical columnar structures [27,28]. This linkage is not possible in the (6′S)-epimer (1) [27-31].

(3R,3′S,5′R,6′R)-Capsanthol [(6′R)-capsanthol (2)], however, forms a left-handed H-type aggregate, which builds up in a time-dependent manner (Fig. 10). A characteristic feature of these two epimers of capsanthol is that a very small structural change (the inversion of one chiral centre) brings about a significant change in the structure of the assembly [27-31].

Four isomers of capsorubin, namely the epimers capsorubin (413) and epicapsorubin (3), and the 2-hydroxy analogues isocapsorubin (4) and epiisocapsorubin (5) (Scheme 1), show different spectroscopic behaviour, depending on the stereochemistry of the OH group. A strong hydrogen bond between the OH proton and the carbonyl lone pair determines the conformation of the (3,6-cis) isomers epicapsorubin (3) and epiisocapsorubin (5) [32].

The aggregation behaviour of these capsorubins is indeed strongly influenced by stereochemical relationships. Intramolecular and intermolecular hydrogen bonds are in free competition in epicapsorubin (3) and epiisocapsorubin (5), but not in the other two isomers which are amenable only to the intermolecular bonding. The spectra of capsorubin (413) and epicapsorubin (3) H-aggregates display notable differences. In capsorubin (413), the hydroxy groups are located *trans* to the keto functions. This structural configuration precludes intramolecular H-bonds but favours the intermolecular counterparts. In contrast to that geometry, the *cis* orientation of the polar OH and CO groups in epicapsorubin (3) allows and promotes intramolecular H-bonding. Since intermolecular and intramolecular interactions compete, the complex stability in epicapsorubin (3) aggregates is reduced; doubled extreme values in the CD spectrum indicate a changed and disturbed structure [18,19]. The spectra of both compounds also show weak CD effects in the range expected of J-aggregation. The H-aggregates also absorb in this region, however, due to the twisted helical geometry, with implications for the selection rules in comparison with idealized card-pack aggregates [19] (Fig. 11).

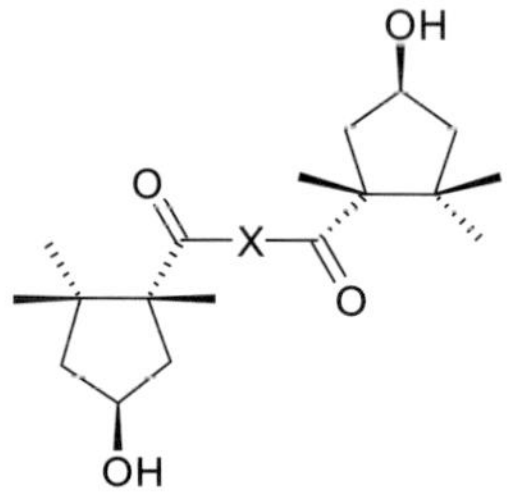

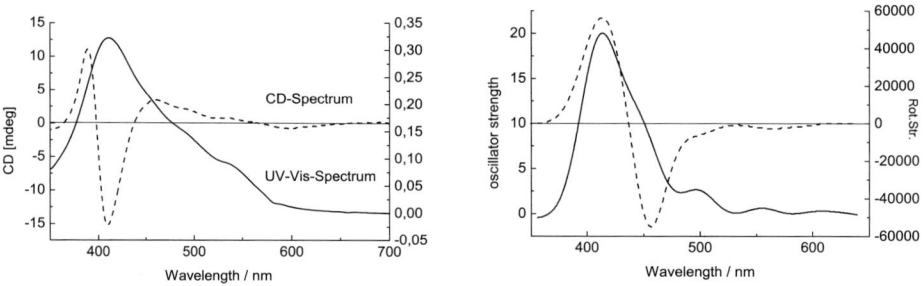

capsorubin (**413**)

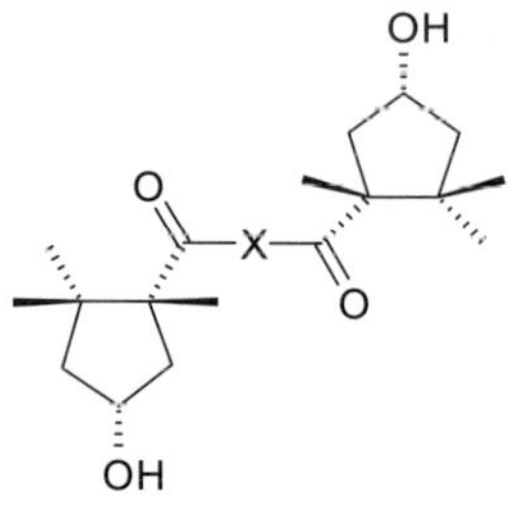

epicapsorubin (*3*)

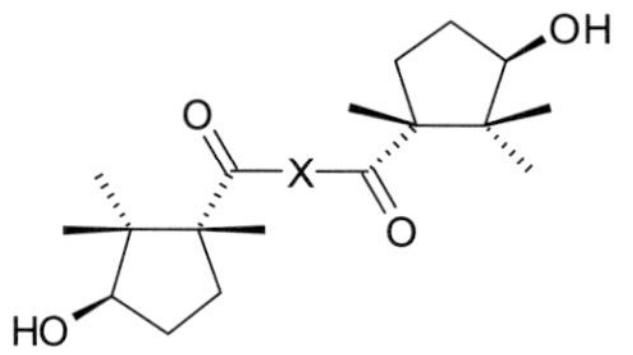

isocapsorubin (*4*)

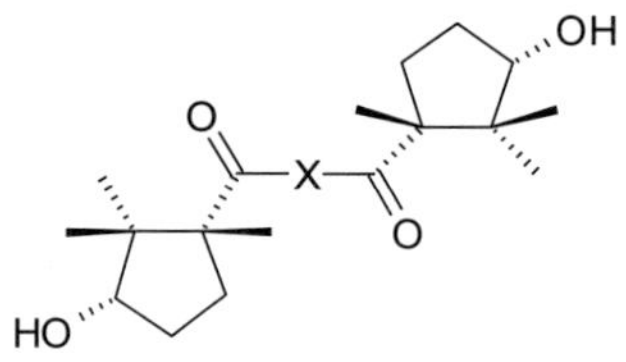

epiisocapsorubin (*5*)

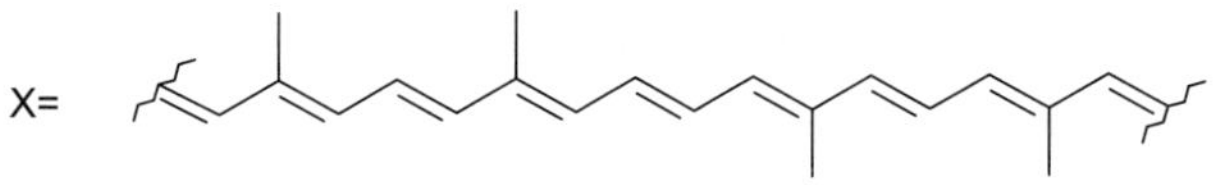

X=

Scheme 1

Fig. 11. Experimental (left) and calculated (right) CD and UV/Vis-spectra of the H-aggregate of capsorubin (**413**) [18,19].

The simulation of structure and spectroscopic properties of aggregates is quite successful. In Fig. 12, the results for a capsorubin (**413**) or epicapsorubin (*3*) dodecamer are illustrated [18]. Since it has been shown that the calculated spectroscopic properties of oligomers converge

when approaching decamers or dodecamers [18], an aggregate number of about 10-12 is usually sufficient to account for experimental data.

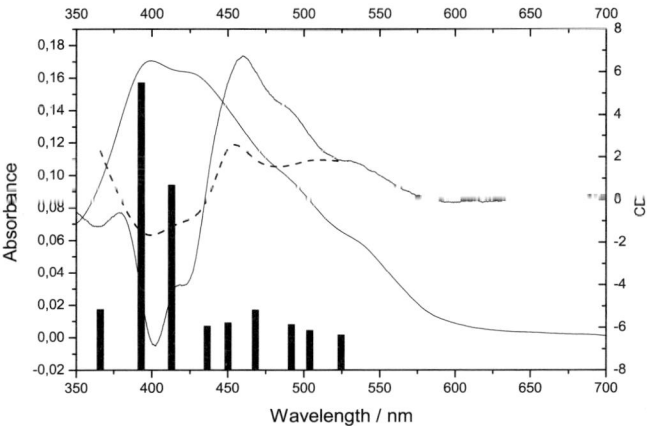

Fig. 12. Calculated and experimental spectra of epicapsorubin (*3*). The dashed curve (CD) and the bars (UV/Vis)
correspond to calculated data; the solid curves represent the experimental results [18,19].

The chirality of aggregates of lutein and zeaxanthin has been investigated [10,14,16,18,33-
37]. Because of its diverse importance in photosynthesis and in the macula of human eyes,
(3R,3'R,6'R)-lutein (**133**), which contains three centres of chirality, and its diacetate were chosen for UV/Vis and CD spectroscopic characterization of their supramolecular self-assemblies
[36] (Fig. 13). In solution and thin film, free lutein forms a typical H-aggregate in which the
molecules are closely packed and in a nearly parallel orientation. Simple exciton calculation,
based on the point-dipole approximation, showed that, in a hypothetical lutein dimer, the
molecules are separated by *ca.* 5.5 Å. The chiral arrangement of these units leads to the observed strong excitonic CD activity with opposite handedness in solution and thin film. In
contrast, lutein diacetate, which lacks the opportunity to form intermolecular H-bonds, forms
aggregates of weakly interacting monomers. The moderate bathochromic shift of the absorption spectrum is explained by the existence of a head-to-tail type molecular arrangement. On
the Atomic Force Microscopy (AFM) images, lutein diacetate exhibited beautiful thread-like
morphology, suggesting the existence of a nematic liquid-crystalline mesophase (in which the
molecules are oriented along one direction). Since the CD spectra of the film and the solution
were very similar, it is highly probable that the same mesophase exists in both circumstances.
This means that lutein diacetate is able to form a lyotropic mesophase in aqueous solution
[36]. A model of these mesophases is shown in Fig. 13.

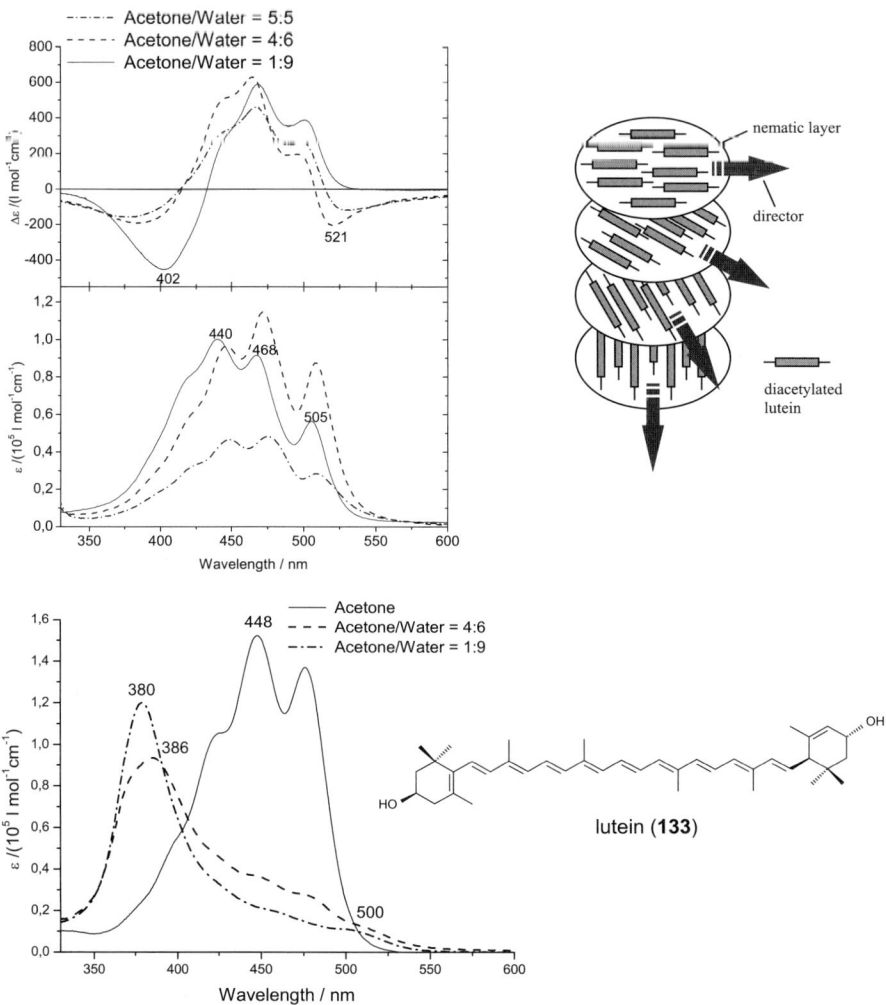

Fig. 13. Left: CD and UV/Vis spectra of lutein (**133**) diacetate (top) and UV/Vis spectra of lutein (bottom), all at 2 mM in acetone/water mixtures. Top right: model of the chiral supramolecular assembly of lutein diacetate.

e) Factors that control J- or H-aggregate formation

In controlling the formation of the two types of zeaxanthin (**119**) aggregates in hydrated etha-nol, namely J-zeaxanthin (characteristic absorption band at 530 nm) and H-zeaxanthin (char-acteristic absorption band at 400 nm) (Fig. 14), three parameters are important: pH, the initial concentration of zeaxanthin, and the ratio of ethanol:water.

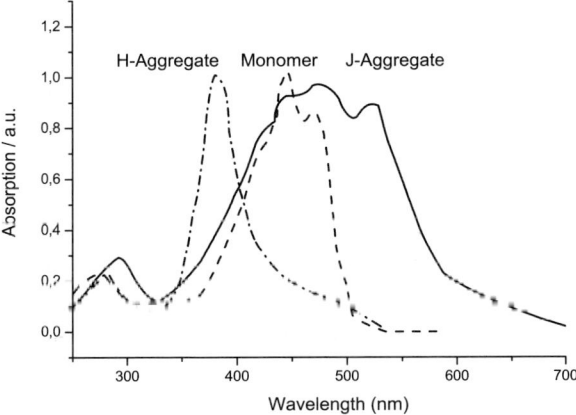

Fig. 14. Steady-state absorption spectra of monomeric zeaxanthin (**119**) (dashed line, 50 μM, ethanol, water concentration 0%), J-zeaxanthin (solid line, 100 μM, water concentration 40%), and H-zeaxanthin (dotted line, 50 μM, water concentration 80%) at 293 K (after [10]).

i) pH. The ability to form hydrogen bonds is decisive for determining whether J-aggregates or H-aggregates are formed. The crucial factor is therefore pH; increasing pH causes deprotonation of the hydroxy groups, so zeaxanthin is not able to create hydrogen bonds. This clearly favours formation of J-zeaxanthin, indicating that the head-to-tail aggregates can be formed only in the absence of hydrogen bonding. On the other hand, the pH dependence supports the earlier hypothesis that the card-pack H-aggregates are held together *via* a hydrogen-bonding network [10,31,32,38]. The inhibition of hydrogen bonding by modification of the molecules (be it by a pH change that can control the H-bonding *via* deprotonation or protonation, a derivatization of OH-groups, or the wrong stereochemistry) will determine the aggregation type of a particular carotenoid. In the simple case of a dimer, hydrogen bonding at both sides of the zeaxanthin molecule helps to keep the two molecules together lying on top of each other with their transition dipoles oriented almost perfectly parallel to each other. Other molecular forces such as π,π-stacking interactions may also contribute significantly to the attractive forces between closely packed carotenoid molecules, however [25] (see below). These attractive forces are stronger when molecules are planar, explaining why the more-or-less planar (3R,3'R)-zeaxanthin (**119**) forms H-aggregates readily. In contrast, weak van der Waals interactions dominate in J-zeaxanthin, resulting in lower stability of the J-aggregates.

ii) Ethanol:water ratio. The different structures of H- and J-aggregates also explain why different ethanol:water ratios are optimal for their formation. Hydrophobic interactions force the conjugated backbone of carotenoids to avoid contact with water molecules. Because of the head-to-tail structure of J-zeaxanthin (slip angle < *ca.* 40°), the conjugated chains are inevitably exposed to solvent. Therefore, rather low water content (optimal ethanol:water ratio of

3:2) is necessary to form and maintain J-zeaxanthin (Fig.14). Increasing the water content destabilizes J-zeaxanthin, as an increasing number of water molecules in the proximity compel the head-to-tail assembly to transform into the card-pack arrangement (slip angle > *ca.* 40°) that pushes the water molecules away from the conjugated chains. The critical water content that initiates transformation of J-zeaxanthin to H-zeaxanthin, however, depends on pH [10].

iii) Initial carotenoid concentration. Another important factor controlling the formation of aggregates is the initial concentration of carotenoid in solution. Apparently, to stimulate J-zeaxanthin formation, the molecules must be close to each other already in ethanol solution. At initial concentrations as high as 100 μM, the proximity of the zeaxanthin molecules in solution can be inferred from a slight loss of vibrational structure of the S_2 state. In such a situation, adding a moderate amount of water will lead to an arrangement of large assemblies and the solution turns slightly opaque upon J-zeaxanthin formation. On the other hand, when the initial concentration drops below a certain level, a moderate water content is not enough to push the molecules close enough to form aggregates, and increasing the water content further will lead only to H-zeaxanthin formation.

The critical values of these factors are dependent on each other, but it is apparent that neutral pH, low zeaxanthin concentration, and low ethanol:water ratio promote formation of H-zeaxanthin, whilst high values of these parameters give rise to J-zeaxanthin [10].

f) Excited-state dynamics

Time-resolved experiments have revealed that excitation of the 530 nm band of J-zeaxanthin produces a relaxation pattern different from that obtained by excitation at 485 and 400 nm, showing that the 530 nm band is not a vibronic band of the S_2 state but a separate excited state resulting from the J-type aggregation. The excited-state dynamics of zeaxanthin aggregates are affected by exciton-exciton annihilation that occurs in both J- and H-aggregates. In H-aggregates, the dominant annihilation component is on the sub-picosecond time scale, while the main annihilation component for the J-aggregate is 5 ps. The S_1 lifetimes of aggregates are longer than for monomers in solution, yielding 20 ps and 30 ps for H- and J-zeaxanthin, respectively. In addition, H-type aggregation promotes a new relaxation channel that forms the zeaxanthin triplet state [10].

Zeaxanthin diacetate, like lutein diacetate as discussed above, is also a good example to demonstrate the breakdown of H-aggregate formation. The modelled structure shows conspicuously the smaller slip angle and accounts well for spectroscopic properties [18].

g) π,π-Stacking interactions

By comparing the three hydrocarbons β-carotene (**3**), lycopene (**31**) and 16,17,18,16′,17′,18′-hexanor-ϕ,ϕ-carotene (*6*), it is easy to demonstrate the importance of the other 'adhesive' groups (planar π-systems) which are composed either of simple olefinic bonds or aromatic rings.

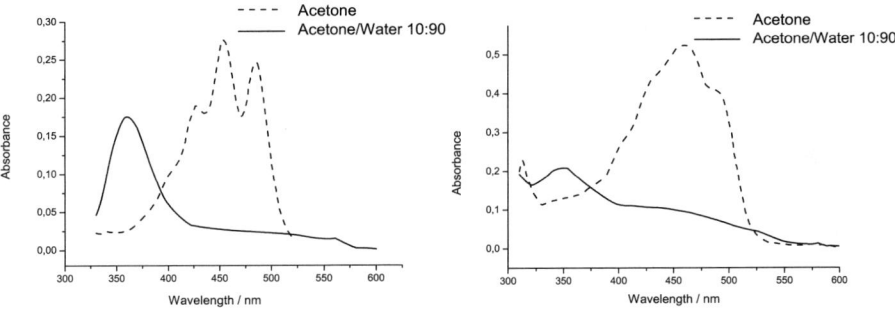

16,17,18,16',17',18'-hexanor-φ,φ-carotene (*6*)

In 45% water/55% acetone, β-carotene (**3**) evolves an aggregate spectrum which is not much different in position from that of the monomer. This may be due to a mixture of H- and J-aggregates, but a single distorted aggregate species may also be responsible for the changes [14,18,39]. Calculations show that dimers with a slip angle around 53° and a distance between the monomers of 3.8 Å would have four strongly allowed UV/Vis transitions at 397 nm, 430 nm, 455 nm and 509 nm, in agreement with the maxima observed for the aggregate [18]. Lycopene (**31**) and 16,17,18,16',17',18'-hexanor-φ,φ-carotene (*6*), however, clearly favour H-aggregates more than the J-type [14,18,40] (Fig. 15).

Fig. 15. Aggregation spectra of (left) lycopene (**31**) and (right) 16,17,18,16',17',18'-hexanor-φ,φ-carotene (*6*).

The reason for this different aggregation behaviour seems to be associated with the adhesive or non-adhesive character of the terminal groups. In β-carotene, the twisted and methylated cyclohexene rings are not conducive to a favourable π,π-stacking interaction whereas the double bonds in the acyclic lycopene and the aromatic rings in 16,17,18,16',17',18'-hexanor-φ,φ-carotene are good candidates for such an architecture and therefore for the building up of H-aggregates. The calculated structures illustrated in Fig. 16 for these H-aggregates support the interpretations impressively [18].

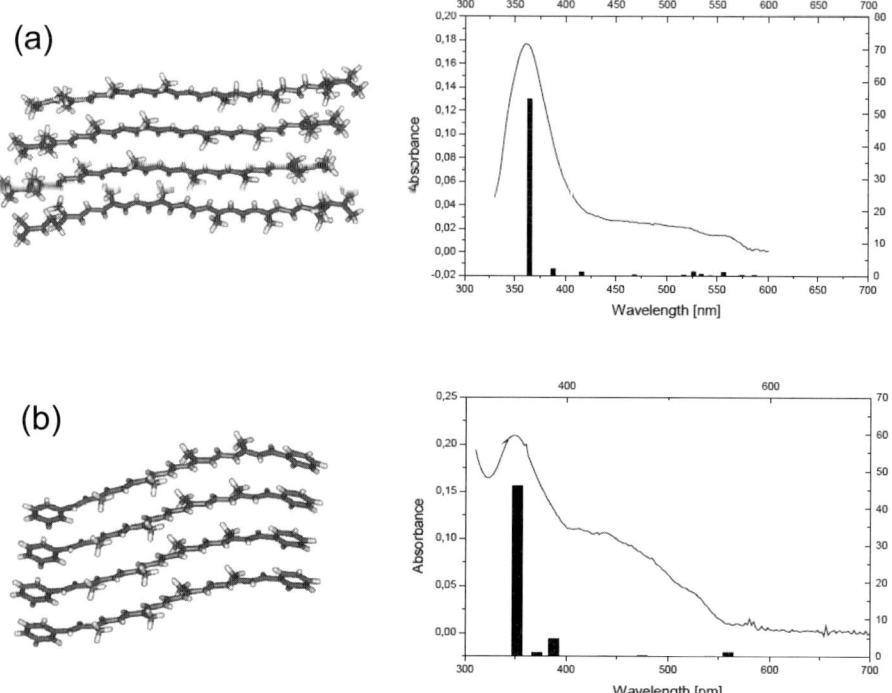

Fig. 16. Calculated spectra and structure of (a) lycopene (**31**) and (b) 16,17,18,16′,17′,18′-hexanor-φ,φ-carotene (**6**) dodecamers (only a tetrameric section is shown). The UV/Vis bars correspond to calculated data for dode-camers, the solid UV/Vis spectrum represents the experimental results [18].

h) Emission spectra

Emission studies have been performed with several aggregated carotenoids. Fluorescence quantum yields for zeaxanthin (**119**) and violaxanthin (**259**) are 4.3 x 10^{-4} and 1.7 x 10^{-3} respectively. The energy of the $^1A_g^*$ state of the aggregates has been measured as 17 860 cm^{-1}. The fluorescence originates from both the ^1B_u singlet state and the forbidden $^1A_g^*$ state. Aggregation increases the probability of electronic transitions to and from the $^1A_g^*$ state [33].

Emission spectra may also be exploited to decide whether the observed band manifolds hide one or more aggregates (Fig. 5). Only the 520 nm J-aggregate band of astaxanthin (**404-406**) gives rise to fluorescence and, therefore, does not belong to the same J-aggregate that gives the 562 nm band [18-20].

2. Light scattering

Light scattering is a useful technique for investigating the size of carotenoid aggregates in dilute solvent/water mixtures. From static light-scattering (SLS) experiments, the radius of gyration of the aggregates, r_G, as well as the second virial coefficient, A_2, can be determined. However, dynamic light scattering (DLS) is even more suited to the study of carotenoid aggregation since not only the size of the aggregate but also the kinetics of aggregate formation can be measured.

In dynamic light scattering [41], a coherent laser beam passes through a transparent sample. Laser light which is being scattered off any particles or any inhomogeneities that give rise to variations in the index of refraction will be detected under a scattering angle θ. The detection optics are set up so that only a small volume of the sample (the 'coherence volume') is imaged onto the detector. Then the light waves scattered from the individual particles within the coherence volume are superimposed on the detector and will give rise to a fluctuating intensity signal, which contains all the information about the kinetics of the system, *i.e.* about the Brownian motion of the particles. The hydrodynamic radius, r_H, and thus the size of the particles, can be calculated by using the Stokes-Einstein equation [41].

The process of carotenoid aggregation can be measured by DLS *in situ* in dilute solvent/water mixtures, with a time resolution of a few seconds. DLS typically covers a length scale from about 5 nm to about 10 µm; the average particle size can be determined unequivocally as an absolute value, *i.e.* without any calibration [41], and the width of polydisperse particle size distributions can be characterized.

Investigation by DLS has shown [19,42] that the size of astaxanthin (**404-406**) aggregates varies strongly with the acetone:water ratio. For an astaxanthin concentration of 3µM in acetone/water (10:90) a particle size of 160 nm was observed. From the corresponding UV/Vis absorption spectra, it is known that H-aggregates are formed under these conditions. Decreasing the water content to a ratio of acetone/water (20:80) again yields H-aggregates, this time with a size of about 275 nm. At an acetone/water ratio of (30:70) the situation is totally different; extremely large aggregates can be observed that exhibit J-aggregate-type absorption spectra and have an aggregate size of up to 7 µm. At an even lower water content (30%), the particle size reduces again, to 70 nm, and, in pure acetone, the astaxanthin molecules are dispersed and dissolved and can no longer be detected by DLS.

The structure of the particles cannot be elucidated by UV/Vis and DLS. It is assumed that primary particles of regularly assembled carotenoid molecules are formed first and then aggregated into the secondary particles that are observed in light scattering experiments [19,42].

3. Raman spectroscopy

The formation of aggregates of organic compounds usually does not affect molecular vibrations significantly, except for some low-energetic modes. This results from the large differences that are seen when the intramolecular force field constants are compared with those originating from intermolecular interactions. However, Raman spectra of assembled carotenoids do show some changes of high-energetic vibrations.

With lutein (**133**), zeaxanthin (**119**) and astaxanthin (**404-406**) [43], the most significant change occurs in the shift of the Raman-active C=C stretching mode. In the Raman spectrum of J-aggregates of astaxanthin, for example, this is recorded at 1516 cm^{-1}, red-shifted by about 11 cm^{-1} compared with that of the monomer species (1527 cm^{-1}). H-Aggregation generally causes a less pronounced red-shift of the C=C stretching mode, to 1522 cm^{-1}. Similar shifts can be observed for carotenoids with longer polyene chains. The Raman spectrum of the aggregates of the C$_{50}$ astaxanthin analogue (*7*) in water/acetone (90:10), for instance, exhibits two new transitions in the region of C=C stretching modes. This may indicate that two different types of aggregate are present simultaneously. As in the case of astaxanthin, one C=C stretching mode is down-shifted (1506 cm^{-1}) compared to the monomer species (1510 cm^{-1}) while the other is found at a higher frequency (1537 cm^{-1}). Shifts of similar size can be observed for other carotenoids [18]. The $\nu_{C=C}$ of aggregates show deviations from the well-known correlation of $\nu_{C=C}$ against $1/\lambda_{max}$ of the polyene [39,44,45].

The alterations of the C=C stretching frequencies after the formation of aggregates may have two sources. The first originates from alterations in the total molecular force field in which the intermolecular force field generated by formation of assembled molecules interacts with the intramolecular force field of monomer units. Second, it is known, from molecular modelling treatments and quantum chemical calculations, that the geometry of carotenoids in aggregates differs considerably from that of molecules which are not aggregated. Classical normal-coordinate treatment shows that this results in different reciprocal masses (1/mass) which may additionally cause a change of vibrational frequencies.

Changes also appear in the range of C-CH$_3$ stretching vibrations. Examples are shifts of 9 cm^{-1} for capsorubin (**413**): (monomer: 975 cm^{-1}, aggregate: 966 cm^{-1}), 19 cm^{-1} for isocapsorubin (*4*): (985 cm^{-1}, 966 cm^{-1}) and 7 cm^{-1} for epicapsorubin (*3*): (977 cm^{-1}, 970 cm^{-1}) [18]. Quantum-mechanically supported vibrational analyses are not available, so these shifts should be related to the changed steric surroundings of methyl groups within molecular complexes.

An interesting example of Raman frequencies and vibronic coupling has been reported for 19,20,19′,20′-tetranor-β,β-carotene (*8*) [46], on aggregation of which the absorption band splits excitonically (400 nm, 500 nm). Four pronounced Raman bands are recorded, denoted as ν_1 and ν_2, ν_a and ν_b. The latter bands ν_a and ν_b are seen only for the aggregate and when excitation in resonance with the 400 nm aggregate band is used. These results are interpreted in terms of vibronic coupling involving the forbidden singlet state $^1A_g^+$ (origin of the *cis*-peak in the near-ultraviolet region).

'C$_{50}$ astaxanthin analogue' (*7*)

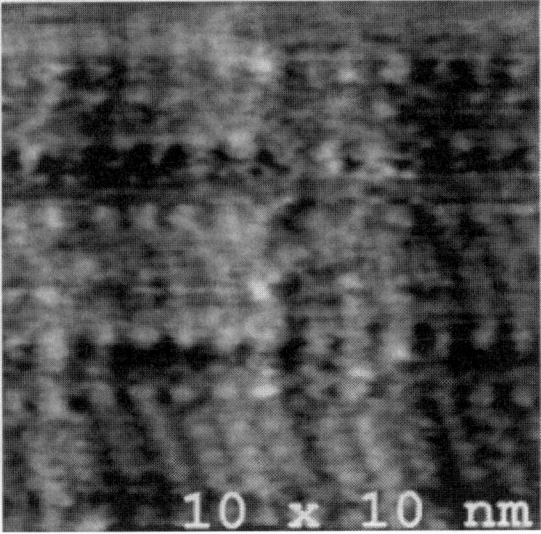

19,20,19',20'-tetranor-β-carotene (*8*)

C. Aggregation in Different Environments

1. Surfaces and interfaces

a) 'Supermicroscopy'

Since the introduction of the first scanning tunnelling microscope (STM) in 1981, 'super-microscopy' has been capable of producing images of individual atoms [47-49]. Currently more than 30 different supermicroscopic techniques are available, a particularly important one being Atomic Force Microscopy (AFM). STM requires electrically conductive surfaces, but is still the best suited for achieving atomic resolution [50], and is widely used to study chirality, dynamics and reactivity in physically adsorbed organic monolayers [51].

Fig. 17. Zoomed STM-images of (3*R*,3'*R*)-astaxanthin (**404**) on graphite (HOPG). Scanning area 10 x 10 nm², bias 800 mV, tunnelling current 2 nA [18,20]

Thin films of carotenoids can be prepared simply, by evaporation of the solvent on a clean surface of highly oriented pyrolytic graphite (HOPG). Figure 17 shows aggregates formed by stacked (3R,3'R)-astaxanthin (**404**) [18,20].

Spontaneously-forming H-type and J-type chiral self-assemblies of lutein (**133**) and lutein diacetate have been studied by UV/Vis and CD spectroscopy and AFM, and the existence of chiral superstructures with opposite or the same handedness in both solution and solid phases demonstrated (Fig. 13). The AFM images revealed definite threadlike morphology in lutein diacetate films, indicating the formation of a nematic liquid crystal phase which probably exists in the aqueous solution as well. The vibronically coupled CD bands of lutein diacetate are therefore attributed to excitonic interaction arising from the twisted nematic layers [18,20].

b) Electrical conductivity of model carotenoids

Due to their rod-like molecular shape, carotenoids are prone to assemble on surfaces or at interfaces if suitable functional groups affect the binding forces to one or both of these phases. Of special interest are aggregated carotenoids which, compared with saturated hydrocarbons, are excellent conductors of electrons, facilitating even measurements of the electrical properties of individual molecules in an insulating matrix. Another important field is the use of carotenoids for molecular recognition in surface monolayers [52-54].

Conducting atomic force microscopy was used to measure the electrical properties of carotenethiol derivatives (*e.g.* *9*), embedded in insulating *n*-alkanethiol self-assembled monolayers and attached to a gold electrode. At a contact force of a few nano-Newtons, the carotenoid molecule is over a million times more conductive than an alkane chain of similar length [52]. Similar studies have been undertaken with a carotenedithiol (*10*) [55,56].

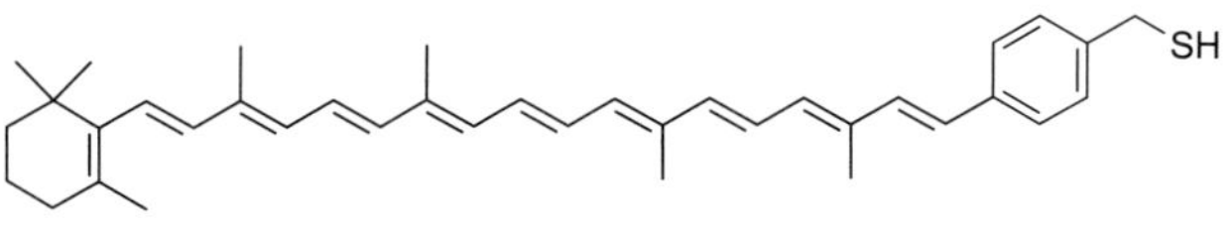

(*9*)

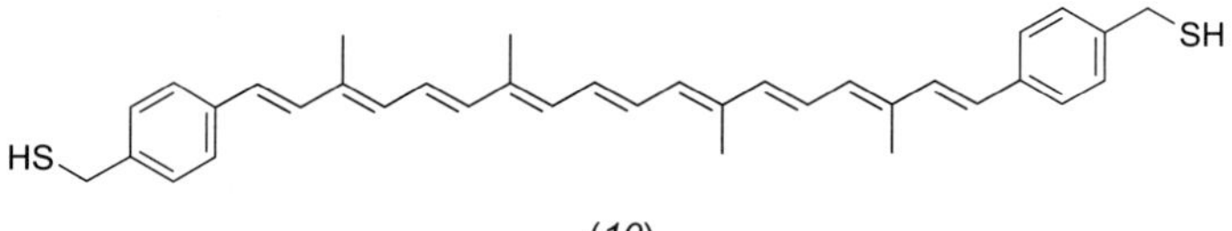

(*10*)

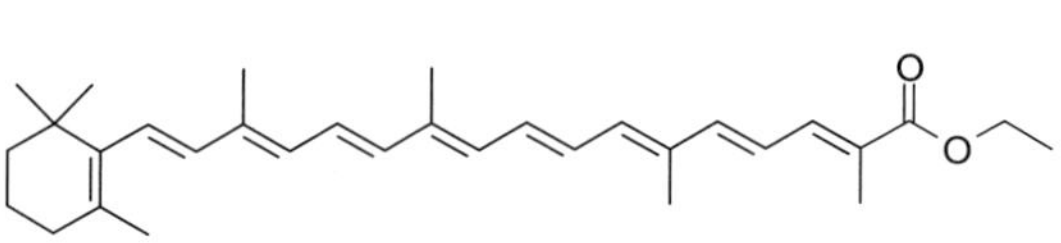

(*11*)

The electrical conductivity of other carotenoid derivatives, including the amphipathic compounds ethyl 8'-apo-β-caroten-8'-oate (*11*) [57] and 7'-apo-7'-(4-carboxyphenyl)-β-carotene (*12*) [58,59], has been studied in LB films, with tetrathiafulvene as the reference compound. For *12*, action spectra implicated the excited state, possibly in an aggregated form, as the photoactive species in the photoinduced electron transfer process. The polyenic carboxylic acid *12* behaves as a photoconductor, in contrast to saturated long-chain acids which act as insulators in similar experiments [58,59]. The electrochemistry of a self-assembled monolayer of 4'-oxo-β,β-carotene-4-thione (*13*) has also been investigated [60].

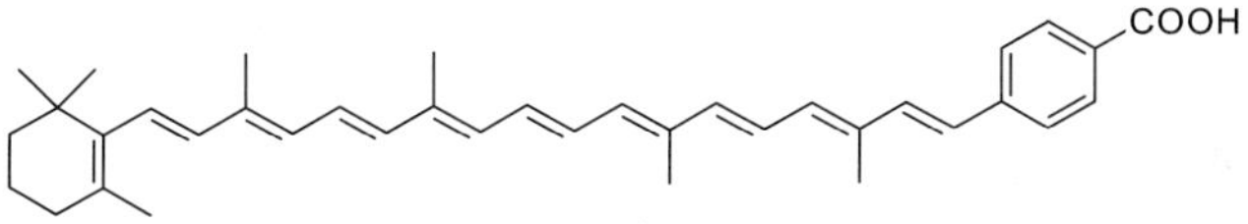

(*12*)

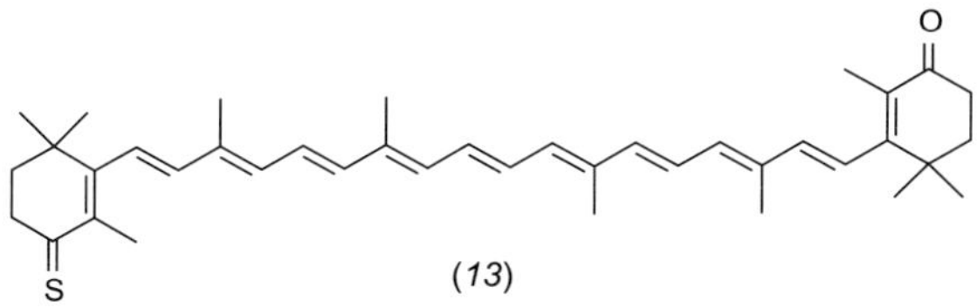

(*13*)

The selenium-containing derivative 1-(8'-apo-β-caroten-8'-oyl)-2-(7-selenaoctanoyl)-glycerol (*14*), bound to a gold electrode surface by the selenium end group, adopted a roughly perpendicular orientation to the electrode surface [61,62].

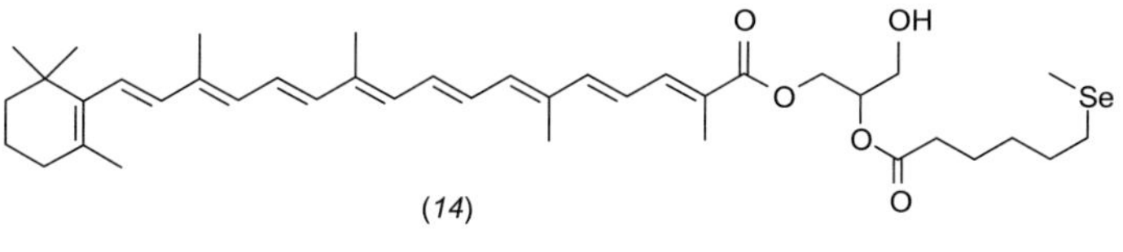

(*14*)

Oleophilic dyes such as β-carotene (**3**) can be co-spread with copoly(γ-n-alkyl L-glutamates) onto the air-water interface from a solvent in which the copolyglutamate assumes a helical conformation. The dichroic absorption of the dye-containing LB-assemblies indicates that the dyes are not isotropically distributed in the layers but located with a preferential orientation of their molecular axes with regard to the director axis of the polypeptide [63].

An electroconductive amphiphilic carotenoid functionalized with a boronic acid, incorporated into an LB film on an electrode, bound a redox-active sugar derivative and allowed the redox current of this to be observed selectively [53].

c) Bola-amphiphiles

Bola-amphiphiles are a special class of amphiphiles in which two polar head groups are present, one on each end of a hydrophobic core [64]. Bixin (**533**) is a natural, asymmetric bola-
amphiphile. The measured surface area (1.0 nm^2/molecule at zero pressure) corresponds to the
area of a bixin molecule lying on the water surface with most methyl groups pointing outside.
If the methyl groups were parallel to the water surface, a molecular area of about 1.7
nm^2/molecule would be occupied [54] (Fig. 18). The *cis* double bond can be 'straightened
out' by iodine-catalysed isomerization to yield (all-*trans*)-bixin, which is bound approximately perpendicular to the surface by only one polar end group.

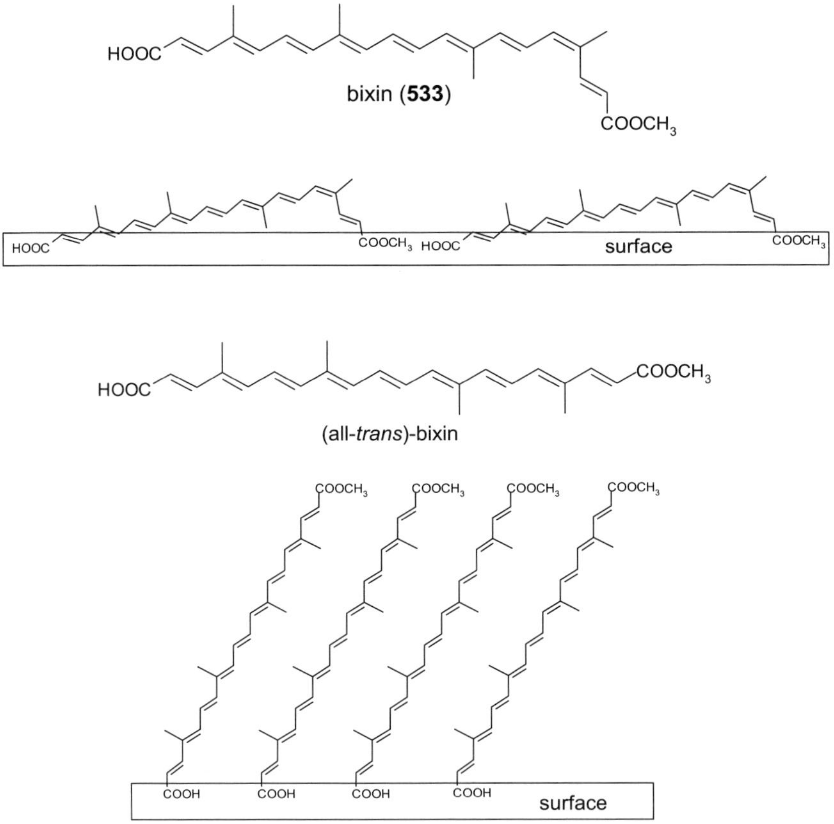

Fig. 18. The bola-amphiphiles bixin (**533**) and (all-*trans*)-bixin and their aggregation on surfaces [54].

The naturally occurring crocin (**545**), the digentiobiosyl ester of crocetin (**538**), should also be classified as a bola-amphiphile. Crocin forms true monomolecular solutions in water; aggregation occurs only at rather high concentrations. Tensiometric data indicate a horizontal orientation of crocin molecules at the water surface [65], similar to that of bixin (Fig. 18).

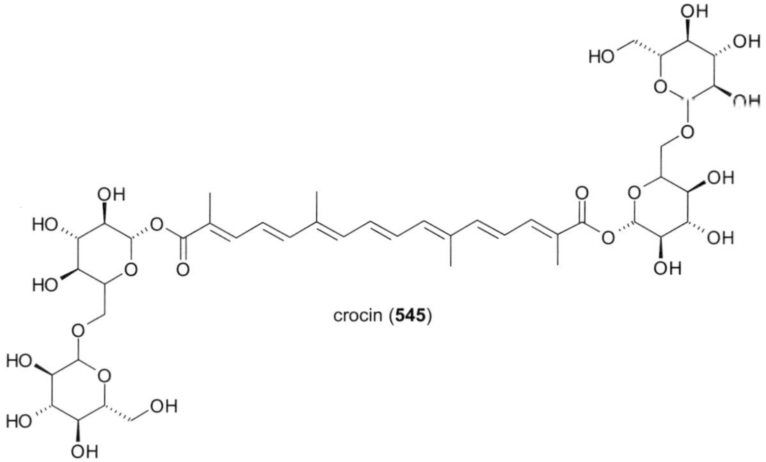

crocin (**545**)

2. Vesicles, lipid bilayers and miscellaneous environments

Membranes of bacteria, plants and animals contain carotenoids as constituents of their lipid phase. An overview of the localization, orientation and solubility of carotenoids in lipid membranes, and their effects on membrane properties is given in *Chapter 10*. Most studies have been carried out with liposomes, artificial vesicle membranes into which carotenoids are incorporated. The following discussion will concentrate on the self-assembly of carotenoids within membranes.

a) Lipid membrane bilayers: phase transitions

At different temperatures, phospholipid membranes can exist in different phases. At the temperature of main phase transitions (T_c), the membrane passes from a tightly ordered gel or solid phase (L_β') (through P_β', a ripple phase) to a liquid-crystalline phase L_α at raised temperatures [66-70]. The physical state of a lipid phase significantly influences the aggregation of carotenoid molecules in membranes. The most representative and convenient technique for observing carotenoid aggregation phenomena is UV/Vis spectroscopy.

l) Xanthophylls. When zeaxanthin (**119**), astaxanthin (**404**) and their C_{50} homologues are incorporated into unilamellar vesicles of dimyristoylphosphatidylcholine (DMPC) at a % molar
ratio *i.e.* carotenoid concentration/(carotenoid concentration + phospholipid concentration)
from 2.5% to 15%, depending on the carotenoid, only the spectrum of the monomer is observed in all cases above the phase transition temperature T_c which, for DMPC, is 23°C.
When the temperature is progressively decreased, no appreciable change is observed until the
phase transition is reached, then a hypsochromic shift of the absorption maximum occurs with
loss of fine structure. This is due to aggregation of carotenoid molecules in a card-packed
manner within the lipid bilayer (intermolecular changes) combined with conformational
change of the carotenoids (intramolecular changes) [71]. A more recent study of these phenomena has compared many different carotenoids [37]; the results for (3*R*,3'*R*)-astaxanthin
(**404**) in dipalmitoylphosphatidylcholine (DPPC) vesicles are shown in Fig. 19.

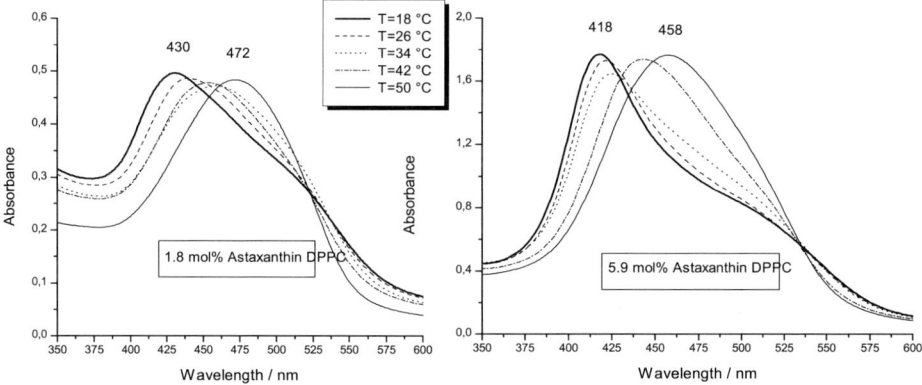

Fig. 19. Temperature-dependent UV/Vis spectra of (3*R*,3'*R*)-astaxanthin (**404**) at two concentrations (1.8 and 5.9
mol%) in unilamellar DPPC vesicles ($T_m = 41.4$ °C) [37].

Lutein (**133**) or zeaxanthin (**119**), incorporated into DPPC unilamellar liposomes, form H-
type molecular aggregates, as indicated by the hypsochromic shift of the main absorption
band of the carotenoids. This aggregation was observed even at relatively low concentrations
in the lipid phase (1-5 mol%). Gaussian analysis of the absorption spectra in terms of the exciton splitting theory revealed the formation of different structures, interpreted as dimers,
trimers, tetramers and large aggregates. Pronounced dissociation of lutein and zeaxanthin to
the monomers was observed to accompany the transition from the P_{β}' phase to the L_{α} phase
of DPPC, mostly at the expense of the trimeric and tetrameric forms. The fraction of lutein as
monomers was always lower by 10-30% than that of zeaxanthin under the same experimental
conditions. The possible physiological significance of different organizational forms of lutein
and zeaxanthin in the eye is discussed in *Volume 5, Chapter 15.*

When zeaxanthin (**119**) is inserted into phospholipid dispersions which are then heated through their gel-liquid crystal phase transitions, large changes are noted in the resonance Raman and absorption spectra of the carotenoid [72]. By analogy with the data from a study of the aggregation of zeaxanthin in acetone-water solutions [43], it is suggested that the carotenoids form aggregates in the phospholipid gel state but dissociate into monomers in liquid crystal phases. The alterations in both the visible absorption and resonance Raman data have been used to monitor phospholipid phase behaviour in dipalmitoylphosphatidylcholine and distearoylphosphatidylcholine one-component systems and binary mixtures. The phase diagram obtained for the binary system, as constructed from visible absorption and resonance Raman data, has been described [72].

ii) β-Carotene and β-cryptoxanthin. β-Carotene (**3**), however, behaves differently when incorporated into multilamellar DPPC liposomes [73-75]. It is proposed that β-carotene is aggregated in both gel and liquid crystal states of the lipid bilayer, as it shows several poorly defined absorption bands in the carotenoid monomer region, plus a J-aggregation band at about 510 nm. This bathochromically shifted new band is also present in the spectra of vesicles charged with β-cryptoxanthin (**55**) below the phase transition temperature but, in fluid membranes, the characteristic carotenoid spectrum is observed, as with zeaxanthin (**119**) preparations [73]. Although the origin of the long-wavelength band can be attributed to small molecular aggregates, another mechanism has been suggested by FE-MO (free electron-molecular orbital) calculations [75], namely that suitably oriented β-carotene (**3**) monomers interact with the neighbouring lipid molecules in the bilayer.

β-cryptoxanthin (**55**)

iii) Violaxanthin. In a study of phosphatidylcholine vesicles injected with a small amount of violaxanthin (**259**) in ethanolic solution [76,77], the UV/Vis spectra obtained were assumed to result from mixing of two forms of spectra. One form, with a blue-shifted maximum at 401 nm, corresponds to H-aggregation of the molecules in the water phase, the other, red-shifted to 449 nm, correlates with the process of pigment adsorption on the surface of phospholipid membranes.

b) CD studies

Important information about the effect of temperature on carotenoid aggregation in membranes and in different environments can be gained by circular dichroism (CD) measurements. (3R,3′R)-Zeaxanthin (**119**), for example, shows optical activity because it possesses

two identical chiral twisted end groups (*Volume 1B, Chapter 3*). When it is incorporated into
vesicles, above the phase transition temperature there is no detectable CD absorption band in
the range of the strong main visible absorption (350-550 nm), as in a pure solvent. Below the
phase transition temperature, however, and at sufficiently high carotenoid concentrations, a
two-wave dichroism signal appears centred near the blue-shifted absorption maximum of the
aggregate (400 nm) and attributed to exciton coupling of chirally arranged chromophores
[71,78]. This CD couplet represents two successive Cotton effects with opposite sign. The CD
curve shows additional signals, in the region of the absorption maximum of the monomer, and
these are also present at lower carotenoid concentration.

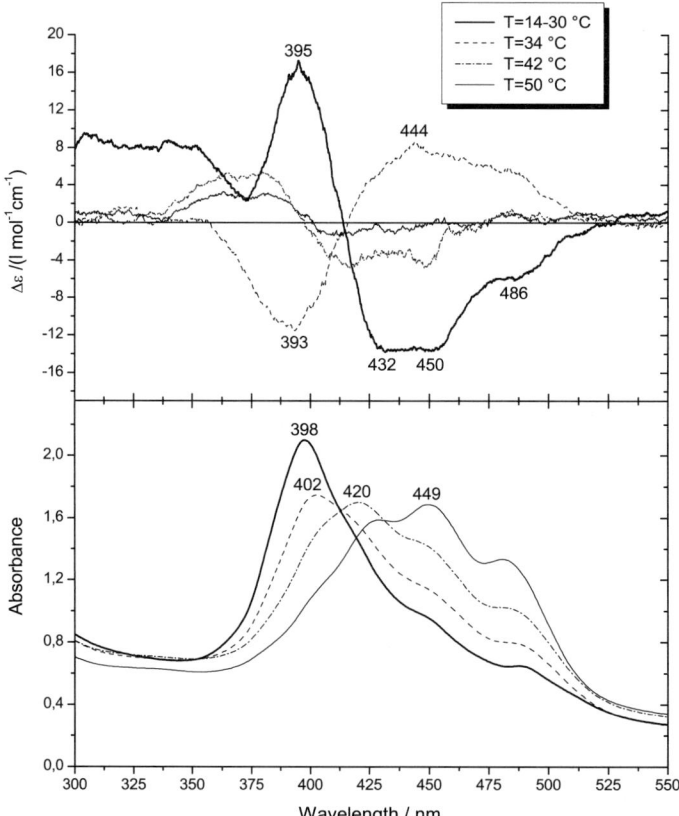

Fig. 20. Temperature-dependent CD (upper) and UV/Vis (lower) spectra of lutein (**133**) (7 mol%) in small
unilamellar DPPC vesicles ($T_m = 41.4°C$) [37].

The results of a recent investigation of the temperature-dependent UV/Vis and CD spectra of
lutein in DPPC vesicles [37] are presented in Fig. 20. Up to 30°C the CD spectrum shows a
negative exciton couplet with extreme values at 432 and 395 nm ($\Delta\varepsilon_{432} - \Delta\varepsilon_{395} = -31$), accom-

panied by two negative maxima at 450 and 486 nm. This pattern indicates that at least tetramers may be present, since helical disorders and other distortions from the ideal card-pack arrangement bestow intensity even on low-energy transitions. At 34°C the CD spectrum is inverted and loses intensity. The resulting positive CD couplet extends from 444 to 393 nm ($\Delta\varepsilon_{444}$ - $\Delta\varepsilon_{393}$ = +20). The zero intersection is retained [37].

Furthermore, some structural investigations were made by electron microscopy of lutein (133) aggregates dispersed in small unilamellar liposomes of phosphatidylcholine and digalactosyldiglyceride [79-81]. A left-handed helical structure of the lutein aggregate, which is expected from the CD pattern of negative exciton chirality, was observed in phosphatidylcholine (PC) liposomes at high pH in the presence of Ca^{2+}. The lutein aggregate in the PC liposome gave fine images which were 20 to 40 nm wide and more than 2000 nm long. The aggregates became longer as the PC concentration increased. Similar helical structures of lutein aggregates were observed in the digalactosyldiglyceride liposomes [79-81].

c) Surfactant solutions

A further possibility to force carotenoid aggregation is in surfactant solutions, as studied in lutein dispersions [79-81]. Lutein (133) acquires a strong CD activity in the visible region when dispersed in an aqueous solution in the presence of sodium dodecyl sulphate (SDS). The CD spectrum of this lutein has positive and negative extrema before and after a crossover at about 390 nm, respectively. The signs of the extrema are inverted when the amount of SDS is increased. Further addition of SDS destroys the CD activity.

These phenomena are suggested to reflect a sequence of events, namely (i) the formation of a helical assembly of lutein molecules; (ii) a large-scale structural change of the assembly resulting in the inversion of its chirality, and (iii) the breakdown of the assembly followed by the inclusion of the lutein molecules into SDS micelles.

As indicated by sedimentation analysis, the size of the lutein aggregate is larger than 450 nm when the ordinary CD pattern remains, and becomes smaller as the ordinary pattern changes to the inverted one [79-81]. The cationic surfactant dodecyltrimethylammonium bromide (DTAB), however, is found to affect the circular dichroism activity of the lutein below the critical micelle concentration in a different way. In this case, no inversion of the CD spectrum occurs. This phenomenon may be interpreted by the card-pack model of the lutein aggregate, in which lutein molecules, slightly shifted with respect to each other, associate body to body. The optical activity abruptly becomes strong just before the critical micelle concentration of DTAB. This seems to correspond to the transition from the polymeric aggregate of lutein to the oligomeric one, which is supported by the binding ratio (surfactant to aggregated lutein) which is eight times higher with DTAB than with SDS.

d) Relationship between structure and formation of H- or J-aggregates

The polar and aromatic end groups illustrated in Scheme 2 promote the formation of H-aggregates of the corresponding symmetrical carotenoids in DPPC membranes [37]. The H-aggregates form in the gel state and dissociate above T_m predominantly into monomers. This transformation is completely reversible, and the carotenoid molecules are localized in the hydrophobic interior, and remain there in the temperature range 18 - 58°C. The molecules tend to integrate into the liquid-crystalline membrane as monomers. Hydroxy substituents are most conducive to this; the 3-hydroxy-ε end group shows particularly high integration ability (due to flexibility) and aggregate diversity. The hypsochromic shifts vary from 2000 to 5000 cm^{-1} [37]. (See also Section **B**).

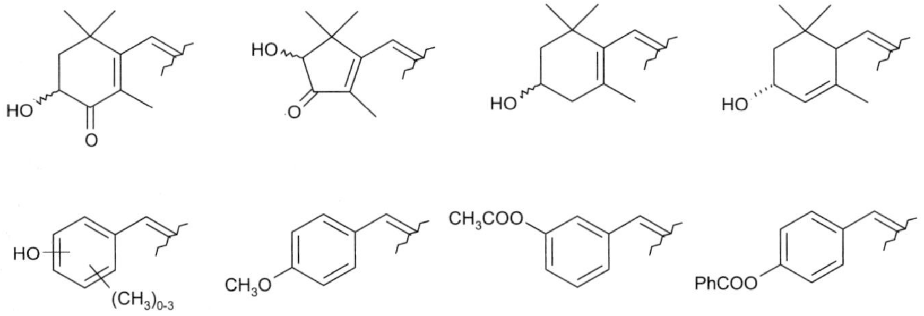

Scheme 2

In contrast, carotenoids containing the end groups shown in Scheme 3 preferentially form J-aggregates in DPPC membranes [37]. These J-aggregates have a tendency not to dissociate into monomers on mild heating.

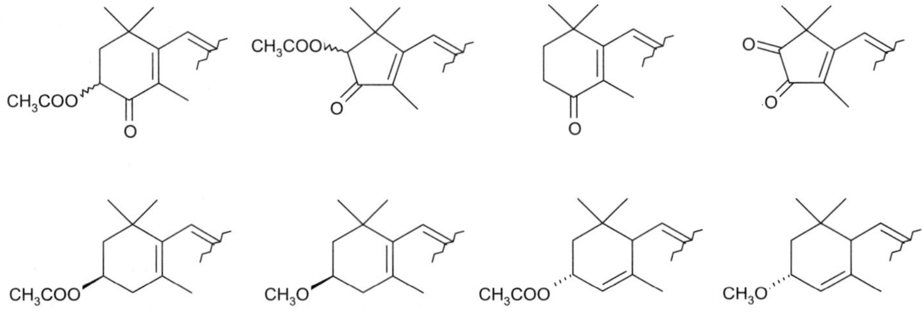

Scheme 3

e) Some natural examples

i) Astaxanthin and carotenoproteins. The CD spectroscopic behaviour of (3*R*,3'*R*)-astaxanthin (**404**) aggregates in membranes [37] (Fig. 21) is interesting because the results allow a comparison with natural (3*S*,3'*S*)-astaxanthin (**406**)-protein complexes (*Chapter 6*). The CD spectrum shows a negative CD couplet from 529 to 434 nm ($\Delta\varepsilon_{529}$ - $\Delta\varepsilon_{434}$ = -38), the minimum of which is located at the bathochromic shoulder of the UV/Vis band. Maxima of both spectra are at the same wavelength. The two Cotton effects coincide with the two UV/Vis transitions. This behaviour indicates the existence of very small H-aggregate oligomers, probably dimers, which have left-handed, helically twisted individual molecules.

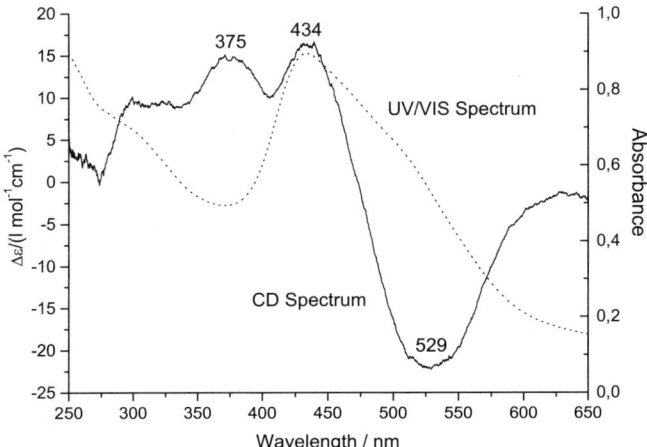

Fig. 21. CD and UV/Vis spectra of (3*R*,3'*R*)-astaxanthin (**404**) (4 mol%) in small unilamellar DPPC vesicles (T_m = 41.4°C) at room temperature [37].

If chiral *L*-DPPC is used for vesicle preparation, no change occurs, which indicates that the origin of the CD lies with chirally twisted carotenoid monomers and not with any possible interaction with a chiral lipid environment [37]. The apparent split attributed to aggregated astaxanthin, ΔN = 4138 cm^{-1} (0.51 eV), and the amplitude A = -38 l mol^{-1} cm^{-1} are typical of tightly packed carotenoid dimers.

Two carotenoid chromophores covalently linked to each other will be in close contact and will mimic a tightly packed intermolecular dimer. Such bichromophoric 'dimers' can therefore serve as good models for non-covalent dimers [82-84]. The C$_{30}$ diamide *15*, for example, in chloroform, gives a split of 3413 cm^{-1} (0.42 eV), with amplitude A= - 27.4 l mol^{-1} cm^{-1}.

It therefore seems justified to take values around 3000 and 4000 cm^{-1} as apparent splittings for tightly bound small oligomers of carotenoids, at van der Waals distances (*ca.* 0.4 nm) and with favourable geometrical alignment.

Carotenoproteins arc discusscd in *Chapter 6*, but it is worthwhile here to discuss exciton splitting in β-crustacyanin in this context. A recent X-ray crystallographic investigation concludes that the exciton coupling due to the proximity of two astaxanthın (**406**) chromophores in β-crustacyanin is large, an exciton splitting of 0.49 eV being calculated by theoretical methods (using experimentally determined geometries of the two astaxanthin molecules) which would mean that interaction between the astaxanthin molecules in β-crustacyanin is approximately as large as in the covalently linked bichromophore *15* [85].

The CD spectrum of β-crustacyanin shows a change of sign at the absorption wavelength; this is attributed to exciton splitting and shows that the chromophores of the two astaxanthin molecules are in proximity and interact energetically [86]. The apparent split for β-crustacyanin amounts to about 2600 cm^{-1} (about 0.32 eV) which is an upper limit to the *Davydov* split. Thus, surprisingly, the astaxanthin chromophores in β-crustacyanin interact energetically as much as the subchromophores of the diiminium dication *16* even though they are considerably further apart (7 Å) than the chromophores of *16* (about 3 Å).

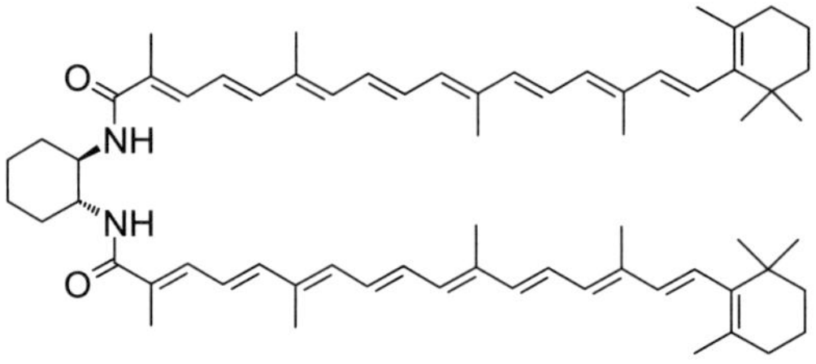

(15) 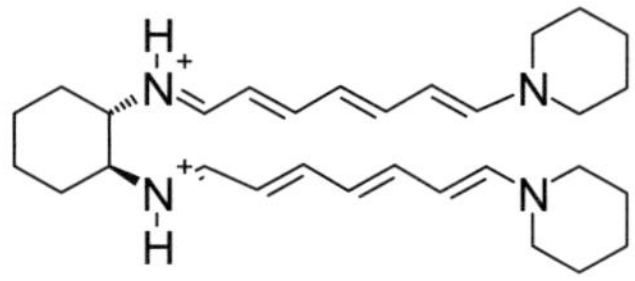   (16)

ii) The macular xanthophyll-binding protein. Recently, a membrane-associated xanthophyll-binding protein (XBP) was purified from human macula and identified as a Pi isoform of human glutathione *S*-transferase (GSTP1). This XBP interacts with (3R,3'S, *meso*)-zeaxanthin (**120**) and dietary (3R,3'R)-zeaxanthin (**119**) but only weakly with (3R,3'R,6'R)-lutein (**133**) [87]. It can integrate up to two molecules of zeaxanthin. From the UV/Vis and CD spectra it is evident that no H-aggregates are being formed, either in GSTP1 or in the detergent CHAPS, but zeaxanthin displays J-type exciton interactions in both of them. Furthermore the UV/Vis and CD spectra of GSTP1-zeaxanthin are almost identical to the UV/Vis and CD spectra of zeaxanthin diacetate (a J-aggregate former) in acetone/water (4:6), or lutein diacetate (a J-aggregate former), in DPPC membranes and are not as bathochromically shifted as the spectra of the zeaxanthin J-aggregate (Fig. 14).

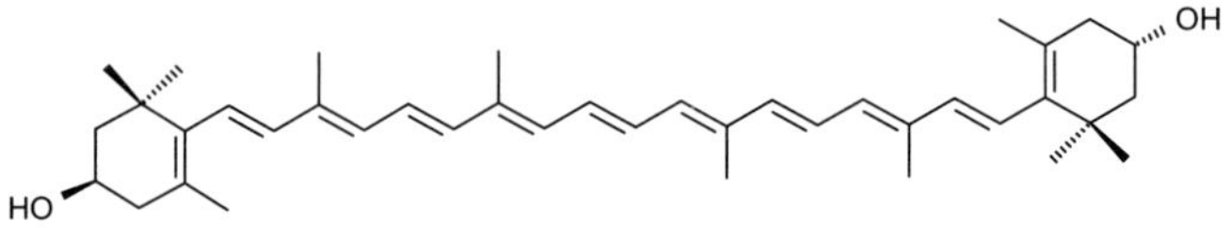

(3R,3'S, *meso*)-zeaxanthin (**120**)

It may be concluded that the two zeaxanthin chromophores are located within GSTP1 in a way completely different from that in DPPC vesicles and held in such a manner that OH-O bridges between zeaxanthin molecules are not possible.

f) New applications

i) Bichromophores. In relation to the design of new carotenoid derivatives for new applications, a dual bixin anilide was shown to enter DPPC membranes in a U-shaped conformation, thereby opening water channels for migration of reactants [54]. Also a membrane-spanning zwitterionic caroviologen, incorporated into vesicle bilayer membranes, functioned as a molecular wire, mediating electron transfer from an external reducing phase to an internal oxidizing phase [88].

ii) Carotenophospholipids. Carotenoid derivatives such as (8'-apo-β-caroten-8'-oyl)-glycerophosphatidylcholine (*17*) show special surface and aggregation effects in the absence of other lipids. In water, the carotenoylphosphatidylcholine forms aggregates with an average hydrodynamic radius r_H of 8 nm within a narrow range of r_H values (6-14 nm), though some larger aggregates with r_H = 40-600 nm are also observed. Single-chain amphiphiles such as this are expected to aggregate in micelles, but the radius of the aggregates of carotenoylphosphatidylcholine points towards another aggregate morphology. In aggregates and monomolecular surface films of saturated or *cis*-unsaturated phospholipids, dense packing of the monomers is hampered by the flexible chain or by the twist of *cis* double bonds. In contrast, carotenoylphosphatidylcholine has a rigid, elongated chain as a result of the all-*trans* configuration of the double bonds, and this favours a close orientation of aggregated and surface-adsorbed molecules [89,90]. The UV/Vis and CD properties of the aggregates can be attributed to an optically active P-oligomer unit, built from eight optically inactive carotenoylphosphatidylcholine monomers [89-91].

(*17*)

iii) Supramolecular assemblies of lycopene. Supramolecular assemblies of lycopene (**31**) in a chitosan matrix have been studied by UV/Vis absorption spectroscopy and transmission electron microscopy (TEM) [92]. As with aqueous acetone solutions (Fig.15) or in Langmuir-Blodgett films [92], aggregate formation was also observed when lycopene-chitosan complexes were dissolved in water containing 1% acetic acid. The main absorption band of the solution, however, was further blue-shifted to 349 nm. The exciton model was used to interpret the spectral shifts of the aggregates. The TEM images showed the dendritic, cubic or triangular shapes of the supramolecular assemblies of lycopene in the chitosan matrix. From this, it is proposed that the supramolecular assemblies form liquid crystals of the nematic type and contain both H-aggregation (more) and J-aggregation (little). The changes in the proportion of H-type or J-type aggregation cause the hypsochromic or bathochromic shift of the absorption spectrum [92].

3. Influence of additives

The aggregation of carotenoid molecules during precipitation in aqueous systems leads to supramolecular complexes with properties that are different from those of the single molecule in, for example, colour strength, colour hue, stability against oxygen and bioavailability.

Conventional UV/Vis spectroscopy is a simple but very effective method to characterize these complexes and can be used to measure the specific colour hue and intensity of a nanoparticle carotenoid dispersion in water. The most important characteristics of the nanoparticles revealed by the intensity and shape of this absorbance curve are:-

(i) Particle size: the smaller the size, the higher is λ_{max} and the lower the 'turbidity', *i.e.* the absorbance in the long wavelength range (>600 nm).

(ii) Molecular packing: the shape of the spectral curve for amorphous nanoparticles is similar to that of dissolved molecules in organic solvent; typical H-aggregates show blue-shifts, typical J-aggregates red-shifts. Figure 22 (left) shows spectra of nanosized β-carotene particles produced by mixing-chamber precipitation [11, 93] under conditions leading to different sizes.

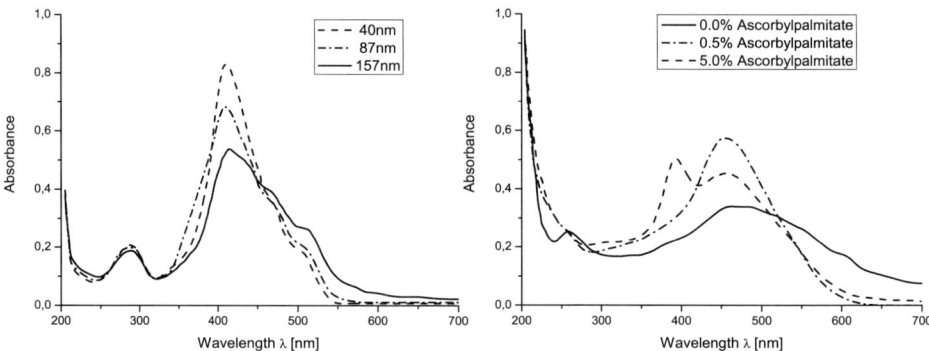

Fig. 22. UV/Vis spectra of (left) nanosized β-carotene (**3**) particles in water, and (right) 5 ppm nanosized astaxanthin (**404**) in water containing different concentrations of ascorbyl palmitate.

By this method, the influence of additives on molecular aggregation can easily be investigated. One example is the precipitation of astaxanthin (**404**) in the presence of various amounts of ascorbyl palmitate, a surface-active substance. As the amount of ascorbyl palmitate is increased, the size of particles is reduced and the degree of H-aggregation is increased (Fig. 22 right).

Another possibility to investigate the influence of additives is molecular modelling [94]. The evolution of a nucleus and its further growth can be simulated stochastically with the software EVOCAP (Excluded Volume Constrained Assembly Packing) [95].

If this technique is applied to astaxanthin in water (two molecules astaxanthin, four molecules water), the H-aggregates (Fig. 2 upper) are shown to be more stable ($\Delta E - $ -58.70 kcal/mol) than the J-aggregates ($\Delta E = $ -66.77 kcal/mol) (Fig. 2 lower). This is confirmed by the experiment. On the other hand the free energy often favours J-aggregates.

4. Natural and artificial nanosized particles containing carotenoids: non-crystalline, crystalline and crystalloidal materials

The terms 'nanosized structures' or 'nano-particles' are applied to any sub-micron structures from *ca.* 50 nm up to *ca.* 1000 nm [96]. There are many examples of natural nano-particles (*ca* 50-500 nm) containing carotenoids, variously described as extra-chloroplastidic lipid globules, cytoplasmic lipid droplets (*Chlamydomonas nivalis* [97,98]), plastoglobuli (*Triticum aestivum* [99,100]), vesicles, lipid granules, globules (*Chlamydomonas reinhardtii* [101-103]), ring vesicles, homogeneous granules, droplets (*Anolis carolinensis* [104], *Cynops pyrrhogaster* [105], *Hyla cinerea* [106]), tubules, plastoglobules, vesicles (*Crocus sativus* [107]), lipid globular droplets, bodies (crystals), crystalloids, plate-like pigment sacs, plastoglobules, membranes, tubules, plastoglobulin-type sacs, tablet-shaped and rod-shaped crystals (*Lycopersicon esculentum* [108-114]), lipid bodies, cytoplasmic lipid vesicles, extraplastidic bodies (*Haematococcus pluvialis* [115-117]), triacylglycerol droplets, interthylakoid globules (*Dunaliella bardawil, D. salina* [118,119]), cytosolic lipid globules (*Xanthophyllomyces dendrorhous*, formerly *Phaffia rhodozyma*, [120-121]).

The following discussion is centred on carotenoid-containing structures in chromoplasts and a comparison between natural particles and artificial, commercial aggregate particles. Other aspects of chromoplast pigmentation are mentioned in *Chapter 10*.

a) Carotenoid aggregates and crystals in chromoplasts

During the maturation of some chromoplasts, the concentration of carotenoids can increase so much that they crystallize. The best known are the large crystals of β-carotene (**3**) in carrot roots and in narcissus flowers, and lycopene (**31**) crystals in tomato fruit. The peculiarity of these crystals is that they develop intra-thylakoidally, *i.e.* inside the lumen of some thylakoids. The crystallization begins even in chloro-chromoplasts. The crystals remain enveloped by a membrane even in mature chromoplasts, which are completely devoid of photosynthetic membranes. In chromoplasts of *Narcissus*, β-carotene crystals are present, in addition to internal membranes. A rarer form of carotene crystals has been found in mature chromoplasts of flowers of the tulip tree (*Liriodendron tulipifera*), where small crystals of β-carotene develop inside the plastoglobules. Structurally somewhat different types of crystals develop inside plastoglobules in the chromoplasts of the fruit of *Cucurbita maxima* cv. *Turbaniformis*; long crystals grow out of plastoglobules, so that they assume the shape of a tadpole [110].

The following terms are used to describe the structures seen [108,112,113].

i) Carotenoid crystals: pure crystalline carotenoid.

ii) Carotenoid bodies: native structures which may occur in plastids or free in cytoplasm, and often remain associated with cytoplasmic residues even after isolation.

iii) Carotenoid crystalloids: similar to carotenoid bodies but membrane-enclosed.

iv) Crystalline thylakoid: initial crystalline structure, apparently consisting of a flat membranous sac.

It is difficult to assess, even approximately, which types of aggregate occur in these carotenoid-containing particles and structures, especially since a variety of pigment-containing structures may exist even in the same chromoplast. Some of these structures are transient, being present only in unripe chromoplasts and disappearring again in mature or senescent chromoplasts; a large variety of carotenoid assemblies play a role for shorter or longer periods.

Isolated carotene crystals from *Narcissus poeticus* are composed of pure (all-*E*)-β-carotene, not a conglomerate of alternating layers of lipoproteins and carotenes [122]. Lycopene bodies and lycopene crystals in tomato are similarly composed of pure (all-*E*)-lycopene [108].

Regardless of the complexity of the final crystal configuration, the initial crystalline structure apparently consists of a 'crystalline thylakoid' [112,113]. During ripening of tomato tissue and chromoplast maturation, active biosynthesis of carotenoids begins and the rapid accumulation of lycopene results in crystallization. During the transformation from chloroplast to chromoplast [111], the crystals remain enveloped by a membrane even after crystallization, with implications for bioavailability (*Volume 5, Chapter 7*). Observations of β-carotene crystals in a high β-carotene variety of tomato by transmission electron microscopy (TEM) indicated that the membranes with which these crystals are associated appear to be different from those associated with lycopene in the normal red variety [112,113]. A cross-sectional view of lycopene crystalloids indicates that they are initiated as plate-like pigment sacs that are formed in association with the thylakoid membrane. The crystalloids increase in size by 'involution in preferred planes' to produce tubular crystals. In the high β-carotene varieties, the greater occurrence of plastoglobuline structures is apparent, and the β-carotene is mainly dissolved in lipid material of the globules. In the red tomato varieties, the relatively small amount of β-carotene that is present is dissolved in lipid material of the globules and does not crystallize whereas lycopene, which is associated with the thylakoid membrane, does [111-114,123].

b) Chiral assemblies

Natural assemblies of carotenoids in tubulous chromoplasts of flowers and fruits have been investigated. Optical activity has been detected by CD spectroscopy in carotenoid-containing living flowers of several species belonging to different families. Under appropriate conditions, natural pure xanthophyll esters give very similar CD spectra *in vitro*, proving the ability of these molecules to form chiral self-assemblies. Circular dichroism and UV/Vis curves of lutein (**133**) diacetate in aqueous ethanol compared excellently with the spectra of a *Chelidonium majus* petal containing a high concentration of lutein esters [9,124]. Apolar carotenoids are able to form a liquid-crystal phase characterized by strong optical anisotropy and are good candidates for building up a nematic phase (Fig. 13).

Another study investigated the CD spectra of carotenoids *in situ* in slices, homogenates and isolated chromoplasts of some fruit, especially red pepper (*Capsicum annuum*) which is known to have tubulous chromoplasts [125]. The results are consistent with a structure in which the carotenoid esters in the chromoplasts form a liquid-crystal phase characterized by strong optical anisotropy. In this, they have no strict positional order but tend to point in the same direction. Because xanthophyll esters are mesogenes containing chiral centres, they form a chiral nematic liquid-crystal phase due to intermolecular forces that favour alignment between molecules at a slight angle to one another. This gives a structure like a stack of two-dimensional layers with the preferred molecular orientation in each layer twisted with respect to those above and below, to form a continuous helical pattern about the line perpendicular to the layers, similar to that illustrated in Fig. 13 [125].

c) Primary aggregates and crystals

In many cases, aggregation is a first stage *en route* to crystallization. Therefore partial and/or incomplete crystal structures or sections from the crystal lattice may be used for building smaller assemblies and, when conditions are appropriate, incipient crystallization will be the favoured process.

The crystal structures of β-carotene (**3**) and some other carotenoids are known [126-128] (see *Chapter 4*). A conspicuous optical feature of crystalline β-carotene is a red shift of the 0-0 absorption band by 2250 cm^{-1} compared with that in hexane [129]. This red shift has been observed in many studies of natural or synthetic β-carotene and rationalized by considering intermolecular interactions in the crystal lattice. The absorption spectrum of β-carotene as a crystal suspension in water, compared with the spectrum in benzene, shows a broad shoulder at *ca.* 540 nm [130].

If polarized light in the visible region is irradiated along the **a** or **b** crystal axis of a single crystal of (all-*E*)-β-carotene at room temperature, an intensity increase is observed for the **b** direction only. Based on symmetry considerations, these results can be explained by assigning the bathochromic absorption to the transition to ^1A_u (parallel to the **b**-axis) and ^1B_u (parallel to the **a**-axis) molecular exciton states with axes as shown in Fig. 23. Changing and modifying the view by taking **a** and **b** axes as viewing directions, one arrives at the two pictures illustrated in Fig. 24.

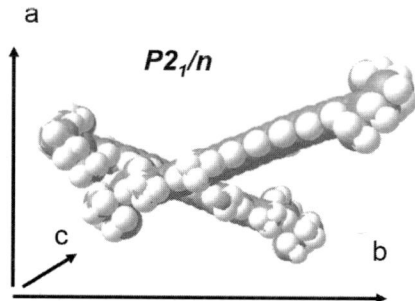

Fig. 23. Projection onto the (001) plane of the 'dimer' of an (all-*E*)-β-carotene single crystal [128]. The transition to the ^1A_u molecular exciton state is polarized parallel to the **b**-axis and the transition to the ^1B_u molecular exciton state is polarized parallel to the **a**-axis. For space group information and unit cell parameters see [128].

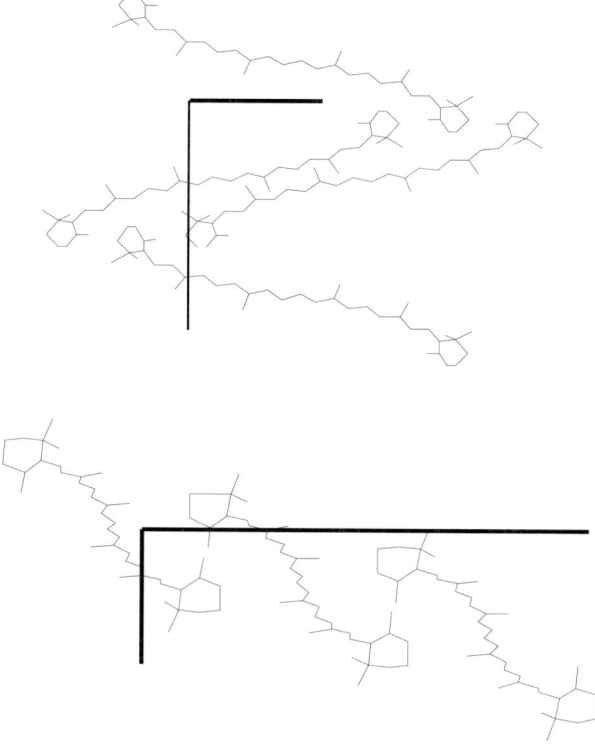

Fig. 24. (Upper) View along axis **a** onto plane (100) of a solid (all-*E*)-β-carotene single crystal. (Lower) View along axis **b**. The fourth carotene is concealed behind the central carotene.

This leads to the hypothesis that if, in the incipient stages of aggregate formation in chromoplasts, growth is not equal in all three dimensions but there is a more linear growth in one or two specific dimensions, according to the limiting conditions set by surrounding molecules and systems, then this would be expected to generate aggregate structures that resemble those produced by experimental procedures with designed boundary conditions and constraints. These 'restricted aggregates' could be among those elusive natural transient assemblies not detected under conventional inspection during chromoplast development.

Violaxanthin (**259**) shows aggregation phenomena that do not fit easily into simple ideas of aggregate structure (*e.g.* head-to-tail or card-pack). When violaxanthin is aggregated either in solution in 2% ethanol in water or on a quartz slide by evaporation from benzene/ethanol (9:1) solution [38], spectra are obtained that show almost identical aggregate forms. Gaussian deconvolution of the absorption spectra of both gives exactly the same maxima: 382, 407, 434, 464 and 500 nm. Such a finding can be taken as conclusive evidence that the same spectral forms are present in both samples, although in different proportions.

If, as a first approximation, the solid structure of violaxanthin is idealized by using the crystal lattice of β-carotene, selecting a number of molecules as points (centres), and allowing transition dipole interactions between four molecules at the indicated lattice positions and with the molecule placed at the lattice point (0,0,1), the calculated wavelengths obtained are 377, 397, 436, 464, 484, 519, 540, 584 nm.

These results are partly satisfactory but partly misleading, especially for the most bathochromically shifted predicted bands. However, if excessive contributions from axis **b** are excluded and the number of molecules interacting along direction **b** is reduced to two, better agreement is obtained. Obviously violaxanthin forms primary aggregates preferentially in the crystallographic plane (010) [38].

Since the findings are close to those obtained from natural situations on cell or membrane surfaces, it is attractive to assume that natural primary aggregates start with two-dimensional sub-lattices of the complete three-dimensional space. Primary aggregates are confined in their growth by caging them in membranes, biomacromolecules or supramolecular frames, and this may lead to observations of otherwise elusive small oligomers. In addition, these aggregates may be considered as low-dimensional precursors of the complete lattice.

A well-known natural example is provided by crustochrin, the yellow protein of the lobster carapace (*Chapter 6*). This contains 20 or so stacked astaxanthin (**406**) molecules which are stabilized by protein, and interact by exciton-exciton interaction, and is thus an excellent example of a natural H-aggregate of astaxanthin units, with the chromophores stacked in more-or-less one dimension and leading to hypsochromically shifted bands.

d) Nanodispersions

It is now possible to comment on data coming from industrial colloid chemistry and the synthesis of nanodispersions of active organic compounds. The increase in solubility and the improvement in biological absorption are achievable only with particle sizes in the middle or lower nanometre range (50-500 nm) [11,42,93].

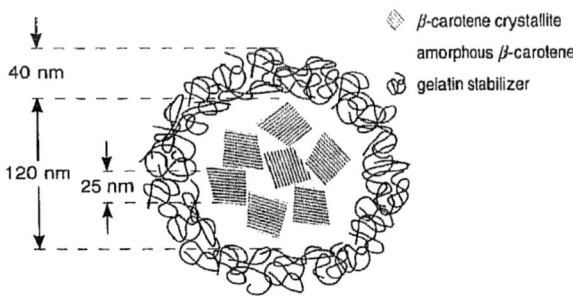

Fig. 25. Core-shell structure of the β-carotene (3) hydrosol particle.

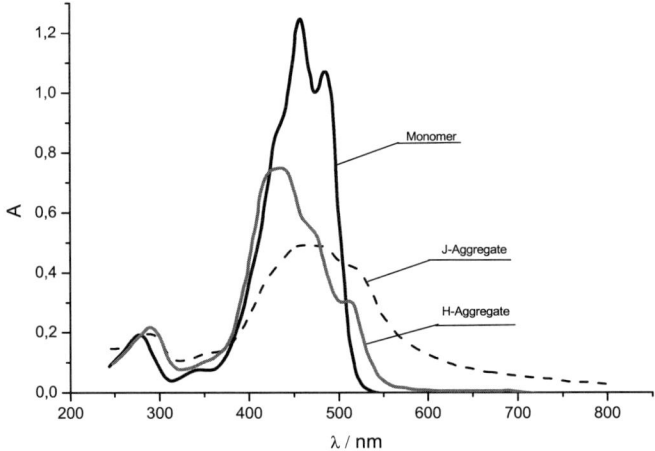

Fig. 26. UV/Vis absorption spectra of a 5 ppm β-carotene (3) solution in cyclohexane (solid black line) and of 5 ppm β-carotene hydrosols. The H-aggregated β-carotene hydrosol (solid grey line) was precipitated under dilute conditions. The J-aggregated β-carotene hydrosol (dashed line) was precipitated at higher concentration and in the presence of corn oil [42].

The simple model in Fig. 25 reflects most of the experimentally determined properties. The spectroscopic data and wide-angle X-ray scattering (WAXS) measurements show the active material core to consist of H- and J-aggregates up to 30 nm in size, which corresponds to a maximum aggregation number of 10,000 molecules [11,42,93]. The spectra of these particles vary according to the kind of aggregate type prevailing in the crystallites (Fig. 26).

e) How natural and artificial assemblies may form

Calculations have been used to explain the origin of the colour and spectral shifts in these core particles. For the treatment of 'real' tetramers, four β-carotene (**3**) molecules (the 'monomers'), structurally optimized to ensure packing effects, are set into two crystallographic unit cells of the crystal lattice of β-carotene. The two cells are joined along the **a** axis (2_1_1), the **b** axis (1_2_1) and the **c** axis (1_1_2) [42].

The resulting structures are illustrated in Fig.27, which shows the view into the lattice that has two unit cells in either the **a**, the **b** or the **c** direction, respectively. The calculated spectral properties of these supercells, listed in Table 1, are consistent with the experimental data. These model aggregates provide a good description of how natural and artificial assemblies form. The behaviour calculated for aggregates (2_1_1) and (1_1_2) is typical of H-aggregates. The fact that there are calculated long-wavelength absorptions of low intensity is normal and does not detract from the H-character. The same is observed with the astaxanthin (**404**) or capsorubin (**413**) dodecamers.

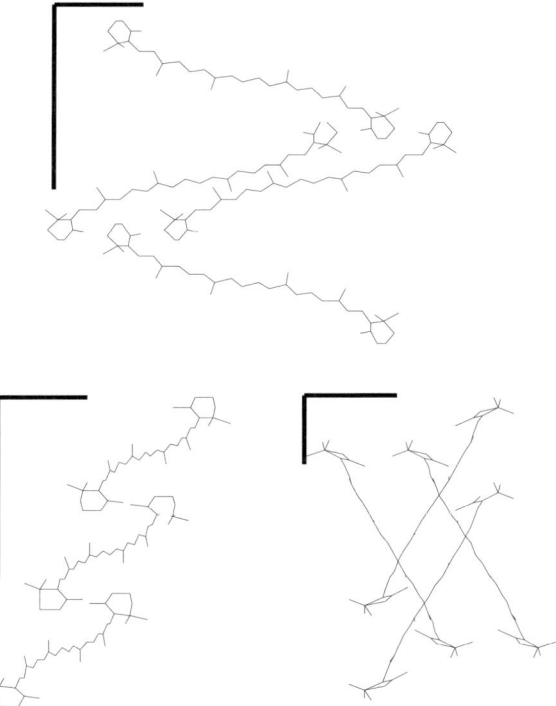

Fig. 27. View into two unit cells of a β-carotene (**3**) crystal. Top: along **a** into **2a**. Bottom left: along **b** into **2b** (the fourth carotene is hidden behind the central one). Bottom right: along **c** into **2c**.

Table 1. UV/Vis transitions and corresponding oscillator strengths of β-carotene, calculated with the CIS model for a monomer, for an ideal parallel oriented tetramer aggregate, and for tetramer supercells as illustrated in Fig. 27 comprising two unit cells along the crystallographic **a**, **b** and **c** axes. The designation of supercell orientation used here avoids confusion with Miller indices or with description of crystal surfaces.

Orientation (**a**, **b**, **c**)	Absorption maximum	Oscillator strength
Monomer	509 nm	3.9
ideal H-Aggregate	465 nm	16.4
(2_1_1)-Aggregate	500 nm	2.3
	469 nm	14.3
(1_2_1)-Aggregate	553 nm	4.0
	502 nm	9.6
(1_1_2)-Aggregate	519 nm	2.1
	487 nm	13.4

These two species, (2_1_1) and (1_1_2), (and their many relatives having larger supercells) can be compared directly with the above-mentioned violaxanthin (**259**) aggregates [38], and with the astaxanthin (**406**) H-aggregate of the yellow lobster protein crustochrin. Both violaxanthin and crustochrin display not only the hypsochromically shifted main absorption band but also the lower intensity absorption around 500 nm. The H-aggregated β-carotene (**3**) hydrosol of Fig. 26, which was precipitated under dilute conditions, can therefore be described well by structures corresponding to sections of the natural crystal lattice of β-carotene, provided that axis **b** is inferior to axes **a** and **c**. The same observation has been made elsewhere [42]. In addition, the analogy of this H-hydrosol with known natural H-aggregates, such as crustochrin, is striking.

Aggregate (1_2_1) serves as a good description of the experimental J-aggregated β-carotene hydrosol, which was precipitated at higher concentration and in the presence of corn oil. Structures like this may also be widely disseminated in Nature and may belong to the plethora of elusive transient species that will finally lead to structures such as carotenoid crystals, carotenoid bodies, carotenoid crystalloids, or crystalline thylakoids. Spectra of solid β-carotene (**3**) allow direct comparison and show that the typical J band above 500 nm originates from excitonic coupling which results in the ^1A_u molecular exciton state polarized parallel to axis **b** [128]. The development of carotenoid assemblies in chromoplasts has mostly been characterized by the later and latest stages in chloro-chromoplast transformations and similar phenomena, but J-aggregates of this hydrosol type presumably represent very early stages of this development.

References

[1]   G. Scheibe, *Angew. Chem., 61*, 300 (1949).

[2]   M. Kasha, *Radiat. Res., 20* (1936).

[3]   R. M. Hochstrasser and M. Kasha, *Photochem. Photobiol., 3*, 317 (1964).

[4]   E. E. Jelly, *Nature,* 1009,**138** (1936).

[5]   E. E. Jelly, *Nature,* **139**, 631 (1937).

[6]   T. M. Raschke, J. Tsai and M. Levitt, *Proc. Natl. Acad. Sci. USA., 98*, 5965 (2001).

[7]   N. Muller, *Acc. Chem. Res., 23*, 23 (1990).

[8]   P. L. Privalov and S. J. Gill, *Pure Appl. Chem., 61*, 1097 (1989).

[9]   J. Derucre, S. Romer, A. Dharlingue, R. A. Backhaus, M. Kuntz and B. Camara, *Plant Cell, 6*, 119 (1994).

[10]   H. H. Billsten, V. Sundström and T. Polivka, *J. Phys. Chem. A, 109*, 1521 (2005).

[11]   D. Horn and J. Rieger, *Angew. Chem. Int. Edn., 40*, 4331 (2001).

[12]   M. Buchwald and W. P. Jencks, *Biochemistry, 7*, 834 (1968).

[13]   M. Buchwald and W. P. Jencks, *Biochemistry, 7*, 844 (1968).

[14]   A. Hager, *Planta, 91*, 38 (1970).

[15]   R. G. Parr and R. G. Pearson, *J. Am. Chem. Soc., 105*, 7512 (1983).

[16]   A. V. Ruban, P. Horton and A. J. Young, *J. Photochem. Photobiol. B, 21*, 229 (1993).

[17]   Y. Mori, K. Yamano and H. Hashimoto, *Chem. Phys. Lett., 254*, 84 (1996).

[18]   C. Köpsel, *Dissertation* , University of Düsseldorf (1999).

[19]   H. Auweter, H. Benade, H. Bettermann, S. Beutner, C. Köpsel, E. Lüddecke, H. D. Martin and B. Mayer, in *Proc. Int. Congr. Pigments in Food Technol.*, (ed. I. M. M. Mosquera, M. J. Galan and D. H. Mendez, p. 197 (Sevilla, 1999).

[20]   C. Köpsel, H. Möltgen, H. Schuch, H. Auweter, K. Kleinermanns, H. D. Martin and H. Bettermann, *J. Mol. Struct., 750*, 109 (2005).

[21]   V. Buss, *Softwareentwicklung in der Chemie,* Springer (1989).

[22]   F. Dietz, *J. Signalauszeichungsmaterialien, 1*, 157 (1973).

[23]   F. Dietz, *J. Signalauszeichungsmaterialien, 4*, 237 (1973).

[24]   F. Dietz and C. Glier, *J. Signalauszeichungsmaterialien, 1*, 221 (1973).

[25]   K. Norland, A. Ames and T. Taylor, *Photographic Sci. Eng., 14*, 295 (1970).

[26]   G. R. Bird, K. S. Norland, A. E. Rosenoff and H. B. Michaud, *Photographic Sci. Eng., 12*, 196 (1968).

[27]   F. Zsila, Z. Bikadi, J. Deli and M. Simonyi, *Chirality, 13*, 446 (2001).

[28]   F. Zsila, J. Deli, Z. Bikadi and M. Simonyi, *Chirality, 13*, 739 (2001).

[29]   F. Zsila, Z. Bikadi, J. Deli and M. Simonyi, *Tetrahedron Lett., 42*, 2561 (2001).

[30]   Z. Bikadi, F. Zsila, J. Deli, G. Mady and M. Simonyi, *Enantiomer, 7*, 67 (2002).

[31]   M. Simonyi, Z. Bikadi, F. Zsila and J. Deli, *Chirality, 15*, 680 (2003).

[32]   H. D. Martin and T. Werner, *J. Mol. Struct., 266*, 91 (1992).

[33]   W. I. Gruszecki, B. Zelent and R. M. Leblanc, *Chem. Phys. Lett., 171*, 563 (1990).

[34]   S. Takagi, T. Yamagami, K. Takeda and T. Takagi, *Agric. Biol. Chem., 51*, 1567 (1987).

[35]   J. Lematre, B. Maudinas and C. Ernst, *Photochem. Photobiol., 31*, 201 (1980).

[36]   F. Zsila, Z. Bikadi, Z. Keresztes, J. Deli and M. Simonyi, *J. Phys. Chem. B, 105*, 9413 (2001).

[37]   M. Korger, *Dissertation*, University of Düsseldorf (2005).

[38]   W. I. Gruszecki, *J. Biol. Phys., 18*, 99 (1991).

[39]   E. Wloch, S. Wieckowski and A. M. Turek, *Photosynthetica, 21*, 2 (1987).

[40]   P. S. Song and T. A. Moore, *Photochem. Photobiol., 19*, 435 (1974).

[41]   B. J. Berne and R. Pecora, *Dynamic Light Scattering: With Applications to Chemistry, Biology, and Physics,* Dover Publications, Inc, New York (2000).

[42]   H. Auweter, H. Haberkorn, W. Heckmann, D. Horn, E. Lüddecke, J. Rieger and H. Weiss, *Angew. Chem. Int. Edn., 38, 2188 (1999).

[43]   V. R. Salares, N. M. Young, P. R. Carey and H. J. Bernstein, *J. Raman Spectrosc., 6, 282 (1977).

[44]   G. Orlandi, F. Zerbetto and M. Z. Zgierski, *Chem. Rev., 91, 867 (1991).

[45]   L. Rimai, M. E. Heyde and D. Gill, *J. Am. Chem. Soc., 95, 4493 (1973).

[46]   H. Okamoto, H. O. Hamaguchi and M. Tasumi, *J. Raman Spectrosc., 20, 751 (1989).

[47]   G. Binnig, H. Rohrer, C. Gerber and E. Weibel, *Phys. Rev. Lett., 49, 57 (1982).

[48]   G. Binnig, H. Rohrer, C. Gerber and E. Weibel, *Appl. Phys. Lett., 40, 178 (1982).

[49]   G. Binnig, C. F. Quate and C. Gerber, *Phys. Rev. Lett., 56, 930 (1986).

[50]   S. N. Maganov and M.-H. Whangho, *Surface Analysis with STM and AFM,* VCH, Weinheim (1996).

[51]   S. De Feyter, A. Gesquiere, M. M. Abdel-Mottaleb, P. C. M. Grim, F. C. De Schryver, C. Meiners, M. Sieffert, S. Valiyaveettil and K. Mullen, *Acc. Chem. Res., 33, 520 (2000).

[52]   G. Leatherman, E. N. Durantini, D. Gust, T. A. Moore, A. L. Moore, S. Stone, Z. Zhou, P. Rez, Y. Z. Liu and S. M. Lindsay, *J. Phys. Chem. B, 103, 4006 (1999).

[53]   T. Miyahara and K. Kurihara, *J. Am. Chem. Soc., 126, 5684 (2004).

[54]   J. H. Fuhrhop, M. Krull, A. Schulz and D. Mobius, *Langmuir, 6, 497 (1990).

[55]   E. Gomar-Nadal, G. K. Ramachandran, F. Chen, T. Burgin, C. Rovira, D. B. Amabilino and S. M. Lindsay, *J. Phys. Chem. B, 108, 7213 (2004).

[56]   G. K. Ramachandran, J. K. Tomfohr, J. Li, O. F. Sankey, X. Zarate, A. Primak, Y. Terazono, T. A. Moore, A. L. Moore, D. Gust, L. A. Nagahara and S. M. Lindsay, *J. Phys. Chem. B, 107, 6162 (2003).

[57]   A. Wegmann, B. Tieke, J. Pfeiffer and B. Hilti, *J. Chem. Soc. Chem. Commun., 586 (1989).

[58]   L. Sereno, J. J. Silber, L. Otero, M. delValle-Bohorquez, A. L. Moore, T. A. Moore and D. Gust, *J. Phys. Chem., 100, 814 (1996).

[59]   A. L. Moore, T. A. Moore, D. Gust, J. J. Silber, L. Sereno, F. Fungo, L. Otero, G. Steinberg-Yfrach, P. A. Liddell, S.-C. Hung, H. Imahori, S. Cardoso, D. Tatman and A. N. Macpherson, *Pure Appl. Chem., 69, 2111 (1997).

[60]   A. Ion, V. Partali, H. R. Sliwka and F. G. Banica, *Electrochem. Commun., 4, 674 (2002).

[61]   B. J. Foss, A. Ion, V. Partali, H. R. Sliwka and F. G. Banica, *J. Electroanal. Chem., 593, 15 (2006).

[62]   B. J. Foss, A. Ion, V. Partali, H. R. Sliwka and F. G. Banica, *Collect. Czech. Chem. Commun., 69, 1971 (2004).

[63]   G. Duda and G. Wegner, *Makromol. Chem. Rapid Commun., 9, 495 (1988).

[64]   J. H. Fuhrhop and J. Köning, *Membranes and Molecular Assemblies: The Synkinetic Approach,* The Royal Society of Chemistry, Cambridge (1994).

[65]   S. N. Naess, A. Elgsaeter, B. J. Foss, B. J. Li, H. R. Sliwka, V. Partali, T. B. Melo and K. R. Naqvi, *Helv. Chim. Acta, 89, 45 (2006).

[66]   R. R. C. New, *Liposmes - A Practical Approach,* Oxford University Press, New York (1997).

[67]   R. B. Gennis, *Biomembranes,* Springer, New York (1989).

[68]   W. Okulski, A. Sujak and W. I. Gruszecki, *Biochim. Biophys. Acta, 1509, 216 (2000).

[69]   A. Sujak and W. I. Gruszecki, *J. Photochem. Photobiol. B, 59, 42 (2000).

[70]   A. Sujak, W. Okulski and W. I. Gruszecki, *Biochim. Biophys. Acta, 1509, 255 (2000).

[71]   A. Milon, G. Wolff, G. Ourisson and Y. Nakatani, *Helv. Chim. Acta, 69, 12 (1986).

[72]   R. Mendelsohn and R. W. Vanholten, *Biophys. J., 27, 221 (1979).

[73]   H. Y. Yamamoto and A. D. Bangham, *Biochim. Biophys. Acta, 507, 119 (1978).

[74]   V. D. Kolev and D. N. Kafalieva, *Photobiochem. Photobiophys., 11, 257 (1986).

[75]   V. D. Kolev, *J. Mol. Struct., 114, 257 (1984).

[76]   W. I. Gruszecki, *Stud. Biophys., 116, 11 (1986).

[77]   W. I. Gruszecki, *Stud. Biophys., 139, 95 (1990).

[78]   M. Cheron and J. Bolard, *C. R. Acad. Sci. Serie III, 292, 1125 (1981).

[79]   S. Takagi, K. Takeda, K. Kameyama and T. Takagi, *Agric. Biol. Chem.*, **46**, 2035 (1982).

[80]   S. Takagi, K. Takeda, and M. Shiroishi, *Agric. Biol. Chem.*, **46**, 2217 (1982).

[81]   S. Takagi and K. Takeda, *Agric. Biol. Chem.*, **47**, 1435 (1983).

[82]   V. Buss, K. Kolster and B. Gors, *Tetrahedron: Asymmetry*, **4**, 1 (1993).

[83]   S. Köhn, *Dissertation*, University of Düsseldorf (2004).

[84]   N. Berova, D. Gargiulo, F. Derguini, K. Nakanishi and N. Harada, *J. Am. Chem. Soc.*, **115**, 4769 (1993).

[85]   A. A. C. van Wijk, A. Spaans, N. Uzunbajakava, C. Otto, H. J. M. de Groot, J. Lugtenburg and F. Buda, *J. Am. Chem. Soc.*, **127**, 1438 (2005).

[86]   R. J. Weesie, R. Verel, F. Jansen, G. Britton, J. Lugtenburg and H. J. M. deGroot, *Pure Appl. Chem.*, **69**, 2085 (1997).

[87]   P. Bhosale, A. J. Larson, J. M. Frederick, K. Southwick, C. D. Thulin and P. S. Bernstein, *J. Biol. Chem.*, **279**, 49447 (2004).

[88]   S. Kugimiya, T. Lazrak, M. Blanchard-desce and J. M. Lehn, *J. Chem. Soc. Chem. Commun.*, 1179 (1991).

[89]   B. J. Foss, S. N. Naess, H. R. Sliwka and V. Partali, *Angew. Chem. Int. Edn.*, **42**, 5237 (2003).

[90]   B. J. Foss, H. R. Sliwka, V. Partali, C. Köpsel, B. Mayer, H. D. Martin, F. Zsila, Z. Bikadi and M. Simonyi, *Chem. Eur. J.*, **11**, 4103 (2005).

[91]   B. J. Foss, H. R. Sliwka, V. Partali, S. N. Naess, A. Elgsaeter, T. B. Melo and K. R. Naqvi, *Chem. Phys. Lipids*, **134**, 85 (2005).

[92]   L. X. Wang, Z. L. Du, R. X. Li and D. C. Wu, *Dyes Pigment.*, **65**, 15 (2005).

[93]   D. Horn and E. Lüddecke, in *Fine Particles Science and Technology* (ed. E. Pelizzetti), p. 761,   Kluwer, Dordrecht (1996).

[94]   E. Hädicke, P. Müller, E. Lüddecke, E. Runge and H. Auweter, in *PARTEC 2001, Int. Congr. Particle Technol.*, Nürnberg, Germany (2001).

[95]   M. Drache, T. Weber and G. Schmidt-Naake, *Angew. Makromol. Chem.*, **273**, 69 (1999).

[96]   H. D. Martin and S. Köhn, "Natural nano-sized particles containing carotenoids or other colorants, BASF-Report", (Ludwigshafen).

[97]   R. R. Bidigare, M. E. Ondrusek, M. C. Kennicutt, R. Iturriaga, H. R. Harvey, R. W. Hoham and S. A. Macko, *J. Phycol.*, **29**, 427 (1993).

[98]   D. Remias, U. Lutz-Meindl and C. Lutz, *Eur. J. Phycol.*, **40**, 259 (2005).

[99]   C. Dahlin and H. Ryberg, *Physiol. Plant.*, **68**, 39 (1986).

[100]   M. D. Smith, D. D. Licatalosi and J. E. Thompson, *Plant Physiol.*, **124**, 211 (2000).

[101]   M. R. Lamb, S. K. Dutcher, C. K. Worley and C. L. Dieckmann, *Genetics*, **153**, 721 (1999).

[102]   D. G. W. Roberts, M. R. Lamb and C. L. Dieckmann, *Genetics*, **158**, 1037 (2001).

[103]   C. L. Dieckmann, *Bioessays*, **25**, 410 (2003).

[104]   N. J. Alexander and W. H. Fahrenbach, *Am. J. Anatomy*, **126**, 41 (1969).

[105]   K. Matsui, J. Marunouchi, and M. Nakamura, *Pigm. Cell. Res.*, **15**, 265 (2002).

[106]   J. T. Bagnara, J. D. Taylor and M. E. Hadley, *J. Cell Biol.*, **38**, 67 (1968).

[107]   M. G. Caiola and A. Canini, *Plant Biosyst.*, **138**, 43 (2004).

[108]   Y. Ben-Shaul and Y. Nafrali, *Protoplasma*, **67**, 333 (1969).

[109]   P. Sitte, H. Falk and B. Liedvogel, in *Pigments in Plants* (ed. F.C.Czygan) p. 117, Fischer, Stuttgart (1980).

[110]   N. Ljubesic, M. Wrischer and Z. Devide, *Int. J. Dev. Biol.*, **35**, 251 (1991).

[111]   M. Nguyen, D. Francis and S. Schwartz, *J. Sci. Food Agric.*, **81**, 910 (2001).

[112]   S. W. Rosso, *J. Ultrastruct. Res.*, **20**, 179 (1967).

[113]   S. W. Rosso, *J. Ultrastruct. Res.*, **25**, 307 (1968).

[114]   W. M. Harris and A. R. Spurr, *Am. J. Bot.*, **56**, 369 (1969).

[115]   Z. Sun, F. X. Cunningham and E. Gantt, *Proc. Natl. Acad. Sci. USA*, **95**, 11482 (1998).

[116]   K. Grunewald, M. Eckert, J. Hirschberg and C. Hagen, *Plant Physiol.,* **122**, 1261 (2000).

[117]   K. Grunewald, J. Hirschberg and C. Hagen, *J. Biol. Chem.,* **276**, 6023 (2001).

[118]   S. Rabbani, P. Beyer, J. von Lintig, P. Hugueney and H. Kleinig, *Plant Physiol.,* **116**, 1239 (1998).

[119]   A. Ben-Amotz, A. Katz and M. Avron, *J. Phycol.,* **18**, 529 (1982).

[120]   P. Lodato, J. Alcaino, S. Barahona, P. Retamales and V. Cifuentes, *Appl. Environ. Microbiol.,* **69**, 4676 (2003).

[121]   A. G. Andrewes and M. P. Starr, *Phytochemistry,* **15**, 1009 (1976).

[122]   H. Kuhn, *J. Ultrastruct. Res.,* **33**, 332 (1970).

[123]   W. M. Harris and A. R. Spurr, *Am. J. Bot.,* **56**, 380 (1969).

[124]   F. Zsila, J. Deli and M. Simonyi, *Planta,* **213**, 937 (2001).

[125]   S. Nechifor, C. Socaciu, F. Zsila and G. Britton, in *Proc. 2nd Int. Congr. Pigments in Food,* p. 155 (Lisbon, 2002).

[126]   M. O. Senge, H. Hope and K. M. Smith, *Z.Naturforsch.C,* **47**, 474 (1992).

[127]   C. Sterling, *Acta Cryst.,* **17**, 1224 (1964).

[128]   H. Hashimoto, Y. Sawahara, Y. Okada, K. Hattori, T. Inoue and R. Matsushima, *Jpn. J. Appl. Phys. Part 1,* **37**, 1911 (1998).

[129]   K. Gaier, A. Angerhofer and H. C. Wolf, *Chem. Phys. Lett.,* **187**, 103 (1991).

[130]   K. Shibata, *Biochim. Biophys. Acta,* **22**, 398 (1956).

Chapter 6

Carotenoid-Protein Interactions

George Britton and John R. Helliwell

A. Introduction: Interactions of Carotenoids with Other Molecules

Chapter 5 shows that the aggregation of carotenoid molecules can have a profound effect on their properties and hence their functioning in biological systems. Another important influence is the interaction between carotenoids and other molecules. The way that interactions of carotenoids with lipid bilayers influence the structure and properties of membranes and membrane-asociated processes is discussed in *Chapter 10*, and the aggregation of carotenoid molecules within the bilayers in *Chapter 5*. Of particular importance, though, are interactions between carotenoids and proteins. These allow the hydrophobic carotenoids to be transported, to exist, and to function in an aqueous environment. In some cases they may modify strongly the light-absorption properties and hence the colour and photochemistry of the carotenoids.

In plants, the importance of chlorophyll-carotenoid-protein complexes in photosynthesis is well known, but other carotenoid-proteins that are not involved in photosynthesis have also been isolated. The animal kingdom provides interesting and functionally significant examples of carotenoid-protein interactions. Notable among these are the blue carotenoproteins of invertebrate animals (Section **D**), but other examples include the lipoproteins which transport carotenoids and other lipids in the blood, and the association of carotenoids with structural proteins, *e.g.* the binding of astaxanthin (**404-406**) to the muscle protein in salmonid fish and of various carotenoids to the structural proteins of bird feathers.

B. Carotenoid-protein Complexes in Plants and Microorganisms

1. Photosynthetic pigment-protein complexes

In the photosynthetic apparatus of bacteria, algae and plants, carotenoids are located in reaction centre and light-harvesting pigment-protein complexes. The precise geometry of binding ensures efficient energy transfer between carotenoids and chlorophylls. The structures of many of these complexes have been determined by X-ray crystallography, and the localization and interactions of the carotenoids with the protein, chlorophylls *etc.* identified. Details of the structures and interactions in relation to the functions of the carotenoids in photosynthesis are described and illustrated in *Chapter 14*.

2. Soluble proteins

Some soluble, carotenoid-containing proteins, which are not directly involved in photo-synthesis, have been isolated from photosynthetic organisms. Notable among these is the 'orange protein' obtained from some cyanobacteria. In many species, *e.g. Arthrospira maxima*, the carotenoid present is 3'-hydroxyechinenone (**295**), whereas in others, *e.g. Anacystis nidulans* and *Lyngbya wholei*, it is zeaxanthin (**119**) [1]. The X-ray structure of the orange protein from *A. maxima* has been determined [2]. The free carotenoid (3'-hydroxyechinenone) is yellow with λ_{max} in organic solvent at around 455 nm. The protein-bound form appears orange with λ_{max} at 495 nm and 465 nm; proteolysis or acid treatment leads to a red form with a further bathochromic shift to 505 nm. The protein crystal structure at 2.1 Å resolution revealed a homodimer with one carotenoid molecule per monomer. The polyene chain is in the all-*E* configuration and curved. The carotenoid-binding site is lined by a striking number (six) of methionine residues. At one end of the carotenoid, the C(4) keto oxygen is hydrogen bonded (2.8 Å) to the NH and OH groups in the side chains of the conserved Trp290 and to Tyr203 (2.7 Å) in the C-terminal domain. At the other end, the C(3') hydroxyl oxygen atom of the carotenoid is hydrogen bonded not to the protein but to a water molecule. The distance between the two carotenoid molecules is >30nm, so there should be no interactions between the chromophores that could affect the spectroscopic properties of the protein-bound form. The mechanism of the colour shift is unknown. The orange proteins are thought to have a photoprotective role (*Chapter 10*).

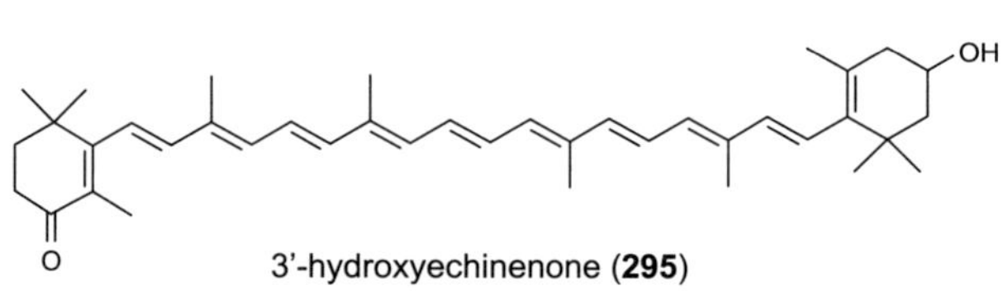

3'-hydroxyechinenone (**295**)

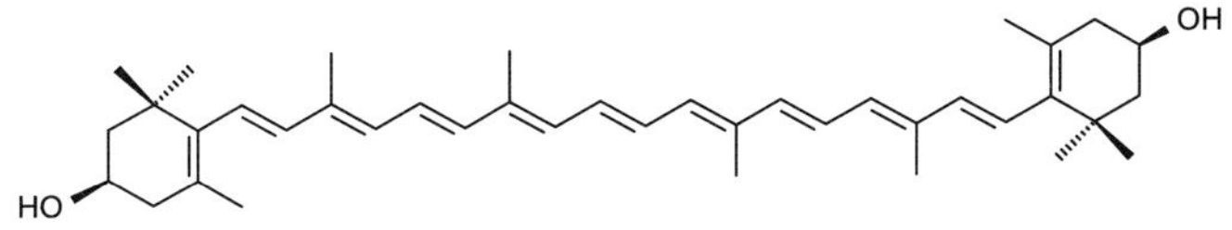

(3R,3'R)-zeaxanthin (**119**)

Soluble carotenoid-containing proteins have been isolated from chromoplast-containing non-photosynthetic tissues of plants. As an example, a protein–β-carotene (**3**) complex from carrot (*Daucus carota*) has been described [3], but no structural studies have been reported.

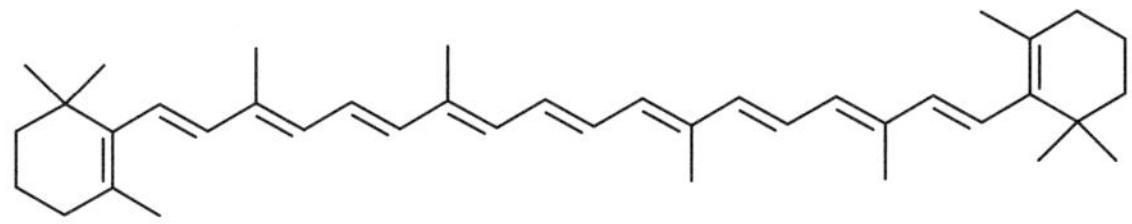

β-carotene (**3**)

C. Carotenoid-Protein Interactions in Animals

1. Blood lipoproteins

In the human body, hydrophobic carotenoids obtained from the diet must be transported to tissues *via* the bloodstream, an aqueous medium. Several different lipoproteins have been identified as the vehicles for this. Details of these transport systems are given in *Volume 5, Chapter 7*. In the various lipoprotein particles, there is some selectivity; carotenes are associated mainly with HDL (high-density lipoprotein) and xanthophylls with LDL (low-density lipoprotein), but there is no specificity of binding [4]. The carotenoids appear simply to be 'dissolved' in the lipid (triacylglycerol) core, and water-solubility is conveyed by the polar lipids and protein that surround this core. Carotenoids are transported on similar lipoproteins in other animals, *e.g.* astaxanthin in salmon (see *Chapter 12*).

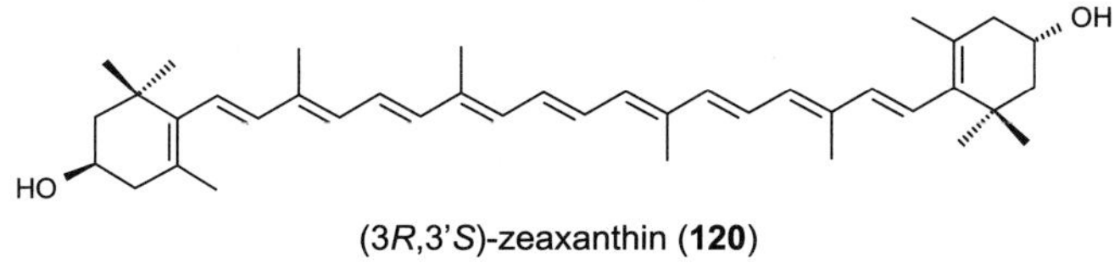

(3R,3'S)-zeaxanthin (**120**)

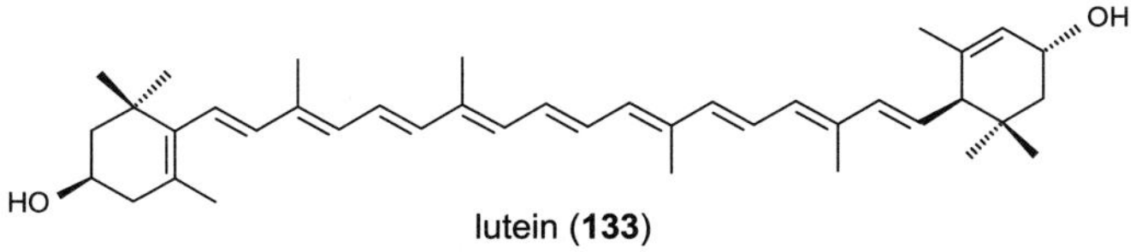

lutein (**133**)

No specific carotenoid binding proteins like plasma retinol binding protein and other retinoid-binding proteins have been identified in the blood. Recently, however, a 'xanthophyll-binding protein' (XBP), which ensures the specific delivery of xanthophylls to the macula, has been found in the retina [5]. It binds two xanthophyll molecules and has highest affinity for (3R,3'R)-zeaxanthin (**119**) and (3R,3'S)-zeaxanthin (**120**); the other macular xanthophyll lutein (**133**) is only weakly bound (see *Chapter 5* and *Volume 5, Chapter 15*).

2. Fish muscle

The carotenoids canthaxanthin (**380**) and especially astaxanthin (**404-406**) in the muscle of salmonid fishes are responsible for the desirable pink-red colour of the flesh. The carotenoid is bound by unspecific hydrogen bonds to the actomyosin complex of the muscle fibres, probably to the α-actinin component [6]. Other aspects of the colouration of the salmon muscle by carotenoids are covered in *Chapter 12*.

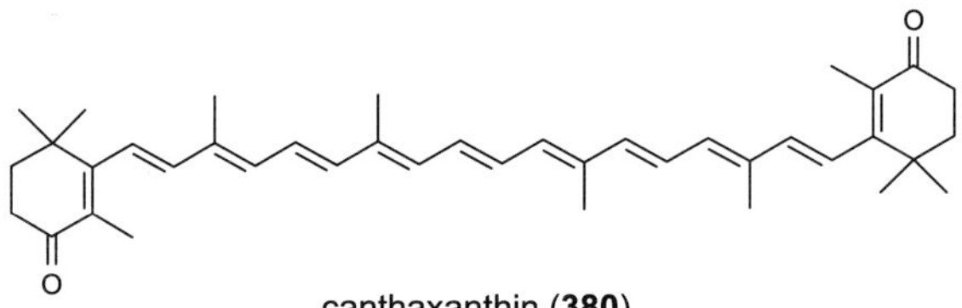

canthaxanthin (**380**)

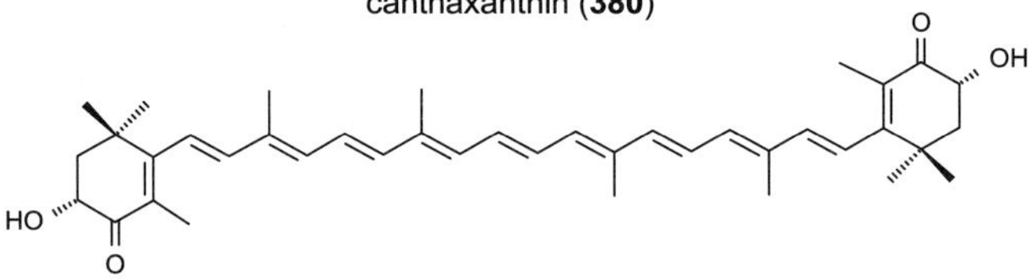

(3R,3'R)-astaxanthin (**404**)

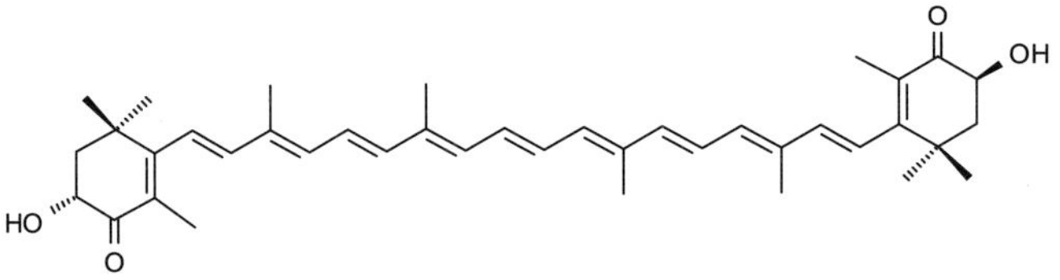

(3R,3'S)-astaxanthin (**405**)

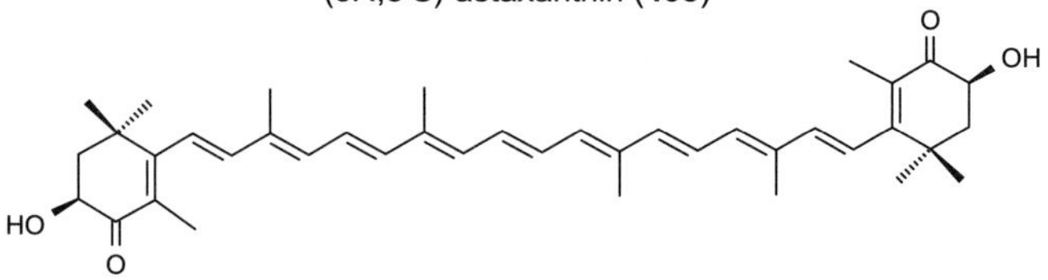

(3S,3'S)-astaxanthin (**406**)

3. Feathers

Although they no longer contain living cells, mature feathers have very complex morphology. The main structural proteins, keratins, are inert and insoluble. Carotenoids are incorporated, with or without structural modification, in the follicles as the feathers develop and they then become strongly bound to the protein filaments of the keratin. This strong binding renders the feather carotenoids difficult to extract. The binding can influence the colour of the carotenoid; the same carotenoid can be yellow, orange or red in feathers of different species or even in different feathers of the same species. For example, ε,ε-carotene-3,3'-dione (**387**, chirality not established) is the main carotenoid in both red and yellow feathers of the goldfinch *Carduelis carduelis* [7]. Details of the carotenoid-keratin interactions have not been elucidated. The functional relevance of carotenoid colours in feathers is discussed in *Chapter 11*.

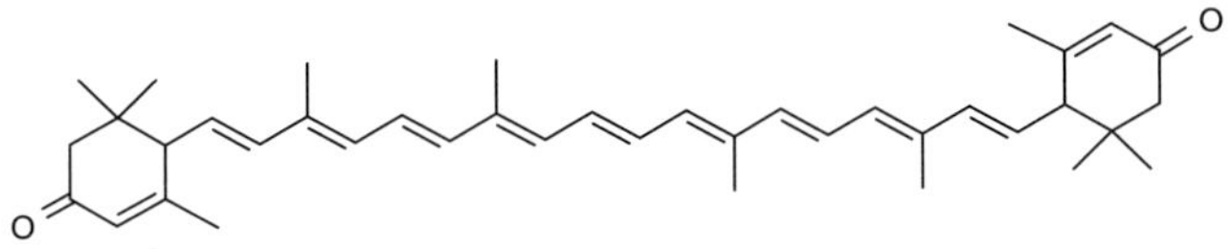

ε,ε-carotene-3,3'-dione (**387**)

D. Carotenoproteins

1. General features

The carotenoproteins of invertebrate animals provide a particularly interesting example of carotenoid-protein interactions because the protein binding has such an enormous effect on the light-absorption properties of the carotenoid, leading to remarkable alterations in colour. Carotenoproteins are widely distributed in many phyla of invertebrate animals, especially ones from a marine environment. They are common in the carapace of crustaceans, giving a grey, black, brown or blue appearance to the live animals; the orange-red colour of the free carotenoid is revealed when the animals are cooked. Some colourless, transparent crustaceans turn red on cooking, implying the presence of colourless carotenoproteins, which may have absorption spectra bathochromically shifted into the near infrared region or hypsochromically shifted below the visible region. With some crustaceans, including the lobsters *Homarus gammarus* and *H. americanus*, and the crayfish *Procambarus clarkii*, which are normally slate blue or even black, bright blue mutants are occasionally found, in which the vivid blue carotenoprotein colour is not modified by the additional presence of free astaxanthin.

Interesting carotenoproteins have also been isolated from several starfish (Asteroidea), notably the purple asteriarubin from *Asterias rubens* [8,9] and the intensely blue linckiacyanin from *Linckia laevigata* [10] (Section **D**.7).

Details of the taxonomic distribution and anatomical location of these complexes can be found in a number of comprehensive reviews [11-14].

The carotenoproteins are specific, stoichiometric combinations of carotenoid and protein. The binding is normally not covalent, but the carotenoid is bound in a precise way and its light absorption properties are drastically altered, leading to substantial bathochromic shifts. The carotenoid component is typically a diketocarotenoid, most commonly astaxanthin (**404-406**).

2. Crustacyanin

The most intensively studied of the carotenoproteins is crustacyanin, the blue (λ_{max} 630 nm) astaxanthin-protein extracted from the carapace of lobsters. Most of the recent work has been done with the European lobster (*H. gammarus*), though the American lobster (*H. americanus*) was used in some of the earlier studies. Scientific study of the blue colouration of the shell of lobsters began at the end of the 19th century [15] and the proposal that the pigment responsible is a combination of carotenoid and protein was first made in 1921 [16]. The nature of the carotenoid-protein interactions and the mechanism of the colour shift in crustacyanin are described in detail below.

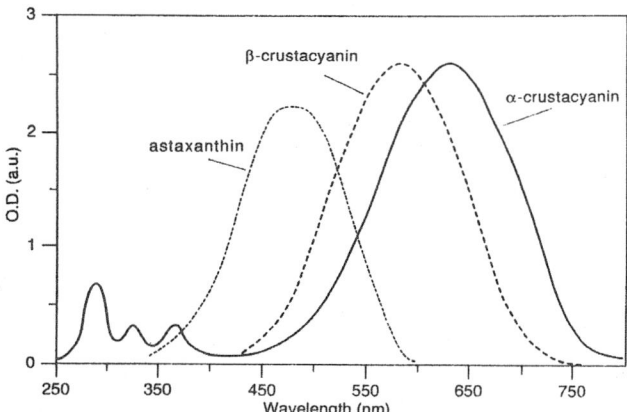

Fig. 1. UV/Vis absorption spectra of α-crustacyanin (solid line), β-crustacyanin (dashed line) and free astaxanthin (dotted line).

A key feature of crustacyanin, and of other carotenoproteins, is that the carotenoid is not linked covalently to the protein. It is therefore possible to remove the carotenoid by solvent extraction and then reconstitute the carotenoprotein by recombining the protein with a selected carotenoid. The reconstituted products appear to be identical to the natural carotenoproteins. This has allowed the structural features of the carotenoid that are important for binding to be identified. Also, the introduction of isotopically labelled carotenoids has made possible detailed NMR and other spectroscopic studies.

Over a period of about 30-40 years, extensive reconstitution and spectroscopic studies have been undertaken in attempts to elucidate details of the carotenoid-protein interactions. These studies have led to several hypotheses to explain the spectral shift. It makes fascinating reading to follow progress in the experiments and development of the arguments, which are summarized in a series of reviews and progress reports [12,17-20]. Recently, the first crystal structure, of a β-crustacyanin dimer, has been determined [21]. Most of the reconstitution and spectroscopic work was done earlier but, rather than presenting a historical, chronological story, it is now appropriate to integrate and discuss all findings, with the structural details revealed by the X-ray study as a focal point. From this a coherent picture is emerging.

The apparently native form is α-crustacyanin, isolated as a large (320 kDa), water-soluble protein containing sixteen astaxanthin molecules. At low ionic strength, α-crustacyanin dissociates irreversibly into the 40 kDa β-crustacyanin, a heterodimer of two closely related monomeric subunits [22,23]. The β-crustacyanin that is isolated is not homogeneous but can be resolved into six components, each representing a different combination of subunits, one of 21 kDa, the other 19 kDa. Five different subunits have been identified, and are classified on the basis of their behaviour on electrophoresis into two groups known as Type 1 (A_1, C_1 and C_2, each *ca.* 21 kDa) and Type 2 (A_2 and A_3, each *ca.* 19 kDa) [24]. Only two genes have been identified, however, one for each group, suggesting that the differences between the members of each group are due to natural or artificial changes that occur post-translation, possibly during isolation. This is supported by the finding that dissociation of α-crustacyanin under different conditions gives different β-crustacyanin patterns [24].

The amino acid sequences of both groups of subunits show significant similarity to those of members of the lipocalin superfamily, which are generally proteins that bind and transport hydrophobic ligands [25,26], an appropriate example being plasma retinol-binding protein, which has the typical lipocalin structure of a β-barrel in which the ligand sits [27].

Analysis of the amino acid sequences and comparison with other lipocalins led to a model [25,26] of β-crustacyanin as a dimeric arrangement in which each protein subunit was proposed to accommodate one astaxanthin (illustrated in [26]), and some overlap of the two astaxanthins was suggested to explain the exciton interaction evident from the CD spectrum [23]. Recent X-ray crystallographic studies have provided precise details of the tertiary and quaternary structures and the location of the astaxanthin ligands within this structure, notably their close proximity, the location of one end ring of each astaxanthin proximal to a histidine provided by respective protein subunits, and end rings coplanar with the polyene chain [21]. The X-ray crystal structure coordinates and structure factor amplitudes of β-crustacyanin (A_1:A_3) at 3.2 Å and of apocrustacyanin A_1 at 1.4 Å are available at the Protein Data Bank (PDB Code 1GKA and 1H91, respectively).

The preparation of crystals of α-crustacyanin that diffract to high resolution has not been possible [28,29]. Determination of the ultrastructure by other biophysical imaging techniques such as solution X-ray scattering and cryo-electron microscopy has been undertaken; for a review see reference [28].

3. The X-ray structure of β-crustacyanin

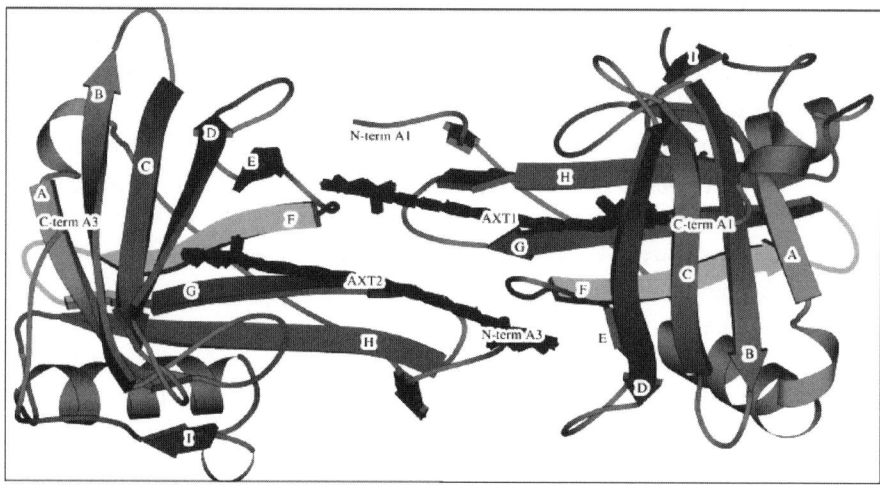

Fig. 2. X-Ray crystal structure of the A_1/A_3 dimeric form of β-crustacyanin, determined at 3.2 Å resolution. The individual β-strands are shown in ribbon form. The two astaxanthin molecules bound are labelled AXT1 and AXT2 [21].

The β-crustacyanin used [21] was a combination of subunits A_1 and A_3, reconstituted with astaxanthin. The classical methods of crystallization repeatedly failed but it was crystallized as blue crystals over a 4-month period by use of a technique of crystal growth under oil, known as 'microbatch' [30]. The structure of the β-crustacyanin was elucidated at 3.2Å by X-ray crystallographic techniques (Fig. 2). A protein structure at 3.2 Å is capable of yielding atomic positions and atomic displacement parameters but it is not optimal for some of the structure interpretation, in particular for elucidating the details of the bound water. First, though, the structure of apocrustacyanin A_1 was established at 1.4 Å resolution [31], and this served as a molecular replacement motif so that the β-crustacyanin structure could be solved in detail. Cross-checks could then be made with the β-crustacyanin studied, which comprised an A_1 protein subunit combined with a sufficiently closely related apocrustacyanin A_3 protein, which has 40% amino acid sequence identity to the A_1 protein [32]. The amino acid sequence identity in the astaxanthin-binding regions was better than the 40% overall value, so that common structural details were available for examination. At the time of that study, no atomic resolution structure of astaxanthin was available, so that of the closely related carotenoid, canthaxanthin (**380**) [33], was used as the basis from which the changes to the astaxanthin conformation due to binding could be elucidated. The assumption that the structures of astaxanthin (**406**) and canthaxanthin (**380**) would be very similar has been confirmed by later X-ray crystallographic studies [34].

Details of the protein structure, including the positioning of individual amino acid residues, are given in the original publication [21]. Here the discussion will concentrate on the salient features and interactions that determine the precise location of the carotenoid molecules and are likely to be relevant to the spectral shift mechanism.

From the crystallographic data, the structure of β-crustacyanin is confirmed as a heterodimer of A_1 and A_3 protomers in the crystal asymmetric unit (Fig. 2). Both subunits have the typical topology of a lipocalin, with a β-barrel made up of two distinct β-sheets, one consisting of β-strands A1, B, C, D, E and F1, the other of β-strands A2, F2, G and H. The A_1 and A_3 subunits interact *via* a loop region that connects strands G and H of each subunit. This differs from the structure of the A_1-A_1 dimer, with no bound carotenoid, in which the interaction is *via* a close contact between the strands F of the two subunits [31].

4. The carotenoid-binding site

The two bound astaxanthins (AXT1 and AXT2) interpenetrate the A_1 to A_3 interface; each subunit contains half of each carotenoid. One astaxanthin (AXT1) has its first end-group ring deep inside the calyx of A_1; the second end group is in a small cavity of the A_3 protein (Fig. 3a). In a similar way, the other astaxanthin (AXT2) has its first end ring inside the calyx of A_3 and its second end ring in a small cavity in A_1 (Fig. 3b). The two (all-*E*)-astaxanthins have very similar linear conformations, with a marked bow shape. The ring-chain conformation in all cases is 6-s-*trans*, which is energetically less favourable than the conformation (6-s-*cis*) normally adopted by free carotenoids, including astaxanthin and canthaxanthin, in solution and crystal structures [34]. The end rings are essentially coplanar with the polyene chain, in contrast to the free carotenoids for which a C(6,7) distortion angle of -40 to -50° is usual. The carotenoids are held so that the central portions of the polyene chains approach within ~7 Å .

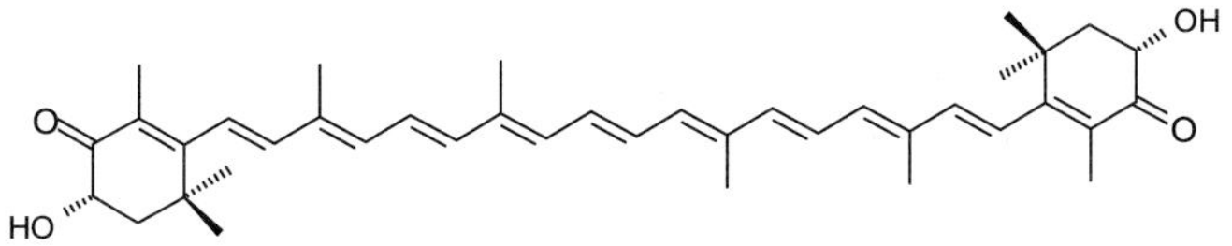

(6,6'-s-*trans*)-astaxanthin

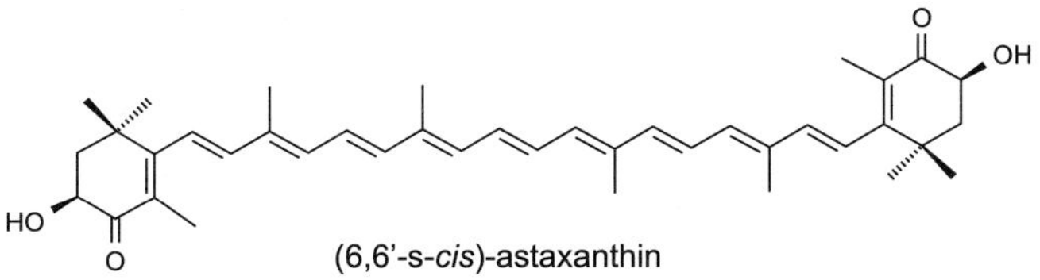

(6,6'-s-*cis*)-astaxanthin

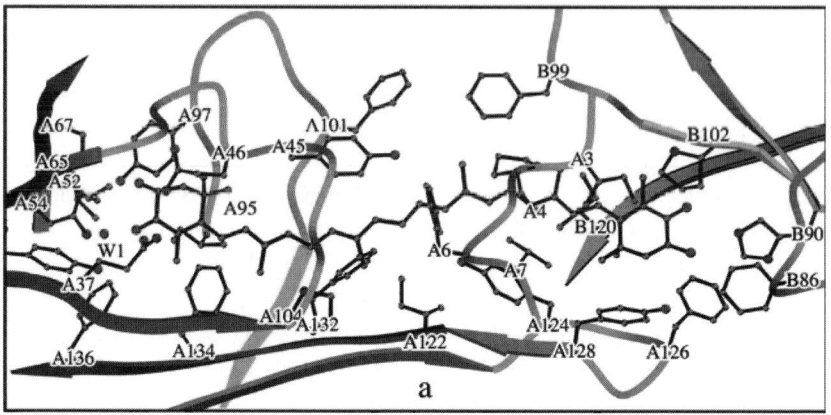

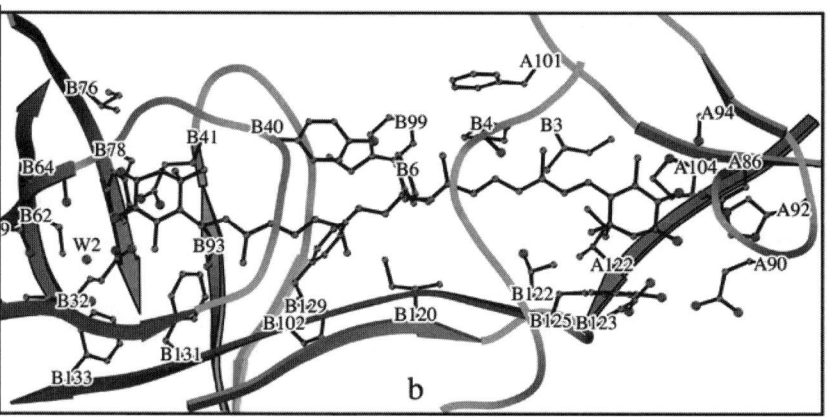

Fig. 3. The binding sites for the two astaxanthin molecules in the A_1/A_3 dimeric form of β-crustacyanin. (a) astaxanthin AXT1; (b) astaxanthin AXT2. From [21] with permission.

5. Carotenoid-protein interactions

The carotenoid-protein interactions are asymmetric; the two end rings of each astaxanthin have dissimilar environments. For each astaxanthin, the first end group [C(1)-C(6)], located within the calyx, has neighbouring hydrophilic residues which coordinate the 3-hydroxy and 4-keto groups, and tether a bound water molecule within very close hydrogen-bonding distance of the 4-keto group (Fig. 4a, 4c). Several hydrophobic residues make the remainder of the cavity hydrophobic. The second end ring [C(1')-C(6')] of AXT1 has the 3'-hydroxy and 4'-keto group close to a His residue in an otherwise largely hydrophobic environment (Fig. 4b). The 3'-hydroxy and 4'-keto groups of AXT2 are also close to a His residue but, in this case, there are additional hydrophilic residues in the vicinity (Fig. 4d).

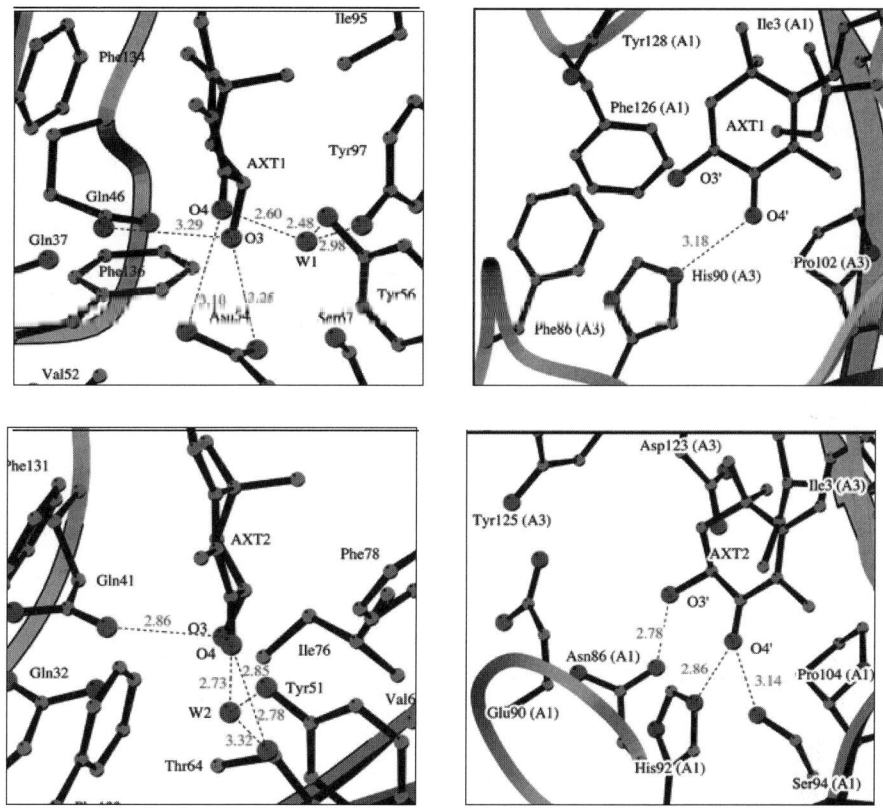

Fig. 4. Details of the interactions of the astaxanthin end groups with nearby aminoacid residues. W1 and W2 indicate water molecules hydrogen-bonding to the astaxanthin oxygen functions. (a) C(1)-C(6) of AXT1; (b) C(1')-C(6') of AXT1; (c) C(1)-C(6) of AXT2; (d) C(1')-C(6') of AXT2.

The X-ray results confirmed the findings of earlier reconstitution experiments [17,20] which showed that only all-*trans* C_{40}-carotenoids could bind and form complexes; slightly longer (C_{42}) or shorter (C_{36}) analogues could not be accommodated, nor could *cis* isomers. Keto groups at C(4) and C(4') were essential for formation of a purple or blue complex, and the presence of hydroxy groups at C(3) and C(3') improved the efficiency of binding; the chirality at C(3) and C(3') had little effect. The position of the keto group in the ring is important. Carotenoids with the keto groups at C(3) and C(3') would not bind. 3,4,3',4'-Tetradehydro-β,β-carotene-2,2'-dione (**377**), bound very efficiently and gave a blue α-crustacyanin complex. The position of the 2-keto group in the 6-s-*cis* conformer of this compound is essentially identical to that of the 4-keto group of astaxanthin in the 6-s-*trans* conformation.

Other alterations to the ring structure, such as changing from a C_6 to a C_5 ring (in actinioerythrol, **552**) or varying the number and positions of the methyl substituents on the ring, had little or no effect on the ability to bind. All these results were confirmed by the crystal structure which shows the defined size and shape of the binding sites, the importance of interactions with the polar substituents of the end groups, and the nature of the hydrophobic environment of the rest of the ring-binding site, including some flexibility of specific residues (see the on-line movie at the *Acta Cryst. D* website associated with [28]).

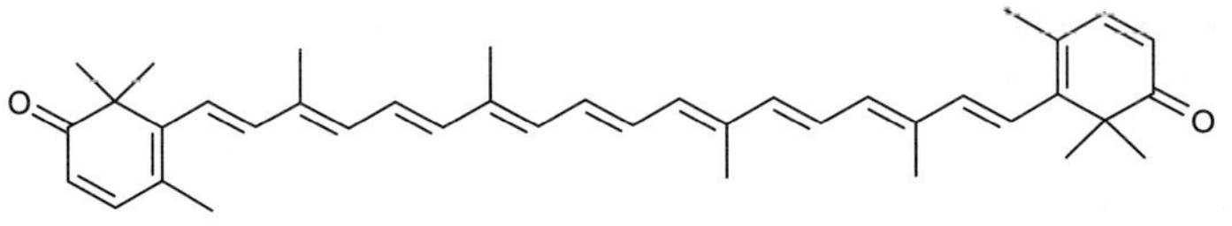

3,4,3',4'-tetrahydro-β,β-carotene-2,2'-dione (**377**)

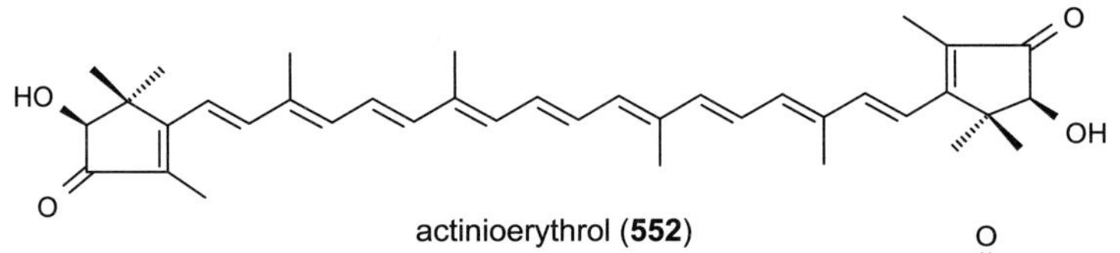

actinioerythrol (**552**)

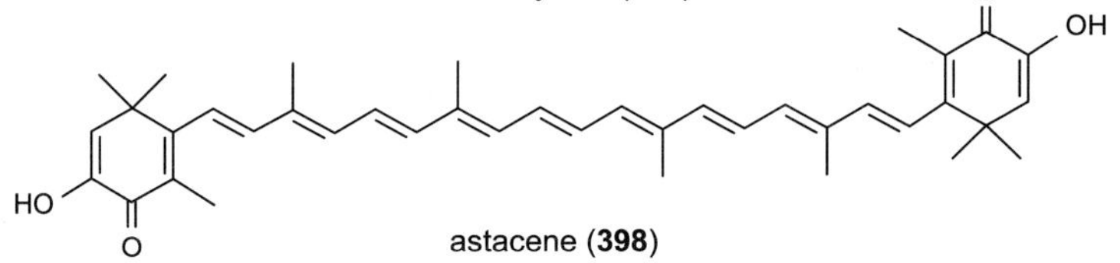

astacene (**398**)

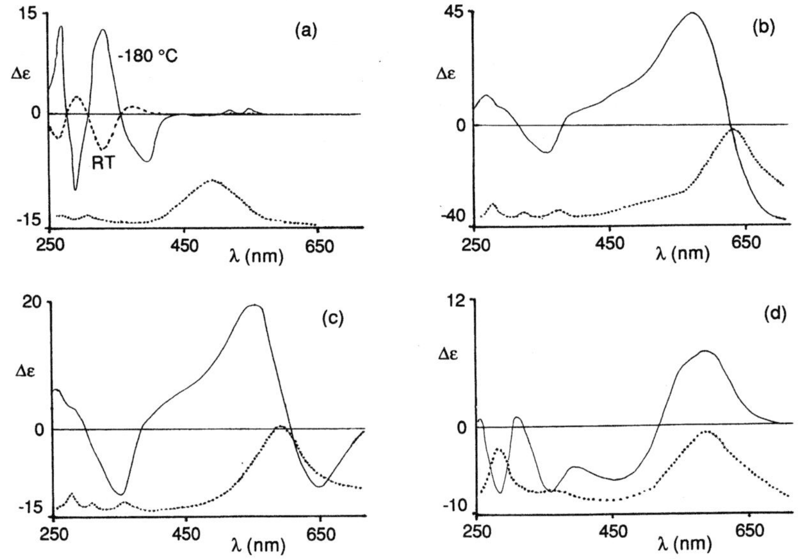

Fig. 5. CD spectra (solid lines) and UV/Vis spectra (dotted lines) of (a) free (3S,3'S)-astaxanthin, (b) α-crustacyanin, (c) β-crustacyanin, (d) asteriarubin [20].

CD spectra of native α-crustacyanin, β-crustacyanin, and reconstitution products revealed that chirality was imposed on the chromophore by the protein binding. Chirality or lack of chirality in the free carotenoid was not an important factor; the different optical isomers of astaxanthin [(3R,3'R), (3R,3'S) and (3S,3'S), **404, 405, 406,** respectively] gave complexes with identical CD spectra, as did achiral carotenoids such as astacene (**398**) and canthaxanthin (**380**) [20,35]. The CD spectra showed strong exciton coupling, consistent with the close proximity of the two astaxanthin chromophores.

The X-ray structure shows that the polyene chains of the astaxanthin molecules are held in place by many hydrophobic residues. A set of hydrophobic residues in protein A_1 makes a central cavity for binding AXT1; an equivalent set of residues in protein A_3 provides a similar cavity for AXT2. Two trios of Phe, Pro and Ile residues, one from A_1, the other from A_3, are sandwiched between the two astaxanthin chains. Binding causes substantial movement of some of these residues, compared with the carotenoid-free A_1-A_1 dimer [28].

These interactions firmly lock the C(20) and C(20') methyl groups of the two astaxanthins in place, an interaction crucial for holding the polyene chains close together. This rationalizes the earlier finding from reconstitution experiments that carotenoid analogues in which these central methyl groups were missing, *e.g.* 20-norastaxanthin (*1*), or displaced from C(13) to C(14), *e.g.* 14,14'-dimethyl-20,20'-dinorcanthaxanthin (*2*), would not bind and form α-crustacyanin or β-crustacyanin complexes [20]. The C(19) and C(19') methyl groups do not experience such interactions. Carotenoid analogues that lack these methyl groups, *e.g.* 19,19'-dinorastacene (*3*), are able to bind and form purple-blue complexes.

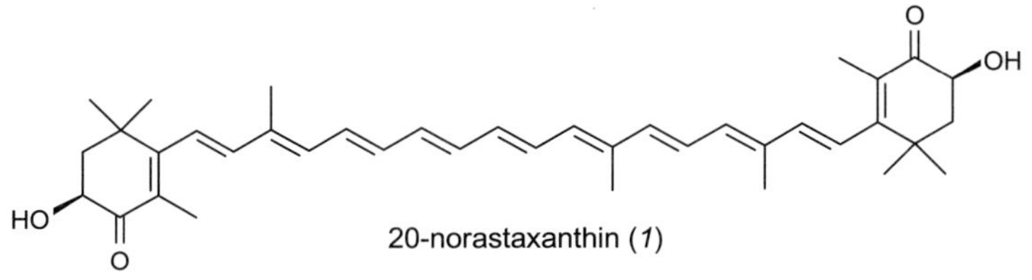

20-norastaxanthin (*1*)

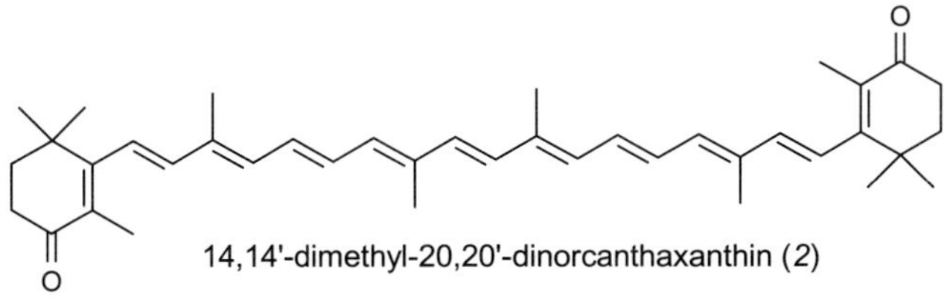

14,14'-dimethyl-20,20'-dinorcanthaxanthin (*2*)

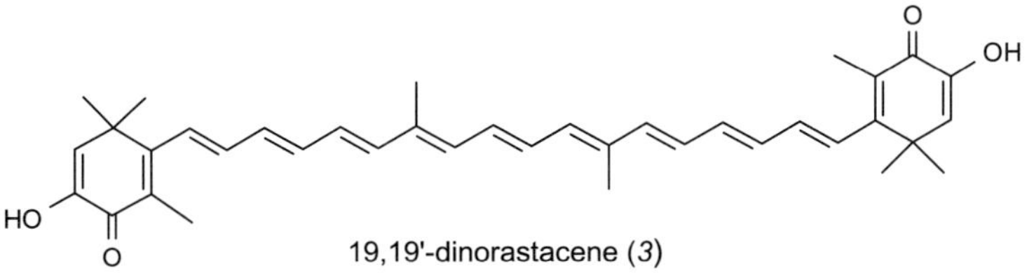

19,19'-dinorastacene (*3*)

6. Mechanism of the spectral shift

From the X-ray structure of β-crustacyanin, a number of features are apparent that could be relevant to the mechanism of the bathochromic shift seen with the bound astaxanthin:

i)   the coplanarity of the rings and the polyene chain, thereby extending and optimizing the conjugation of the chromophore,

ii) the proximity of the astaxanthin keto groups to bound water and histidine, leading to hydrogen bonding and possible electronic polarization effects,

iii) the bowing of the astaxanthin molecules,

iv) the generally hydrophobic environment of the astaxanthin molecules,

v)   the close proximity and orientation of the two astaxanthin chromophores, leading to exciton interaction.

Against this background it is now possible to consider by what mechanism such interactions may produce the large change in electronic and optical properties of astaxanthin. It is now 40 years since the first detailed spectroscopic study of crustacyanin was reported [23]. This gave the first CD evidence of exciton splitting. Several possible mechanisms were suggested to account for the perturbed optical properties of astaxanthin in crustacyanin, including intermolecular interactions of the π-electron systems, distortion such as twisting of double bonds, and polarization of the chromophore. A subsequent early study by resonance Raman spectroscopy showed that there was no distortion caused by twists in the double bonds of the conjugated stystem [36]. The exciton interaction mechanism hypothesis was also discarded [36]. The conclusion was that the bathochromic shift is caused by a charge polarization mechanism induced by charged amino acid residues and/or by hydrogen bonding in the binding site. Extensive investigations since then by various spectroscopic techniques yielded a large amount of information but failed to resolve the question of the mechanism.

The synthesis of a range of forms of astaxanthin enriched with ^{13}C at selected positions has now made possible a detailed study by NMR spectroscopy, complemented by resonance Raman spectroscopy and quantum mechanical calculations [20,37-40].

In the first investigations, the labelled species used were [12,12'-^{13}C$_2$]-, [13,13'-^{13}C$_2$]-, [14,14'-^{13}C$_2$]-, [15,15'-^{13}C$_2$]- and [20,20'-^{13}C$_2$]-astaxanthins. Each was incorporated into reconstituted α-crustacyanin complexes and examined by solid-state magic-angle spinning (MAS) ^{13}C-NMR spectroscopy [20,37,38]. Comparison with free astaxanthin showed significant downfield shifts for the signals for C(12,12') and C(14,14') in the complex, and small upfield shifts for the signals for C(13,13'), C(15,15') and C(20,20'). These results indicate significant differences in electronic charge density in this central part of the molecule, and an alternating effect along the polyene chain, reversed at the centre [38]. This is consistent with polarization from both ends of the molecule, and calculations showed that, when possible exciton interactions were not considered, the results were compatible with a structure in which the keto groups at C(4) and C(4') are both protonated [38]. The signals for C(12,12') and C(14,14') each split into two components, showing that these carbons were experiencing different environments, *e.g.* that the proposed polarization effect from the two ends would be quantitatively different. This model predicts that the magnitude of the changes in electronic charge density would increase towards the ends of the molecule, *i.e.* in the C(5)-C(10) region.

Recently, astaxanthin species enriched with ^{13}C in this part of the molecule have been synthesized and used to extend the NMR study into the C(6)-C(11) and C(6')-C(11') region [40,41]. The results for C(10,10') continued the trend seen in the earlier study, with a significant shift difference of +9 ppm due to binding. The shift differences for C(6) and C(8), however, predicted by the polarization model to be even larger, were in fact negligible. The results from this and subsequent studies are therefore not consistent with protonation of the C(4) and C(4') keto groups [42], so this hypothesis can be discarded as an explanation for the optical properties of crustacyanin.

The X-ray crystal structure of β-crustacyanin had revealed three features that might lead to alterations of the electronic structure in the bound astaxanthin, namely extension of the chromophore due to the planar 6-s-*trans* conformation, hydrogen bonding between the keto groups and water and histidine residues, and the close approach of the two astaxanthin molecules leading to excitonic interactions between the two chromophores (Fig. 6).

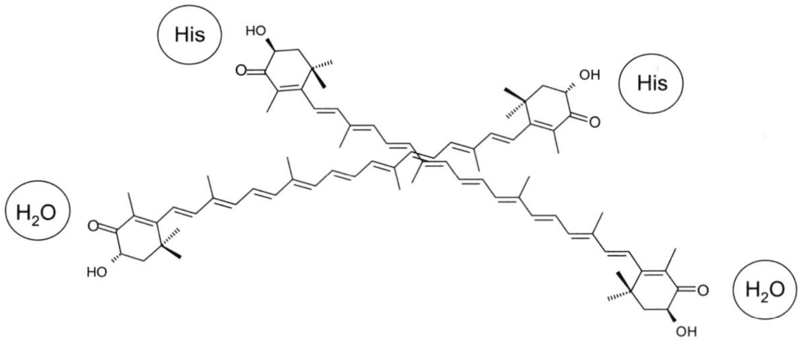

Fig. 6. View from above of the two astaxanthin molecules in β-crustacyanin. The bowing of the polyene chains is not shown. The approximate positions of the histidine residues and the hydrogen-bonding water molecules are indicated. When viewed from the side (not illustrated) the molecules are approximately parallel and are *ca.* 7 nm apart at the crossing point [21].

In an evaluation of these possibilities [43-45], it was proposed that the polarization effect is caused by hydrogen-bonding of the C(4) keto group, particularly to the His residue, suggested to be protonated. The bowing of the bound astaxanthin molecule was considered to have a negligible effect on the absorption spectrum, and in the unbound astaxanthin there is already some partial conjugation of the end rings with the polyene chain.

The coordinates defined by the crystal structure were used in quantum mechanical calculations to evaluate the effects of each of these features on the optical properties and light absorption maximum [41]. All gave rise to a bathochromic effect, but that due to the first two features was relatively small, generating in total a shift of only 39 nm (0.19 eV), to 517 nm.

By far the greatest effect could be attributed to the interaction between the two astaxanthin molecules, even at a minimum distance of 7 nm between the chromophores (Fig. 6). With the long axes of the molecules forming an angle of about 120°, the transition dipole is computed to be very intense (*ca.* 19 Debye), in agreement with results of a study of α-crustacyanin by Stark spectroscopy [46]. The calculation gives an exciton splitting of 0.49 eV, corresponding to a large and decisive contribution to the bathochromic shift.

Resonance Raman spectroscopic studies were also performed with crustacyanins prepared by reconstitution with the ^{13}C-enriched astaxanthins. The results gave a direct indication that the electronic excited state undergoes delocalization upon binding and supported a charge displacement from the centre of the chromophore towards the end rings upon excitation as indicated by calculations of the HOMO and LUMO orbitals [40].

The NMR and resonance Raman studies were performed with α-crustacyanin, whereas the computational analysis was based on the structure determined for β-crustacyanin. The assumption that the chromophore-protein interactions and the effects responsible for the bathochromic shift are qualitatively similar in α- and β-crustacyanins is supported by the similar patterns seen in the UV/Vis, ORD, Raman and CD spectra of the two forms, and the fact that α-crustacyanin dissociates into eight β-crustacyanin units. The clear conclusion, therefore, is that the bathochromic shift in the spectrum of crustacyanin is caused by a combination of effects; 'on-site' conformational changes and polarization effects contribute about one-third, but the aggregation effects causing exciton coupling are the major factor.

7. Other carotenoproteins

Crustacyanin is just one example of the many carotenoproteins that have been found in a variety of invertebrate animals. All would be interesting to study in detail but there are some with properties that seem to reveal features that are not consistent with the crustacyanin mechanism. Detailed crystallographic and NMR studies such as those undertaken with crustacyanin would be interesting and informative.

a) Asteriarubin

This purple carotenoprotein, λ_{max} 555-570 nm, isolated from the dorsal skin of the common starfish *Asterias rubens*, is a 43 kDa, 4-subunit protein [8,9] that has as its prosthetic group a mixture of carotenoids, mainly astaxanthin (**406**) and its acetylenic analogues 7,8-didehydroastaxanthin (**402**) and 7,8,7',8'-tetradehydroastaxanthin (**400**). The amino acid sequence shows no homology with the crustacyanin subunits and indeed seems not to be related to the sequences of any other proteins in the data banks. Reconstitution studies show similarities between the binding requirements of asteriarubin and crustacyanin, but the tetrameric asteriarubin contains only one carotenoid molecule and the CD spectrum shows no exciton splitting [17,20]. In this case, therefore, molecular aggregation and chromophore-

chromophore interactions are unlikely to be a factor in determining the bathochromic shift. The same is true of alloporin, a carotenoprotein (λ_{max} 545nm) from the soft coral *Allopora californica* [47].

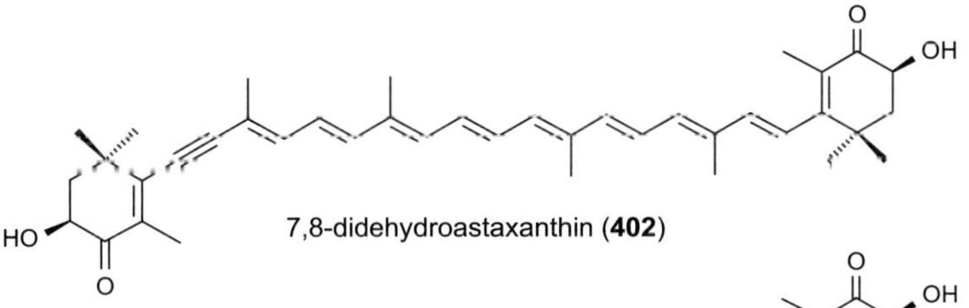

7,8-didehydroastaxanthin (**402**)

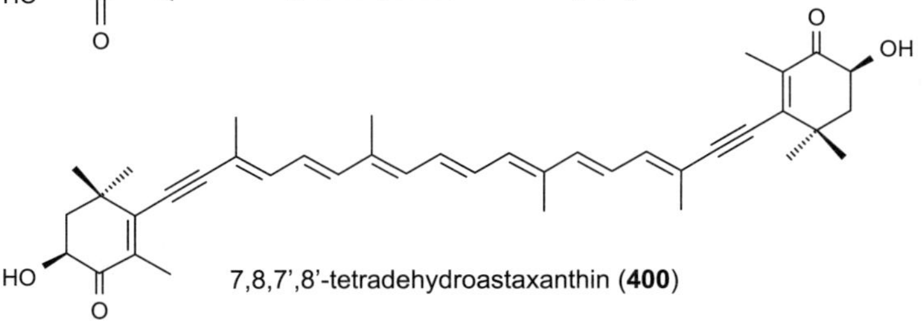

7,8,7',8'-tetradehydroastaxanthin (**400**)

b) Linckiacyanin

The bright blue skin of the calcified starfish *Linckia laevigata*, is coloured by the carotenoprotein linckiacyanin, which has λ_{max} 395, 612 nm [10,48]. The main carotenoid present is astaxanthin, mainly the (3*S*,3'*S*) isomer (**406**), but significant amounts of other carotenoids are also present, notably zeaxanthin (**119**) and the novel hydroxyclathriaxanthin (**335.1**), which is not fully characterized but has only one 3-hydroxy-4-keto-β end group and one aromatic (χ) end group containing a CH$_2$OH substituent. The predominant glycoprotein subunit is small (6 kDa) but the native linckiacyanin is large (>1000 kDa), with a complex quaternary structure and containing at least 200 carotenoid molecules. The CD spectrum shows strong exciton interaction between the carotenoid chromophores, and this could, as with crustacyanin, be largely responsible for the bathochromic shift. The unusual primary and quaternary structures and the presence of both diketocarotenoids and monoketocarotenoids make linckiacyanin also an interesting example for further study.

c) The carotenoprotein of the Western Rock Lobster

The Australian Western Rock Lobster (*Panulirus cygnus*) has an astaxanthin-binding protein similar in size, composition and multimeric structure to α-crustacyanin [49,50], but the UV/Vis spectrum of astaxanthin is not altered significantly by protein binding, and the complex is red. The relative amounts of the apoproteins expressed from the two genes are different, however, from those in crustacyanin [50] A comparison of the tertiary and quaternary structure of this carotenoid-binding protein with those of crustacyanin from *H. gammarus* should be informative.

d) Crustochrin

Another astaxanthin-protein can be isolated from the lobsters *H. gammarus* and *H. americanus* [23,36,51]. This yellow complex, known as crustochrin, has a protein structure rather similar to crustacyanin but has a hypsochromically shifted spectrum with λ_{max} 400-410nm. About 20 astaxanthin molecules are present and exhibit exciton-exciton interactions typical of a natural H-aggregate (*Chapter 5*) of astaxanthin units with their chromophores stacked in a pack-of-cards manner [23,36,52].

e) Ovoverdin

The dark green ovoverdin is an astaxanthin lipoglycoprotein (lipovitellin) present in lobster ovaries and eggs. Its absorption spectrum shows two main bands in the visible region, at *ca.* 460 and 640 nm, which are considered to arise from carotenoid molecules in two different environments, one a non-stoichiometric, unspecific dispersion in lipid of the lipoprotein (460nm) and the other a true carotenoid-protein interaction (640 nm) [35]. No detailed spectroscopic studies have been reported.

f) Ovorubin

Isolated from the eggs of a river snail *Pomacea canaliculata*, ovorubin is a large (330 kDa), red astaxanthin glycoprotein [53]. Binding leads to only a small bathochromic shift (20-30 nm) to ~510 nm, but has the interesting effect of increasing the spectral vibronic fine structure [17]. It is not known what interactions are responsible for this effect. Reconstitution with carotenoids with keto groups at C(4) and C(4') always resulted in increased fine stucture. Carotenoids without these keto groups but with hydroxy groups at C(3) and C(3') gave red complexes but the spectral fine structure was lost [17].

g) Velellacyanin

Velellacyanin is a blue astaxanthin-protein isolated from the 'By-the-wind-sailor' jellyfish (*Velella velella*). Its absorption maximum is similar to that of α-crustacyanin but, whereas α-crustacyanin has an invariant number of protein subunits, velellacyanin has a helical structure of varying lengths, implying a different supramolecular arrangement [54].

8. Conclusions - future prospects

Much has been achieved in recent years in eludiating details of the structure and interactions in crustacyanin, but much remains to be done. X-Ray structures and further spectroscopic work for α-crustacyanin are required and cloning of the apoprotein genes, the use of recombinant proteins and site-directed mutagenesis would be particularly useful to investigate interactions with specific amino acid side chains in crustacyanin and other carotenoproteins. Understanding the interactions and mechanism of the the spectral shift may enhance the prospects for commercially attractive green, purple and blue carotenoid based products and formulations.

References

[1]   C. A. Kerfeld, *Photosynth. Res.*, **81**, 215 (2004).
[2]   C. A. Kerfeld, M. R. Sawaya, V. Brahmandam, D. Cascio, K. K. Ho, C. C. Trevithick-Sutton, D. W. Krogmann and T. O. Yeates, *Structure*, **11**, 55 (2003).
[3]   J. C. G. Milicua, J. L. Juarros, J. de las Rivas, J. Ibarrondo and R. Gomez, *Phytochemistry*, **30**, 1535 (1991).
[4]   B. A. Clevidence and J. G. Bieri, *Meth. Enzymol.*, **214**, 3 (1993).
[5]   P. Bhosale, A. J. Larson, J. M. Frederick, K. Southwick, C. D. Thulin and P. S. Bernstein, *J. Biol. Chem.*, **279**, 49447 (2004).
[6]   S. J. Mathews, N. W. Ross, R. E. Olsen and S. P. Lall, *Comp. Biochem. Physiol.*, **B144**, 206 (2006).
[7]   R. Stradi, *The Colour of Flight*, Solei Press, Milan (1998).
[8]   A. Elgsaeter, J. D. Tauber and S. Liaaen-Jensen, *Biochim. Biophys. Acta*, **B530**, 402 (1978).
[9]   C. C. Shone, G. Britton and T. W. Goodwin, *Comp. Biochem. Physiol.*, **B62**, 507 (1979).
[10] P. F. Zagalsky, F. Haxo, S. Hertzberg and S. Liaaen-Jensen, *Comp. Biochem. Physiol.*, **B93**, 339 (1989).
[11] D. L. Cheesman, W. L. Lee and P. F. Zagalsky, *Biol. Rev.*, **42**, 132 (1967).
[12] P. F. Zagalsky, *Pure Appl. Chem.*, **47**, 103 (1976).
[13] W. L. Lee (ed.), *Carotenoproteins in Animal Coloration*, Dowden, Hutchinson and Ross, Stroudsberg, PA (1977).
[14] T. W. Goodwin, *The Biochemistry of the Carotenoids, Vol. II: Animals*, p. 1, Chapman and Hall, London (1984).
[15] M. I. Newbigin, *J. Physiol.*, **21**, 237 (1897).
[16] J. Verne, *Bull. Soc. Zool. Fr.*, **46**, 61 (1921).
[17] G. Britton, G. M. Armitt, S. Y. M. Lau, A. K. Patel and C. C. Shone, in *Carotenoid Chemistry and Biochemistry* (ed. G. Britton and T. W. Goodwin), p. 237, IUPAC-Pergamon, Oxford (1982).
[18] J. B. C. Findlay, D. J. C. Pappin, M. Brett and P. F. Zagalsky, in *Carotenoids: Chemistry and Biology* (ed. N. I. Krinsky, M. M. Mathews-Roth and R. F. Taylor), p. 75, Plenum Press, New York (1990).
[19] P. F. Zagalsky, *Pure Appl. Chem.*, **66**, 973 (1994).
[20] G. Britton, R. J. Weesie, D. Askin, J. D. Warburton, L. Gallardo-Guerrero, F. J. Jansen, H. J. M. de Groot, J. Lugtenburg, J.-P. Cornard and J.-C. Merlin, *Pure Appl. Chem.*, **69**, 2075 (1997).
[21] M. Cianci, P. J. Rizkallah, A. Olczak, J. Raftery, N. E. Chayen, P. F. Zagalsky and J. R. Helliwell, *Proc. Natl. Acad. Sci. USA*, **99**, 9795 (2002).
[22] D. F. Cheesman, P. F. Zagalsky and H. J. Ceccaldi, *Proc. Roy. Soc. London, Ser. B*, **164**, 130 (1966).

[23] M. Buchwald and W. P. Jencks, *Biochemistry*, **7**, 844 (1968).

[24] R. Quarmby, D. A. Norden, P. F. Zagalsky, II. J. Ceccaldi and R. Daumas, *Comp. Biochem. Physiol.*, **B56**, 55 (1977).

[25] J. N. Keen, I. Caceres, E. E. Eliopoulos, P. F. Zagalsky and J. B. C. Findlay, *Eur. J. Biochem.*, **197**, 407 (1991).

[26] J. N. Keen, I. Caceres, E. E. Eliopoulos, P. F. Zagalsky and J. B. C. Findlay, *Eur. J. Biochem.*, **202**, 31 (1991).

[27] M. E. Newcomer and E. O. Ong, *Biochim. Biophys. Acta*, **1482**, 57 (2000).

[28] N. E. Chayen, M. Cianci, J. G. Grossmann, J. Habash, J. R. Helliwell, G. A. Nneji, J. Raftery, P. J. Rizkallah and P. F. Zagalsky, *Acta Cryst.*, **D59**, 2072 (2003).

[29] G. A. Nneji and N. E. Chayen, *J. Appl. Cryst.*, **37**, 502 (2004).

[30] N. E. Chayen, P. D. Shaw Stewart, D. L. Maeder and D. M. Blow, *J. Appl. Cryst.*, **23**, 297 (1990).

[31] M. Cianci, P. J. Rizkallah, A. Olczak, J. Raftery, N. E. Chayen, P. F. Zagalsky and J. R. Helliwell, *Acta Cryst.*, **D57**, 1219 (2001).

[32] P. F. Zagalsky, E. E. Eliopoulos and J. B. C. Findlay, *Comp. Biochem. Physiol.*, **B97**, 1 (1990).

[33] J. C. J. Bart and C. H. McGillavry, *Acta Cryst.*, **B24**, 1587 (1968).

[34] G. Bartalucci, J. Coppin, S. Fisher, G. Hall, J. R. Helliwell, M. Helliwell and S. Liaaen-Jensen, *Acta Cryst.*, **B63**, 328 (2007).

[35] B. Renstrøm, H. Rønneberg, G. Borch and S. Liaaen-Jensen, *Comp. Biochem. Physiol.*, **B71**, 249 (1982).

[36] V. R. Salares, N. M. Young, H. J. Bernstein and P. R. Carey, *Biochim. Biophys. Acta*, **576**, 176 (1979).

[37] R. J. Weesie, R. Verel, F. J. H. M. Jansen, G. Britton, J. Lugtenburg and H. J. M. de Groot, *Pure Appl. Chem.*, **69**, 2085 (1997).

[38] R. J. Weesie, F. J. H. M. Jansen, J.-C. Merlin, J. Lugtenburg, G. Britton and H. J. M. de Groot, *Biochemistry*, **36**, 7288 (1997).

[39] R. J. Weesie, J.-C. Merlin, J. Lugtenburg, G. Britton, F. J. H. M. Jansen and J.-P. Cornard, *Biospectroscopy*, **5**, 19 (1999).

[40] A. A. C. van Wijk, A. Spaans, N. Uzunbajakava, C. Otto, H. J. M. de Groot, J. Lugtenburg and F. Buda, *J. Am. Chem. Soc.*, **127**, 1438 (2005).

[41] A. van Wijk, *Doctoral Thesis*, Leiden University (2005).

[42] G. Kildahl-Andersen, B. F. Lutnaes and S. Liaaen-Jensen, *Org. Biomol. Chem.*, **2**, 489 (2004).

[43] B. Durbeej and L. A. Eriksson, *Chem. Phys. Lett.*, **375**, 30 (2003).

[44] B. Durbeej and L. A. Eriksson, *Phys. Chem. Chem. Phys.*, **6**, 4190 (2004).

[45] B. Durbeej and L. A. Eriksson, *Phys. Chem. Chem. Phys.*, **8**, 4053 (2006).

[46] S. Krawczyk and G. Britton, *Biochim. Biophys. Acta*, **1544**, 301 (2001).

[47] H. Rønneberg, G. Borch, D. L. Fox and S. Liaaen-Jensen, *Comp. Biochem. Physiol.*, **B62**, 309 (1979).

[48] R. J. H. Clarke, G. A. Rodley, A. F. Drake, R. A. Church and P. F. Zagalsky, *Comp. Biochem. Physiol.*, **B95**, 847 (1990).

[49] N. Wade, K. C. Goulter, K. J. Wilson, M. R. Hall and B. M. Degnan, *Comp. Biochem. Physiol.*, **B141**, 307 (2005).

[50] N. Wade, *Ph.D. Thesis*, University of Queensland (2005).

[51] W. P. Jencks and P. Buten, *Arch. Biochem. Biophys.*, **107**, 511 (1964).

[52] V. R. Salares, N. M. Young, H. J. Bernstein and P. R. Carey, *Biochemistry*, **16**, 4751 (1979).

[53] D. F. Cheesman, *Proc. Roy. Soc. London, Ser. B*, **149**, 571 (1958).

[54] P. F. Zagalsky and R. Jones, *Comp. Biochem. Physiol.*, **B21**, 237 (1982).

Carotenoids
Volume 4: Natural Functions
© 2008 Birkhäuser Verlag Basel

Chapter 7

Carotenoid Radicals and Radical Ions

Ali El-Agamey and David J McGarvey

A. Introduction

1. Definitions

Various types of carotenoid-derived ions, radicals and radical ions are referred to in this Chapter and elsewhere in this Volume.The different species are defined below and their relationship to the parent carotenoid is illustrated by the example of β-carotene (**3**, $C_{40}H_{56}$).

Carotenoid radical anion ($CAR^{\bullet-}$)

$$C_{40}H_{56} \quad + \; e^- \quad \longrightarrow \quad C_{40}H_{56}{}^{\bullet-}$$

Carotenoid radical cation ($CAR^{\bullet+}$)

$$C_{40}H_{56} \quad - \; e^- \quad \longrightarrow \quad C_{40}H_{56}{}^{\bullet+}$$

(Carotenoid – H) cation (${}^{\#}CAR^{+}$)

$$C_{40}H_{56} \quad - \; H^{\bullet} \; - \; e^- \quad \longrightarrow \quad C_{40}H_{55}{}^{+}$$

Carotenoid dication (CAR^{2+})

$$C_{40}H_{56}^{\bullet +} - e^- \longrightarrow C_{40}H_{56}^{2+}$$

Carotenoid neutral radical (CAR$^\bullet$ or $^{\#}$CAR$^\bullet$)

$$C_{40}H_{56} \quad - H^\circ \longrightarrow C_{40}H_{55}^\bullet$$

$$C_{40}H_{56}^{\bullet +} - H^+ \longrightarrow C_{40}H_{55}^\bullet$$

$$C_{40}H_{55}^+ + e^- \longrightarrow C_{40}H_{55}^\bullet$$

2. The roles of carotenoid radicals

Carotenoids are susceptible to oxidation but relatively resistant to reduction. Consequently, carotenoid radicals can arise from their reductive quenching of excited states and from free-radical scavenging. The widespread natural occurrence of carotenoids in oxidizing and photo-oxidizing environments reflects these properties. Carotenoid radicals are often produced as intermediates in oxidizing environments, although the extent to which such species arise in biological systems under oxidative stress is largely unknown, as are any biological effects these species may have. The associations between the incidence of certain degenerative diseases and dietary intake of carotenoids (*Volume 5*) may be linked to the ability of carotenoids to scavenge free radicals, but the significance of this remains unclear.

Carotenoids have been shown to inhibit free-radical mediated oxidation of oxidizable substrates [1-6]. However, the oxidation chemistry of carotenoids is complex, as reflected in the variety of carotenoid oxidation products that are produced and the dependence of the product profile on the structure of the carotenoid under study and the nature of the oxidizing environment [7-11]. One of the main challenges in the study of the chemistry of carotenoid oxidation is to characterize fully the pathways from initial free-radical attack to oxidation products and subsequently to reconcile these data with the antioxidant/pro-oxidant properties of the carotenoids [8].

Carotenoid radicals also arise in photosynthetic systems (see *Chapter 14*). A specific example is the generation in photosystem II of carotenoid radical cations (CAR$^{\bullet +}$) [12-15], possibly formed by reaction of the carotenoid with the most strongly oxidizing species in photosystem II, P$^{\bullet +}_{680}$. This reaction (Scheme 1) has been proposed to be part of a sequential electron transfer from cytochrome b$_{559}$ (cyt b$_{559}$) to P$^{\bullet +}_{680}$ *via* accessory chlorophylls (chl$_z$) and carotenoids (CAR) [12].

$$\text{cyt b}_{559} \longrightarrow \text{chl}_z \longrightarrow \text{CAR} \longrightarrow \text{P}_{680}$$

Scheme 1

Carotenoid radical cations have also been observed in the light-harvesting complex 2 (LH2), from *Rhodobacter sphaeroides* into which different carotenoids were incorporated [16]. The generation of $CAR^{\bullet+}$ is attributed to electron transfer from the S_2 excited state of the carotenoid to bacteriochlorophyll a (BChl) to form $CAR^{\bullet+}$ and $BChl^{\bullet-}$ (Scheme 2).

$$*CAR\ (S_2)\ +\ BChl\ \longrightarrow\ [*CAR\ (S_2)...\ BChl]\ \longrightarrow\ CAR^{\bullet+}\ +\ BChl^{\bullet-}$$

Scheme 2

Carotenoids can scavenge oxidizing free radicals *via* at least three primary reactions (equations 1-3), namely electron transfer, addition and hydrogen atom transfer [17,18].

$$CAR\ +\ R^{\bullet}\ \longrightarrow\ CAR^{\bullet+}\ +\ R^{-}\qquad \text{Electron transfer}\qquad (1)$$

$$CAR\ +\ R^{\bullet}\ \longrightarrow\ CAR^{\bullet}\ +\ RH\qquad \text{Hydrogen atom transfer}\quad (2)$$

$$CAR\ +\ R^{\bullet}\ \longrightarrow\ RCAR^{\bullet}\qquad\qquad \text{Addition}\qquad\qquad (3)$$

However, secondary reactions are possible that can lead to different carotenoid radicals from the one formed in the initial step.

   Progress has been made in understanding the free-radical chemistry of carotenoids, although there are many questions that remain unanswered. Carotenoid radicals are commonly detected *via* their UV/Vis/NIR optical absorption. Neutral radicals generally absorb in the same spectral region as the parent compound, though at somewhat longer wavelength. Radical anions and radical cations absorb at much longer wavelength (by several hundred nm), λ_{max} commonly being in the NIR region in the range 800-1100 nm (see Fig. 1, *Chapter 8*). Radical cations generally absorb at somewhat longer wavelength than the corresponding radical anion. The spectra of all carotenoid radicals show little or no fine structure, and the λ_{max} are strongly influenced by solvent/medium and experimental conditions (see Table 1). One area where there has been relatively little activity, however, is in the use of theoretical calculations to predict radical structures and physicochemical properties. Such work could prove invaluable in aiding assignment of the various carotenoid radicals that can be observed *via* UV/Vis/NIR spectrophotometric measurements.

It is therefore important that:
   i) The routes by which carotenoid radicals are formed are understood.
  ii) Carotenoid radicals are characterized structurally and spectroscopically.
 iii) The chemistry of carotenoid radicals, in terms of the rates and mechanisms of intra-
      molecular and intermolecular chemical reactions, is fully understood.
  iv) The pathways from carotenoid radical intermediates to final oxidation products are fully
      elucidated.

B. Radical Ions

1. Formation and detection of carotenoid radical ions

a) Pulse radiolysis

In pulse radiolysis, the sample is irradiated by a pulsed beam of high-energy electrons. The energy is absorbed by the solvent, resulting in the production of excited states and radicals of the solvent molecules. Scavenging/quenching of these solvent species by the solute(s) results in formation of solute excited states and radicals, which are normally monitored *via* UV/Vis/NIR absorption spectroscopy. The radiation chemistry is not the same for all solvents, so different solute-derived transient species are obtained in different media, although the use of various chemical additives allows some selectivity in terms of the solute transient species produced.

   In media of low polarity, such as argon-saturated hexane [19,20] and benzene [17,20,21], carotenoid radical ions can be generated *via* scavenging of solvent radical cations and electrons (Schemes 3 and 4).

$$C_6H_{14} \rightsquigarrow C_6H_{14}{}^{\bullet+} + e^-$$

$$e^- + CAR \longrightarrow CAR^{\bullet-}$$

$$C_6H_{14}{}^{\bullet+} + CAR \longrightarrow CAR^{\bullet+} + C_6H_{14}$$

Scheme 3

$$C_6H_6 \rightsquigarrow C_6H_6{}^{\bullet+} + e^-$$

$$e^- + CAR \longrightarrow CAR^{\bullet-}$$

$$C_6H_6{}^{\bullet+} + CAR \longrightarrow C_6H_6 + CAR^{\bullet+}$$

Scheme 4

In the presence of oxygen, only $CAR^{\bullet+}$ is observed, due to scavenging of the electron and $CAR^{\bullet-}$ by oxygen.

   In more polar solvents, *e.g.* argon-saturated methanol [19,20], only carotenoid radical anions ($CAR^{\bullet-}$) are generated by the reaction of the solvated electron with carotenoids (Scheme 5).

$$CH_3OH \xrightarrow{\hspace{1cm}} {}^\bullet CH_2OH + e^-_{MeOH}$$

$$e^-_{MeOH} + CAR \xrightarrow{\hspace{1cm}} CAR^{\bullet-}$$

Scheme 5

In aqueous environments [20,22,23], both oxidizing species ($^\bullet$OH) and reducing species (e^-_{aq} and H$^\bullet$) are formed (equation 4) in the radiolysis of water. In argon-saturated solution, a reducing environment is afforded by the use of additives such as formate ion (HCO$_2^-$), which reacts with $^\bullet$OH to form CO$_2^{\bullet-}$. Under such conditions CAR$^{\bullet-}$ may be generated (Scheme 6).

$$H_2O \xrightarrow{\hspace{1cm}} e^-_{aq} + {}^\bullet OH + H^\bullet + H_2 + H_2O_2 \qquad (4)$$

$${}^\bullet OH + HCO_2^- \xrightarrow{\hspace{1cm}} H_2O + CO_2^{\bullet-}$$

$$e^-_{aq} + CAR \xrightarrow{\hspace{1cm}} CAR^{\bullet-}$$

Scheme 6

An oxidizing environment is afforded by saturation of the solution with N$_2$O, which scavenges e^-_{aq} leading to the formation of the highly oxidizing hydroxyl radical ($^\bullet$OH). Milder oxidizing species may be formed by reaction of hydroxyl radical with halide ions such as bromide, leading to formation of Br$_2^{\bullet-}$ and generation of CAR$^{\bullet+}$ (Scheme 7).

$$e^-_{aq} + H_2O + N_2O \xrightarrow{\hspace{1cm}} N_2 + OH^- + {}^\bullet OH$$

$${}^\bullet OH + 2Br^- \xrightarrow{\hspace{1cm}} OH^- + Br_2^{\bullet-}$$

$$Br_2^{\bullet-} + CAR \xrightarrow{\hspace{1cm}} CAR^{\bullet+} + 2Br^-$$

Scheme 7

b) Laser flash photolysis

Laser flash photolysis involves irradiation of the sample with a pulse (usually nanoseconds) of monochromatic laser light. This technique has been used extensively for generating free radicals [24-29] (see also section **D**) and excited states of carotenoids. There are no reports of the use of laser flash photolysis for the generation of carotenoid radical anions, however.

Carotenoid radical cations (CAR$^{\bullet+}$) can be formed *via* reductive quenching of the triplet excited states of certain photosensitizers. For example, CAR$^{\bullet+}$ is produced from the reaction of CAR with the triplet excited state of toluidine blue, a reaction which involves competitive energy-transfer and electron-transfer quenching pathways [30]. Similar behaviour is observed for the reaction of CAR with triplet excited nitronaphthalene (^{3}NN) [31] (Scheme 8).

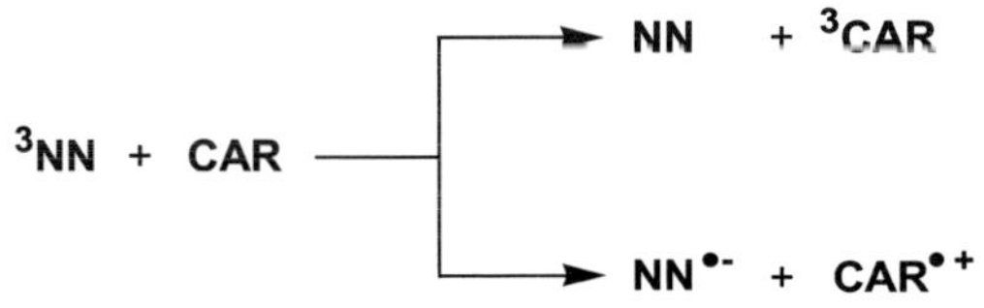

$$^3\text{NN + CAR} \longrightarrow \begin{cases} \text{NN} \quad + \ ^3\text{CAR} \\ \\ \text{NN}^{\bullet-} + \ \text{CAR}^{\bullet+} \end{cases}$$

Scheme 8

c) Chemical methods

The carotenoid radical cation $CAR^{\bullet+}$ can be produced by the reactions of carotenoids with strong oxidizing agents. For example, reaction of carotenoids with $FeCl_3$ in CH_2Cl_2 gives rise to the formation of carotenoid radical cations and dications ($CAR^{\bullet+}$ and CAR^{2+} respectively) (equation 5) [32]. $CAR^{\bullet+}$ is also produced from the reactions of carotenoids with quinones in dichloromethane or with iodine (equation 6; *cf Chapter 8*) [33,34]. More recent studies of the latter reaction and characterization of the complex as $CAR.I_4$ are described in *Chapter 8*.

$$\text{CAR} \xrightarrow{\text{FeCl}_3} \text{CAR}^{2+} \xrightarrow{\text{CAR}} 2\ \text{CAR}^{\bullet+} \tag{5}$$

$$\text{CAR} + 3\ \text{I}_2 \rightleftharpoons 2\left(\text{CAR}^{\bullet+}...\ \text{I}_3^-\right) \tag{6}$$

Reactions of carotenoids with Metal-MCM-41 mesoporous molecular sieves [35-38], where the metal is Fe^{3+}, Cu^{2+}, Ni^{2+} or Al^{3+}, also give rise to the corresponding $CAR^{\bullet+}$.

d) Electrochemical methods

Electrochemical methods [32,39-48] have been used to generate and investigate carotenoid radical cations, carotenoid radical anions [49] and carotenoid neutral radicals (Section **B**.3).

2. Structural and spectroscopic properties of carotenoid radical ions

a) Vis/NIR spectroscopy

An extensive list of the λ_{max} and molar absorption coefficient (ε) data of the radical ions $CAR^{\bullet+}$ and $CAR^{\bullet-}$ for a range of carotenoids with different chromophore lengths in different solvents is given in Table 1.

Table 1. The λ_{max} and molar absorption coefficients (ε) of the radical anions (CAR$^{\bullet-}$) and radical cations (CAR$^{\bullet+}$) of some carotenoids and related compounds in different solvents [50].

| Compound | cdb[a] | Parent[b] | Spectroscopic data. Solvent: λ_{max} (ε x 10^{-5}) [Ref] | |
			Radical anion	Radical cation
β-Cyclocitral (*1*)	1		M: ~315 [51]	
			T: 370 [51]	
β-Ionone (*2*)	2		M: 350 [51]	II: 375 [52]
			T: 380 [51]	DCE: 385 [51]
			H: 385 [52]	
			AM: 300 (0.12) [52]	
β-Ionylideneacetaldehyde (*3*)	3		H: 460 [52]	H: 470 [52]
			AM: 350 (0.24) [52]	
(15*Z*)-Phytoene (**44**)	3	H: 286	H: 470 [19]	H: 470 [19]
(all-*E*)-Phytoene (**44**)	3	H: 286	H: 460 [19]	H: 460 [19]
C_{17}-Aldehyde (*4*)	4		M: 395 [51]	A: 515 [51]
Retinol (*5*)	5	E: 326	M: 370 [53]	A: 585 [54]
			I: 370 [53]	DCE: 600 [54]
				H: 600 [54]
				TX: 590 [55]
Retinal (*6*)	5	E: 370	A: 495 [56]	A: 585 [54]
			H: 580 [56]	DCE: 595 [54]
			M: 445 [56]	H: 590 [52]
			E: 458 [56]	ACN: 585 [56]
			I: 460 [53]	BZN: 595 [56]
			T: 530 [51]	B: 600 [56]
			AM: 405 (0.79) [52]	TX: 590 [55]
			ACN:520 [56]	
			BZN: 550 [56]	
			B: 570 [56]	
			TX10:460 [56]	
			CTAB: 448 [56]	
Retinal *n*-butylamine Schiff base (*7*)	5		M: 430 [53]	A: 615 [54]
			I: 435 [53]	H: 635 [54]
Retinyl acetate (*8*)	5		M: 390 [53]	A: 580 [54]
			I: 390 [53]	DCE: 595 [54]
				H: 590 (1.0) [54,57]
Retinoic acid (*9*)	5		H: 510 [57]	A: 575 [54]
			M: 480 (1.2) [53,57]	DCE: 585 [54]
			I: 505 [53]	H: 590 [54]
				M: 590 (0.7) [57]
				TX: 590 [55]
Methyl retinoate (*10*)	5		M: 480 [53]	A: 575 [54]
			I: 510 [53]	DCE: 585 [54]
				H: 590 [54]
14'-Apo-β-caroten-14'-al (**513**)	6		M: 465 [51]	A: 640 [51]
C_{24}-Aldehyde (*11*)	7		M: 490 [51]	A: 700 [51]

Table 1 continued

Compound	cdb[a]	Parent[b]	Radical anion	Radical cation
		Spectroscopic data. Solvent: λ_{max} (ε x 10^{-5}) [Ref]		
ζ-Carotene (**38**)	7	E: 399		M: 740 [24]
7,7'-Dihydro-β-carotene (**49**)	8	H: 402 [21]	H: 785 [19] M: 705 [19]	H: 830 [19] M: 770 [24] TX· 770 [58]
Heptapreno-β-carotene (*12*)	9	H: 414 [21]	H: 785 [19] M: 705 [19] TX: 760 [23]	H: 915 [19] M: 820 [24] TX: 850 (0.37) [22]
Violaxanthin (**259**)	9	E: 440		E: 830 [59]
8'-Apo-β-caroten-8'-al (**482**)	9	E: 452	H: 840 [52] M: 555 [51] B: 820 (3.41) [60] AM: 500 (0.94) [52] T: 725 [51] TX: 590-600 [62] TX40: 590 [51]	H: 890 [22] M: 820 [21] B: 880 [17] C: 855 [28] DCM: 848 [61]
Lutein (**133**)	10	E: 445	H: 855 [23] B: 870 (3.320) [60] TX: 840 [23]	H: 973 [22] B: 950 [17] M: 865 [21] E: 880 [59] C: 940 [28] TX: 900 (0.67) [22]
β-Carotene (**3**)	11	E: 450	H: 880 (4.42) [63] B: 880 (3.27) [60] M: 800 [19] Cy: 900 (3.24) [63] TX10: 865 [64] Mix 3: 850 [65]	H: 1040 (2.18) [19,63] B: 1020 (1.24) [17,21] M: 900 [24] E: 910 (1.30) [30] Cy: 1050 (2.02) [63] C: 1000 [26] DCM: 970 [32] DMSO: 942 (0.16) [66] TX: 936 (0.87) [22] Mix1: 920 [67] Mix2: 910 (0.94) [68] ME: 840 [69]
(15Z)-β-Carotene (**3**)	11		H: 900 (2.51) [63] M: 820 [63]	H: 1050 (1.54) [63]
Lycopene (**31**)	11	E: 472	H: 950 (4.1) [19,63] B: 950 (3.71) [60] M: 870 [19]	H: 1070 (3.25) [19,63] B: 1050 (1.58) [17,21] M: 950 [21] C: 1040 [28] TX Mix: 970 [58]
Zeaxanthin (**119**)	11	E: 450	H: 900 [23] B: 880 (2.78) [60] M: 820 [60] TX: 861 [23]	H: 1040 [22] B: 1000 [17] M: 910 [21] E: 890 [30] C: 980 [28] TX: 936 (0.41) [22]

Table 1 continued

| Compound | cdb[a] | Parent[b] | Spectroscopic data. Solvent: λ_{max} (ε x 10^{-5}) [Ref] | |
			Radical anion	Radical cation
(*meso*)-Zeaxanthin (**120**)	11	E: 450	B: 900 (2.48) [60]	B: 1020 [21]
				H: 1040 [21]
				M: 900 [21]
Canthaxanthin (**380**)	11	E: 474	H: 1150 [19]	H: 960 [19]
			B: ≥1100 [60]	B: 940 [17]
			M: 600-610 [62]	M: 840 [21]
			TX: ~720 (0.21) [62]	C: 900 [28]
				DCM: 887 [32]
				TX: 862 (0.67) [22]
Astaxanthin (**406**)	11	E: 478	H: 1120 [70]	H: 940 [22]
			B: ≥1100 [60]	B: 920 [17]
			M: 610-620 [62]	M: 840 [21]
			TX: ~720 (0.53) [62]	TX: 875 (0.30) [22]
Torularhodinaldehyde (**271**)	13	E: 507	H: 1130 [52]	H: 1130 [52]
			AM: 600 [52]	
Decapreno-β-carotene (*13*)	15	P: 495 [19]	H: 1130 [19]	H: 1250 [19]
			M: 1010 [19]	
Dodecapreno-β-carotene (*14*)	19	P: 534 [19]	H: 1355 [19]	H: 1480 [19]
			M: 1140 [19]	

Solvents: A, acetone; ACN, acetonitrile; B, benzene; C, chloroform; E, ethanol; M, methanol; P, petroleum ether; BZN, benzonitrile; Cy, cyclohexane; H, hexane; DCE, 1, 2-dichloroethane; DCM, dichloromethane; AM, alkaline methanol (methanol containing 0.01 M NaOH); I, isopropyl alcohol; T, tetrahydrofuran; TX, 2% (w/v) aq. Triton X-100; TX10, 10 mM aq. Triton X-100; TX40, 40mM aq. Triton X-100; TX Mix, 3% (w/v) Triton X-405 and 1% (w/v) Triton X-100; Mix1, di-*t*-butylperoxide/cyclohexane (7:3 v/v); Mix2, *t*-butanol/water (1:1 v/v); Mix3, ethanol/water (86:14 v/v); CTAB, 18mM cetyltrimethylammonium bromide; ME, a microemulsion of sodium lauryl sulphate (3.23% w/v), cyclohexane (75%v/v), water (6.45% v/v) and pentan-1-ol (15.32% v/v).

[a] cdb is the number of conjugated double bonds (excluding carbonyl groups where relevant).

[b] The parent carotenoid. λ_{max} values are taken from *Vol. 1B, Chapter 2*, unless otherwise stated.

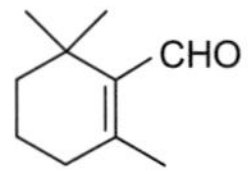

(*1*) β-cyclocitral

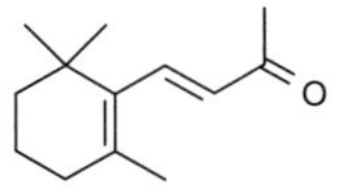

(*2*) β-ionone

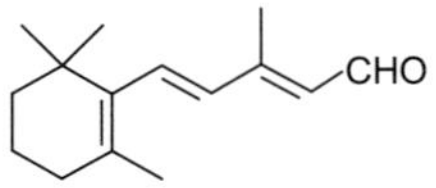

(*3*) β-ionylideneacetaldehyde

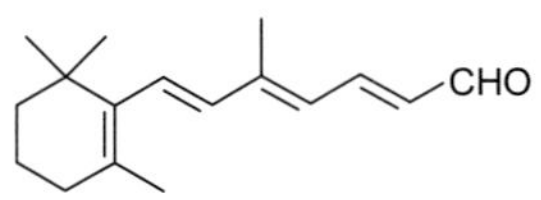

(*4*) 'C$_{17}$-aldehyde'

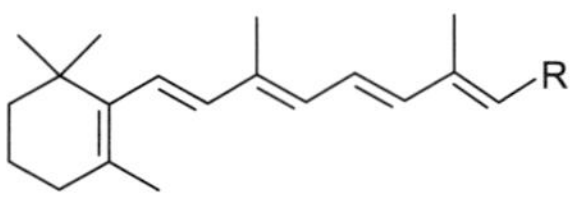

(*5*) R = CH$_2$OH: retinol
(*6*) R = CHO: retinal
(*7*) R = CH=N(CH$_2$)$_3$CH$_3$: retinal *n*-butylamine Schiff base
(*8*) R = CH$_2$OCOCH$_3$: retinyl acetate
(*9*) R = COOH: retinoic acid
(*10*) R = COOCH$_3$: methyl retinoate

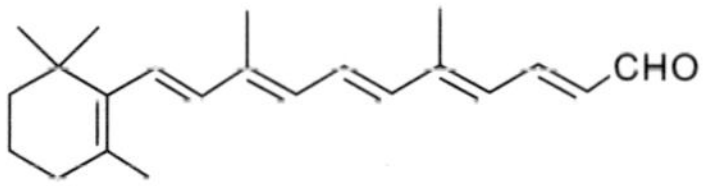

(**513**) 14'-apo-β-caroten-14'-al

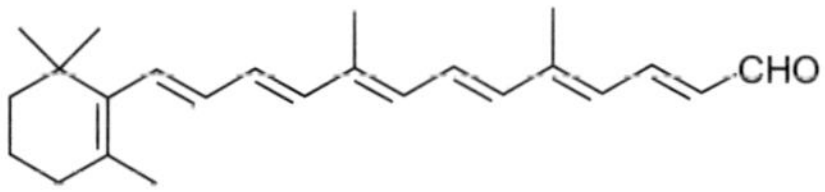

(**11**) 'C$_{24}$-aldehyde'

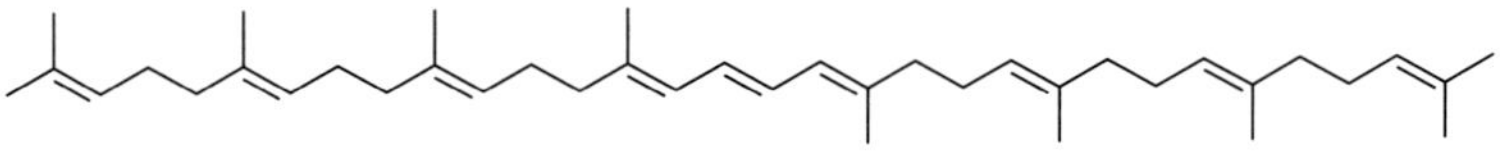

(**44**) phytoene

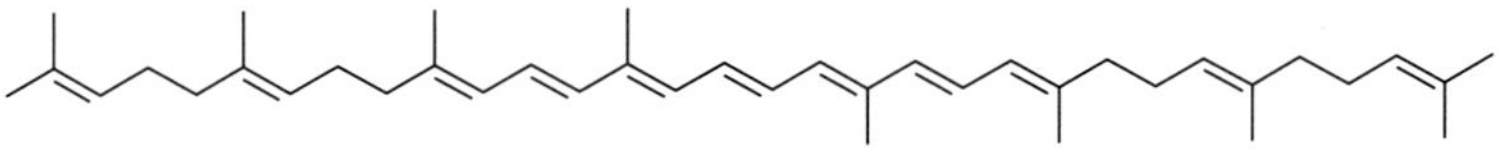

(**42**) phytofluene

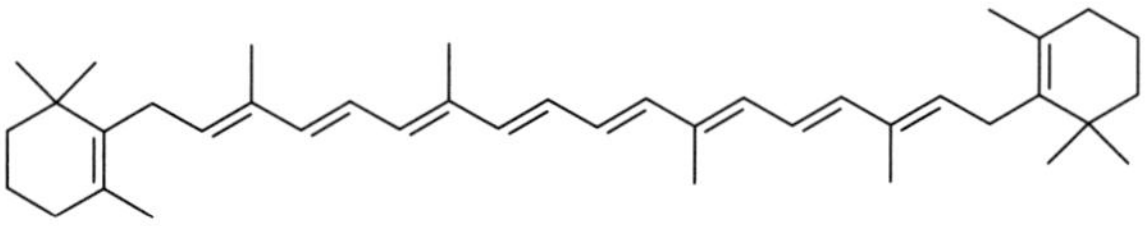

(**49**) 7,7'-dihydro-β-carotene (77DH)

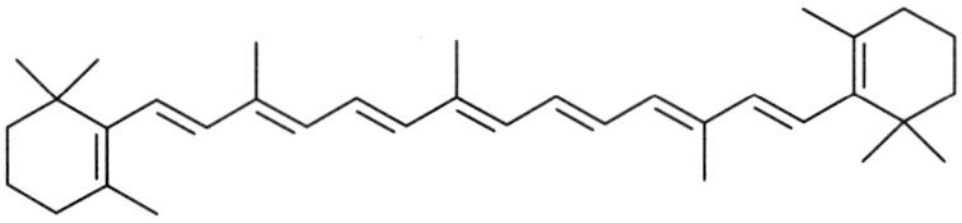

(**12**) heptapreno-β-carotene

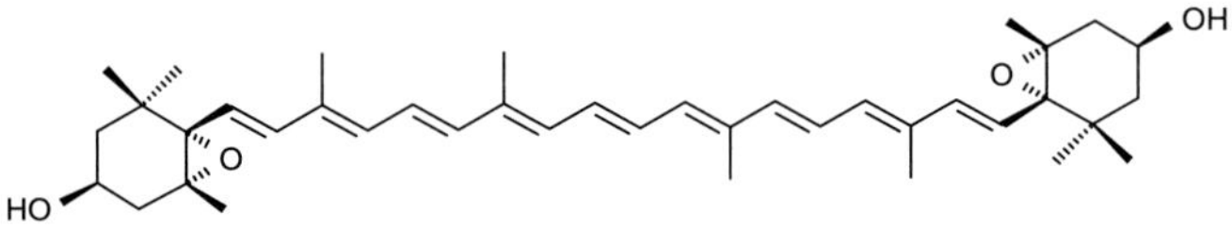

(**259**) violaxanthin

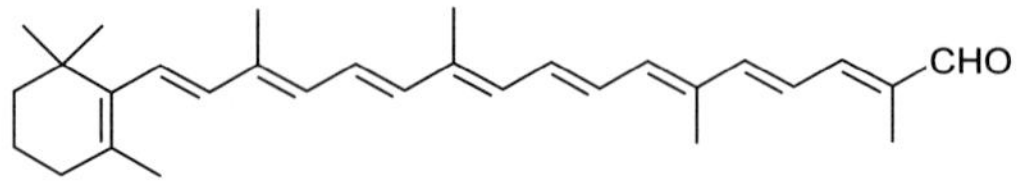

(**482**) 8'-apo-β-caroten-8'-al

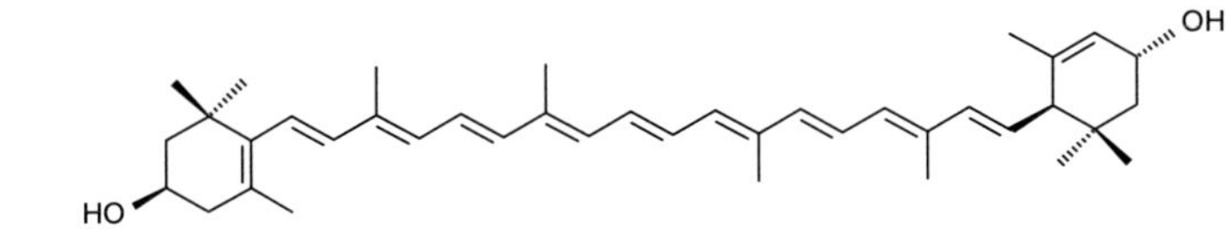

(**133**) lutein

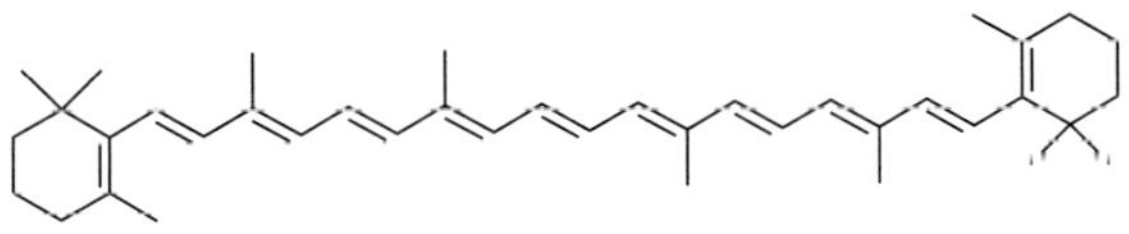

(**3**) β-carotene

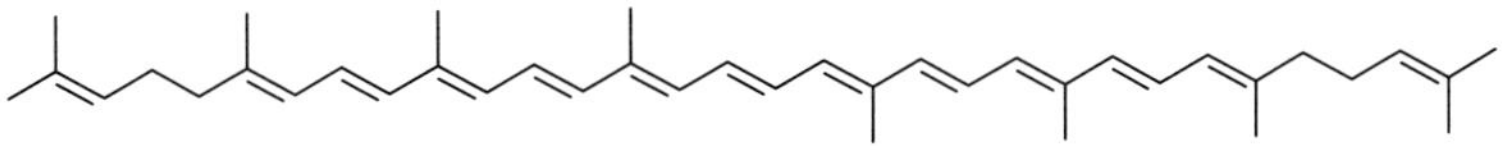

(**31**) lycopene

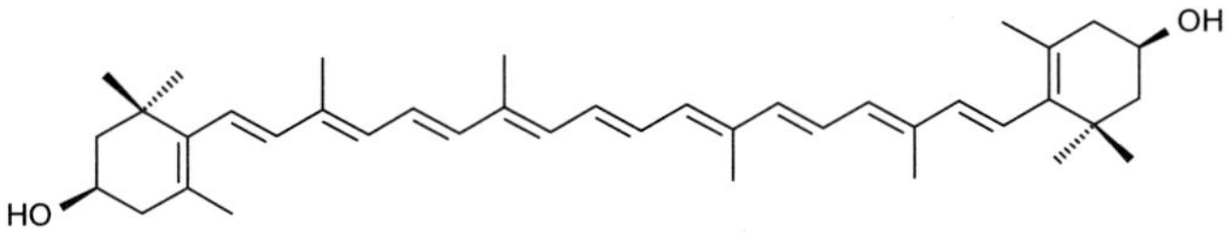

(**119**) zeaxanthin

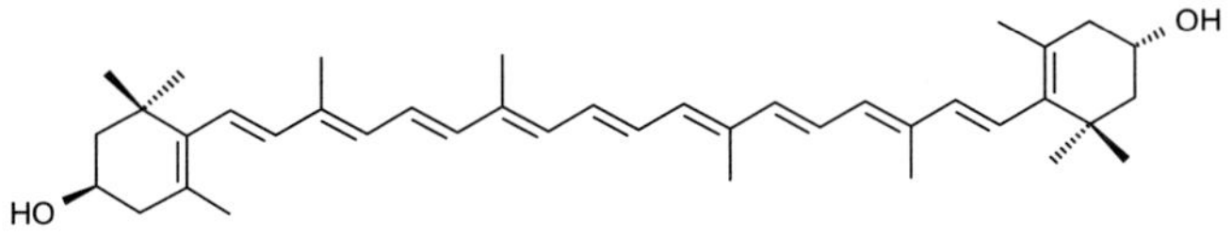

(**120**) *meso*-zeaxanthin

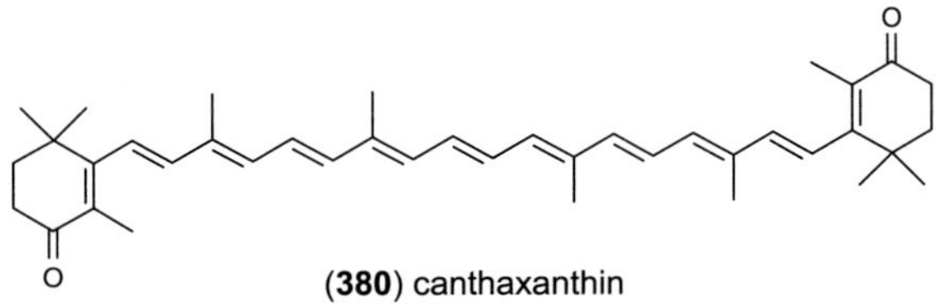

(**380**) canthaxanthin

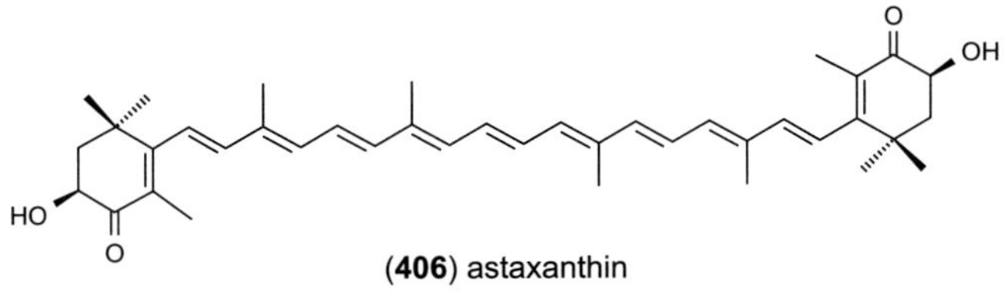

(**406**) astaxanthin

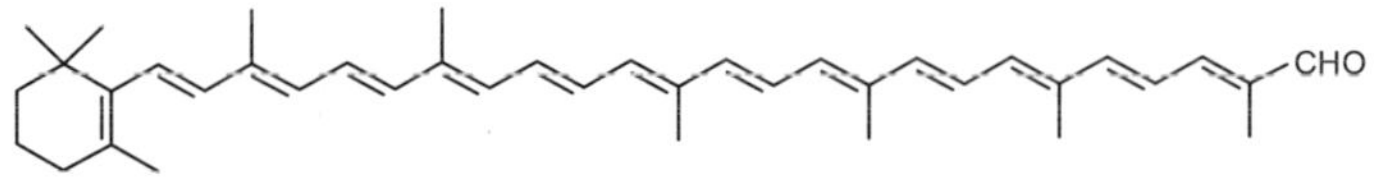

(**271**) torularhodinaldehyde

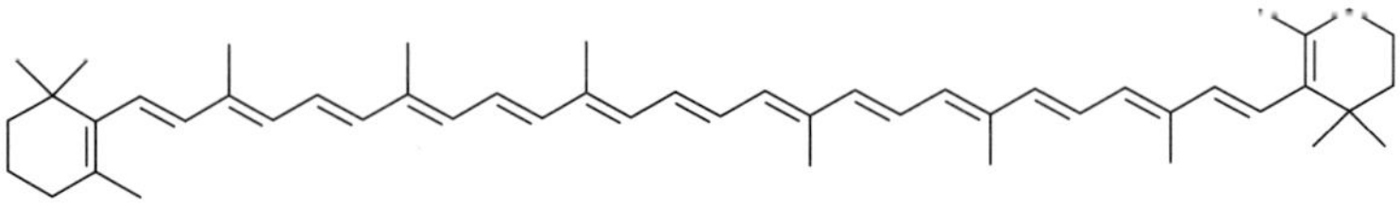

(**13**) decapreno-β-carotene

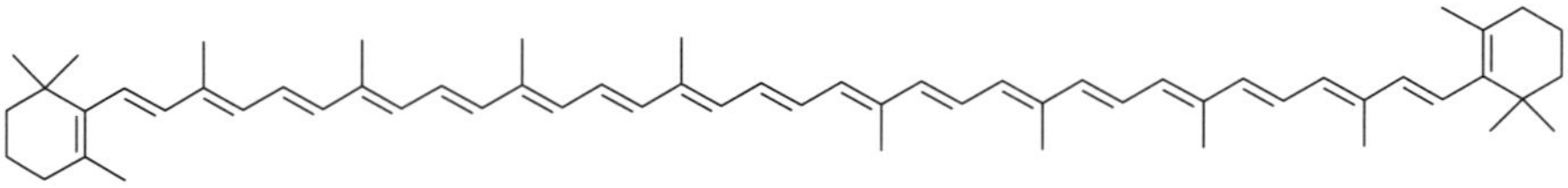

(**14**) dodecapreno-β-carotene

Generally, the λ_{max} of CAR$^{\bullet+}$ is at longer wavelength than that of CAR$^{\bullet-}$ [19] and, as the number of double bonds increases, the λ_{max} of CAR$^{\bullet+}$ and CAR$^{\bullet-}$ increases [52]. The spectral band positions for the radical ions are strongly dependent on solvent. The data show that the λ_{max} of CAR$^{\bullet-}$ is blue-shifted (to shorter wavelength) on changing the solvent from hexane to methanol. This is attributed to a less uniform charge distribution in the ground state of CAR$^{\bullet-}$ than in the excited state. Therefore, in methanol, the ground state is stabilized relative to the excited state [19].

More recently, theoretical and experimental work [71] has revealed the presence of an additional weak absorption band for polyene radical cations at longer wavelengths, attributed to the $D_0 \rightarrow D_1$ transition ('D' signifies 'doublet-state' in the same way that 'S' and 'T' signify 'singlet-state' and 'triplet-state' respectively). The more familiar intense band at shorter wavelengths is assigned to the $D_0 \rightarrow D_2$ transition (Table 2).

Table 2. The λ_{max} and molar absorption coefficients of the $D_0 \rightarrow D_1$ and $D_0 \rightarrow D_2$ transitions of CAR$^{\bullet+}$ in CH$_2$Cl$_2$ [32,72].

CAR$^{\bullet+}$	λ_{max} of $D_0 \rightarrow D_1$ / nm (log ε)	λ_{max} of $D_0 \rightarrow D_2$ / nm (log ε)
β-CAR$^{\bullet+}$	1425 (3.3)	970 (4.8)
CAN$^{\bullet+}$	1310 (4.6)	887 (5.4)

Abbreviations. β-CAR: β-carotene (**3**); CAN: canthaxanthin (**380**)

b) Electron paramagnetic resonance (EPR)

Vis/NIR spectra are used extensively for characterizing radical ions, but the presence of unpaired electrons is not proved by this method only. The absorption of carotenoid cations that have no unpaired electrons occurs within the same spectral region [73] (see *Chapter 8*). Electron paramagnetic resonance, which measures transitions between spin states of an unpaired electron, is a general method for demonstrating the presence of unpaired electrons [74] and is thus used as a tool for identifying and studying carotenoid radicals, including triplet states [34,42,44,75-84]. Most spectra are recorded in first-derivative form, as illustrated by the representative EPR spectrum of the radical cation of β-carotene (**3**), shown in Fig. 1. The single signal spectrum is characteristic of carotenoid radicals [86]. The spin properties of the electron (g-factor, hyperfine coupling) are influenced by nearby nuclei with non-zero nuclear spin, allowing structural information to be deduced.

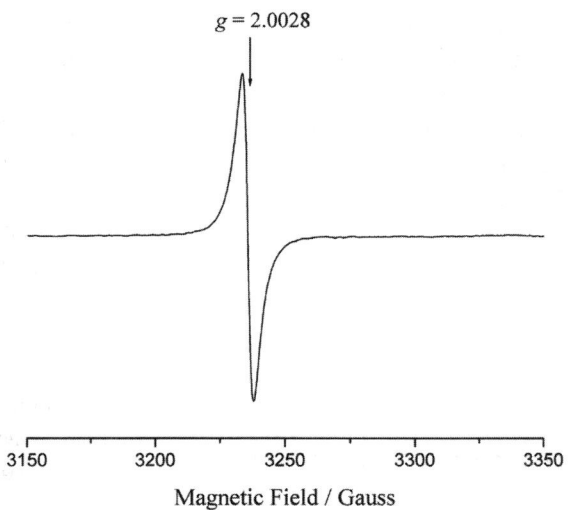

Fig. 1. EPR spectrum, measured at 207K, of β-carotene radical cation CAR$^{•+}$ generated by treatment of β-carotene with antimony trichloride [85].

c) Electron nuclear double resonance (ENDOR)

The more recent development, ENDOR, which employs both nuclear magnetic and electron spin resonance, is used in modern studies and gives greatly enhanced high-resolution EPR spectra. Pulsed ENDOR spectra of the radical cations of β-carotene (**3**) and canthaxanthin (**380**) have been reported [79,82].

d) Resonance Raman spectroscopy

Resonance Raman spectroscopy [31,43] of $CAR^{\bullet+}$, together with AM1 [42] and gas-phase B3LYP [87] calculations, suggest the delocalization of the unpaired electron density over the entire carbon backbone. Also, earlier calculations [42,52] and resonance Raman studies [31,43] indicated that the bond lengths in $CAR^{\bullet+}$ and $CAR^{\bullet-}$ have intermediate values between the single-bond and double-bond values of parent carotenoids (see also *Chapter 8*).

3. Reduction (redox) potentials for carotenoid radical ions

Most redox potential data for carotenoids [44-48,61,88] have been determined by electrochemical methods [47,49]. Reduction potentials [in CH_2Cl_2, *versus* the saturated calomel electrode (SCE)] involving carotenoid radical cations and neutral radicals are given in Table 3. The more positive the E° value, the more strongly oxidizing is the species on the left-hand-side of the reduction half-reaction. In relative terms, a low E° value means that it is easier, in principle, to oxidize the species on the right-hand-side of the reduction half-reaction to form the species on the left-hand side.

In the equations given in Scheme 9, $CAR^{\bullet+}$, CAR^{2+}, $^{\#}CAR^{+}$ and $^{\#}CAR^{\bullet}$ represent respectively the carotenoid radical cation, carotenoid dication, [carotenoid – H] cation and carotenoid neutral radical (see Section A.1)..

$$CAR^{\bullet+} + e^- \; \underset{}{\overset{E°_1}{\rightleftharpoons}} \; CAR$$

$$CAR^{2+} + e^- \; \underset{}{\overset{E°_2}{\rightleftharpoons}} \; CAR^{\bullet+}$$

$$^{\#}CAR^{+} + e^- \; \underset{}{\overset{E°_3}{\rightleftharpoons}} \; {}^{\#}CAR^{\bullet}$$

Scheme 9

Moreover, reactive species such as CAR^{2+} and $CAR^{\bullet+}$ can undergo various reactions (see equations 7-9).

$$CAR^{2+} + CAR \; \rightleftharpoons \; 2\,CAR^{\bullet+} \qquad\qquad (7)$$

$$CAR^{2+} \; \rightleftharpoons \; {}^{\#}CAR^{+} + H^+ \qquad\qquad (8)$$

$$CAR^{\bullet+} \; \rightleftharpoons \; {}^{\#}CAR^{\bullet} + H^+ \qquad\qquad (9)$$

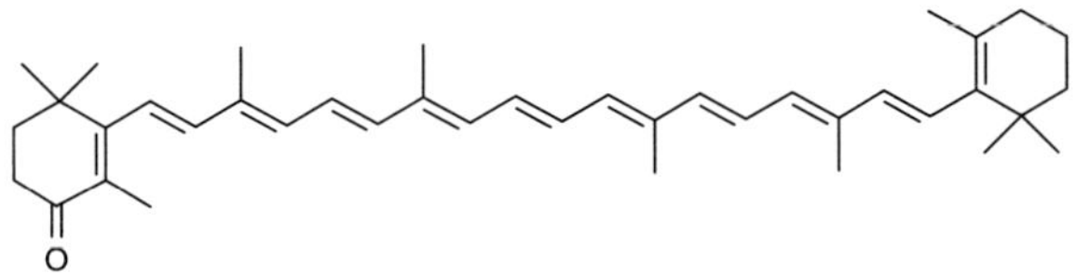

(**283**) echinenone

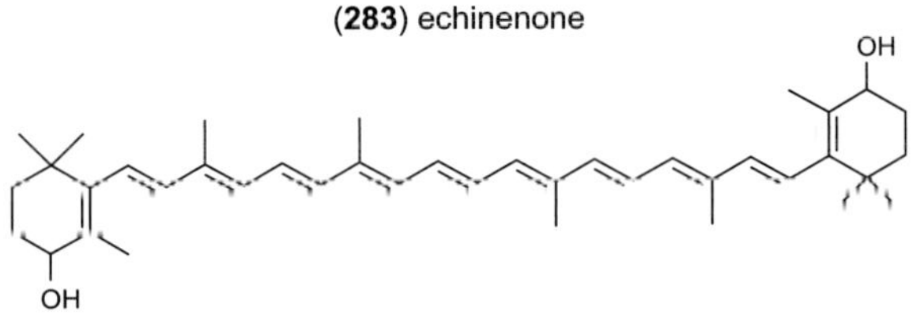

(**129**) isozeaxanthin

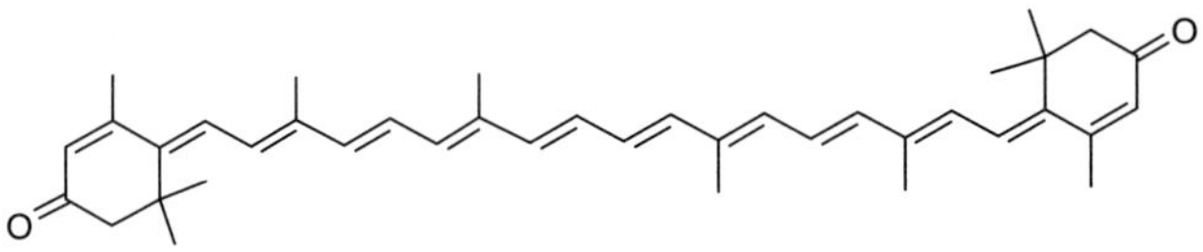

(**424**) rhodoxanthin

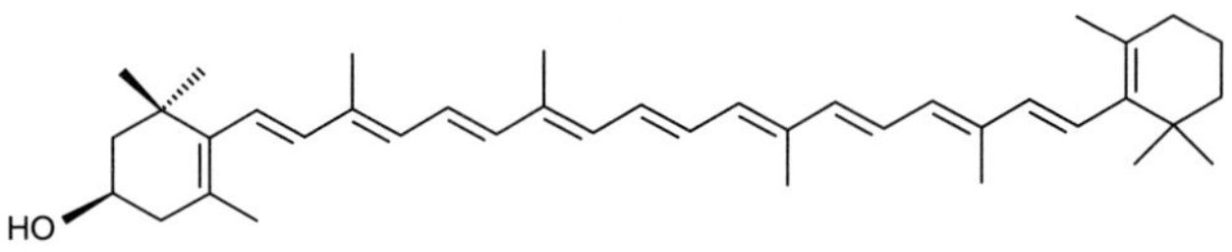

(**55**) β-cryptoxanthin

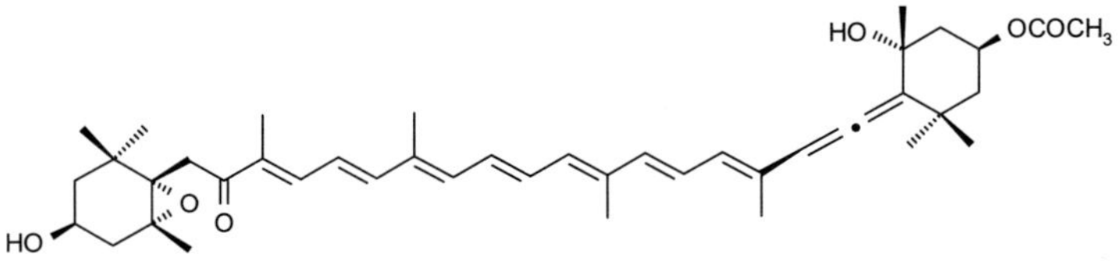

(**369**) fucoxanthin

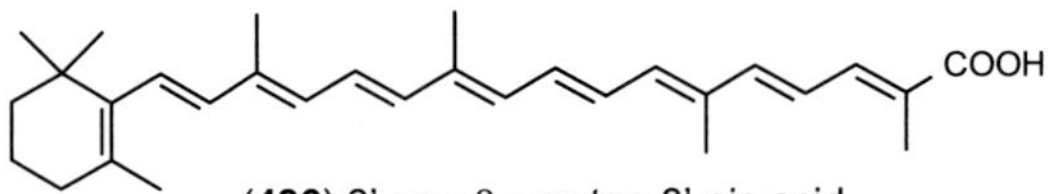

(**486**) 8'-apo-β-caroten-8'-oic acid

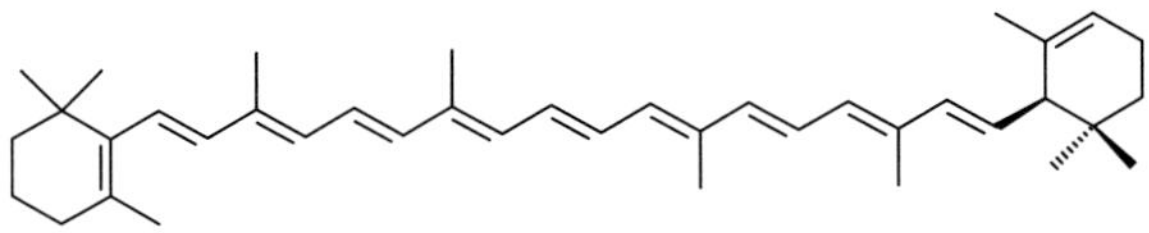

(**7**) α-carotene

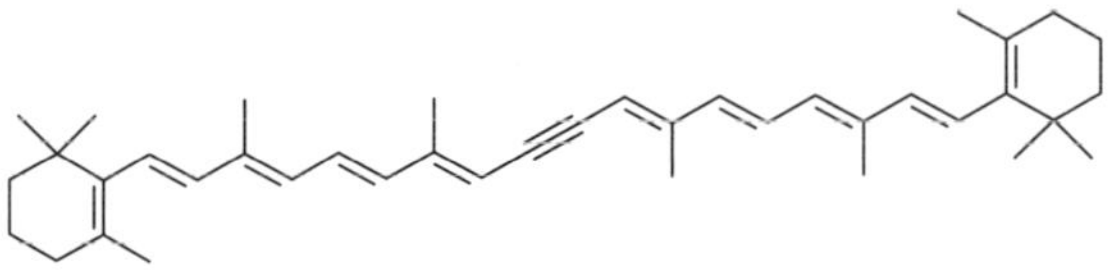

(15) 15,15'-didehydro-β-carotene

Table 3. Reduction potentials (E°_1) *versus* saturated calomel electrode (SCE) (± 2 mV, unless otherwise stated) for the carotenoid radicals as described in Scheme 9 [47]; in CH_2Cl_2 unless otherwise stated[a].

Carotenoid	E°_1 (mV)	E°_2 (mV)	E°_3 (mV)	Ref
β-Carotene (**3**)	567 ± 4	-	-	[45]
	540	545	35	[46]
	530	560	45	[44]
	510[a]			[49]
Canthaxanthin (**380**)	689	894	264	[46]
	705	945	250	[44]
Echinenone (**283**)	590	690	110	[44]
Isozeaxanthin (**129**)	570	550	80	[44]
Rhodoxanthin (**424**)	655	900	250	[44]
8'-Apo-β-caroten-8'-al (**482**)	720 ± 10	865 ± 25	75 ± 25	[61]
β-Cryptoxanthin (**55**)	560	570	60	[47]
Zeaxanthin (**119**)	571 ± 11	-	-	[45]
	530	550	90	[47]
Fucoxanthin (**369**)	790	820	240	[47]
8'-Apo-β-caroten-8'-oic acid (**486**)	686 ± 4	846 ± 8	263 ± 30	[47]
α-Carotene (**7**)	596 ± 4	623 ± 10	66 ± 16	[47]
	570[a]			[49]
Lycopene (**31**)	507 ± 6	524 ± 6	51 ± 28	[47]
	480[a]			[49]
Violaxanthin (**259**)	681 ± 14	-	-	[45]
15,15'-Didehydro-β-carotene (*15*)	763	766	240	[88]
	680[a]			[49]

[a]In acetonitrile/benzene (2:1) containing 0.027 M Et_4NClO_4.

Pulse radiolysis has been used to determine the reduction potentials of a variety of carotenoids in micelles by monitoring the reversible electron transfer between tryptophan radical cation (TrpH[•+]) and the carotenoid at different pH values. Since the reduction potentials of TrpH[•+]/ TrpH at different pH values are known, the reduction potentials of CAR[•+]/ CAR can be estimated (Table 4) [15,89].

Table 4. Reduction potentials (E°_1) ($\pm$ 25 mV) *versus* SCE for carotenoid radical cations in micelles (converted by adding 242 mV to the original data [*versus* normal hydrogen electrode (NHE)] [15,89].

Carotenoid	Micelle	E°_1 (mV)
β-Carotene (**3**)	TX-100[a]	818
β-Carotene (**3**)	TX-405/ TX-100[b]	786
Canthaxanthin (**380**)	TX-100	799
Zeaxanthin (**119**)	TX-100	789
β-Cryptoxanthin (**55**)	TX-100	786 [90]
Astaxanthin (**406**)	TX-100	788
Lycopene (**31**)	TX-405/ TX-100[b]	738

[a]TX-100: 2% (w/v) aqueous Triton X-100.
[b]TX-405/ TX-100: 3% (w/v) Triton X-405 and 1% (w/v) Triton X-100.

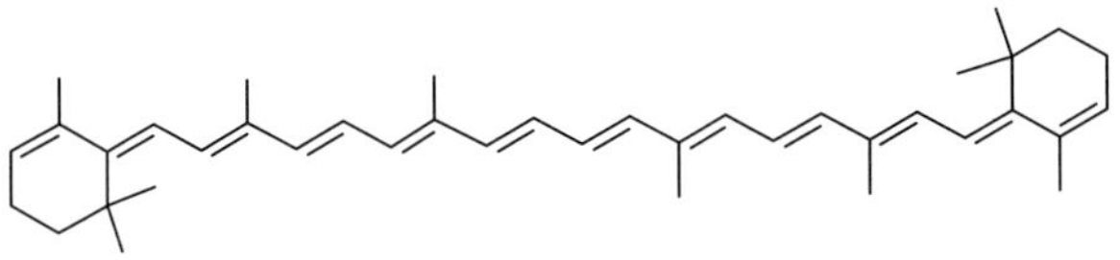

(*48*) retrodehydro-β-carotene

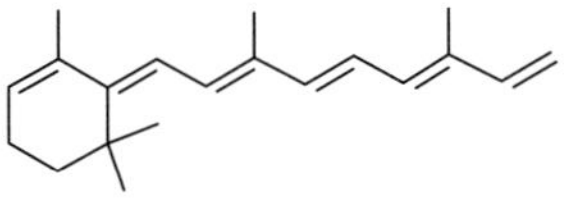

(*16*) anhydrovitamin A

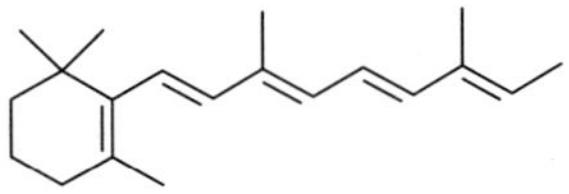

(17) axerophtene

The one-electron reduction potentials (*v.* SCE) of β-carotene (**3**), lycopene (**31**), α-carotene (**7**), neo-A-*retro*dehydro-β-carotene (**48**) 15,15'-didehydro-β-carotene (*15*), anhydrovitamin A (*16*) and axerophtene (*17*) (in dimethylformamide/benzene (2:1, v/v) containing 0.027 M tetraethyl ammonium iodide) have been reported as -1.68, -1.65, -1.70, -1.65, -1.69, -1.97 and -2.32 V, respectively [49]. The relative order of reduction potentials for β-carotene and lycopene is in agreement with the observed reaction of β-carotene[•−] with lycopene in hexane ($k \sim 1.4 \times 10^{10}$ M^{-1} s^{-1}) and benzene ($k \sim 6.2 \times 10^9$ M^{-1} s^{-1}) (equation 10) [21].

$$\text{β-CAR}^{\bullet-} + \text{LYCO} \longrightarrow \text{β-CAR} + \text{LYCO}^{\bullet-} \qquad (10)$$

C. Carotenoid Neutral Radicals

1. Formation and detection of carotenoid neutral radicals

There are several potential routes to the formation of carotenoid neutral radicals, including radical addition to the polyene backbone (giving a carotenoid addition radical), allylic hydrogen abstraction and de-protonation/protonation of radical ions (equations 11 and 12). Carotenoid addition radicals and other neutral radicals appear to absorb light in a similar spectral region to the parent carotenoid and, in some cases, there is a spectrally resolved absorption band on the red edge (long wavelength) of the parent carotenoid absorption [24,27,51-53,56,68,91,92]. These experimental observations are consistent with theoretical ZINDO/S calculations [93] of the absorption maxima of carotenoid neutral radicals, which are predicted to absorb close to the region where the parent carotenoid absorbs.

An early report [50] showed that the radical anions of carbonyl-containing carotenoids can be protonated in methanol to form the corresponding α-hydroxy radical derivatives (equation 11). In alkaline methanolic solutions, the reaction shifts toward the left and only carotenoid radical anions were observed. Similar observations were reported in different protic solvents and in micelles [51,53,56]. For example, the λ_{max} of the corresponding α-hydroxy radical derivatives of astaxanthin (406), canthaxanthin (380), 8'-apo-β-caroten-8'-al (482) and retinal (6) in aqueous Triton X-100 are 570, 580, 510 and 405 nm, respectively [56,62].

$$\underset{\text{CAR}}{\overset{\text{O}^-}{\underset{\bullet}{\parallel}}}\text{H} + \text{CH}_3\text{OH} \rightleftharpoons \underset{\text{CAR}}{\overset{\text{OH}}{\underset{\bullet}{\parallel}}}\text{H} + \text{CH}_3\text{O}^- \qquad (11)$$

$$\text{CAR}^{\bullet+} \rightleftharpoons {}^{\#}\text{CAR}^{\bullet} + \text{H}^+ \qquad (12)$$

When carotenoid addition radicals were first observed spectroscopically, the assignment at the time was uncertain. The reaction of glutathione thiyl free radicals (GS$^{\bullet}$) with retinol (5) was studied in aqueous methanol [91] and a strongly absorbing species was observed (λ_{max} 380 nm, $\varepsilon_{380\ nm} = 4.0 \times 10^4\ M^{-1}\ cm^{-1}$) that was attributed to either retinol$^{\bullet+}$ or [GS-retinol]$^{\bullet}$. However, retinol$^{\bullet+}$ is known to absorb around 585 nm [54] so the species absorbing at 380 nm is almost certainly [GS-retinol]$^{\bullet}$ (equation 13).

$$\text{Retinol} + \text{GS}^{\bullet} \longrightarrow [\text{GS- Retinol}]^{\bullet} \qquad (13)$$

Addition radicals were shown to be formed by reaction of β-carotene (3) with thiyl radicals and sulphonyl radicals (which also yield carotenoid radical cations under the experimental conditions used) [68,92]. This work was subsequently extended to the study of other carotenoids [astaxanthin (**406**), canthaxanthin (**380**), zeaxanthin (**119**), lutein (**133**), lycopene (**31**)] [94].

The *retro*-carotenoid 7,7'-dihydro-β-carotene (7,8-dihydro-8,7'-*retro*-β,β-carotene, **49**) is distinctive in that addition radicals derived from it display exceptionally intense absorption bands on the red edge of the parent absorption [24,27]. The spectral profile of addition radicals derived from 7,7'-dihydro-β-carotene is not strongly affected by the nature of the scavenged radical, although the chemical properties of the addition radicals vary considerably (see Section **E.2**).

D. Unidentified Carotenoid Radicals

The reactions of various radicals, including $CCl_3O_2{}^{\bullet}$, acylperoxyl radicals, sulphonyl radicals and phenoxyl radicals, with different carotenoids in polar environments leads to formation of carotenoid radicals that absorb in the near infrared, but at shorter wavelengths than the corresponding radical cations [22,25,26,28,67,94]. The decay of these carotenoid radical species appears to lead to formation of $CAR^{\bullet+}$ [22]. Speculation surrounding the assignment of these NIR absorption features initially included carotenoid radical adducts and other neutral radicals [22,25,26,28,67,95] and ion pairs [28,29,94]. Subsequent work on carotenoid addition radicals in polar and apolar environments, however, appears to preclude the assignment of these species as carotenoid neutral radicals, which would absorb in the visible region close to the absorption of the parent carotenoid [27,52,56,68,91]. More recently, in the light of evidence that these carotenoid radical species are not formed directly, but from ionic dissociation of carotenoid addition radicals, the possibility that these species are geometrical (*i.e. cis-trans*) isomers of the carotenoid radical cation has been suggested [24].

E. Reaction of Carotenoids with Oxidizing Free Radicals

1. Factors that influence the mechanism of reactions of free radicals with carotenoids

The mode of reaction of carotenoids with free radicals [96-100] is dependent upon the nature of the free radical (see sections **E.3.a** and **E.3.c**), solvent polarity and also the nature of the carotenoid *i.e.* number of conjugated double bonds, presence and type of oxygen functions (see section **E.3.d**).

2. Free radical scavenging mechanisms in environments of low polarity

The reactions of carotenoids with oxidizing free radicals in polar environments often proceeds *via* electron transfer leading to formation of the carotenoid radical cation. In apolar environments, however, this is generally not possible, as charge separation is not supported in such environments. Given that carotenoids are likely to be found in apolar environments within biological systems, it is surprising that comparatively few studies of free-radical reactions with carotenoids in apolar environments have been carried out. Nevertheless, even in apolar environments, the reactions of carotenoids with certain free radicals are extremely rapid [24,101]. Thus, the reaction of carotenoids with phenylthiyl and acylperoxyl radicals proceeds *via* radical addition [24,27,101] with rate constants in the region of 10^9 $M^{-1}s^{-1}$. Alternative scavenging mechanisms such as H-atom abstraction have been ruled out on the basis that the reactions of different radicals (*e.g.* acylperoxyl and phenylthiyl radicals) with 7,7'-dihydro-β-carotene (**49**) lead to carotenoid neutral radicals with very similar spectra, but very different kinetic behaviour [24,27]; reaction with acylperoxyl radicals is 100-1000 times more rapid. Less reactive peroxyl radicals such as benzylperoxyl radical ($C_6H_5CH_2OO^{\bullet}$) also react with carotenoids in apolar environments, probably by radical addition but at a much lower rate [102]. Recent EPR studies of the reaction of $NO_2^{\bullet}$ with β-carotene in chloroform [103] suggest an addition mechanism.

3. Free-radical scavenging mechanisms in polar and heterogeneous environments

In polar and heterogeneous environments (*e.g.* micelles and microemulsions), the mechanisms by which carotenoids scavenge free radicals are more complex and more varied. The behaviour with different classes of free radical is summarized below.

a) Thiyl radicals ($RS^{\bullet}$)

In polar (and low polarity) solvents, thiyl radicals react with carotenoids [27,68,94] exclusively *via* radical addition (see equation 13 and Scheme 14) and there is no evidence for formation of $CAR^{\bullet+}$ in these reactions. This reflects the relatively low reduction potentials of these radicals (*e.g.* for 2-mercaptoethanol thiyl radical, $E° \sim 0.78$ V *versus* NHE) [104] compared with, for example, acylperoxyl radicals ($E° \sim 1.12$ V *versus* NHE) [105].

b) Sulphonyl radicals ($RSO_2^{\bullet}$)

The reactions of $RSO_2^{\bullet}$ with carotenoids [68,94] in polar environments (*t*-butanol/water mixtures) appear to proceed *via* radical addition (RSO_2-$CAR^{\bullet}$) and electron transfer to produce the carotenoid radical cation (CAR•+). However, formation of CAR•+ is preceded (at least in part) by formation of an intermediate species that absorbs at shorter wavelengths, and the possibility that both these species are preceded by the addition radical cannot be rigorously excluded. The reactions of sulphonyl radicals with carotenoids in apolar environments have not been reported.

c) Peroxyl radicals (RO$_2$•)

From studies of the reactions of acylperoxyl radicals with carotenoids in polar environments there is some evidence that radical addition is the initial step and that the addition radical (RO$_2$-CAR•) subsequently undergoes an ionic dissociation *via* an intermediate (NIR1) to form CAR•+ in a consecutive process (Pathway A in Scheme 10) [24]. This process probably operates in competition with the unimolecular S$_H$i process that prevails in environments of low polarity. Whether this pathway operates more generally remains unresolved and direct formation of NIR1 (Pathway B in Scheme 10) cannot be excluded [22]. Also, for less strongly oxidizing peroxyl radicals (*e.g.* alkylperoxyl radicals), the formation of CAR•+ may not be feasible thermodynamically when the respective redox potentials are taken into account [106].

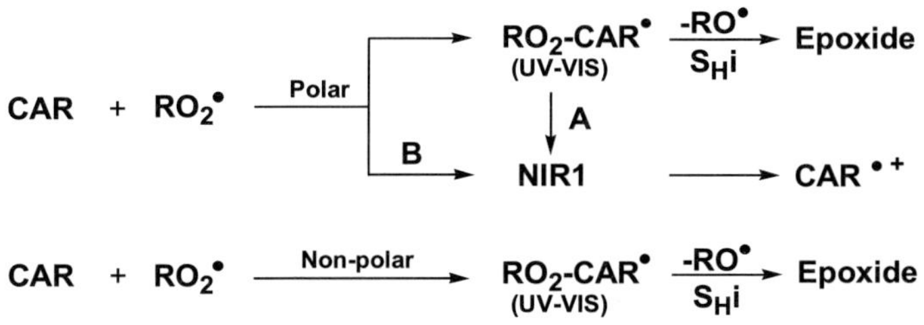

Scheme 10

d) Phenoxyl radicals (PhO•)

In di-*t*-butyl peroxide/benzene (70:30), PhO• reacts with β-carotene (**3**) to form the carotenoid radical cation *via* a NIR1 species [25]. It is possible that addition precedes NIR1, but further work is required to establish whether this is so. For canthaxanthin (**380**), the reaction with PhO• is very much slower than that with β-carotene, and no reaction was observed at all with astaxanthin (**406**), possibly reflecting the higher reduction potentials of the xanthophylls [25]. The reactions of phenoxyl and alkoxyl radicals with carotenoids in low polarity environments have not been studied.

e) Other radicals

The reactions of carotenoids with $CCl_3^\bullet$ [22], $NO_2^\bullet$ [68,92,94,107,108] and with $Br_2^{\bullet-}$ and $(SCN)_2^{\bullet-}$ in polar environments [109,110] appear to proceed by direct electron transfer to produce $CAR^{\bullet+}$. There is no evidence for the formation of addition radicals or NIR1 species.

F. Reactions of Carotenoid Radicals

1. Carotenoid radical cations ($CAR^{\bullet+}$)

a) Reactions with nucleophiles

The presence of the positive charge means that carotenoid radical cations may be susceptible to nucleophilic attack. These reactions have been observed for the radical cations of shorter carotenoid-like polyenes such as retinoids. The rate constants for the reactions of these radical cations with water, triethylamine and bromide ion in acetone are shown in Table 5. From this, it is clear that the reaction rate decreases as the length of the conjugated polyene chain increases [51,54].

Table 5. The rate constants k_q ($\pm$ 20%) for the reaction of the radical cations of some retinoids and related compounds with nucleophiles [water, triethylamine (TEA), bromide ion] in oxygen-saturated acetone [51,54].

Compound	cdb[a]	k_q/ M^{-1} s^{-1}		
		Water/10^5	TEA/10^8	Br^-/10^9
C_{17}-Aldehyde (4)	4	1.9	21	56
Retinal (6)	5	0.84	1.8	53
Retinol (5)	5	1.9	0.27	41
Retinyl acetate (8)	5	1.3	0.39	31
Retinoic acid (9)	5	3.5	1.9	39
Methyl retinoate (10)	5	1.6	1.2	37
Retinal Schiff base (7)	5	1.5	0.39	31
14'-Apo-β-caroten-14'-al (513)	6	0.45	0.56	41
8'-Apo-β-caroten-8'-al (482)	9	< 0.1	0.056	0.12

[a] cdb is the number of conjugated double bonds (carbonyl group not included).

In the absence of any other reactants, carotenoid radical cations disproportionate to give carotenoid dication and the parent carotenoid (equation 14) [32]. The equilibrium constants for these reactions have been compiled [88].

$$2CAR^{\bullet+} \rightleftharpoons CAR + CAR^{2+} \qquad (14)$$

b) De-protonation

The equilibrium constants (K'_{dp}) and forward rate constants (k'_f) for the de-protonation of carotenoid radical cations (equation 15) in dichloromethane have been reported [88].

$$CAR^{\bullet +} \rightleftharpoons {}^{\#}CAR^{\bullet} + H^+ \qquad (15)$$

c) Reactions with amino acids and peptides

Various carotenoid radical cations are able to oxidize both tyrosine (TyrOH) and cysteine (CySH) ($k \sim 10^4$ and 10^6 M^{-1} s^{-1} respectively) (equations 16 and 17) [15,58,89]. Also, some carotenoid radical cations react reversibly with tryptophan (TrpH) to form the tryptophan radical cation (equation 18) in a pH-dependent process [15,89].

$$CAR^{\bullet +} + TyrOH \longrightarrow CAR + TyrO^{\bullet} + H^+ \qquad (16)$$
$$CAR^{\bullet +} + CysH \longrightarrow CAR + Cys^{\bullet} + H^+ \qquad (17)$$
$$CAR^{\bullet +} + TrpH \rightleftharpoons CAR + Trp^{\bullet +} + H^+ \qquad (18)$$

d) Reactions with porphyrins

Some porphyrin derivatives (POR) such as chlorophyll a can be oxidized by $CAR^{\bullet +}$ (equation 19) [111-113].

$$CAR^{\bullet +} + POR \longrightarrow CAR + POR^{\bullet +} \qquad (19)$$

e) Reactions with other carotenoids

The relative reduction potentials of a variety of carotenoid radical cations have been established by monitoring the reaction of the radical cation of one carotenoid ($CAR1^{\bullet +}$) with another carotenoid (CAR2) in benzene (equation 20 and Table 6) [17]. For example, astaxanthin$^{\bullet +}$ can react with lycopene to form lycopene$^{\bullet +}$, therefore the reduction potential of astaxanthin$^{\bullet +}$ ($E^\circ_{ASTA^{\bullet +}/ASTA}$) is greater than that of lycopene ($E^\circ_{LYCO^{\bullet +}/LYCO}$, equation 21). By studying various pairs of carotenoids, the relative reduction potentials of $CAR^{\bullet +}/CAR$ were established [17]. This study reveals that $E^\circ_{ASTA^{\bullet +}/ASTA} > E^\circ_{APO^{\bullet +}/APO} > E^\circ_{CAN^{\bullet +}/CAN} > E^\circ_{LUT^{\bullet +}/LUT} > E^\circ_{ZEA^{\bullet +}/ZEA} \approx E^\circ_{MZEA^{\bullet +}/MZEA} > E^\circ_{\beta-CAR^{\bullet +}/\beta-CAR} > E^\circ_{LYCO^{\bullet +}/LYCO}$. The relative order of the reduction potentials is similar to the order reported in other studies [25,47,114,115] (see Table 6 for abbreviations).

$$CAR1^{\bullet +} + CAR2 \longrightarrow CAR1 + CAR2^{\bullet +} \qquad (20)$$

$$ASTA^{\bullet +} + LYCO \longrightarrow ASTA + LYCO^{\bullet +} \qquad (21)$$

Table 6. Rate constants k_q ($\pm$ 10%) for electron transfer between a carotenoid (CAR2) and a carotenoid radical cation (CAR1$^{\bullet +}$) (Equation 20) [17].

CAR$^{\bullet +}$	$k_q / 10^9$ M^{-1} s^{-1}		
	Lycopene (31)	β-Carotene (3)	Zeaxanthin (119)
ASTA$^{\bullet +}$	9	8	5
APO$^{\bullet +}$	11	6	8
CAN$^{\bullet +}$	8	5	< 1
LUT$^{\bullet +}$	5	<1	<1
MZEA$^{\bullet +}$	7.8 [21]	< 1 [21]	-
ZEA$^{\bullet +}$	7	< 1	-

Abbreviations. ZEA: zeaxanthin (119); ASTA: astaxanthin (406); APO: 8'-apo-β-caroten-8'-al (482); CAN: canthaxanthin (380); LUT: lutein (133); MZEA: (meso)-zeaxanthin (120).

Recently [106], the reactivity of the radical cation of the retinoid product A2E$^+$ (Scheme 11) with carotenoids has been investigated. Oxidation of A2E$^+$ by Br$_2^{\bullet -}$ in 2% (w/v) aqueous Triton X-100 (Scheme 11) gives A2E$^{\bullet 2+}$ (λ_{max} = 590 nm; ε_{590} = 8400 $\pm$ 500 M^{-1} cm^{-1}), which subsequently reacts with β-carotene to form β-carotene$^{\bullet +}$ (k ~2 × 10^9 M^{-1} s^{-1}).

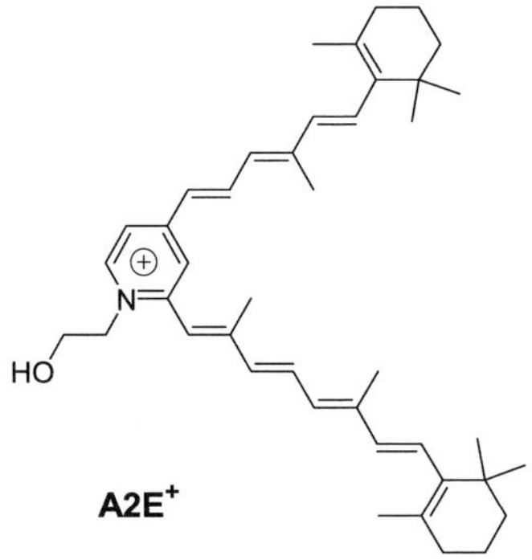

$$A2E^+ + Br_2^{\bullet -} \longrightarrow A2E^{\bullet 2+} + 2Br^-$$

$$A2E^{\bullet 2+} + \beta\text{-CAR} \longrightarrow \beta\text{-CAR}^{\bullet +} + A2E^+$$

A2E$^+$

Scheme 11

f) Association with the parent carotenoid

Radical cations of retinal (6), retinoic acid (9) and methyl retinoate (10) in acetone have been observed to undergo association with the parent compound at high concentration [117]. This behaviour has also been observed for the radical cations of canthaxanthin (380) in CH$_2$Cl$_2$ [118] and β-carotene (3) in DMSO (k = 2.5 × 10^7 M^{-1} s^{-1}) [66,119] (equation 22). Carotenoid radical cation dimer $(CAR_2^{\bullet+})$ can deprotonate to give the carotenoid radical dimer $(^\#CAR_2^{\bullet})$ (equation 23) [118].

$$CAR^{\bullet+} + CAR \rightleftharpoons CAR_2^{\bullet+} \qquad (22)$$

$$CAR_2^{\bullet+} \rightleftharpoons {}^\#CAR_2^{\bullet} + H^+ \qquad (23)$$

g) Reactions with oxygen

In fast, time-resolved studies of carotenoid radical cations (e.g. by pulse radiolysis), no evidence has been found for reaction with oxygen on the timescales employed [63]. In a recent investigation [120], however, mechanistic arguments were presented to suggest that carotenoid radical cations do react with oxygen to form $CARO_2^{\bullet+}$ (Scheme 12).

$$CAR^{\bullet+} + O_2 \longrightarrow CARO_2^{\bullet\,+}$$

$$CARO_2^{\bullet\,+} + CAR \ (or \ Fe^{2+}) \rightleftharpoons CARO_2 + CAR^{\bullet+} \ (or \ Fe^{3+})$$

Scheme 12

h) Miscellaneous reactions

In aqueous medium [2% (w/v) Triton X-100], carotenoid radical cations $CAR^{\bullet+}$ can be quenched by dopamelanin ($k \sim 10^6$ M^{-1} s^{-1}) [121], cysteinyldopamelanin ($k \sim 10^6$ M^{-1} s^{-1}) [121], trolox ($k \sim 10^8$ M^{-1} s^{-1}) [58], uric acid ($k \sim 10^7$ M^{-1} s^{-1}) [58] and ferulic acid ($k \sim 10^6$ M^{-1} s^{-1}) [58]. Moreover, carotenoid radical cations react with ascorbic acid (AscH, equation 24) via electron-transfer reactions {$k \sim 10^8$-10^9, $\sim 10^7$-10^8 and $\sim 10^7$ M^{-1} s^{-1} in methanol [21,122], 2% (w/v) Triton X-100 [55,58,122] and unilamellar dipalmitoylphosphatidylcholine (DPPC) vesicles [109], respectively}. Also, some tocopherols (TOH), e.g. α-tocopherol, are able to reduce various carotenoid radical cations [29,115] (equation 25).

$$CAR^{\bullet+} + AscH \longrightarrow CAR + Asc^{\bullet} + H^+ \qquad (24)$$

$$CAR^{\bullet+} + TOH \longrightarrow CAR + TO^{\bullet} + H^+ \qquad (25)$$

2. Carotenoid radical anions (CAR$^{\bullet-}$)

a) Reactions with oxygen

Carotenoid radical anions are readily scavenged by oxygen *via* an electron-transfer process to produce superoxide ion, $O_2^{\bullet-}$ (equation 26) [123]. The rate constant of this reaction decreases as the carotenoid chain length increases (Table 7).

$$CAR^{\bullet-} + O_2 \longrightarrow CAR + O_2^{\bullet-} \qquad (26)$$

Table 7. Rate constants k_q for the reaction of the radical anions (CAR$^{\bullet-}$) of some carotenoids with oxygen in hexane (Equation 26) [123].

Carotenoid	cdb[a]	$k_q/10^8$ M^{-1} s^{-1}
β-Carotene (**3**)	11	25 ± 5
(15Z)-β-Carotene (**3**)	11	20 ± 5
(9Z)-β-Carotene (**3**)	11	20 ± 5
Lycopene (**31**)	11	2 ± 1
Decapreno-β-carotene (*13*)	15	1 ± 0.5

[a] cdb is the number of conjugated double bonds.

Also, the rate constant of the reaction of O_2 with A2E$^{\bullet}$ ($\lambda_{max} \sim 500$ nm, $\varepsilon_{500} = 25100 \pm 1200$ M^{-1} cm^{-1}), formed by reduction of A2E$^+$ by NAD$^{\bullet}$ or $CO_2^{\bullet-}$ in 2% (w/v) aqueous Triton X-100, has been determined as k $\sim3 \pm 1 \times 10^8$ M^{-1} s^{-1} [116].

b) Reactions with porphyrins

Carotenoid radical anions react very efficiently ($\sim10^9$-10^{10} M^{-1} s^{-1}) with porphyrins (POR) *via* electron transfer (Table 8, equation 27) [111-113].

$$CAR^{\bullet-} + POR \longrightarrow CAR + POR^{\bullet-} \qquad (27)$$

Table 8. Rate constants k_q for the reaction of the radical anion (CAR$^{\bullet-}$) of some carotenoids with porphyrins in hexane (Equation 27) [111].

	$k_q/10^{10}$ M^{-1} s^{-1}		
Carotenoid	Chlorophyll a	Phaeophytin a	Phaeophytin b
Decapreno-β-carotene (*13*)	0.54	0.61	1.14
Lycopene (**31**)	0.70	0.99	1.49
β-Carotene (**3**)	0.85	1.93	2.45

c) Other carotenoids

A similar approach to that described in Section **F.1.e** has allowed the relative reduction potentials of a variety of carotenoids to be established by monitoring the reaction of carotenoid radical cation (CAR1$^{\bullet+}$) with another carotenoid (CAR2) in hexane and benzene (equation 28) [60]. For example, β-CAR$^{\bullet+}$ can react with LYCO to form LYCO$^{\bullet+}$, therefore the reduction potential of LYCO ($E^0_{LYCO/LYCO^{\bullet+}}$) is greater than $E^0_{\beta-CAR/\beta-CAR^{\bullet+}}$ (equation 29). By repeating the same reaction with different pairs of carotenoids, the relative reduction potentials of CAR/CAR$^{\bullet+}$ have been established [63]. In hexane, the order is $E^0_{DECA/DECA^{\bullet+}}$ > $E^0_{LYCO/LYCO^{\bullet+}}$ > $E^0_{\beta-CAR/\beta-CAR^{\bullet+}}$ > $E^0_{HEPT/HEPT^{\bullet+}}$. In benzene, the order is $E^0_{ASTA/ASTA^{\bullet+}}$ > $E^0_{CAN/CAN^{\bullet+}}$ ≈ $E^0_{APO/APO^{\bullet+}}$ > $E^0_{LYCO/LYCO^{\bullet+}}$ > $E^0_{LUT/LUT^{\bullet+}}$ ≈ $E^0_{\beta-CAR/\beta-CAR^{\bullet+}}$ > $E^0_{ZEA/ZEA^{\bullet+}}$. The rate constants for the reaction of CAR$^{\bullet+}$ with oxygen and porphyrins, in hexane, are in agreement with the observed reduction potential order (see Tables 7 and 8) [111-113,123]. Also, the rate constants for the reaction of CAR1$^{\bullet+}$ with CAR2 in hexane and benzene have been estimated (Tables 9 and 10) [60].

$$\text{CAR1}^{\bullet-} + \text{CAR2} \longrightarrow \text{CAR1} + \text{CAR2}^{\bullet-} \qquad (28)$$

$$\beta\text{-CAR}^{\bullet-} + \text{LYCO} \longrightarrow \beta\text{-CAR} + \text{LYCO}^{\bullet-} \qquad (29)$$

Table 9. Rate constants k_q (±10 %) for electron transfer between a carotenoid (CAR2) and a carotenoid radical anion (CAR1$^{\bullet-}$) in argon-saturated hexane (Equation 28) [60].

	$k_q/ 10^9 M^{-1} s^{-1}$		
CAR$^{\bullet-}$	Decapreno-β-carotene (*13*)	Lycopene (**31**)	β-Carotene (**3**)
HEPT$^{\bullet-}$	63	12	20
β-CAR$^{\bullet-}$	11	14	-

Abbreviations: β-CAR: β-carotene (**3**); HEPT: heptapreno-β-carotene (*12*).

Table 10. Rate constants k_q (± 10%) for electron transfer between between a carotenoid (CAR2) and a carotenoid radical anion (CAR1$^{\bullet-}$) in argon-saturated benzene (Equation 28) [60].

	$k_q/ 10^9 M^{-1} s^{-1}$					
	Astaxanthin	Canthaxanthin	8'-Apo-β-carot-	Lycopene	Lutein	β-Carotene
CAR$^{\bullet-}$	(**406**)	(**380**)	en-8'-al (**482**)	(**31**)	(**133**)	(**3**)
ZEA$^{\bullet-}$	15	15	10	3.0	3.8	3.7
β-CAR$^{\bullet-}$	14	7.7	13	6.2	≤ 0.5	
LUT$^{\bullet-}$	13	7.5	10	2.5		
LYCO$^{\bullet-}$	12	10	10			
APO$^{\bullet-}$	1.1	≤ 0.2				
CAN$^{\bullet-}$	1.9					

Abbreviations. ZEA: zeaxanthin (**119**); β-CAR: β-carotene (**3**); LUT: lutein (**133**); LYCO: lycopene (**31**); APO: 8'-apo-β-caroten-8'-al (**482**); CAN: canthaxanthin (**380**).

3. Carotenoid neutral radicals

a) Unimolecular fragmentation reactions

Unimolecular fragmentation reactions have been observed for addition radicals derived from reactions of 77DH (**49**) with acylperoxyl radicals in non-polar environments [24,124]. These radicals absorb strongly in the 450-470 nm region and the relatively fast first order decay is attributed to an S_Hi mechanism leading to epoxide formation (see Scheme 10). The direct observation of similar processes for other carotenoids was precluded by kinetic and/or spectroscopic resolution factors.

b) Ionic dissociation

This type of reaction may be relevant to the behaviour of carotenoid addition radicals when these encounter the lipid/aqueous interface in biological systems. Good evidence for this type of reaction has been found for carotenoid addition radicals derived from reaction with acylperoxyl radicals in polar environments (Scheme 13) [24]. In these systems, the addition radical RO_2-CAR$^{\bullet}$ that is formed initially decays to the carotenoid radical cation *via* an intermediate (NIR1). Addition radicals derived from reactions of carotenoids with other peroxyl radicals (*e.g.* $CCl_3O_2{}^{\bullet}$) or phenoxyl radicals may undergo similar reactions in polar environments.

$$\textbf{RO}_2^{\bullet} \; + \; \textbf{CAR} \longrightarrow \textbf{RO}_2\text{-}\textbf{CAR}^{\bullet} \longrightarrow \textbf{NIR1} \longrightarrow \textbf{RO}_2^{\bullet-} \; + \; \textbf{CAR}^{\bullet+}$$

Scheme 13

c) Reactions with oxygen

The reversible reaction of carotenoid neutral radicals with oxygen has been proposed to explain the dependence of the antioxidant behaviour of carotenoids on oxygen concentration [125]. However, it is only recently [27] that such reactions have been directly observed in real time *via* laser flash photolysis. Carotenoid addition radicals derived from the reaction of phenylthiyl radicals (PhS$^{\bullet}$) with 7,7'-dihydro-β-carotene (**49**) and β-carotene (**3**) (*i.e.* PhS-77DH$^{\bullet}$ and PhS-β-CAR$^{\bullet}$) (Scheme 14) are exceptionally long-lived (tens of milliseconds) and such radicals are ideal candidates for detecting any reaction with oxygen. These radicals are observed to react reversibly with O_2 probably *via* formation of carotenoid peroxyl radicals (PhS-CARO$_2{}^{\bullet}$) (Scheme 14) [27]. The rate constants for oxygen addition to PhS-77DH$^{\bullet}$ and PhS-β-CAR$^{\bullet}$ are $(4.3 \pm 0.07) \times 10^4$ and $(0.64 \pm 0.09) \times 10^4$ M^{-1} s^{-1}, respectively, and these data suggest that the oxygen addition rate constant decreases as the conjugated chain length increases. Recently, this has been confirmed in a study extended to other carotenoids with shorter chromophores, e.g. the heptaene ζ-carotene (**38**). The results are presented in Table 11 [126].

$$PhS^\bullet + CAR \longrightarrow PhS\text{-}CAR^\bullet$$

$$PhS\text{-}CAR^\bullet + O_2 \rightleftharpoons PhS\text{-}CARO_2^\bullet$$

Scheme 14

Table 11. Rate constants k_q for the reaction of carotenoid neutral radical (PhS-CAR$^\bullet$) with oxygen (Scheme 14) and the λ_{max} of PhS-CAR$^\bullet$ in benzene [126].

Carotenoid	cdb[a]	$k_q / 10^4\ M^{-1}\ s^{-1}$	λ_{max} (nm)
ζ-Carotene (**38**)	7	3.44 ± 0.4	455
7,7'-Dihydro-β-carotene (**49**)	8	4.3 ± 0.07	470
Heptapreno-β-carotene (*12*)	9	3.5 ± 1.5	500
β-Carotene (**3**)	11	0.64 ± 0.09	540
Lycopene (**31**)	11	0.32 ± 0.03	545
Zeaxanthin (**119**)	11	1.1 ± 0.5	540

[a] cdb is the number of conjugated double bonds.

G. Antioxidant and Pro-oxidant Properties

Carotenoids are commonly included amongst the lipid-soluble components of the vast array of dietary antioxidants. Working towards an understanding of the impact of carotenoids on biological systems under oxidative stress requires a detailed understanding of their free radical chemistry, ideally within environments that mimic biological environments as closely as possible. The role of carotenoids in free-radical processes within lipophilic environments and at lipid/water interfaces has been the subject of much work [22,58,109,121,122,127] but many of the mechanistic and kinetic details of the processes involved remain to be resolved. Figure 2 illustrates some of the free-radical processes that may be relevant for the participation of carotenoids in peroxyl radical mediated oxidation within a biological system (see *Volume 5, Chapter 12*).

Based on the results of recent studies of free-radical scavenging by carotenoids in apolar media [24,124] the mechanism by which carotenoids scavenge peroxyl radicals in the lipophilic compartment within the cell is likely to be *via* a radical addition reaction (Scheme 10, Fig. 2). Ionic dissociation of the addition radical is possible at the interface between the lipophilic and hydrophilic compartments and gives rise to the carotenoid radical cation.

Evidence for ionic dissociation reactions of addition radicals derived from the addition of acylperoxyl radicals to 7,7'-dihydro-β-carotene (49) in polar solvents has recently been reported [24]. Although there were some differences in interpretation at the time of publication, the kinetics of carotenoid radical cation formation in the scavenging of acylperoxyl radicals in polar solvents may be taken to suggest a preceding ionic dissocation process [24].

Moreover, there is the possibility of epoxide formation from ROO-CAR$^{\bullet}$ in competition with ionic dissociation as well as addition of oxygen to ROO-CAR$^{\bullet}$ to produce a carotenoid-derived peroxyl radical ROO-CARO$_2^{\bullet}$ (Fig. 2). It is only recently that the rate constants for oxygen addition to carotenoid neutral radicals have been reported. It has been shown that the rate constant of the oxygen addition reaction displays a moderate dependence on the chain length of the carotenoid [27,126]. In the literature, there are numerous reports concerning the influence of oxygen concentration on the antioxidant properties of carotenoids [1,3,5,6,27,96,125,128]. This influence is attributed to the reversible addition of oxygen to ROO-CAR$^{\bullet}$ to form ROO-CARO$_2^{\bullet}$ (Fig. 2). At low oxygen concentrations, the equilibrium is positioned toward ROO-CAR$^{\bullet}$ whilst at high oxygen concentrations the equilibrium is positioned toward the ROO-CARO$_2^{\bullet}$, a peroxyl radical that may contribute to the propagation of the lipid peroxidation process and hence inhibit the antioxidant potency of the carotenoid.

There are many factors, besides the oxygen concentration, that may influence the antioxidant properties of carotenoids. The structure of the carotenoid can play a major role in its orientation within biological membranes and thereby on its ability to scavenge free radicals. For example, zeaxanthin (119) is able to scavenge both lipid and aqueous phase radicals since it spans the lipid membrane in a way that brings it into contact with both aqueous and lipid media. β-Carotene (3), however, is only able to scavenge lipid phase radicals due to its limited exposure to aqueous media [18,129,130]. Carotenoid concentration can influence antioxidant properties; decreased antioxidant effects are attributed to carotenoid aggregation [18,131].

Moreover, the presence of other antioxidants can reduce the damage induced by harmful species that are formed by the reactions of free radicals with carotenoids. For example, CAR$^{\bullet+}$, generated from free radical oxidation of the carotenoid, can oxidize amino acids (equations 16-18) [15,89,95]. Should such reactions occur *in vivo*, they may lead to structural modifications within proteins, with consequent effects upon the functions of the proteins [15,89]. The long lifetimes of carotenoid radical cations, as observed in micelles and liposomes [58,109,122], increase the chance of their interaction with biological molecules. However, in the presence of other antioxidants such as ascorbic acid (AscH) and vitamin E (TOH), CAR$^{\bullet+}$ can be recycled (equations 24 and 25) [18,29,115,122].

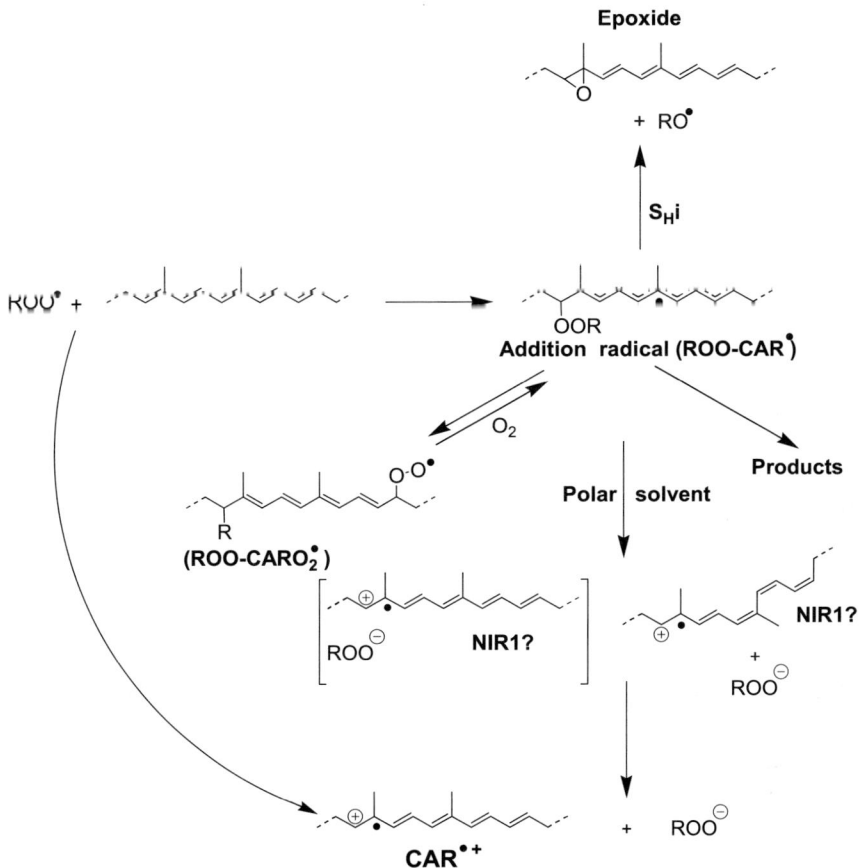

Fig. 2. Scheme summarizing the reactions of a carotenoid polyene chain with peroxyl radicals and the subsequent reactions of the addition radical product. NIR1: an uncharacterized transient intermediate detected by its NIR absorption properties. The significance of these reactions for the antioxidant properties of carotenoids in biological systems is discussed in *Vol. 5, Chapter 12.*

To conclude this section, it is important to discuss the use of azo-compounds as free radical initiators since many studies have employed these compounds to investigate the antioxidant properties of carotenoids [1,3,4,6,7,10,11,125,129,132]. Scheme 15 illustrates the mechanism for generation of free radicals by use of azo-initiators such as 2,2′-azobis-(2,4-dimethylvaleronitrile) (AMVN) and 2,2′-azobis-(2-amidinopropane) dihydrochloride (AAPH) [133]. When the temperature is raised, azo-initiators undergo fragmentation reactions to give alkyl free radicals ($R^•$). In the presence of oxygen, these react with oxygen ($k \sim 10^9$ M^{-1} s^{-1}) to form alkylperoxyl radicals ($RO_2^•$), which can initiate lipid peroxidation. In Scheme 15, Lipid-H represents peroxidizable lipid.

$$R-N=N-R \longrightarrow 2R^\bullet + N_2$$

$$R^\bullet + O_2 \longrightarrow ROO^\bullet$$

$$ROO^\bullet + \text{Lipid-H} \longrightarrow ROOH + \text{Lipid}^\bullet$$

Scheme 15

H. Conclusion

The influence of the structural characteristics of the carotenoid on the propensity to participate in the free-radical processes discussed above is an area that requires further work and may shed some light on the differences in antioxidant behaviour between the hydrocarbon carotenes and the xanthophylls, for example. In addition, the behaviour, in heterogeneous systems, of carotenoid radicals derived from reaction with radicals other than peroxyl radicals is an area that warrants further study.

Future work concerned with an understanding of the free-radical chemistry of carotenoids necessitates an integrated approach that combines direct observations of individual reactions with product analyses and theoretical calculations. The potential of such an approach has been shown [93] and this work has contributed significantly to our understanding of the electrochemical, spectroscopic and acid-base properties of carotenoid radicals and radical ions, as well as other carotenoid species. A similar approach is required in order to interpret fully the results obtained from fast time-resolved studies of short-lived carotenoid radical intermediates.

With regard to the interpretation of antioxidant/pro-oxidant behaviour of carotenoids, the influence of the carotenoid oxidation products is an area that requires further study [134-137]. In addition, the system used to investigate the antioxidant activity of carotenoids has a significant effect on the evaluation. Therefore, such an evaluation must be assessed after conducting many experiments *in vitro* and *in vivo*, exploring the effectiveness in a range of different environments [138].

References

[1]   P. Palozza, C. Luberto, G. Calviello, P. Ricci and G. M. Bartoli, *Free Rad. Biol. Med.,* **22**, 1065 (1997).
[2]   H.-D. Martin, C. Jäger, C. Ruck, M. Schmidt, R. Walsh and J. Paust, *J. Prakt. Chem.,* **341**, 302 (1999).
[3]   H. Tsuchihashi, M. Kigoshi, M. Iwatsuki and E. Niki, *Arch. Biochem. Biophys.,* **323**, 137 (1995).
[4]   L. Tesoriere, D. D`Arpa, R. Re and M. A. Livrea, *Arch. Biochem. Biophys.,* **343**, 13 (1997).
[5]   G. F. Vile and C. C. Winterbourn, *FEBS Lett.,* **238**, 353 (1988).

[6]   T. A. Kennedy and D. C. Liebler, *J. Biol. Chem.*, **267**, 4658 (1992).

[7]   T. A. Kennedy and D. C. Liebler, *Chem. Res. Toxicol.*, **4**, 290 (1991).

[8]   D. L. Baker, E. S. Krol, N. Jacobsen and D. C. Liebler, *Chem. Res. Toxicol.*, **12**, 535 (1999).

[9]   R. C. Mordi, J. C. Walton, G. W. Burton, L. Hughes, K. U. Ingold, D. A. Lindsay and D. J. Moffatt, *Tetrahedron*, **49**, 911 (1993).

[10]   R. Yamauchi, N. Miyake, H. Inoue and K. Kato, *J. Agric. Food Chem.*, **41**, 708 (1993).

[11]   G. J. Handelman, F. J. G. M. Van Kuijk, A. Chatterjee and N. I. Krinsky, *Free Rad. Biol. Med.*, **10**, 427 (1991).

[12]   J. S. Vrettos, D. H. Stewart, J. C. De Paula and G. W. Brudvig, *J. Phys. Chem. B*, **103**, 6403 (1999).

[13]   P. Mathis and A. W. Rutherford, *Biochim. Biophys. Acta*, **767**, 217 (1984).

[14]   C. C. Schenck, B. Diner, P. Mathis and K. Satoh, *Biochim. Biophys. Acta*, **680**, 216 (1982).

[15]   R. Edge, E. J. Land, D. J. McGarvey, M. Burke and T. G. Truscott, *FEBS Lett.*, **471**, 125 (2000).

[16]   T. Polivka, T. Pullerits, H. A. Frank, R. J. Cogdell and V. Sundstrom, *J. Phys. Chem. B*, **108**, 15398 (2004).

[17]   R. Edge, E. J. Land, D. J. McGarvey, L. Mulroy and T. G. Truscott, *J. Am. Chem. Soc.*, **120**, 4087 (1998).

[18]   A. J. Young and G. M. Lowe, *Arch. Biochem. Biophys.*, **385**, 20 (2001).

[19]   J. Lafferty, A. C. Roach, R. S. Sinclair, T. G. Truscott and E. J. Land, *J. Chem. Soc. Faraday Trans. 1*, **73**, 416 (1977).

[20]   R. V. Bensasson, E. J. Land and T. G. Truscott, *Flash Photolysis and Pulse Radiolysis: Contributions to the Chemistry of Biology and Medicine, 1st Ed.*, p.1 and p.67, Pergamon Press, Oxford (1983).

[21]   R. Edge, PhD Thesis, Keele University (1998).

[22]   T. J. Hill, E. J. Land, D. J. McGarvey, W. Schalch, J. H. Tinkler and T. G. Truscott, *J. Am. Chem. Soc.*, **117**, 8322 (1995).

[23]   T. J. Hill, PhD Thesis, Keele University (1994).

[24]   A. El-Agamey and D. J. McGarvey, *J. Am. Chem. Soc.*, **125**, 3330 (2003).

[25]   A. Mortensen and L. H. Skibsted, *J. Agric. Food Chem.*, **45**, 2970 (1997).

[26]   A. Mortensen and L. H. Skibsted, *Free Rad. Res.*, **25**, 355 (1996).

[27]   A. El-Agamey and D. J. McGarvey, *Org. Lett.*, **7**, 3957 (2005).

[28]   A. Mortensen and L. H. Skibsted, *Free Rad. Res.*, **26**, 549 (1997).

[29]   A. Mortensen and L. H. Skibsted, *Free Rad. Res.*, **27**, 229 (1997).

[30]   P. Mathis and A. Vermeglio, *Photochem. Photobiol.*, **15**, 157 (1972).

[31]   J. H. Tinkler, S. M. Tavender, A. W. Parker, D. J. McGarvey, L. Mulroy and T. G. Truscott, *J. Am. Chem. Soc.*, **118**, 1756 (1996).

[32]   J. A. Jeevarajan, C. C. Wei, A. S. Jeevarajan and L. D. Kispert, *J. Phys. Chem.*, **100**, 5637 (1996).

[33]   R. Ding, J. L. Grant, R. M. Metzger and L. D. Kispert, *J. Phys. Chem.*, **92**, 4600 (1988).

[34]   N. E. Polyakov, V. V. Konovalov, T. V. Leshina, O. A. Luzina, N. F. Salakhutdinov, T. A. Konovalova and L. D. Kispert, *J. Photochem. Photobiol. A: Chem.*, **141**, 117 (2001).

[35]   T. A. Konovalova, Y. Gao, L. D. Kispert, J. van Tol and L.-C. Brunel, *J. Phys. Chem. B*, **107**, 1006 (2003).

[36]   Y. Gao, T. A. Konovalova, J. N. Lawrence, M. A. Smitha, J. Nunley, R. Schad and L. D. Kispert, *J. Phys. Chem. B*, **107**, 2459 (2003).

[37]   T. A. Konovalova, Y. Gao, R. Schad, L. D. Kispert, C. A. Saylor and L.-C. Brunel, *J. Phys. Chem. B*, **105**, 7459 (2001).

[38]   T. A. Konovalova, S. A. Dikanov, M. K. Bowman and L. D. Kispert, *J. Phys. Chem. B*, **105**, 8361 (2001).

[39]   J. L. Grant, V. J. Kramer, R. Ding and L. D. Kispert, *J. Am. Chem. Soc.*, **110**, 2151 (1988).

[40]   M. Khaled, A. Hadjipetrou, L. D. Kispert and R. D. Allendoerfer, *J. Phys. Chem.*, **95**, 2438 (1991).

[41]   D. Liu and L. D. Kispert, *J. Phys. Chem. B*, **105**, 975 (2001).

[42]   G. Gao, C. C. Wei, A. S. Jeevarajan and L. D. Kispert, *J. Phys. Chem.*, **100**, 5362 (1996).

[43]   A. S. Jeevarajan, L. D. Kispert, G. Chumanov, C. Zhou and T. M. Cotton, *Chem. Phys. Lett.*, **259**, 515 (1996).

[44]   A. S. Jeevarajan, M. Khaled and L. D. Kispert, *Chem. Phys. Lett.*, **225**, 340 (1994).

[45]   D. Niedzwiedzki, J. F. Rusling and H. A. Frank, *Chem. Phys. Lett.*, **415**, 308 (2005).

[46]   J. A. Jeevarajan and L. D. Kispert, *J. Electroanal. Chem.*, **411**, 57 (1996).

[47]   D. Liu, Y. Gao and L. D. Kispert, *J. Electroanal. Chem.*, **488**, 140 (2000).

[48]   L. D. Kispert, T. Konovalova and Y. Gao, *Arch. Biochem. Biophys.*, **430**, 49 (2004).

[49]   V. G. Mairanovsky, A. A. Engovatov, N. T. Ioffe and G. I. Samokhvalov, *J. Electroanal. Chem.*, **66**, 123 (1975).

[50]   R. V. Bensasson, E. J. Land and T. G. Truscott, *Excited States and Free Radicals in Biology and Medicine*, p.201, Oxford University Press, Oxford (1993).

[51]   K. Bobrowski and P. K. Das, *J. Phys. Chem.*, **91**, 1210 (1987).

[52]   E. J. Land, J. Lafferty, R. S. Sinclair and T. G. Truscott, *J. Chem. Soc. Faraday Trans. I*, **74**, 538 (1978).

[53]   N. V. Raghavan, P. K. Das and K. Bobrowski, *J. Am. Chem. Soc.*, **103**, 4569 (1981).

[54]   K. Bobrowski and P. K. Das, *J. Phys. Chem.*, **89**, 5079 (1985).

[55]   M. Rozanowska, A. Cantrell, R. Edge, E. J. Land, T. Sarna and T. G. Truscott, *Free Rad. Biol. Med.*, **39**, 1399 (2005).

[56]   K. Bobrowski and P. K. Das, *J. Phys. Chem.*, **89**, 5733 (1985).

[57]   K. K. N. Lo, E. J. Land and T. G. Truscott, *Photochem. Photobiol.*, **36**, 139 (1982).

[58]   M. Burke, PhD Thesis, Keele University (2001).

[59]   S. Amarie, J. Standfuss, T. Barros, W. Kuhlbrandt, A. Dreuw and J. Wachtveitl, *J. Phys. Chem. B*, **111**, 3481 (2007).

[60]   R. Edge, A. El-Agamey, E. J. Land, S. Navaratnam and T. G. Truscott, *Arch. Biochem. Biophys.*, **458**, 104 (2007).

[61]   Y. Deng, G. Gao, Z. He and L. D. Kispert, *J. Phys. Chem. B*, **104**, 5651 (2000).

[62]   A. El-Agamey, R. Edge, S. Navaratnam, E. J. Land and T. G. Truscott, *Org. Lett.*, **8**, 4255 (2006).

[63]   E. A. Dawe and E. J. Land, *J. Chem. Soc. Faraday Trans. I*, **71**, 2162 (1975).

[64]   M. Almgren and J. K. Thomas, *Photochem. Photobiol.*, **31**, 329 (1980).

[65]   N. Getoff, *Radiat. Phys. Chem.*, **55**, 395 (1999).

[66]   N. Getoff, *Radiat. Res.*, **154**, 692 (2000).

[67]   A. Mortensen and L. H. Skibsted, *Free Rad. Res.*, **25**, 515 (1996).

[68]   S. A. Everett, M. F. Dennis, K. B. Patel, S. Maddix, S. C. Kundu and R. L. Willson, *J. Biol. Chem.*, **271**, 3988 (1996).

[69]   S. Adhikari, S. Kapoor, S. Chattopadhyay and T. Mukherjee, *Biophys. Chem.*, **88**, 111 (2000).

[70]   J. H. Tinkler, PhD Thesis, Keele University (1995).

[71]   T. Bally, K. Roth, W. Tang, R. R. Schrock, K. Knoll and L. Y. Park, *J. Am. Chem. Soc.*, **114**, 2440 (1992).

[72]   A. S. Jeevarajan, L. D. Kispert and X. Wu, *Chem. Phys. Lett.*, **219**, 427 (1994).

[73]   J. A. Jeevarajan, C. C. Wei, A. S. Jeevarajan and L. D. Kispert, *J. Phys. Chem.*, **100**, 5637 (1996).

[74]   C. P. Poole and H. A. Farach, *Handbook of Electron Spin Resonance: Data Sources, Computer Technology, Relaxation, and ENDOR*, AIP Press, New York (1994).

[75]   K. V. Lakshmi, M. J. Reifler, G. W. Brudvig, O. G. Poluektov, A. M. Wagner and M. C. Thurnauer, *J. Phys. Chem. B*, **104**, 10445 (2000).

[76]   L. Piekara-Sady, M. M. Khaled, E. Bradford, L. D. Kispert and M. Plato, *Chem. Phys. Lett.*, **186**, 143 (1991).

[77]   Y. Gao, T. A. Konovalova, T. Xu and L. D. Kispert, *J. Phys. Chem. B*, **106**, 10808 (2002).

[78]   T. A. Konovalova, J. Krzystek, P. J. Bratt, J. van Tol, L.-C. Brunel and L. D. Kispert, *J. Phys. Chem. B*, **103**, 5782 (1999).

[79]   L. Piekara-Sady and L. D. Kispert, *Mol. Phys. Rep.*, **6**, 220 (1994).

[80] T. A. Konovalova, L. D. Kispert and V. V. Konovalov, *J. Phys. Chem. B*, **101**, 7858 (1997).

[81] A. S. Jeevaran, M. Khaled and L. D. Kispert, *J. Phys. Chem.*, **98**, 7777 (1994).

[82] L. Piekara-Sacy, A. S. Jeevarajan and L. D. Kispert, *Chem. Phys. Lett.*, **207**, 173 (1993).

[83] T. A. Konovalova, L. D. Kispert, N. E. Polyakov and T. V. Leshina, *Free Rad. Biol. Med.*, **28**, 1030 (2000).

[84] J. A. Bautista, V. Chynwat, A. Cua, F. J. Jansen, J. Lugtenburg, D. Gosztola, M. R. Wasielewski and H. A. Frank, *Photosynth. Res.*, **55**, 49 (1998).

[85] G. Kildahl-Anderson, Dissertation, NTNU, Trondheim (2007).

[86] R. Ding, J. L. Grant, M. Metzger and L. D. Kispert, *J. Phys. Chem.*, **92**, 4600 (1988).

[87] P. Hapiot, L. D. Kispert, V. V. Konovalov and J.-M. Saveant, *J. Am. Chem. Soc.*, **123**, 6669 (2001).

[88] D. Liu and L. D. Kispert, *Rec. Res. Devel. Electrochem.*, **2**, 139 (1999).

[89] M. Burke, R. Edge, E. J. Land, D. J. McGarvey and T. G. Truscott, *FEBS Lett.*, **500**, 132 (2001).

[90] A. El-Agamey, A. Cantrell, E. J. Land, D. J. McGarvey and T. G. Truscott, *Photochem. Photobiol. Sci.*, **3**, 802 (2004).

[91] M. D`Aquino, C. Dunster and R. L. Willson, *Biochem. Biophys. Res. Commun.*, **161**, 1199 (1989).

[92] S. A. Everett, S. C. Kundu, S. Maddix and R. L. Willson, *Biochem. Soc. Trans.*, **23**, 230S (1995).

[93] Y. Gao, S. Webb and L. D. Kispert, *J. Phys. Chem. B*, **107**, 13237 (2003).

[94] A. Mortensen, L. H. Skibsted, J. Sampson, C. Rice-Evans and S. A. Everett, *FEBS Lett.*, **418**, 91 (1997).

[95] A. Mortensen, L. H. Skibsted and T. G. Truscott, *Arch. Biochem. Biophys.*, **385**, 13 (2001).

[96] C. A. Rice-Evans, J. Sampson, P. M. Bramley and D. E. Holloway, *Free Rad. Res.*, **26**, 381 (1997).

[97] R. Edge and T. G. Truscott, *The Spectrum*, **13**, 12 (2000).

[98] N. I. Krinsky and K.-J. Yeum, *Biochem. Biophys. Res. Commun.*, **305**, 754 (2003).

[99] R. Edge, D. J. McGarvey and T. G. Truscott, *J. Photochem. Photobiol. B: Biol.*, **41**, 189 (1997).

[100] A. El-Agamey, G. M. Lowe, D. J. McGarvey, A. Mortensen, D. M. Phillip, T. G. Truscott and A. J. Young, *Arch. Biochem. Biophys.*, **430**, 37 (2004).

[101] A. Mortensen, *J. Photochem. Photobiol. B: Biol.*, **61**, 62 (2001).

[102] A. Mortensen, *Free Rad. Res.*, **36**, 211 (2002).

[103] S. M. Khopde, K. I. Priyadarsini, M. K. Bhide, M. D. Sastry and T. Mukherjee, *Res. Chem. Intermed.*, **29**, 495 (2003).

[104] D. A. Armstrong, in *The Chemistry of Free Radicals: S-Centered Radicals, 1st Ed.*, (ed. Z. B. Alfassi), p.27, Wiley, Chichester (1999).

[105] G. Merényi, J. Lind and L. Engman, *J. Chem. Soc. Perkin Trans. 2*, 2551 (1994).

[106] A. Mortensen and L. H. Skibsted, *FEBS Lett.*, **426**, 392 (1998).

[107] F. Böhm, J. H. Tinkler and T. G. Truscott, *Nature Med.*, **1**, 98 (1995).

[108] F. Böhm, R. Edge, D. J. McGarvey and T. G. Truscott, *FEBS Lett.*, **436**, 387 (1998).

[109] M. Burke, R. Edge, E. J. Land and T. G. Truscott, *J. Photochem. Photobiol. B: Biol.*, **60**, 1 (2001).

[110] J.-P. Chauvet, R. Viovy, E. J. Land, R. Santus and T. G. Truscott, *J. Phys. Chem.*, **87**, 592 (1983).

[111] J. McVie, R. S. Sinclair, D. Tait, T. G. Truscott and E. J. Land, *J. Chem. Soc. Faraday Trans. 1*, **75**, 2869 (1979).

[112] J. Lafferty, T. G. Truscott and E. J. Land, *J. Chem. Soc. Faraday Trans. 1*, **74**, 2760 (1978).

[113] J. Lafferty, E. J. Land and T. G. Truscott, *J. Chem. Soc. Chem. Commun.*, 70 (1976).

[114] N. J. Miller, J. Sampson, L. P. Candeias, P. M. Bramley and C. A. Rice-Evans, *FEBS Lett.*, **384**, 240 (1996).

[115] A. Mortensen and L. H. Skibsted, *FEBS Lett.*, **417**, 261 (1997).

[116] A. Broniec, A. Pawlak, T. Sarna, A. Wielgus, J. E. Roberts, E. J. Land, T. G. Truscott, R. Edge and S. Navaratnam, *Free Rad. Biol. Med.*, **38**, 1037 (2005).

[117] K. Bobrowski and P. K. Das, *J. Phys. Chem.*, **90**, 927 (1986).

[118] G. Gao, Y. Deng and L. D. Kispert, *J. Phys. Chem. B*, **101**, 7844 (1997).

[119] N. Getoff, I. Platzer and C. Winkelbauer, *Radiat. Phys. Chem.*, **55**, 699 (1999).

[120] Y. Gao and L. D. Kispert, *J. Phys. Chem. B*, **107**, 5333 (2003).

[121] R. Edge, E. J. Land, M. Rozanowska, T. Sarna and T. G. Truscott, *J. Phys. Chem. B*, **104**, 7193 (2000).

[122] F. Böhm, R. Edge, E. J. Land, D. J. McGarvey and T. G. Truscott, *J. Am. Chem. Soc.*, **119**, 621 (1997).

[123] P. F. Conn, C. Lambert, E. J. Land, W. Schalch and T. G. Truscott, *Free Rad. Res. Commun.*, **16**, 401 (1992).

[124] A. El-Agamey and D. J. McGarvey, *Free Rad. Res.*, **36**, 337 (2002).

[125] G. W. Burton and K. U. Ingold, *Science*, **224**, 569 (1984).

[126] A. El-Agamey and D. J. McGarvey, *Free Rad. Res.*, **41**, 295 (2007).

[127] K.-J. Yeum, R. M. Russell, N. I. Krinsky and G. Aldini, *Arch. Biochem. Biophys.*, **430**, 97 (2004).

[128] P. Palozza, S. Serini, S. Trombino, L. Lauriola, F. O. Ranelletti and G. Calviello, *Carcinogenesis*, **27**, 2383 (2006).

[129] A. A. Woodall, G. Britton and M. J. Jackson, *Biochim. Biophys. Acta*, **1336**, 575 (1997).

[130] G. Britton, *FASEB J.*, **9**, 1551 (1995).

[131] P. Palozza, *Nutr. Rev.*, **56**, 257 (1998).

[132] P. Zhang and S. T. Omaye, *Toxicology*, **146**, 37 (2000).

[133] M. C. Hanlon and D. W. Seybert, *Free Rad. Biol. Med.*, **23**, 712 (1997).

[134] J. S.Hurst, M. K. Saini, G.-F. Jin, Y. C. Awasthi and F. J. G. M. van Kuijk, *Exp. Eye Res.*, **81**, 239 (2005).

[135] W. Siems, I. Wiswedel, C. Salerno, C. Crifo, W. Augustin, L. Schild, C.-D. Langhans and O. Sommerburg, *J. Nutr. Biochem.*, **16**, 385 (2005).

[136] C. Salerno, C. Crifo, E. Capuozzo, O. Sommerburg, C.-D. Langhans and W. Siems, *Biofactors*, **24**, 185 (2005).

[137] S.-L. Yeh, M.-L. Hu and C.-S. Huang, *Eur. J. Nutr.*, **44**, 365 (2005).

[138] L. Packer, in *Carotenoids in Human Health*, (ed. L. M. Canfield, N. I. Krinsky and J. A. Olson), *Ann. NY Acad. Sci.*, **691**, 48, The New York Academy of Sciences, New York, (1993).

Carotenoids
Volume 4: Natural Functions
© 2008 Birkhäuser Verlag Basel

Chapter 8

Structure and Properties of Carotenoid Cations

Synnøve Liaaen-Jensen and Bjart Frode Lutnæs

A. Introduction

Naturally occurring carotenoids are familiar as yellow, orange or red pigments with λ_{max} below 600 nm in organic solvents. Carotenoids with short polyene chains, such as the triene phytoene (**44**) and the pentaene phytofluene (**42**), absorb light only in the UV region and are colourless. Compounds that absorb light in the near infrared (NIR) region above *ca.* 900 nm are also colourless, unless they also absorb light in the 600-900 nm region, in which case they appear blue. True carotenoproteins (*Chapter 6*) are purple-blue and absorb light above 600 nm.

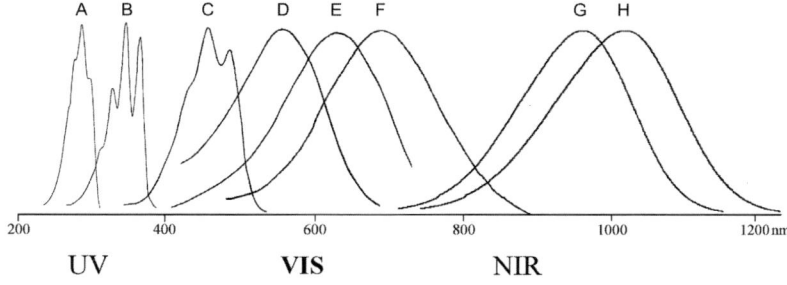

Fig. 1. UV/Vis/NIR absorption spectra of selected carotenoids: (A) phytoene (**44**); (B) phytofluene (**42**); (C) β,β-carotene (**3**); (D) violerythrin (**553**); (E) crustacyanin; (F) fucoxanthin (**369**) oxonium ion; (G) β,β-carotene dication (CAR^{2+}) (*1, 2*); (H) β,β-carotene-iodine complex (*8, 9*).

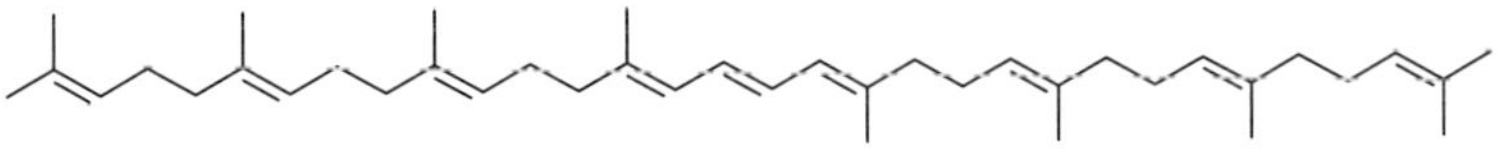

phytoene (**44**)

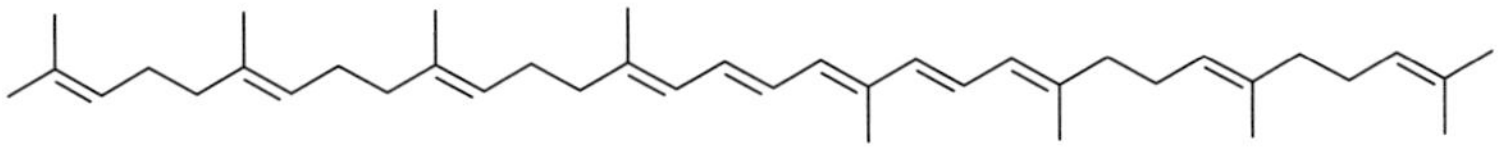

phytofluene (**42**)

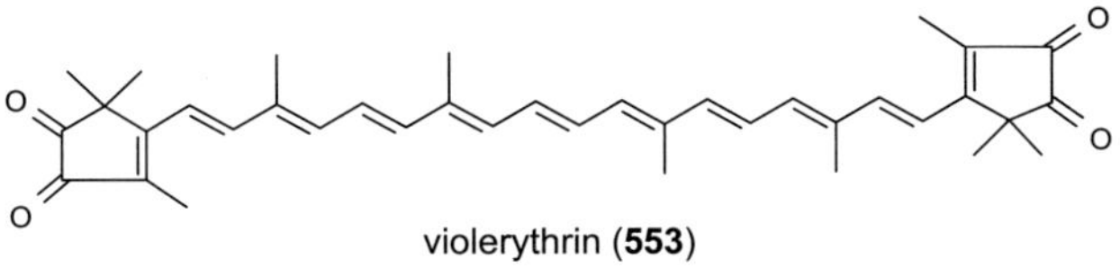

violerythrin (**553**)

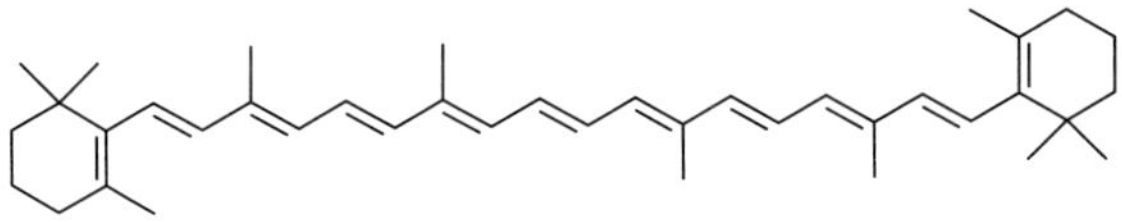

β-carotene (**3**)

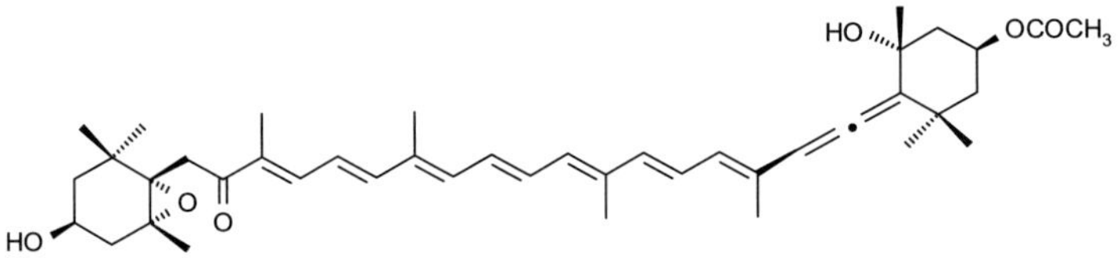

fucoxanthin (**369**)

This Chapter describes the structural evidence for and properties of unstable carotenoid species such as oxonium ions, monocarbocations, dicarbocations, radical cations and iodine complexes, which exhibit absorption in the 600-1100 nm range (Fig. 1). Carotenoid anions, which have been much less studied [1], are not treated here. General features of carotenoid radicals, including radical ions, are treated more thoroughly in *Chapter 7*.

Because of the long conjugated polyene chain of the carotenoids, these electron-deficient and positively charged species have strongly delocalized π-electrons, resulting in higher stability than for other carbocations. Hence their excitation energy is much lower than that of neutral carotenoids, and corresponds to light absorption in the NIR region.

B. Carotenoid Oxonium Ions

1. Preparation and chemistry

A diagnostic feature used to identify carotenoids containing epoxide groups is the blue colour produced on treatment with strong protic acids. These blue products, obtained by treatment of an ethereal solution of the carotenoid with concentrated HCl or trifluoroacetic acid, have been identified as carotenoid oxonium ions with positive charge on oxygen. Detailed experimental procedures are available [2,3].

Treatment of carotenoid 5,6-epoxides with weak acid leads to 5,8-furanoxides and a hypso-chromic shift (*Volume 1A, Chapter 4*). The 5,8-furanoxides are probable intermediates in the formation of the oxonium ion, by reactions that require a hydride transfer, as shown in Scheme 1. On treatment with KOH, the blue, unstable oxonium ions provide yellow hemiketals which, in contrast to the oxonium ions, are stable enough to be characterized fully [2,3].

Resonance considerations suggest a certain delocalization of the charge of the oxonium ions into the polyene chain, but NMR spectra of sufficient quality to evaluate the extent of charge delocalization have not been achieved.

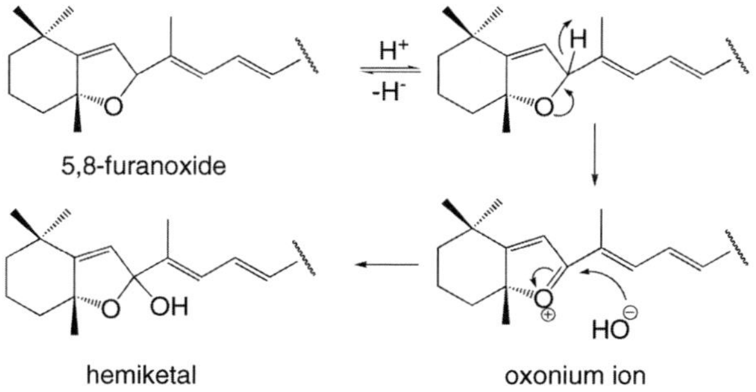

5,8-furanoxide

hemiketal

oxonium ion

Scheme 1

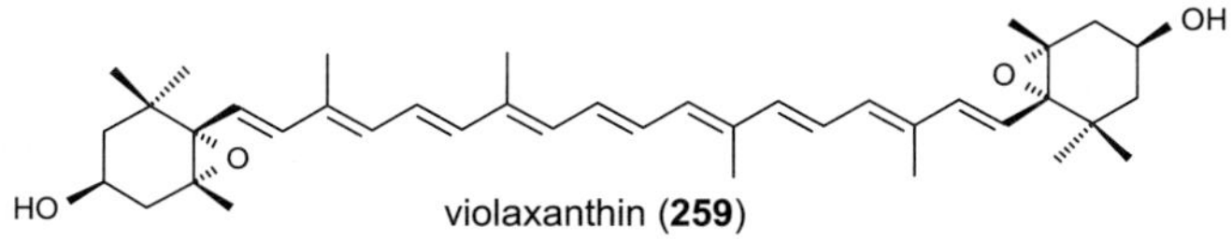

violaxanthin (**259**)

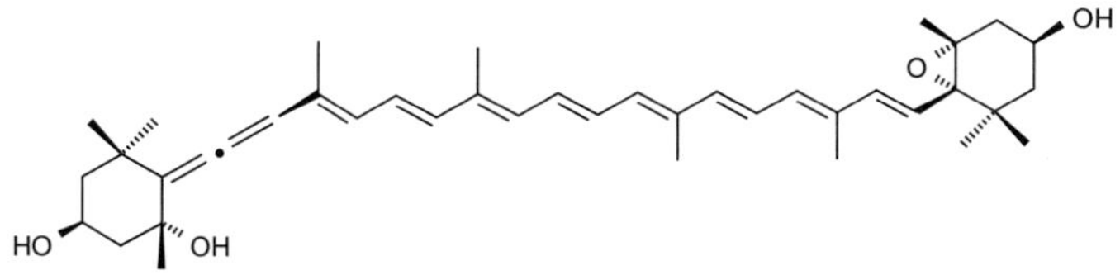

neoxanthin (**234**)

2. Vis/NIR spectra

Blue oxonium ions have been characterized for fucoxanthin (**369**), neoxanthin (**234**) and violaxanthin (**259**), with λ_{max} of the **369** oxonium ion 686 nm (HCl-MeOH), as shown in Fig. 2, 720 nm (AcOH), 740 nm (CF$_3$COOH), of the **234** oxonium ion 745 nm (CF$_3$COOH), and of the **259** oxonium ion 700 nm (CF$_3$COOH).

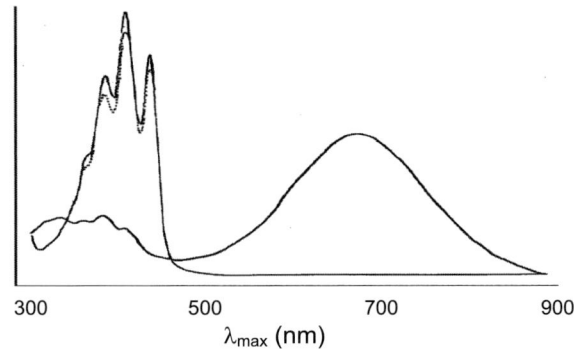

Fig. 2. The almost identical absorption spectra of (solid line) fucoxanthin (**369**) hemiketal and (dotted line) fucoxanthinol (**368**) hemiketal formed by treatment with base of the blue oxonium ion produced by treatment with acid (λ_{max} 690 nm).

fucoxanthinol (**368**)

C. Delocalized Carotenoid Monocarbocations and Dicarbocations

1. Preparation

The following radical cation, monocation and dications, formally derived from β,β-carotene (**3**), have been prepared and characterized:

i)   β,β-carotene (**3**) – 1 e$^-$ →   radical cation CAR$^{\bullet+}$ (C$_{40}$H$_{56}$, 21 π e$^-$)
ii) β,β-carotene (**3**) – 2 e$^-$ →   dication CAR^{2+} (C$_{40}$H$_{56}$, 20 π e$^-$)
iii) β,β-carotene-4-ol (**57**) + H$^+$ →   monocation ‡CAR$^+$ (C$_{40}$H$_{55}$, 22 π e$^-$) + H$_2$O
iv) β,β-carotene-4,4'-diol (**129**) + 2H$^+$ →   dication ‡CAR^{2+} (C$_{40}$H$_{54}$, 22 π e$^-$) + 2 H$_2$O

In principle, electron-deficient Lewis acids (non-protic acids that can accept an electron pair) are suitable electrophiles for abstracting π-electrons from the polyene chain. Those mainly used are BF_3-diethyl etherate, BF_3-dimethyl etherate and BF_3-tetrahydrofuran etherate. Stepwise abstraction, *via* the radical cation $CAR^{•+}$ to the dication CAR^{2+} takes place [4].

To obtain the monocation $^{‡}CAR^{+}$ and dication $^{‡}CAR^{2+}$ from the respective allylic carotenols **57** and **129**, trifluoroacetic acid is the preferred reagent, because HCl results in allylic elimination [5,6]. Other cations have recently been prepared from canthaxanthin (**380**). By treatment with protic acids (CF_3COOH and CF_3SO_3H), the C(5) protonated, C(7) protonated, enolised O(4) protonated and O(4,4'),C(7) triprotonated cations of canthaxanthin (**380**) have been prepared [7].

Reaction temperatures of -15 to -20°C are recommended.

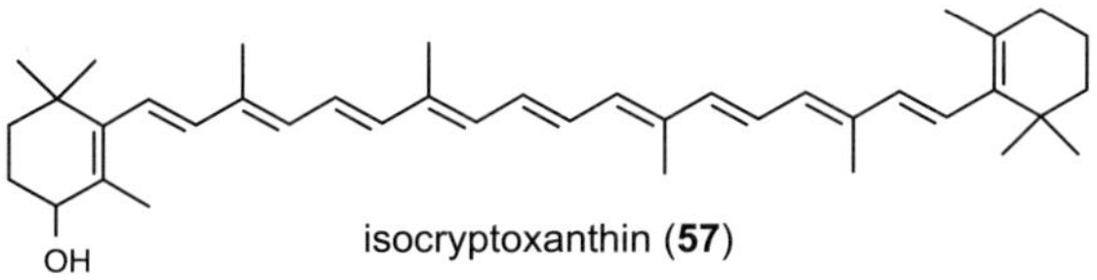

isocryptoxanthin (**57**)

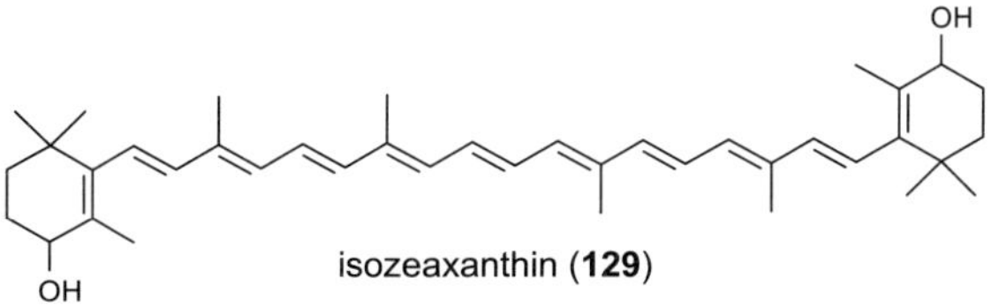

isozeaxanthin (**129**)

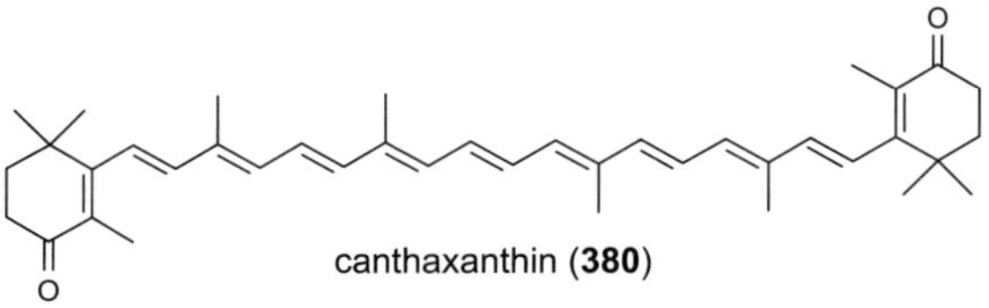

canthaxanthin (**380**)

2. Vis/NIR spectra and stability

Absorption maxima have been reported [4-6] for the monocation $^{‡}CAR^{+}$ [1020 nm (CH_2Cl_2, -15°C)], the dication CAR^{2+} [960 nm ($CHCl_3$, RT); 925 nm ($CHCl_3$, -20°C)] and the dication $^{‡}CAR^{2+}$ [1020 nm ($CHCl_3$, -15°C)].

The stability of the cations is readily monitored by Vis/NIR spectroscopy. The β,β-carotene dication CAR^{2+} decays by *ca* 20% in the first 30 minutes then more slowly until *ca.* 70% remains after 3 hours. The stability at -20°C is higher. The stability appears to be higher when evaluated by Vis/NIR than by NMR spectroscopy.

3. Cyclic voltammetry

Carotenoid dications have been characterized by electrochemical methods. Cyclic voltammetry is powerful for determining formal redox potentials and evaluating electron transfer
kinetics. Studies of cyclic voltammetry of carotenoids have been published since 1990 [8].

4. AM1 calculations

For the doubly charged ion state, information on bond lengths, charge distribution and orbital
energies has been obtained by semi-empirical Austin Model 1 (AM1) calculations [9]. The
dication CAR^{2+} of β,β-carotene (**3**) was depicted (Scheme 2) as a pair of charged species (*1*,
2). In the centre of the molecule, strong bond alternations and a reversal of single and double
bonds relative to the parent polyene should be noted [9]. The rotation of the end groups upon
cation formation was not considered.

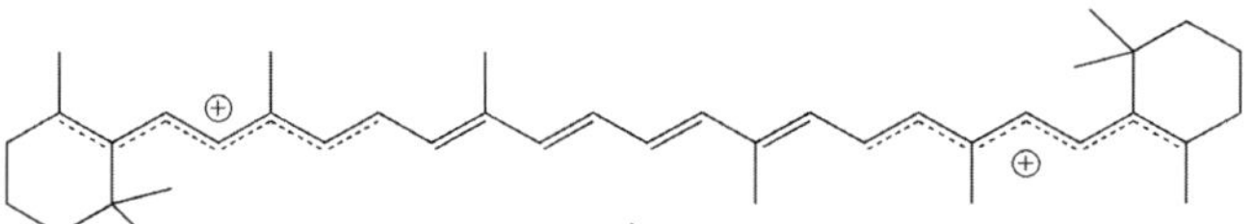

Scheme 2

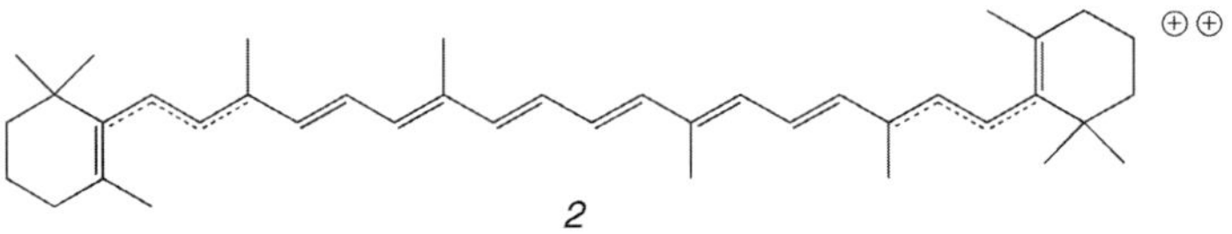

5. NMR spectra and structure

Modern 2D NMR spectroscopy at high field (500-600 MHz) and low temperature (-15°C) is
prerequisite for structure determination of carotenoid cations. In the electron-deficient cation
the nuclei are more naked and downfield shifts (to higher δ-values) are observed for protons
and especially carbon relative to the neutral parent carotenoid. Chemical shifts must, therefore,
be assigned for each proton and carbon atom in both the parent carotenoid and the cation.
The following 2D NMR methods are employed for this purpose:
- The homonuclear correlation spectroscopy (COSY) technique establishes through-bond
correlations between protons bound to two neighbouring carbon atoms.
- The rotating frame Overhauser enhancement spectroscopy (ROESY) technique
determines through-space correlations between two protons close in space, and is important
for establishing *E/Z* isomeric form and the configuration/conformation of the end groups.

•   Heteronuclear single quantum coherence (HSQC) experiments provide correlation between a proton and the carbon atom it is attached to. The method is much more sensitive than 1D ^{13}C NMR, and assignment of the ^{13}C chemical shift is obtained by the proton correlation.

•   Heteronuclear multiple bond correlation (HMBC) experiments also provide correlation between proton and carbon, but to the second and third neighbouring carbon atoms. This method allows assignment of ^{13}C chemical shifts for quaternary carbon atoms. Moreover, the spin systems of the polyene chain may be assigned by correlations from the methyl protons.

Below is outlined a procedure for structure elucidation based on these techniques, that allows:

i)   determination of the total positive charge (monocation, dication, trication),

ii) detailed determination of the delocalization of the charge in the polyene system,

iii) determination of carbon-carbon bond types: *i.e.* single, double or intermediate.

The total charge is determined from the difference in the sum of all downfield ^{13}C shifts for the cation and the 'parent' carotenoid. This difference, $\Sigma\Delta\delta_C$, is *ca.* 250 ppm per positive charge (*ca.* 500 ppm for a dication).

The distribution of the charge follows from the difference in ^{13}C shift, $\Delta\delta_C$, for each individual carbon in the cation and the neutral parent carotenoid. Larger $\Delta\delta_C$ means higher partial positive charge.

Finally, the size of the vicinal ^{1}H,^{1}H coupling constant $^3J_{H,H}$ between protons on the polyene chain is used to determine the bond type. It is well established that $J \sim 14$ Hz for *trans* double bonds and $J \sim 10$ Hz for s-*trans* single bonds. Intermediate values are interpreted as indicating intermediate bonds.

The structures determined by NMR for the monocation ‡CAR$^+$ (*3*) [5] and the dications CAR^{2+} (*4*) [4] and ‡CAR^{2+} (*5*) [6] are depicted in Fig. 3.

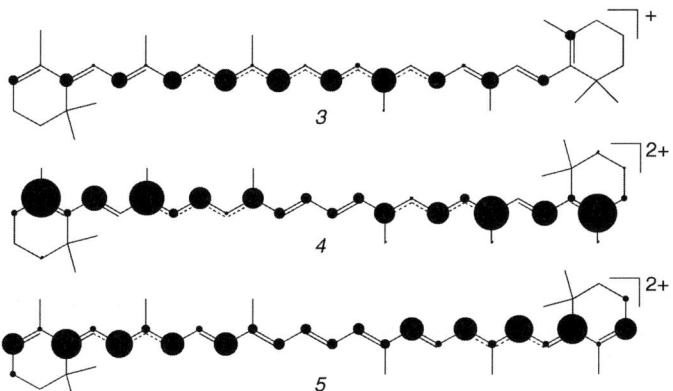

Fig. 3. Structures determined by NMR for the monocation ‡CAR$^+$ (*3*), dication CAR^{2+} (*4*) and dication ‡CAR^{2+} (*5*) of β,β-carotene (*3*). The relative magnitude of the partial charge on individual carbon atoms is indicated by the diameter of the filled circles.

For the monocation ‡CAR$^+$ (*3*), the single charge is located in the central part of the molecule, where intermediate bonds and bond reversal are clearly identified.

In the symmetrical dication *4*, containing 20 π-electrons, the charge is distributed near the ends of the polyene system with bond reversal in the centre, and intermediate bonds in the C(9)-C(12) and C(9')-C(12') regions, in agreement with the results based on AM1 calculations (Scheme 2).

The symmetrical dication ‡CAR²⁺ with 22 π-electrons (*5*), also has charge localization towards the ends of the polyene system. Here, the position of the double bonds in the central part is opposite to that of *4*, and bond alternation takes place in the C(6)-C(10) and C(6')-C(10') regions, which possess intermediate bonds.

In order to illustrate the strong potential of the NMR method, the total composition, including four stereoisomers, determined [5] for the monocation *3* is shown in Scheme 3.

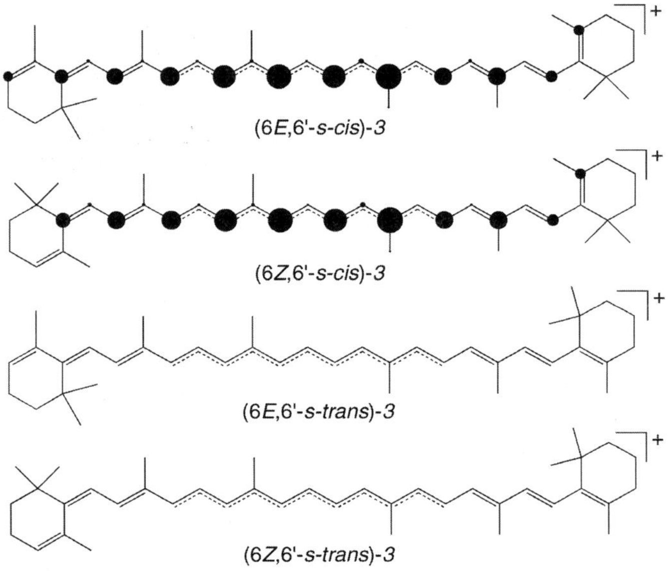

(6E,6'-s-cis)-3

(6Z,6'-s-cis)-3

(6E,6'-s-trans)-3

(6Z,6'-s-trans)-3

Scheme 3

6. Relation to the soliton model

The soliton model provides a mathematical description of how the charge is distributed in a charged polyene. For shorter π-electron systems, the charge distribution may be reasonably well described in terms of resonance, assuming the same contribution from each resonance structure. For polyenes, however, this approach fails as the charge is not evenly distributed along the polyene, but localized in a sine-shaped charge-wave, called a soliton. A positive soliton is electron deficient, *i.e.* a cation, whilst a negative soliton is electron rich, *i.e.* an anion. Structures (6E,6'-s-*cis*)-*3* and (6Z,6'-s-*cis*)-*3* (Scheme 3) illustrate a positive soliton as part of a sine-shaped wave, whereas the end groups affect the charge distribution in *4* and *5* (Fig. 3).

The width of this charge-wave is expected to approach a maximum value when the polyene system in question is sufficiently long to avoid effects from the termini of the polyene chain. Cations prepared from carotenoids have given the first experimental value for the width of a positive soliton [6]. Thus, for the monocation *3* with the charge delocalized over 23 carbon atoms, the soliton half-width was found to be 7.8 carbon atoms, in good agreement with previous AM1 calculations for a free soliton [6,10].

7. Reactions with nucleophiles

In order to obtain support for the cationic structures assigned, the reactions with suitable O, N and S nucleophiles have been investigated, Scheme 4 [4,11]. It may be predicted that carbon atoms with high partial positive charge are preferentially attacked. However, product stability is also an important factor. Product analysis has been carried out by HPLC, UV/Vis and MS.

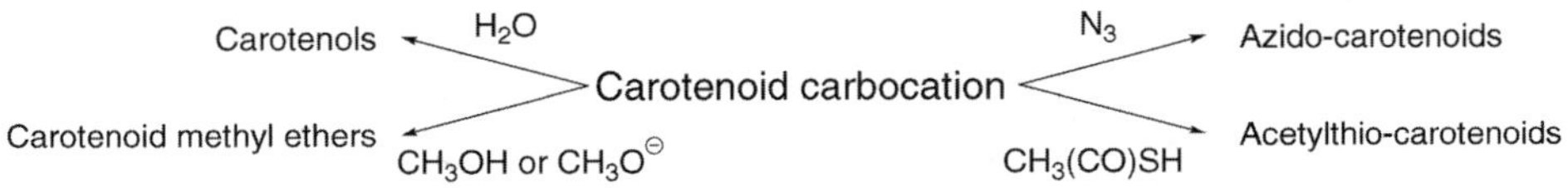

Carotenols $\xleftarrow{H_2O}$ Carotenoid carbocation $\xrightarrow{N_3}$ Azido-carotenoids

Carotenoid methyl ethers $\xleftarrow{CH_3OH \text{ or } CH_3O^{\ominus}}$ $\xrightarrow{CH_3(CO)SH}$ Acetylthio-carotenoids

Scheme 4

In addition to the nucleophilic addition products mentioned above, dehydrogenated products were obtained due to competing elimination reactions. The structures of the isolated products were consistent with the cationic structure of the carotenoid cation.

D. Carotenoid Radical Cations

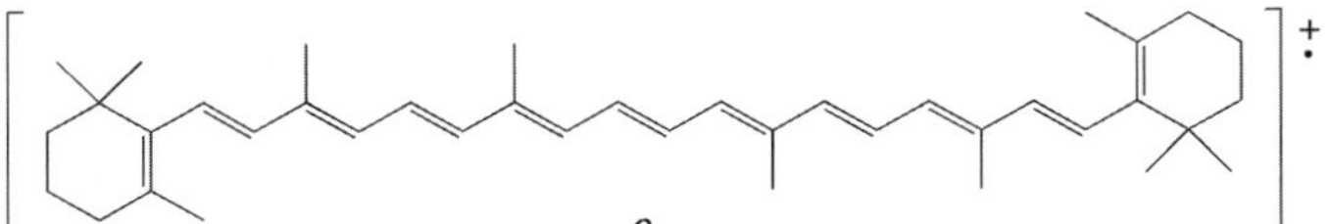

6

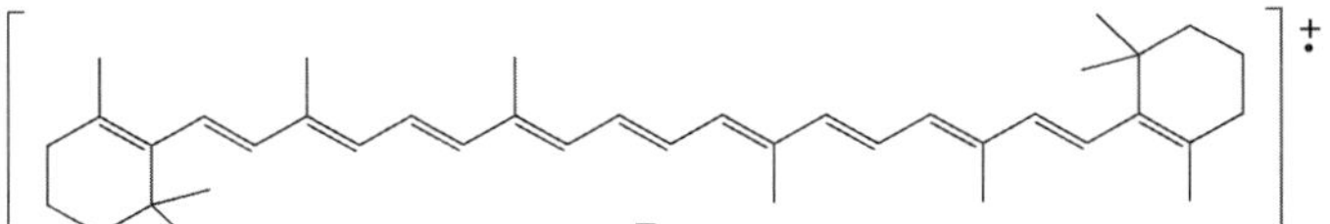

7

Scheme 5

In principle, a carotenoid radical cation (*Chapter 7*) is formed when one π-electron from the polyene chain is removed, *cf.* Section C.1. The current picture of the β,β-carotene radical cation, based on theoretical calculations [8], shows delocalization of the unpaired electron over the polyene system in a structure called a polaron (Scheme 5). Contributions of both the species *6* and *7*, with the 6-s-*cis* and 6-s-*trans* conformation, respectively, are indicated [12,13].

Carotenoid radical cations absorb light in the NIR region, and may produce colourless solutions. From the NIR spectra alone, carotenoid radical cations may be difficult to differentiate from the monocations and dications. For the β,β-carotene radical cation CAR$^{•+}$, the λ$_{max}$: 942 nm (DMSO); 970, 1425 nm (CH$_2$Cl$_2$) (Fig 4); 975 nm (hexane) are reported [8].

They may be prepared by electrochemical oxidation, *via* excited states or by treatment with Lewis acids. Spectroscopic methods used for characterization are NIR, electron paramagnetic resonance (EPR) spectroscopy for detecting free radicals, electron nuclear double resonance (ENDOR) spectroscopy and resonance Raman spectroscopy [8]. Radical cations cannot be studied by NMR because of the rapid relaxation induced by the free radical. Carotenoid radical cations are less stable than cationic species that contain no unpaired electrons.

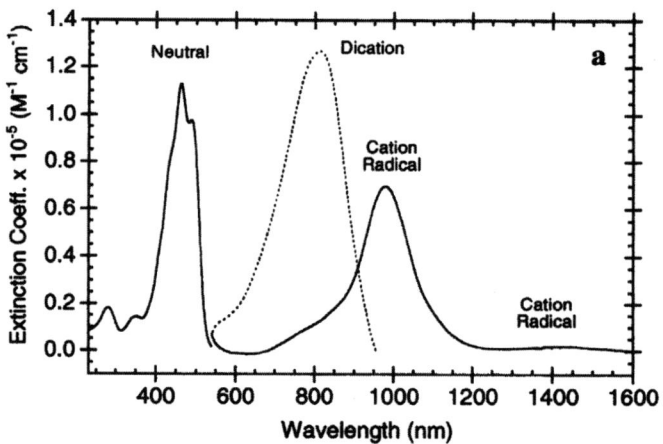

Fig. 4. Absorption spectra of neutral β,β-carotene (**3**), β,β-carotene radical cation CAR$^{•+}$ and β,β-carotene dication CAR^{2+} in CH$_2$Cl$_2$. Reprinted from [16] with permission. Copyright 1996 American Chemical Society.

E. Carotenoid-Iodine Complexes

Although carotenoid-iodine complexes have been studied since 1886, it was not until recently that a detailed structure, expressed as a π-complex with both cationic and radical cationic properties, was published [14]. Methods including Vis/NIR, IR, MS, EPR, ENDOR and NMR (COSY, TOCSY, 2D ROESY, HSQC and HMBC) spectroscopy, as well as chemical reactions, have been used to give a structure for the β,β-carotene (**3**)-iodine complex (C$_{40}$H$_{56}$·4I), as a π-complex with the contributing structures *8* and *9* (Scheme 6).

The Vis/NIR spectrum of the β,β-carotene (**3**)-iodine complex at room temperature in CH₂Cl₂ is shown in Fig. 5. The stability during 17 hours was comparable with that of the β,β-carotene dication CAR^{2+} (Section **C**. 2).

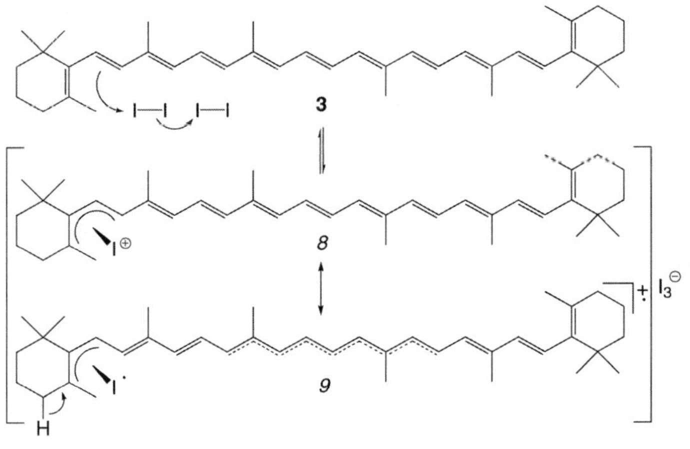

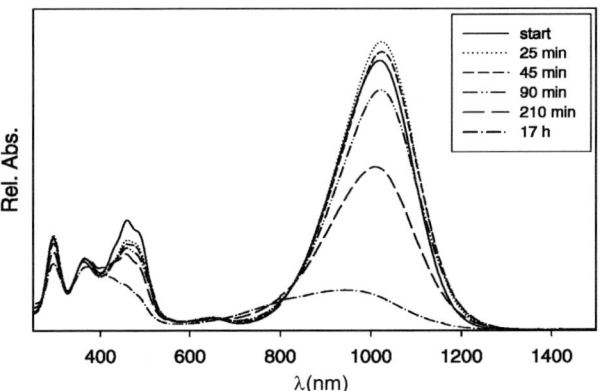

Scheme 6

Fig. 5. UV/Vis/NIR spectra and stability of the β,β-carotene–iodine complex (**3** + 4I; *8,9*) in CH₂Cl₂ at room temperature.

F. Biological Relevance

Of the positively charged carotenoid species, only radical cations have been detected in biological systems, though the presence of carotenoid dications in photosynthetic reaction centres has been suggested [15]. The presence of β,β-carotene radical cation in the reaction centre of photosystem 2 has been determined by spectroscopic methods such as IR, resonance Raman, EPR and ENDOR. For details of its role in photosynthetsis, see *Chapter 14*. Carotenoid radical cations are considered as unstable intermediates in reactions between carotenoids and reactive oxygen species. This topic is discussed in *Volume 5, Chapter 12*.

An intermediate role of radical cations in *E/Z* isomerization *in vitro* has been suggested, because AM1 molecular orbital calculations have shown that the energy barrier for the *E* to *Z* isomerization is much lower in cation radicals (*ca.* 20 kcal/mol) and dications (*ca.* 0 kcal/mol) than in neutral carotenoids [16]. *E/Z* (*cis-trans*) isomerization is treated in detail in *Chapter 3*.

True carotenoproteins, such as α-crustacyanin, also exhibit λ_{max} above 600 nm. The interactions responsible for the bathochromic shift include cationic features in the prosthetic group astaxanthin (**406**) attributed to hydrogen bonding (see *Chapter 6*).

(3*S*,3'*S*)-astaxanthin (**406**)

References

[1]   G. Kildahl-Andersen, B. F. Lutnaes and S. Liaaen-Jensen, *Abstr. 15th Int. Symp. Carotenoids*, 56 (2005).

[2]   J. A. Haugan and S. Liaaen-Jensen, *Acta Chem. Scand.*, **48**, 68 (1994).

[3]   J. A. Haugan and S. Liaaen-Jensen, *Acta Chem. Scand.*, **48**, 152 (1994).

[4]   B. F. Lutnaes, L. Bruas, G. Kildahl-Andersen, J. Krane and S. Liaaen-Jensen, *Org. Biomol. Chem.*, **1**, 4064 (2003).

[5]   G. Kildahl-Andersen, B. F. Lutnaes, J. Krane and S. Liaaen-Jensen, *Org. Lett.*, **5**, 2675 (2003).

[6]   B. F. Lutnaes, G. Kildahl-Andersen, J. Krane and S. Liaaen-Jensen, *J. Am. Chem. Soc.*, **126**, 8981 (2004).

[7]   G. Kildahl-Andersen, B. F. Lutnaes and S. Liaaen-Jensen, *Org. Biomol. Chem.*, **2**, 489 (2004).

[8]   S. Liaaen-Jensen and B. F. Lutnaes, in *Natural Products Research, Vol 30: Bioactive Natural Products* (ed. Atta-ur-Rahman), p. 515, Elsevier, Amsterdam (2005).

[9]   E. Ehrenfreund, D. Moses, K. Lee, A. J. Heeger, J. Cornil and J. L. Bredas, *Synth. Metals*, **57**, 4707 (1993).

[10]   J. R. Reimers, J. S. Craw and N. S. Hush, *J. Phys. Chem.*, **97**, 2778 (1993).

[11]   G. Kildahl-Andersen, L. Bruas, B. F. Lutnaes and S. Liaaen-Jensen, *Org. Biomol. Chem.*, **2**, 2496 (2004).

[12]   P. Hapiot, L. D. Kispert, V. V. Konovalov and J.-M. Saveant, *J. Am. Chem. Soc.*, **123**, 6669 (2001).

[13]   J.-D. Guo, Y. Luo and F. Himo, *Chem. Phys. Lett.*, **366**, 73 (2002).

[14]   B. F. Lutnaes, J. Krane and S. Liaaen-Jensen, *Org. Biomol. Chem.*, **2**, 2821 (2004).

[15]   Z. He and L. D. Kispert, *J. Phys. Chem. B*, **103**, 10524 (1999).

[16]   J. A. Jeevarajan, C. C. Wei, A. S. Jeevarajan and L. D. Kispert, *J. Phys. Chem.*, **100**, 5637 (1996).

Carotenoids
Volume 4: Natural Functions
© 2008 Birkhäuser Verlag Basel

Chapter 9

Excited Electronic States, Photochemistry and Photophysics of Carotenoids

Harry A. Frank and Ronald L. Christensen

A. Introduction

The most striking characteristic of carotenoids is their palette of colours. Absorption of light in the visible region of the electromagnetic spectrum by molecules such as β-carotene (3) and lycopene (31) not only readily accounts for their colours but also signals the ability of these long-chain polyenes to serve as antenna pigments in diverse photosynthetic systems [1-4].

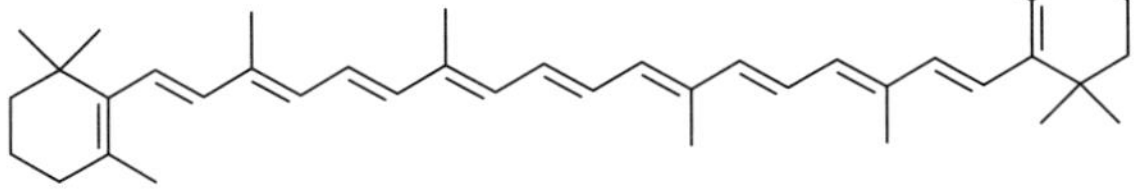

(**3**) β-carotene

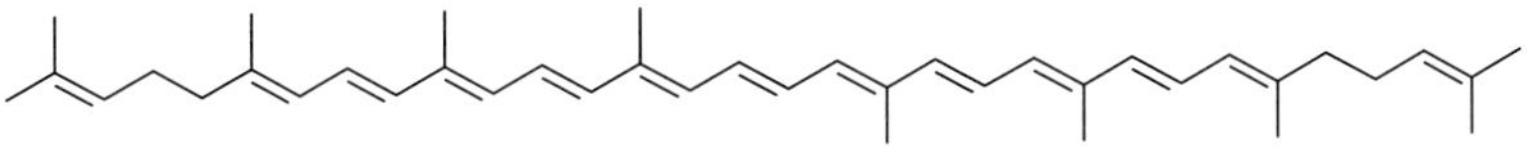

(**31**) lycopene

The absorption spectra of some carotenoids and related compounds undergo significant energy shifts upon binding to proteins and form the basis for biological colouration, *e.g.*

astaxanthin (**406**) in the lobster carapace (α-crustacyanin, *Chapter 6*) and retinal in vision (rhodopsin *etc., Chapter 15*).

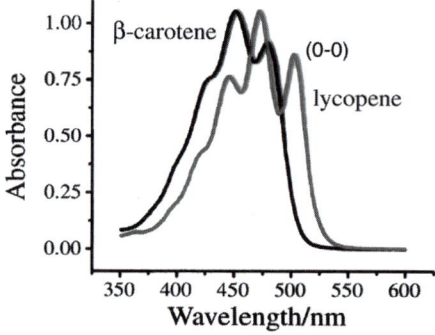

(**406**) astaxanthin

The spectra of β-carotene and lycopene given in Fig. 1 illustrate the light absorption properties of carotenoids. The distinctive, strongly allowed absorption band that appears between 400 and 550 nm is responsible for the characteristic yellow, orange and red hues of carotenoids in solution and in biological matrices. These absorptions provide an important diagnostic for probing carotenoid photophysics and photochemistry.

Fig 1. Room temperature absorption spectra of β-carotene (**3**) and lycopene (**31**). The electronic origin of lycopene is labelled (0-0).

Photosynthesis and other biological functions of carotenoids that are based on light absorption and energy transfer and transduction are covered in *Chapters 10* and *14* and in *Volume 5, Chapters 15* and *16*. The focus of this Chapter is on the absorption and fluorescence properties of carotenoids, including practical considerations central to the successful implementation of the methodology, and also on the fundamental photochemistry and photophysics that underlie the natural functions.

B. Conceptual Background and Terminology

In order to understand the molecular details of the photochemistry and photophysics of carotenoids, it is important to establish the basic principles and terminology. A brief introduction is provided here. For a more comprehensive discussion of these terms, the reader should consult any of a number of available textbooks on molecular photochemistry (*e.g.* [5]). The electronic structure of carotenoids and other polyenes is treated in detail in *Volume 1B, Chapter 1*, and the application of UV/Vis spectroscopy and spectrophotometry for the identification and quantitative analysis of carotenoids, with emphasis on the relationship between chromophore structure and absorption spectrum, in *Volume 1B, Chapter 2*.

1. Electronic structure and electronic states

(i) The **electronic structure** is a description of the distribution of electrons in a molecule (see *Volume 1B, Chapter 1*). Electrons in low-energy (π) and high-energy (π^*) delocalized orbitals formed from a combination of $2p_z$ atomic orbitals are important in determining the photochemical properties of polyenes and carotenoids.

(ii) The **ground electronic state** is the electronic structure that gives the lowest possible electronic energy of a molecule. For carotenoids this is a singlet state, designated S_0.

(iii) An **excited electronic state** is a higher energy electronic state of a molecule. For carotenoids this occurs by promotion of an electron from a low-energy molecular orbital to a higher one, by absorption of light. The excited states thus produced are normally also singlet states (S_1, S_2, *etc*).

(iv) A **singlet state** is an electronic state of a molecule in which all the electronic spins are paired so that no net spin angular momentum exists.

(v) A **triplet state** is an electronic state that has two unpaired parallel electronic spins. A triplet state thereby has biradical character and may be detected and studied by EPR (Electron Paramagnetic Resonance) techniques. It is difficult to produce the lowest energy triplet state, T_1, of a carotenoid by direct absorption of light into that state. Instead, it is usually formed by energy transfer from another triplet species, *e.g.* triplet chlorophyll, or by intersystem crossing, the radiationless interconversion of singlet states and triplet states.

(vi) The energy levels of a molecule may change in response to the application of an external electric field. This is known as an **electrochromic** or **Stark effect**.

2. Electronic transitions

(i) A change in the electronic state of a molecule is known as an **electronic transition**. It is typically brought about by the radiative processes of absorption or emission of light, but can also occur by non-radiative thermal processes. In carotenoids, a transition typically involves the promotion of an electron from a low-energy π orbital to a higher energy π^* orbital, or *vice versa*. The lowest energy excited configuration is generated from the ground state by promoting one electron from the highest-energy occupied molecular orbital (HOMO) to the lowest-energy unoccupied molecular orbital (LUMO), i.e. a **HOMO to LUMO transition**, subject to selection rules.

(ii) The transitions are governed by selection rules, determined by symmetry considerations. The overall symmetry of all-*trans* carotenoids, characterized by a collection of **symmetry elements**, places them in the C_{2h} point group. The electronic state of a molecule can be classified as A or B, according to elements of the symmetry of its electronic distribution. A states show rotational symmetry, B states do not.

(iii) **Selection rules**: for most practical purposes, excitation by light absorption is possible if the transition involves a change in the symmetry designation g $\leftrightarrow$ u and a change in the pseudoparity sign - $\leftrightarrow$ +. Transitions which do not involve such changes are forbidden. As discussed in Section **C** below, the transition from the ground state S_0 ($1^1A_g^-$) to the lowest energy excited state S_1 ($2^1A_g^-$) does not involve a change in symmetry nor in pseudoparity and is forbidden. The characteristic strong absorption in the visible region arises from a strongly allowed transition to the second excited state S_2 ($1^1B_u^+$).

(iv) The **electronic origin**, also referred to as the (0-0) band, indicates the transition between electronic states in their lowest energy vibrational states (zero-point vibrational levels). The electronic origin is the lowest energy band in absorption spectra and the highest energy band in emission spectra.

(v) The **oscillator strength** expresses the strength of light absorption by a molecule from the integration of the experimental absorption spectrum. It is proportional to the square of the magnitude of the **transition dipole moment**, *i.e.* the dipole moment of the molecule induced by its interaction with the electric field of the incident radiation. For the very strong $1^1A_g^-$ (S_0) $\rightarrow$ $1^1B_u^+$ (S_2) transition, the oscillator strength is *ca.* 1, and increases with extension of the π-electron conjugation.

(vi) The energy change associated with a transition between states is described by Planck's law, $\Delta E = h\nu$, where ν is the frequency of the light emitted or absorbed. Because $\nu = c/\lambda$, where c is the speed of light and λ is the wavelength, there is an inverse relationship between energy and wavelength.

(vii) The term **absorption cross-section** refers to a quantity that is proportional to the probability of light absorption by a molecule. It is used to indicate this probability.

(viii) The **potential energy surface** describes the energy of an electronic state as a function of nuclear coordinates. Potential energy surfaces are extremely useful for visualizing how energy is interconverted in carotenoids.

(ix) A **red shift** (bathochromic shift) is a shift of a spectral feature to longer wavelength. A shift to shorter wavelength is termed a **blue shift** (hypsochromic shift).

3. Interconversions and loss of excitation energy

(i) Excited electronic states may lose their excitation energy in a number of ways.

(ii) They may revert to the ground state by releasing the energy as radiation (emission). Emission from a singlet excited state is known as **fluorescence**, whereas **phosphorescence** is emission from an excited triplet state.

(iii) The radiationless deactivation of an excited state is known as **internal conversion**. This is the primary decay pathway of carotenoids from their excited states back to the ground state. The energy is released as heat.

4. Exciton interactions

When two or more chromophores are close together, their excited electronic states may interact. The excitation energy may be visualized as hopping from molecule to molecule, and this requires rather close spacing between donor and acceptor. When the inter-molecular distance becomes very close, *e.g.* as the two astaxanthins in crustacyanin, the energy levels of the individual molecules are perturbed and the spectral properties may be altered. Such systems involving pairs of molecules are termed '**exciton-coupled dimers**'.

5. Energy transfer

Especially significant in photosynthesis (*Chapter 14*), energy transfer processes are ones in which the excited state energy of one molecule is passed on to another. The lifetimes of the excited species are important for this. Typically for carotenoids, the lifetimes of the S_1, S_2 and T_1 states are in the picosecond, femtosecond and microsecond range, respectively. Rapid radiationless decay (or rapid internal conversion) and the short singlet lifetimes of carotenoids present significant challenges for efficient energy transfer to chlorophyll. Donor and acceptor molecules must be close together, and the rate of energy transfer depends on the mechanism and interactions that control the process. In a dipolar (Förster) mechanism [6] the probability of transfer follows a $1/r^6$ dependence, where r is the distance between the donor and acceptor molecules. In the exchange (Dexter) mechanism [7] the probability follows an exponential dependence on r. The relative geometry between the donor and acceptor molecules is also a factor, and varies depending on the mechanism. Triplet energy from chlorophyll triplet excited states can be transferred in the other direction to form carotenoid triplet states.

6. Quantum yield

The term **quantum yield** is used as a measure of the efficiency of a photo-induced process. A process that is 100% efficient in terms of photons absorbed *versus* products formed has a quantum yield of unity.

C. Absorption Spectroscopy

The absorption of ultraviolet or visible light involves an electronic transition from a low-lying state, usually the ground state, to an excited state. A simple energy level diagram that describes the singlet-state energies and many of the photochemical properties of polyenes and carotenoids is shown in Fig. 2. The ground state, S_0, and first excited singlet state, S_1, have A_g symmetry elements in the idealized C_{2h} point group of all-*trans* configurations. The second excited state, S_2, is a more energetic, excited electronic state of B_u symmetry. Higher excited singlet states, S_3, S_4, *etc.* are not indicated in Fig. 2, though it is important to point out that, in addition to the S_1 state, there may be other low-energy singlet states that are not easily detected by standard absorption measurements [8,9]. Symmetry labels for carotenoid electronic states are based on the C_{2h} geometry of undistorted (all-*trans*)-polyenes. These designations and their implications for transition intensities, vibronic interactions, and radiative lifetimes also work remarkably well for a wide range of unsymmetrical *cis* and *trans* polyenes and carotenoids. Thus, the A_g and B_u labels are used throughout this Chapter.

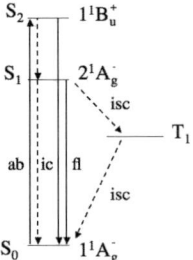

Fig. 2. Energy level diagram used to describe many of the photoprocesses that carotenoids and polyenes undergo, involving their singlet and triplet states. ab, absorption; ic, internal conversion; fl, fluorescence; isc, intersystem crossing.

As mentioned above, the electronic states S_0 and S_1 are both A_g states, so transitions between them are forbidden by the powerful $g \leftrightarrow u$ selection rule for electronic transitions. Also, + and - signs frequently accompany the group theoretical labels of the electronic states of polyenes and carotenoids. These designate pseudoparity elements derived from orbital pairing relationships that emerge when high-level computations involving configuration interaction among singly excited configurations are done [10-12]. At this level of description, the ground state of a polyene belonging to the C_{2h} point group is represented as 1^1A_g, and one-photon transitions between states having the same pseudoparity element are forbidden. So, for example, a transition from $1^1A_g^-$ (ground state) $\rightarrow 1^1B_u^-$ (excited state) is forbidden by pseudoparity selection rules that supplement the $g \leftrightarrow u$ selection rule derived from group theory. The strongly-allowed absorptions of polyenes and carotenoids (e.g. Fig. 1) cor-respond to $1^1A_g^-$ (S_0) $\rightarrow 1^1B_u^+$ (S_2) transitions that involve both $g \leftrightarrow u$ and $- \leftrightarrow +$ changes.

The emission of light by large organic molecules in condensed phases typically occurs from the lowest energy excited singlet or triplet electronic state following rapid radiationless decay from higher excited electronic states [5]. Longer polyenes and carotenoids, however, e.g. those with more than five conjugated double bonds, often show $1^1B_u^+$ (S_2) $\rightarrow 1^1A_g^-$ (S_0) emission, with the ratio of ($S_2 \rightarrow S_0$):($S_1 \rightarrow S_0$) emission increasing with the length of conjugation. For carotenoids such as β-carotene and lycopene, the $S_1 \rightarrow S_0$ emission yields are almost negligible ($<10^{-5}$). Another important characteristic of the photophysics of carotenoids and polyenes is the apparent absence of reports of radiative decay (phosphorescence) from their lowest triplet states. The rate of the spin-forbidden phosphorescence apparently cannot compete with rapid, nonradiative $T_1 \rightarrow S_0$ relaxation from the low energy triplet states.

From an orbital standpoint, the absorption spectra shown in Fig. 1 are due to $^1\pi\pi^*$ transitions of the conjugated π-system. Figure 3 shows the π and π* molecular orbitals of butadiene (1) as a simple illustration.

The energies of $\pi\pi^*$ transitions can be rationalized either by molecular orbital theory or by the free-electron model ('particle-in-a-box') [13]. The simplest versions of these models explain why the strongly-allowed, low-energy transition [$1^1A_g^-$ (S_0) $\rightarrow 1^1B_u^+$ (S_2) in Fig. 2] shifts to longer wavelength with increasing conjugation length. These models also predict that the energy of the $S_0 \rightarrow S_2$ transition can be approximated by $\Delta E = A + B/N$, where N is the effective number of conjugated double bonds and is proportional to the conjugation length. The experimental data suggest an asymptotic limit of ≈700 nm ($A \approx 14{,}000$ cm^{-1}) for the $S_0 \rightarrow S_2$ absorptions of infinite polyenes and carotenoids with alternating C-C and C=C bond lengths [13,14]. This stands in contrast to $A \approx 0$ for cyanine dyes and other linearly conjugated systems that lack alternation of the carbon-carbon bond lengths.

Energy

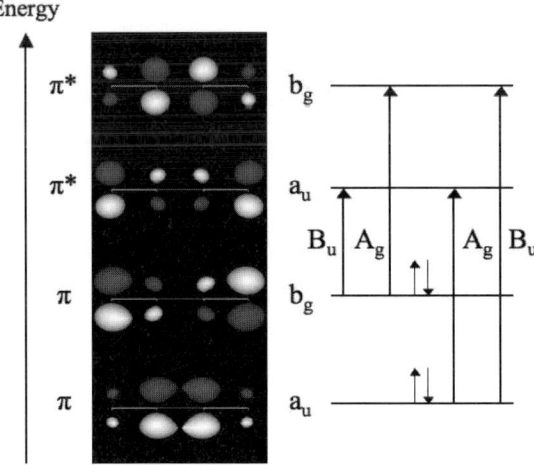

Fig. 3. The π molecular orbitals and $\pi\pi^*$ states in butadiene (*1*).

There is now considerable experimental evidence that the transition energies of other electronic transitions ($S_0 \rightarrow S_1$, $S_0 \rightarrow S_3$, $S_0 \rightarrow S_4$, *etc.*) also exhibit 1/N approaches to their asymptotic limits [13,15,16]. However, unlike the $S_0 \rightarrow S_2$ transition, there is little theoretical basis for this behaviour, particularly in the long polyene limit. Nevertheless, these empirical relationships prove extremely useful for describing how the excited state energies of intermediate length polyenes and carotenoids ($5 \leq N \leq 15$) vary with conjugation length.

The S_2 ($1^1B_u^+$) states are well described in simple molecular orbital or free-electron treatments as being of HOMO $\rightarrow$ LUMO parentage ($a_u \rightarrow b_g$ or $b_g \rightarrow a_u$ where a_u and b_g refer to the irreducible representations of the π-electron molecular orbitals in Fig. 3), and the symmetry-allowed, S_0 ($1^1A_g^-$) $\rightarrow$ S_2 ($1^1B_u^+$) transitions have long been appreciated to be responsible for the colours of long polyenes and carotenoids. However, satisfactory descriptions of the S_1 ($2^1A_g^-$) and other low-energy excited states demand a detailed consideration of the correlation of electron-electron interactions. Generally, this is approached by using configuration interaction (CI), which preserves the orbital description of the π electrons. Even a qualitative understanding of the experimentally observed ordering of the lowest lying excited states, $E(2^1A_g^-) < E(1^1B_u^+)$, requires the interaction of both singly excited, *e.g.* HOMO-1 $\rightarrow$ LUMO and HOMO $\rightarrow$ LUMO+1 ($b_g \rightarrow b_g$ or $a_u \rightarrow a_u$) and doubly excited (*e.g.* HOMO, HOMO $\rightarrow$ LUMO, LUMO) electronic configurations. A more quantitative agreement between theoretical and experimental $2^1A_g^-$ energies requires extensive configuration interaction. For example, the inclusion of all singly up to quadruply excited configurations explains the slight increase in the energy gap, $E(1^1B_u^+) - E(2^1A_g^-)$, as a function of increasing polyene length [15,17]. However, the computational effort for calculations at this level of CI increases exponentially with the number of π-electrons, and high level CI calculations for polyenes with more than five or six double bonds have proven prohibitive [15].

In addition to the semi-empirical, multi-reference CI calculations [15,17], *ab-initio* electronic structure calculations on polyenes of moderate size are now feasible [18-20]. However, calculating accurate energies for the $2^1A_g^-$ and $1^1B_u^+$ states remains a very challenging problem, especially for linearly conjugated systems that are comparable to the carotenoids.

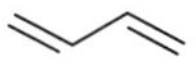

(1) butadiene (N = 2)

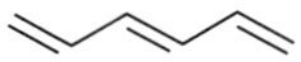

(2) hexatriene (N = 3)

(3) octatetraene (N = 4)

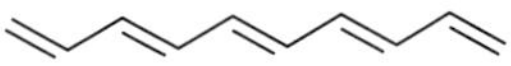

(4) decapentaene (N = 5)

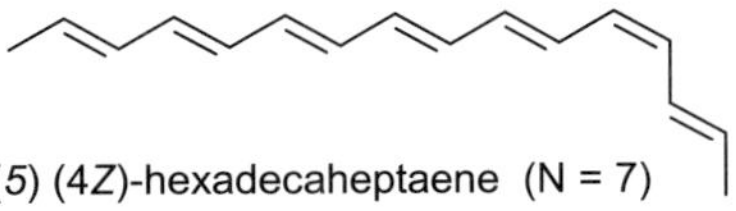

(5) (4Z)-hexadecaheptaene (N = 7)

Time-dependent density functional theory (TDDFT) has been applied [20-22] to the simple, unsubstituted polyenes butadiene (*1*, N = 2), hexatriene (*2*, N = 3), octatetraene (*3*, N = 4), and decapentaene (*4*, N = 5). The $2^1A_g^-$ energies were described quite well. The results were considerably poorer for the $1^1B_u^+$ state, however, with theory underestimating the experimental excitation energies by ~4000-6000 cm^{-1}, a value which is comparable to the $1^1B_u^+ - 2^1A_g^-$ energy differences in these molecules. Although the TDDFT approaches and other *ab-initio* and empirical treatments account for the $1^1B_u^+$ and $2^1A_g^-$ state orderings and the general trends of decreasing excitation energies with increasing conjugation lengths, it is likely that our understanding of carotenoid excited states will rely on experimental data for the foreseeable future.

D. Fluorescence Spectroscopy

Absorption spectroscopy is not entirely adequate to resolve the complex character of the excited states of carotenoid molecules. Because the $S_0 \rightarrow S_1$ transition is forbidden, this has hindered the direct observation of this transition in an absorption spectroscopic experiment. In a small number of cases, *e.g.* for short, simple polyenes in solvent environments and at the low temperatures that lead to well-resolved spectra, the S_0 $(1^1A_g^-) \rightarrow S_1$ $(2^1A_g^-)$ transition can be observed in absorption [23-26]. However, even for *cis* polyenes and distorted *trans* isomers, this symmetry-forbidden, vibronically induced transition is very weak; its molar absorptivities are only $\sim 10^{-2}$-10^{-3} that of the allowed $S_0 \rightarrow S_2$ transition, which has ε_{max} values of $\sim 10^5$ L/mol cm.

The inherent sensitivity of fluorescence spectroscopy provides a useful alternative for probing the energies and dynamics of the S_1 states in carotenoids, yet there are many technical impediments that prohibit the observation of fluorescence, including difficulties in obtaining samples free of fluorescing impurities, and the inherently low quantum yields of emission of longer polyenes and carotenoids. However, by combining HPLC to obtain ultra-pure samples, laser excitation for efficient and stable optical pumping, and photon counting to enhance the sensitivity of the detection of the weak emission, emission from carotenoids can be observed [13,27-29]. Although fluorescence can be detected readily, the carotenoid fluorescence bands often are broad and featureless, precluding an unambiguous assignment of the spectral origins.

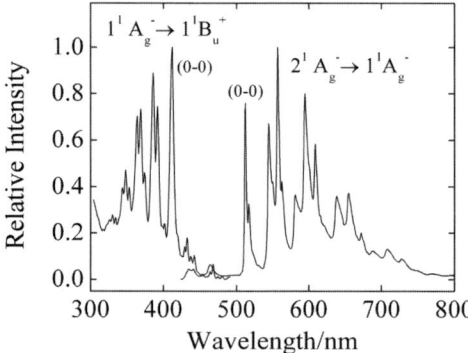

Fig. 4. Excitation and emission spectra of (4Z)-hexadecaheptaene (*5*) in *n*-pentadecane at 77 K. The spectra were normalized to their λ_{max} values.

Detection of the low-lying S_1 state, especially in shorter polyenes and carotenoids, is illustrated in the low-temperature absorption and fluorescence spectra of (4Z)-hexadeca-heptaene (*5*) (Fig. 4). The vibronic structure exhibited in these spectra is broadened for carotenoids of photobiological interest, particularly for dicyclic molecules such as β-carotene (3) where non-planarity between the central polyene chain and terminal cyclohexenylidene rings results in a distribution of absorbing and emitting species [30,31]. The well-resolved spectra of unsubstituted, model polyenes at low temperature facilitates the unambiguous identification of electronic origins [(0-0) bands] and the precise measurement of $1^1B_u^+$ and $2^1A_g^-$ electronic energies. Figure 4 shows the characteristic gap between the onsets [(0-0)s] of the strongly allowed $1^1A_g^- \rightarrow 1^1B_u^+$ absorption and the $2^1A_g^- \rightarrow 1^1A_g^-$ emission. The fluorescence yields for S_1 $(2^1A_g^-) \rightarrow S_0$ $(1^1A_g^-)$ decrease steadily with increasing conjugation length, ranging from ~1 for octatetraene (3) at low temperature [32] to <1 x 10^{-5} for β-carotene [33,34] and most other carotenoids of biological interest [35-40].

The vibronic features of absorption and emission spectra of polyenes are worth noting. Higher-resolution versions of the spectra of (4Z)-hexadecaheptaene (5), obtained in low-temperature, mixed crystals, have been discussed [23]. The electronic spectra are dominated by combinations of totally symmetric (a_g) C-C and C=C stretching modes of ~1200 cm^{-1} and ~1600 cm^{-1}, with frequencies decreasing (as ~ 1/N) with increasing conjugation. These details are easily identified in the S_1 $(2^1A_g^-) \rightarrow S_0$ $(1^1A_g^-)$ fluorescence spectrum, which shows Franck-Condon maxima that characteristically involve at least one quantum of the double-bond stretch. In the broader, less well resolved absorption and emission spectra of carotenoids, the vibronic features corresponding to single-bond and double-bond stretches often coalesce into progressions of what appears to be a single, intermediate frequency of 1300-1400 cm^{-1}. Note that the absorption spectrum of the heptaene 5 in Fig. 4 has its maximum intensity in the (0-0) band; it is likely that this reflects the relatively small geometry change experienced in the S_0 $(1^1A_g^-) \rightarrow S_2$ $(1^1B_u^+)$ transition. This is consistent with theory, which predicts a more significant transposition of the ground state π-bond orders in the $2^1A_g^-$ state [41]. The vibronic features in Fig. 4, those seen in low-temperature spectra of mixed crystals [23,25,26,42,43], and the highly detailed vibronic development observed in high-resolution spectra of isolated tetraenes in supersonic expansions [44] all are consistent with planar $2^1A_g^-$ and $1^1B_u^+$ excited states in longer polyenes. With long polyenes and carotenoids, there is no evidence for the substantial deviations from planarity experienced by the excited states of dienes and trienes [45-47].

Carotenoids with eight or fewer carbon-carbon double bonds exhibit fluorescence bands associated with the $S_1 \rightarrow S_0$ transition. In longer-chromophore carotenoids, the fluorescence is weak but dominated by the $S_2 \rightarrow S_0$ transition. The crossover from S_1 to S_2 fluorescence can be explained by increases in the rates of $S_1 \rightarrow S_0$ non-radiative decay due to a combination of smaller S_1 - S_0 energy gaps and the increased density of S_0-accepting modes in the longer molecules [48]. This leads to the disappearance of $S_1 \rightarrow S_0$ fluorescence, allowing the weaker, residual $S_2 \rightarrow S_0$ fluorescence to dominate the emissions of longer conjugated systems. This idea is supported by the significant decreases in $S_1 \rightarrow S_0$ lifetimes and quantum yields that are observed as the extent of conjugation increases. The $S_2 \to S_1$ internal conversion rates and fluorescence yields are relatively constant for carotenoids [9,13]. The abrupt decrease in $S_1 \to S_0$ fluorescence with increasing conjugated chain length accounts for the crossover from S_1 to S_2 fluorescence. Fluorescence is clearly the most direct method by which the forbidden $S_1 \to S_0$ transition may be observed and the S_1 state energy assigned.

E. Other Optical Techniques and Aspects

1. Pump-probe spectroscopy

The $S_0 \leftrightarrow S_1$ absorption and emission transitions are forbidden by symmetry and are difficult to detect, especially in longer polyenes and carotenoids. However, the $S_1 \to S_2$ transition is symmetry allowed and exhibits a substantial oscillator strength. Femtosecond lasers have been used [49] to excite carotenoids from S_0 to S_2, which then relaxes to the zero-point vibrational level of S_1. The $S_1 \to S_2$ absorption spectrum then can be detected by means of an infrared probe laser. Subtraction of the energy of the spectral origin of the $S_1 \to S_2$ transition from the energy of the spectral origin of the $S_0 \to S_2$ transition gives the S_1 energy of the carotenoid. This approach was applied to violaxanthin (**259**), zeaxanthin (**119**) [49], and spheroidene (**97**) [50,51].

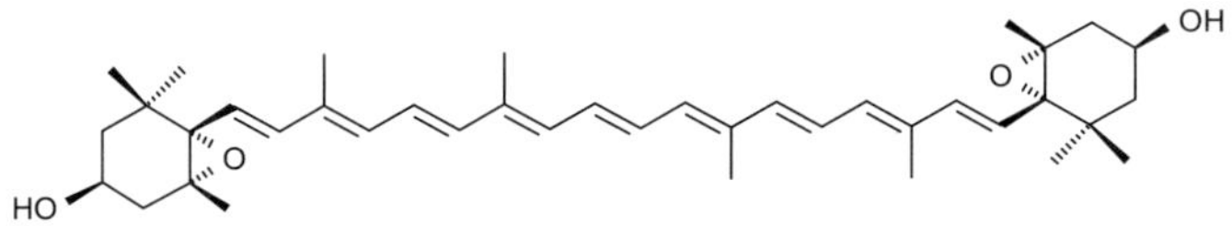

(**259**) violaxanthin

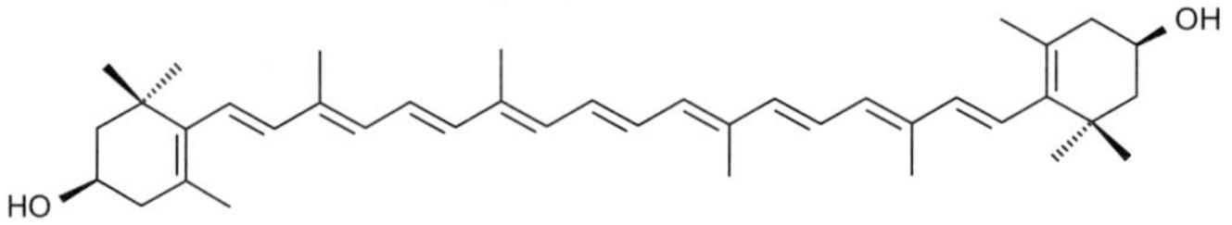

(**119**) zeaxanthin

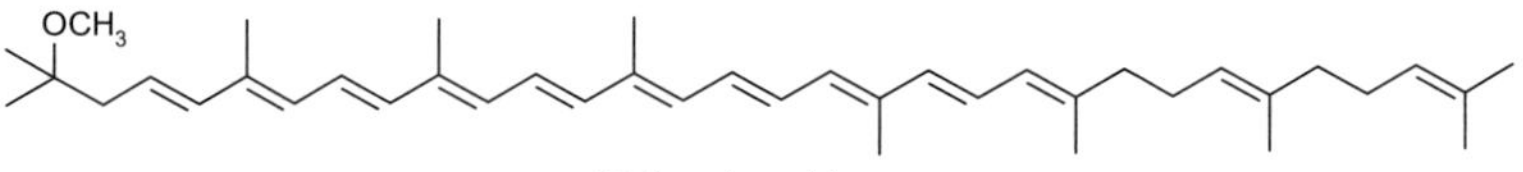

(**97**) spheroidene

Semi-empirical and *ab-initio* quantum calculations [15,17,21,52,53] have suggested that additional excited singlet states may lie in the vicinity of S_1 or between S_1 and S_2. These are states ($^1A_g^-$ and $^1B_u^-$) into which absorption from the $^1A_g^-$ ground state is symmetry forbidden. Different variations on ultra-fast laser technology have been used to seek evidence for these states. For example, ultra-fast pump-probe spectroscopy was used [54] to obtain results on several carotenoids including lycopene (31) and β-carotene (3) and, from this, the existence of an intermediate singlet state (S_x), which facilitates internal conversion between the S_2 ($1^1B_u^+$) and S_1 ($2^1A_g^-$) states, was postulated. A wavelength dependence of the dynamics of spirilloxanthin (166) was observed [55] and interpreted in terms of a singlet electronic state, denoted S*. Recent experiments on several cyclic xanthophylls and open-chain carotenoids have suggested that S* is associated with twisted conformations of carotenoids (see Section G). Application of pump–probe optical techniques to β carotene [56] suggested yet another carotenoid excited state, referred to as $S^\ddagger$, formed directly from S_2 ($1^1B_u^+$). Spectroscopic features have been assigned to a low-lying $1^1B_u^-$ state [16], and the presence of an intramolecular charge transfer state, S_{ICT}, has been invoked [57,58] to explain spectroscopic observations on peridinin (558). Details of these results and other experiments are summarized in a recent review [9].

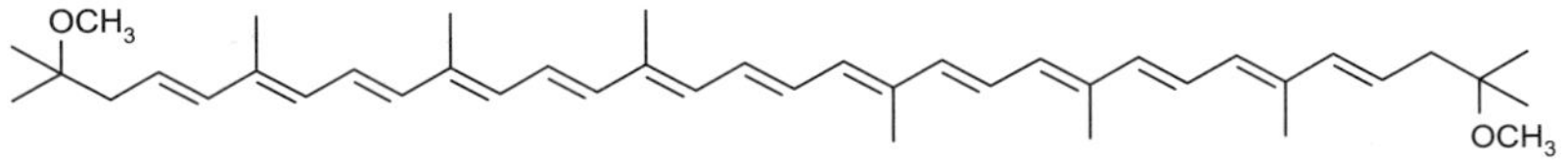

(166) spirilloxanthin

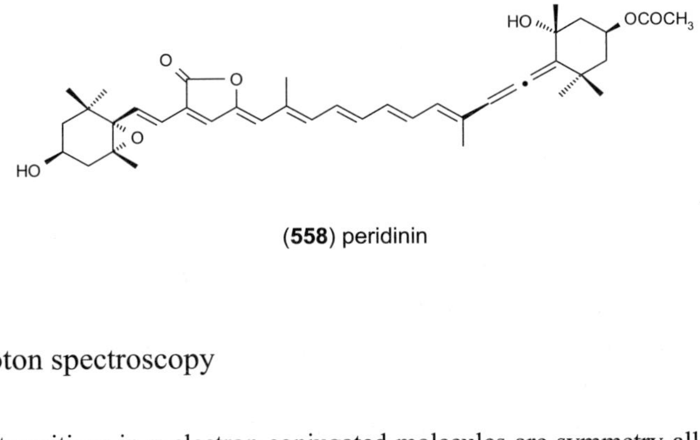

(558) peridinin

2. Two-photon spectroscopy

Two-photon transitions in π-electron conjugated molecules are symmetry-allowed between states that have the same parity [10]. Therefore, in principle, the $S_0 \rightarrow S_1$ ($1^1A_g^- \rightarrow 2^1A_g^-$) transition should be observable by, for example, fluorescence-detected, two-photon excitation. This technique has been exploited in high-resolution optical studies of short polyenes in mixed crystals [59] and as cold, isolated molecules in supersonic jets [44]. However, the very low quantum yields of S_1 emission in longer polyenes have precluded broad application of the technique to purified carotenoids in solution. The $S_0 \rightarrow S_1$ transition

$(1^1A_g^- \rightarrow 2^1A_g^-)$ has been observed for carotenoids bound in pigment-protein complexes from photosynthetic organisms [60-67] where the two-photon excitation profile can be monitored by using the emission from the highly fluorescent bound chlorophylls to which carotenoids typically transfer energy. These studies are important because they provide information on the energies and dynamics of the S_1 states of carotenoids bound in these pigment-protein complexes.

3. Resonance Raman spectroscopy

Resonance Raman spectroscopy is a powerful technique for analysing the electronic structures and dynamics of excited carotenoids in solution as well as those bound to proteins [68]. Both steady-state and time-resolved methods have been used extensively to probe carotenoids in their ground and excited singlet and triplet states. A variation of the technique, resonance Raman excitation spectroscopy, monitors resonance Raman line intensities as a function of the excitation energy of the photons used to induce scattering. The technique has been applied to several carotenoids including β-carotene (**3**), lycopene (**31**), spheroidene (**97**), anhydrorhodovibrin (**91**), and spirilloxanthin (**166**) [16,69-72].

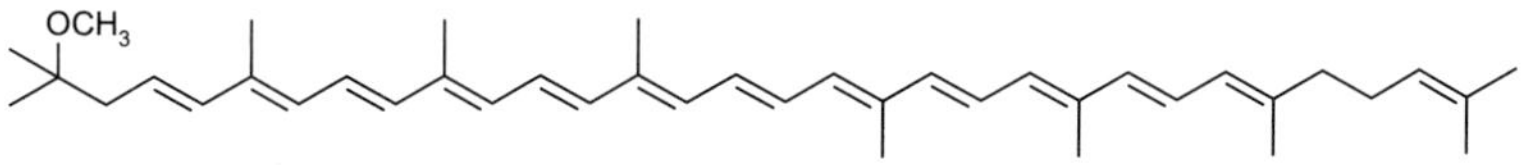

OCH$_3$

(**91**) anhydrorhodovibrin

The experiments employ crystalline carotenoids to induce strong self-absorption of the resonance Raman lines associated with the strongly allowed $S_0 \rightarrow S_2$ absorption. In this manner, signals associated with optically forbidden transitions may be revealed. The excitation profiles are strongly dependent on the concentration of the sample and the geometry of the optical set-up, and assigning the vibrational progressions to particular electronic states is difficult. Nevertheless, signals have been assigned to the $1^1B_u^+$, $1^1B_u^-$ and $2^1A_g^-$ states of several carotenoids, and the energies of the $2^1A_g^-$ states, based on resonance Raman excitation profiles, are in reasonable agreement with the values obtained from steady-state fluorescence methods.

The experiments and computations described above have suggested a much more complex energy level diagram (Fig. 5). At this stage, it is not clear how many, if any, distinct electronic states lie between $1^1B_u^+$ (S_2) and $2^1A_g^-$ (S_1), or what their significance may be for biological functions of carotenoids, but this clearly will remain an active area of research.

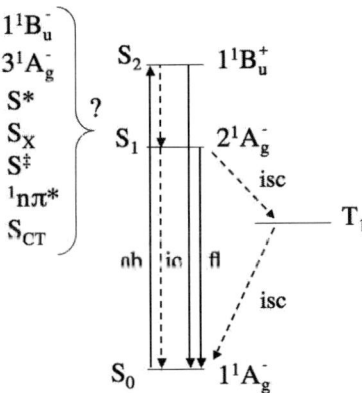

Fig. 5. Energy level diagram depicting the notation of various additional excited singlet states under investigation.

F. Experimental Considerations

1. Purity of carotenoids

For the spectroscopic experiments described above, it is imperative that carotenoid samples be of high purity. Fortunately, for most optical experiments, analytical-scale chromatography yields sufficient material. Purification of carotenoids generally uses combinations of column chromatography, TLC and/or recrystallization, followed by HPLC, as treated in *Volume 1A*.

In a typical procedure used by the authors [40], carotenoids were extracted from anaerobically grown cells of *Rhodobacter sphaeroides* wild-type strain 2.4.1 with methanol/acetone and fractionated by column chromatography on alumina. The fraction containing spheroidene (**97**) was then subjected to reversed-phase HPLC, either on a NovaPak C_{18} column with gradient elution or a YMC C_{30} column with isocratic elution (methyl *t*-butyl ether/methanol, 11:89). (all-*E*)-Spheroidene and several *Z*-isomers were resolved, and the isomerically pure (all-*E*)-spheroidene, which had the longest retention time on the C_{30} column, was collected for use in spectroscopy experiments.

2. Fluorescence spectral corrections

Fluorescence spectra should be corrected for the wavelength dependence of the optical components (including lenses, gratings, and photomultipliers) by using correction factors traceable to NIST or comparable illumination standards and, normally, a quartz-halogen tungsten coiled filament lamp.

3. Band-pass corrections

When emission spectra are displayed on a wavenumber ($\tilde{\nu}$) scale, it is important to recognize that, although wavelengths may be converted into wavenumbers by taking the reciprocal, $\tilde{\nu} = 1/\lambda$, the bandpass on a wavelength scale ($\Delta\lambda$) is not simply the inverse of the bandpass on a wavenumber scale ($\Delta\tilde{\nu}$) [73]. For an emission spectrum obtained with a fixed wave-length bandpass, the usual experimental method, the bandpass on a wavenumber scale decreases with wavelength. from the relationships: $d\tilde{\nu} = -d\lambda/\lambda^2$ or $|\tilde{\nu}| = |\Delta\lambda|/\lambda^2$. For an emission spectrum obtained as photons/sec nm, $I(\lambda)/\Delta\lambda$, conversion into emission intensity in photons/sec cm^{-1}, $I(\tilde{\nu})/\Delta\tilde{\nu}$, requires multiplying each value of the intensity by the square of the detection wavelength [73]. Thus, $I(\tilde{\nu}) = \lambda^2 I(\lambda)$ gives the corrected spectral response for the conversion of an emission spectrum from wavelength to wavenumber scales. This correction does not apply to absorption or fluorescence excitation spectra. In these cases the signals are related to **ratios** of intensities, and $I(\tilde{\nu})/I_0(\tilde{\nu}) = I(\lambda)/I_0(\lambda)$ where $I(\tilde{\nu})$, $I(\lambda)$, etc. represent the average of signals (e.g. photons/s) integrated over the band pass.

4. Gaussian deconvolution

Absorption and fluorescence spectra may be deconvoluted into vibronic components by fitting the lineshape to a sum of Gaussian functions. Given the fundamental importance of energy in quantum mechanics and spectroscopy, spectra are typically converted to wave-number or other energy scales before spectral fitting and analysis of vibronic spacings and intensities. 'Origin' software [74], which employs a Levenberg-Marquardt curve-fitting algorithm, is a typical package for accomplishing this. The frequencies (in cm^{-1}) of the vibronic bands are initially estimated from the locations of the peaks and shoulders in the experimental spectra, and the frequencies, widths, and amplitudes can then be allowed to vary through several iterations of the non-linear least squares program. In this manner, best-fit parameters for each of the Gaussian bands comprising the spectra are obtained. These parameters should not depend on the initial estimates of band positions and their intensities.

5. Excitation spectra

Fluorescence excitation spectra should be corrected for the intensity of light incident on the sample. Most spectrometers accomplish this by splitting the excitation beam and detecting the incident light by means of a quantum counter, for example Rhodamine 610 in ethylene glycol (0.3 g/100 mL) [75], or a calibrated photodiode. The sample emission is then divided by the incident light response, to yield a corrected excitation spectrum. The quantum counter technique is limited by the absorption cut-off (~600 nm for Rhodamine) of the fluorescent dye. The use of a calibrated photodiode can extend the correction range into the infrared.

6. Correlation of absorption and excitation spectra

Fluorescence excitation spectra can be used to identify the emitting molecular species and, by comparing the excitation and absorption spectra, to determine the efficiencies of energy transfer between molecules, but caution should be exercised in making these comparisons. Fluorescence intensity (F) as a function of the excitation wavelength (λ) is given by

$$F(\lambda) = I_0(1 - 10^{-\varepsilon bC})Q_f = I_0(1 - 10^{-A})Q_f = I_0(1 - T)Q_f$$
$$(1)$$

where I_0 is the intensity of incident light, εbC is the sample absorbance (A) at the wavelength of excitation, $T (= I/I_0)$ is the transmission of the sample, and Q_f is the quantum yield of the fluorescence. If $A = \varepsilon bC$ is small, then the series

$$10^{-A} = 1 - A(2.302) + A^2[(2.302^2)/2!] - A^3[(2.302^3)/3!] + ...$$
$$(2)$$

rapidly converges. Keeping only the first order term in A gives

$$F(\lambda) = I_0[A(2.302)]Q_f$$
$$(3)$$

For samples with low absorbance, fluorescence intensities thus will be linearly proportional to A, and the fluorescence excitation spectrum should be superimposable with the absorption spectrum [A(λ)]. At higher concentrations, the series convergence is not rapid and the fluorescence excitation spectra then should be compared with (1 - T) not Absorbance (A). For example, comparing $F = I_0[A(2.302)]Q_f$ with $F = I_0(1 - 10^{-\epsilon bC})Q_f$ at A = 0.03 yields an error of approximately 4% which increases to 12% for A = 0.1. This will result in an apparent distortion of relative intensities in the fluorescence excitation spectrum, if compared with A(λ), rather than [1-T(λ)].

7. Quantum yields

The fluorescence quantum yields of carotenoids can be measured by using standards such as Rhodamine 590 in methanol ($\phi_F = 0.95$) [76]. The evaluation of the quantum yields is based on equation 4 [77]:

$$\phi_c = \phi_r \left(\frac{1-10^{-A}r\lambda}{1-10^{-A}r\lambda} \right) \left(\frac{I_{r\lambda}}{I_{c\lambda}} \right) \left(\frac{n_c^2}{n_r^2} \right) \left(\frac{D_c}{D_r} \right) \tag{4}$$

where ϕ_c and ϕ_r are the quantum yields of the carotenoid and reference (standard) solutions, respectively. $I_{c\lambda}$ and $I_{r\lambda}$ are the relative intensities of the excitation light at wavelength λ for the carotenoid and standard solutions. $A_{c\lambda}$ and $A_{r\lambda}$ are the absorbances of the carotenoid and standard solutions at λ. n_c is the refractive index of the solvent used for the carotenoid, and n_r is the refractive index of the solvent used for the Rhodamine 590 standard. In quantum yield experiments, the solvent for the carotenoid and standard solutions should preferably be the same, so the ratio of the indices of refraction can be set to unity. The ratio of the excitation light intensities can be obtained by use of a reference photomultiplier. D_c and D_r are the integrated areas of the corrected emission spectra of the carotenoid and standard solutions and should be obtained under identical instrumental conditions of slit width, gain, etc.

8. Solvents

The absorption spectrum associated with the $S_0 \rightarrow S_2$ ($1^1A_g^- \rightarrow 1^1B_u^+$) transition of carotenoids is influenced by solvent. Dispersion interactions between the solvent environment and the large transition dipole moment shift the spectral profiles [78-81]. For apolar carotenoids in apolar solvents, the magnitude of this effect depends linearly on solvent polarizability given by $R(n) = (n^2-1)/(n^2+2)$ where n is the refractive index of the solvent [79]. The energies of the $S_1 \leftrightarrow S_0$ ($2^1A_g^- \leftrightarrow 1^1A_g^-$) transitions of carotenoids are not as strongly affected by changes in the solvent due to the much smaller dipole moment associated with this transition [82]. Solvent shifts of the $S_1 \leftrightarrow S_0$ ($2^1A_g^- \leftrightarrow 1^1A_g^-$) transition are $<ca.$ 10% of those associated with the $S_0 \leftrightarrow S_2$ ($1^1A_g^- \leftrightarrow 1^1B_u^+$) transition. An exception is peridinin (**558**), which exhibits a pronounced solvent effect on the wavelength of the $S_1 \rightarrow S_0$ ($2^1A_g^- \rightarrow 1^1A_g^-$) fluorescence and on the lifetime of its lowest excited singlet state [57]. An examination of the spectroscopic behaviour and dynamics of peridinin and other carbonyl-containing carotenoids revealed that the lifetime of the lowest excited singlet state of these molecules is strongly dependent on solvent polarity, $P(\varepsilon) = (\varepsilon-1)/(\varepsilon+2)$, where ε is the dielectric constant of the solvent, In general, carotenoids containing carbonyl groups also have complex transient absorption spectra and show a pronounced dependence of the excited singlet state lifetime on solvent environment. These effects have been related to the presence of an intramolecular charge transfer state strongly coupled to the S_1 ($2^1A_g^-$) excited singlet state.

G. Recent Developments

1. Geometrical isomerization

A recent report of spectroscopic studies on *cis* and all-*trans* isomers of the simple polyene, hexadecaheptaene (**5**), indicates a low barrier for conversion between *cis* and all-*trans* isomers in the $2^1A_g^-$ state [24]. In solutions at room temperature, the essentially non-fluorescent *trans* isomer is in equilibrium, on the $2^1A_g^-$ potential surface, with non-symmetric, fluorescent *cis* isomers. These experiments suggest: (i) that the S_1 states of longer polyenes and carotenoids have local energy minima corresponding to a range of conformations and isomers; and (ii) that these minima are connected by relatively low energy barriers. Steady-state and time-resolved optical measurements on the S_1 states thus sample a distribution of conformers and geometrical isomers, even for molecules represented by a single, dominant ground-state structure. Complex S_1 potential energy surfaces may help to explain the complicated $S_2 \rightarrow S_1$ relaxation kinetics of carotenoids and reduce the need to invoke many of the intermediate electronic states indicated in Fig. 5. These recent experiments highlight the limitations in using fluorescence techniques to study the S_1 ($2^1A_g^-$) states of longer all-*trans* polyenes and carotenoids. Previous reports of fluorescence from these symmetrical systems can probably be attributed to distorted *trans* isomers and/or *cis* impurities. Excited state absorption experiments [49,51] thus have the important advantage of exploiting an allowed electronic transition to probe the energy levels of all-*trans* samples with signals that reflect accurately the distribution of symmetrical and asymmetrical species.

Other recent work [83,84] supports the essential features of the three-level energy level diagram given in Fig. 2. At least some of the spectroscopic transients observed in pump-probe experiments on carotenoids and assigned to the states included in Fig. 5 can be attributed to two-photon processes induced by the high light intensities of these pulsed experiments. Also, recent ultrafast, time-resolved experiments on xanthophylls [85] and open-chain carotenoids [86], with increasing number of π-electron carbon-carbon double bonds, have focused on elucidating the nature of the state denoted S* implicated as an intermediate in the depopulation of S_2 ($1^1B_u^+$) and as a pathway for the formation of carotenoid triplet states in light-harvesting complexes [55]. The experimental data are supported by quantum computations which suggest that S* is simply an S_1 state with a twisted conformational structure, the yield of which increases as the π-electron conjugated chain length of the molecule increases. Thus, upon photo-excitation of the carotenoids into the S_2 ($1^1B_u^+$) state, relaxation into the S_1 ($2^1A_g^-$) state is accompanied by conformational twisting leading to branched decay pathways of the molecules that populate various conformers with different spectroscopic features. These then decay independently at different rates back to the ground state. Further work will resolve the precise structures of these intermediates in the excited-state decay pathways, thereby increasing our understanding of their role in controlling the important photochemical processes that carotenoids undergo in biological systems.

References

[1]   O. Isler (ed.), *Carotenoids*, Birkhauser, Basel (1971).
[2]   A. Young and G. Britton (eds), *Carotenoids in Photosynthesis*, Kluwer Academic, London (1993).
[3]   H. A. Frank and R. J. Cogdell, *Photochem. Photobiol.*, **63**, 257 (1996).
[4]   H. A. Frank, A. J. Young, G. Britton and R. J. Cogdell (eds), *The Photochemistry of Carotenoids*, Kluwer Academic Publishers, Dordrecht (1999).
[5]   N. J. Turro, *Modern Molecular Photochemistry*, University Science Books, California (1991).
[6]   T. Förster, *Ann. Phys.*, **2**, 55 (1968).
[7]   D. L. Dexter, *J. Chem. Phys.*, **21**, 836 (1953).
[8]   R. L. Christensen, E. A. Barney, R. D. Broene, M. G. I. Galinato and H. A. Frank, *Arch. Biochem. Biophys.*, **430**, 30 (2004).
[9]   T. Polívka and V. Sundström, *Chem. Rev.*, **104**, 2021 (2004).
[10]   R. R. Birge, *Acc. Chem. Res.*, **19**, 138 (1986).
[11]   R. Pariser, *J. Chem. Phys.*, **24**, 250 (1955).
[12]   P. R. Callis, T. W. Scott and A. C. Albrecht, *J. Chem. Phys.*, **78**, 16 (1983).
[13]   R. L. Christensen, in *The Photochemistry of Carotenoids* (ed. H. A. Frank, A. J. Young, G. Britton and R. J. Cogdell), p. 137, Kluwer Academic Publishers, Dordrecht (1999).
[14]   R. L. Christensen, A. Faksh, J. A. Meyers, I. D. W. Samuel, P. Wood, R. R. Schrock and K. C. Hultzsch, *J. Phys. Chem. A*, **108**, 8229 (2004).
[15]   P. Tavan and K. Schulten, *J. Chem. Phys.*, **85**, 6602 (1986).
[16]   Y. Koyama, F. S. Rondonuwu, R. Fujii, and Y. Watanabe, *Biopolymers*, **74**, 2 (2004).
[17]   P. Tavan, and K. Schulten, *J. Chem. Phys.*, 70, 5407 (1979).
[18]   L. Serrano-Andrés, R. Lindh, B. O. Roos, and M. Merchán, *J. Phys. Chem.*, **97**, 9360 (1993).
[19]   L. Serrano-Andrés, M. Merchán, I. Nebot-Gil, R. Lindh and B. O. Roos, *J. Chem. Phys.*, **98**, 3151 (1993).
[20]   K. Nakayama, H. Nakano and K. Hirao, *Int. J. Quantum Chem.*, **66**, 157 (1998).
[21]   C.-P. H. Hsu, S. Hirata and M. Head-Gordon, *J. Phys. Chem. A*, **105**, 451 (2001).
[22]   J. H. Starcke, M. Wormit, J. Schirmer and A. Dreuw, *Chem. Phys. Lett.*, **329**, 39 (2006).

[23]   J. H. Simpson, L. McLaughlin, D. S. Smith and R. L. Christensen, *J. Chem. Phys.,* **87**, 3360 (1987).
[24]   R. L. Christensen, M. G. I. Galinato, E. F. Chu, R. Fujii, H. Hashimoto and H. A. Frank, *J. Am. Chem. Soc.,* **129**, 1769 (2007).
[25]   R. L. Christensen and B. E. Kohler, *J. Chem. Phys.,* **63**, 1837 (1975).
[26]   B. E. Kohler, C. Spangler and C. Westerfield, *J. Chem. Phys.,* **89**, 5422 (1988).
[27]   R. Fujii, K. Onaka, M. Kuki, Y. Koyama and Y. Watanabe, *Chem. Phys. Lett.,* **288**, 847 (1998).
[28]   Y. Koyama and R. Fujii, in *The Photochemistry of Carotenoids* (ed. H. A. Frank, A. J. Young, G. Britton and R. J. Cogdell), p. 161, Kluwer Academic Publishers, Dordrecht (1999).
[29]   H. A. Frank, J. A. Bautista, J. S. Josue and A. J. Young, *Biochemistry,* **39**, 2831 (2000).
[30]   R. L. Christensen and B. E. Kohler, *Photochem. Photobiol.,* **18**, 293 (1973).
[31]   R. Hemley and B. Kohler, *Biophys. J.,* **20**, 377 (1977).
[32]   R. M. Gavin, C. Weisman, J. K. McVey and S. A. Rice, *J. Chem. Phys.,* **68**, 522 (1978).
[33]   S. L. Bondarev and V. N. Knyukshto, *Chem. Phys. Lett.,* **225**, 346 (1994).
[34]   P. O. Andersson and T. Gillbro, *J. Chem. Phys.,* **103**, 2509 (1995).
[35]   H. Petek, A. J. Bell, R. L. Christensen and K. Yoshihara, *J. Chem. Phys.,* **96**, 2412 (1992).
[36]   S. L. Bondarev, S. M. Bachilo, S. S. Dvornikov and S. A. Tikhomirov, *J. Photochem. Photobiol. A: Chemistry,* **46**, 315 (1989).
[37]   S. L. Bondarev, S. S. Dvornikov and S. M. Bachilo, *Opt. Spectrosc. (USSR),* **64**, 268 (1988).
[38]   T. Gillbro and R. J. Cogdell, *Chem. Phys. Lett.,* **158**, 312 (1989).
[39]   S. A. Cosgrove, M. A. Guite, T. B. Burnell and R. L. Christensen, *J. Phys. Chem.,* **94**, 8118 (1990).
[40]   H. A. Frank, R. Z. B. Desamero, V. Chynwat, R. Gebhard, I. van der Hoef, F. J. Jansen, J. Lugtenburg, D. Gosztola and M. R. Wasielewski, *J. Phys. Chem. A,* **101**, 149 (1997).
[41]   K. Schulten, I. Ohmine and M. Karplus, *J. Chem. Phys.,* **64**, 4422 (1976).
[42]   R. A. Auerbach, R. L. Christensen, M. F. Granville and B. E. Kohler, *J. Chem. Phys.,* **74**, 4 (1981).
[43]   J. R. Andrews and B. S. Hudson, *J. Chem. Phys.,* **68**, 4587 (1978).
[44]   H. Petek, A. J. Bell, Y. S. Choi, K. Yoshihara, B. A. Tounge and R. L. Christensen, *J. Chem. Phys.,* **102**, 4726 (1995).
[45]   G. Orlandi, F. Zerbetto and M. Z. Zgierski, *Chem. Rev.,* **91**, 867 (1991).
[46]   G. J. M. Dormans, G. C. Groenenboom and H. M. Buck, *J. Chem. Phys.,* **86**, 4895 (1987).
[47]   F. Zerbetto and M. Z. Zgierski, *J. Chem. Phys.,* **93**, 1235 (1990).
[48]   H. A. Frank, J. S. Josue, J. A. Bautista, I. van der Hoef, F. J. Jansen, J. Lugtenburg, G. Wiederrecht and R. L. Christensen, *J. Phys. Chem. B,* **106**, 2083 (2002).
[49]   T. Polívka, J. L. Herek, D. Zigmantas, H. E. Akerlund and V. Sundström, *Proc. Natl. Acad. Sci. USA,* **96**, 4914 (1999).
[50]   T. Polívka, D. Zigmantas, H. A. Frank, J. A. Bautista, J. L. Herek, Y. Koyama, R. Fujii and V. Sundström, *J. Phys. Chem. B,* **105**, 1072 (2001).
[51]   T. Polívka, D. Zigmantas, J. L. Herek, J. A. Bautista, H. A. Frank and V. Sundström, *Springer Ser. Chem. Phys.,* **66**, 668 (2001).
[52]   P. Tavan and K. Schulten, *Phys. Rev. B: Condens. Matter,* **36**, 4337 (1987).
[53]   A. Dreuw, *J. Phys. Chem. A,* **110**, 4592 (2006).
[54]   G. Cerullo, D. Polli, G. Lanzani, S. De Silvestri, H. Hashimoto and R. J. Cogdell, *Science,* **298**, 2395 (2002).
[55]   C. C. Gradinaru, J. T. M. Kennis, E. Papagiannakis, I. H. M. van Stokkum, R. J. Cogdell, G. R. Fleming, R. A. Niederman and R. van Grondelle, *Proc. Natl. Acad. Sci. USA,* **98**, 2364 (2001).
[56]   D. S. Larsen, E. Papagiannakis, I. H. M. van Stokkum, M. Vengris, J. T. M. Kennis and R. van Grondelle, *Chem. Phys. Lett.,* **381**, 733 (2003).
[57]   J. A. Bautista, R. E. Connors, B. B. Raju, R. G. Hiller, F. P. Sharples, D. Gosztola, M. R. Wasielewski and H. A. Frank, *J. Phys. Chem. B,* **103**, 8751 (1999).
[58]   H. A. Frank, J. A. Bautista, J. Josue, Z. Pendon, R. G. Hiller, F. P. Sharples, D. Gosztola and M. R. Wasielewski, *J. Phys. Chem. B,* **104**, 4569 (2000).
[59]   M. F. Granville, G. R. Holtom, B. E. Kohler, R. L. Christensen and K. L. D'Amico, *J. Chem. Phys.,* **70**, 593 (1979).
[60]   A. P. Shreve, J. K. Trautman, T. G. Owens and A. C. Albrecht, *Chem. Phys. Lett.,* **170**, 51 (1990).
[61]   B. P. Krueger, J. Yom, P. J. Walla and G. R. Fleming, *Chem. Phys. Lett.,* **310**, 57 (1999).
[62]   P. J. Walla, J. Yom, B. P. Krueger and G. R. Fleming, *J. Phys. Chem. B,* **104**, 4799 (2000).
[63]   P. J. Walla, P. A. Linden, C.-P. Hsu, G. D. Scholes and G. R. Fleming, *Proc. Natl. Acad. Sci. USA,* **97**, 10808 (2000).

[64]   S. Shima, R. P. Ilagan, N. Gillespie, B. J. Sommer, R. G. Hiller, F. P. Sharples, H. A. Frank and R. R.
        Birge, *J. Phys. Chem. A,* **107,** 8052 (2003).
[65]   R. P. Ilagan, S. Shima, A. Melkozernov, S. Lin, R. E. Blankenship, F. P. Sharples, R. G. Hiller, R. R.
        Birge and H. A. Frank, *Biochemistry,* **43,** 1478 (2004).
[66]   M. Hilbert, A. Wehling, E. Schlodder and P. J. Walla, *J. Phys. Chem. B,* **108,** 13022 (2004).
[67]   A. Wehling and P. J. Walla, *J. Phys. Chem. B,* **109,** 24510 (2005).
[68]   B. Robert, in *The Photochemistry of Carotenoids* (ed. H. A. Frank, A. J. Young, G. Britton and R. J.
        Cogdell), p. 189, Kluwer Academic Publishers, Dordrecht (1999).
[69]   K. Gaier, A. Angerhofer and H. C. Wolf, *Chem. Phys. Lett.,* **187,** 103 (1991).
[70]   T. Sashima, M. Shiba, H. Hashimoto, H. Nagae and Y. Koyama, *Chem. Phys. Lett.,* **290,** 36 (1998).
[71]   T. Sashima, H. Nagae, M. Kuki and Y. Koyama, *Chem. Phys. Lett.,* **299,** 187 (1999).
[72]   T. Sashima, Y. Koyama, T. Yamada and H. Hashimoto, *J. Phys. Chem. B,* **104,** 5011 (2000).
[73]   J. R. Lakowicz, *Principles of Fluorescence Spectroscopy, 2nd ed.,* Kluwer Academic, Plenum
        Publishers, New York (1999).
[74]   Microcal Software, Inc., Northampton, MA, (1999).
[75]   W. H. Melhuish, *J. Opt. Soc. Am.,* **52,** 1256 (1962).
[76]   D. B. Benfey, D. C. Brown, S. J. Davis, L. G. Piper and F. R. Fouter, *Appl. Opt.,* **31,** 7034 (1992).
[77]   J. N. Demas, and G. A. Crosby, *J. Phys. Chem.,* **75,** 991 (1971).
[78]   A. L. LeRosen and C. E. Reid, *J. Chem. Phys.,* **20,** 233 (1952).
[79]   P. O. Andersson, T. Gillbro, L. Ferguson and R. J. Cogdell, *Photochem. Photobiol.,* **54,** 353 (1991).
[80]   H. Nagae, M. Kuki, R. J. Cogdell and Y. Koyama, *J. Chem. Phys.,* **101,** 67 (1994).
[81]   S. Basu, *Adv. Quantum Chem.,* **1,** 145 (1964).
[82]   B. S. Hudson, B. E. Kohler, and K. Schulten, in *Excited States* (ed. E. D. Lim), p. 1, Academic Press,
        New York (1982).
[83]   P. Kukura, D. W. McCamant and R. A. Mathies, *J. Phys. Chem. A,* **108** (2004).
[84]   D. Kosumi, M. Komukai, H. Hashimoto and M. Yoshizawa, *Phys. Rev. Lett.,* **95,** 213601 (2005).
[85]   D. M. Niedzwiedzki, J. O. Sullivan, T. Polivka, R. R. Birge and H. A. Frank, *J. Phys. Chem. B,* **110,**
        22872 (2006).
[86]   D. Niedzwiedzki, J. F. Koscielecki, H. Cong, J. O. Sullivan, G. N. Gibson, R. R. Birge and H. A.
        Frank, *J. Phys. Chem. B,* **111,** 5984 (2007).

Carotenoids
Volume 4: Natural Functions
© 2008 Birkhäuser Verlag Basel

Chapter 10

Functions of Intact Carotenoids

George Britton

A. Functions, Actions and Associations

The traditional view that carotenoids are a class of plant pigments does not do justice to their versatility. This versatility will become clear from the overview of the biological roles of carotenoids, in animals and microorganisms as well as in plants, that is given in this Chapter. It has become customary and convenient to differentiate biological effects of carotenoids into functions, actions and associations [1]. 'Functions' have been defined as effects or properties that are essential for the normal well-being of the organism. Biological responses that follow the administration of carotenoids in the diet or as supplements are considered as 'actions'. When an effect is seen but a causal relationship to the carotenoid has not been demonstrated, this is described as an 'association'. The line between these is often not clear, however.

The most extensively studied functions of carotenoids, namely in photosynthesis, colouration and signalling, are discussed in detail in *Chapters 11-14*. Other functions, which are not described elsewhere, are treated in this Chapter. Proven and proposed actions of carotenoids in human health, including an evaluation of antioxidant activity, are described in *Volume 5*.

B. Overview: Diversity of Functions

1. Functions involving interactions with light

An enormous amount of energy from the sun reaches the surface of the Earth. This light is the primary source of energy for life on our planet but this energy needs to be harvested and used to drive biochemical processes. Plants, algae and some bacteria do this by photo-synthesis and they, in turn, serve as an energy source ('food') for other organisms. This light energy can be very dangerous to life, however, and can cause damage and destruction to cells and tissues unless the organism has some means of protecting against this. Light also forms the basis of systems of communication between organisms, provided they have suitable sensitive photoreceptors capable of using and detecting these signals. Carotenoids play essential roles in all these diverse areas of interactions between life and light [2,3].

The characteristic feature of carotenoid structures is the long conjugated double-bond system which absorbs light in the high-energy part of the visible region of the electromagnetic spectrum (400-500 nm). The conjugated polyene system of the carotenoids gives them the special photochemical properties that form the basis of the various biological interactions with light. A good understanding of the principles of the photochemistry and photophysics of carotenoids is an essential foundation for understanding their various photofunctions. This topic is presented in *Chapter 9*.

It is likely that the first functional role of carotenoids was their contribution to the harvesting of light energy by absorbing light and passing excitation energy on to chlorophylls, thereby extending the wavelength range of light that could be harvested, and improving the efficiency of photosynthesis.

The conjugated double-bond system also proved to have other properties that living organisms could use to good effect in other ways. With the development of oxygen-evolving photosynthesis, the environment became aerobic, and living organisms were faced with a new and critical challenge, to combat the lethal combination of light and oxygen. The photosynthetic apparatus is a perfect system for generating the very reactive and damaging species singlet oxygen, $^1O_2$. A battery of enzymic and other defences against photooxidation is now available, but carotenoids were already in place and provided highly effective protection against the formation and effects of $^1O_2$. Without carotenoids, life may not have survived on Earth in its aerobic atmosphere. This and other kinds of protection by carotenoids now extend way beyond photosynthetic systems.

Living organisms move. There is an advantage of being able to move towards or away from light, and of adopting the most favourable orientation with respect to the incident light. Light can be a signal, but the organism needs a sensitive photoreceptor to detect the light and initiate the response. Some very sophisticated mechanisms have now been developed, notably vision. Photoreceptors generally are sensitive to a particular narrow range of wavelengths. When several photoreceptors are employed, each sensitive to a different wavelength range, this allows wavelength discrimination, leading to sophisticated colour vision. Colours can be distinguished, so colour becomes an extremely important sensory feature and signal. Carotenoids are among the natural substances ('pigments', 'biochromes') that absorb light in some part of the visible spectrum, and can thus be perceived as coloured. This is another highly significant natural function in relation to recognition of species and individuals.

2. Functions not involving interactions with light

The conjugated double-bond system of carotenoids has other properties that lead to biological effects that do not necessarily involve any interaction with light, so extending further the functional versatility of the carotenoids [4]. The lipophilic nature of the carotenoids and the rigid rod-like structure that allows them to fit perfectly into photosynthetic pigment-protein complexes also allow them to fit into or span across the lipid bilayer of membranes where they can influence aspects of the structure and properties of the membrane.

The electron-rich conjugated double-bond system is highly reactive towards oxidizing agents and oxidizing free radicals. This gives carotenoids the ability to protect other sensitive systems against oxidative damage, and forms the basis of their proposed antioxidant action. Products of the enzymic or non-enzymic oxidative breakdown of carotenoids, have many important biological actions. Vitamin A, retinol (*1*) is a notable example.

Carotenoids occur naturally in food, especially fruit and vegetables. They are thus normal components of the diet for animals, including humans. They are absorbed, transported and deposited in tissues, where several actions have been reported, and associations have been identified between the presence of carotenoids and health.

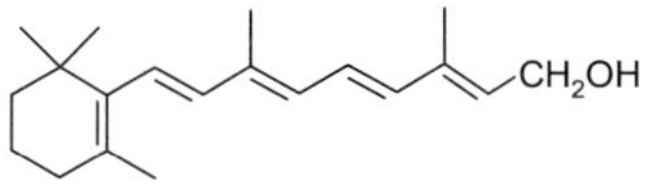

(*1*) retinol 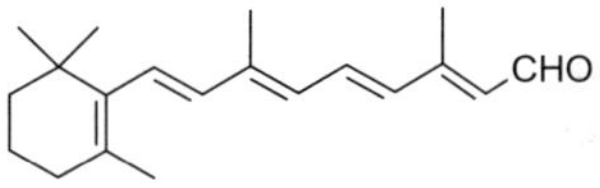   (*2*) retinal

C. Light harvesting

1. Photosynthesis

a) Carotenoids and light harvesting

Carotenoids are essential components of the photosynthetic apparatus of plants, algae and phototrophic bacteria, and the functioning of carotenoids in photosynthesis is of paramount importance. The carotenoids are located, in association with (bacterio)chlorophylls, in specific pigment-protein complexes. Some carotenoid, in plants mainly β-carotene (**3**), is present in the reaction centre complexes but the bulk, especially xanthophylls, is in the light-harvesting antenna complexes. With very few exceptions, the chloroplasts of leaves and all green tissues of higher plants contain the same collection of main carotenoids in approximately the same proportion [5], namely β-carotene, 20-25%, lutein (**133**), 40-45%, violaxanthin (**259**), 10-15%, and neoxanthin (**234**), 10-15%. Darker green leaves contain more chloroplasts, and consequently more carotenoid. The lipophilic nature of the carotenoids facilitates the necessary close interactions between them and chlorophylls which are the primary light-harvesting pigments in photosynthesis. The main light-harvesting pigment is chlorophyll, but carotenoids serve as accessory light-harvesting pigments. Light energy absorbed by carotenoids is passed on from the excited singlet state of the carotenoid to chlorophyll by singlet-singlet energy transfer. Light of wavelengths absorbed by carotenoids but not by chlorophyll can thus be used as an energy source for photosynthesis.

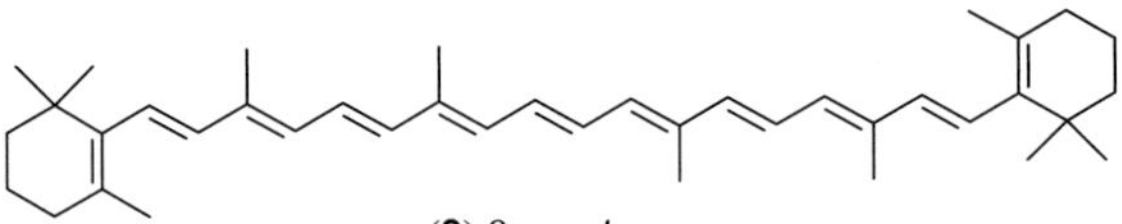

(**3**) β-carotene

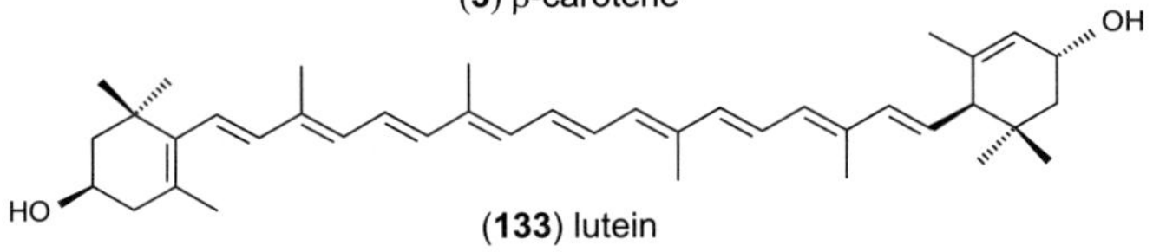

(**133**) lutein

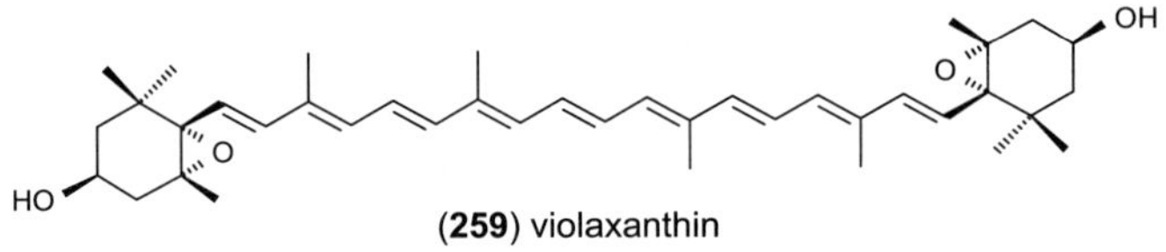

(**259**) violaxanthin

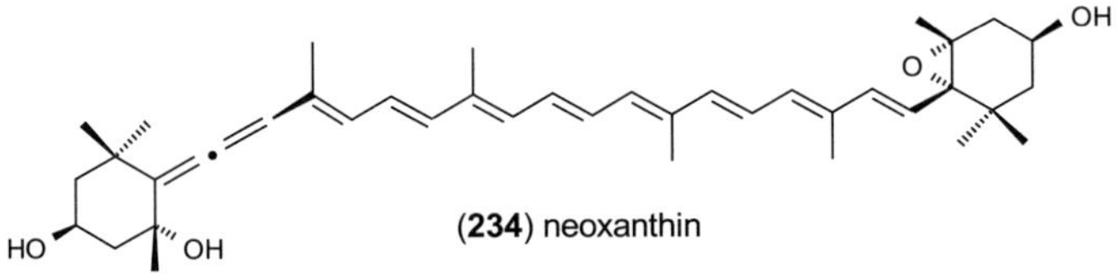

(**234**) neoxanthin

Algae and phototrophic bacteria contain different collections of carotenoids, but these play similar roles in photosynthesis. In many cases, the contribution of carotenoids to light harvesting is greater than with higher plants because less light reaches the environment in which the organisms live. Specialized carotenoid-rich pigment-protein complexes, *e.g.* the fucoxanthin (**369**)-chlorophyll-proteins of diatoms, have developed to allow some algae to make best use of the available light in the under-sea environment [6]. The concentration of the carotenoid may be higher than that of chlorophyll. Comprehensive information about the structure of these carotenoid-containing pigment-protein complexes and the significance of carotenoids in light harvesting is given in *Chapter 14*.

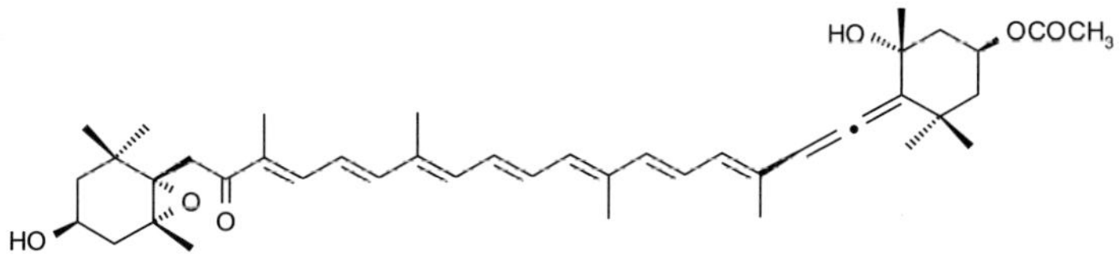

(**369**) fucoxanthin

b) Protection against excess light energy

Light harvesting is not the only role of carotenoids in photosynthesis. Fundamental to the survival of the plant and of life itself, are the actions of carotenoids in protecting against damaging and potentially lethal effects of excess light energy. Light harvesting in the antenna complexes is very efficient, even at relatively low light intensity. At high intensity, energy is absorbed at a greater rate than it can be used in photosynthesis. This can lead to loss of photosynthetic efficiency (photoinhibition) and to the formation of the destructive species singlet oxygen ($^1O_2$). Carotenoids play vital roles in quenching the excess energy and in preventing the formation and reactions of $^1O_2$, by a triplet-state energy quenching mechanism. Again this is discussed in detail in *Chapter 14*. Some of the basic features of mechanisms of photoprotection in this and other systems are given in Section **D**.1.

2. Other light-harvesting roles

Some members of the Archaea, notably *Halobacterium* species, can live by harvesting light energy and using this to generate biochemical energy as ATP *via* a proton-pumping mechanism [7]. The light-harvesting pigment, bacteriorhodopsin, has as its chromophore the carotenoid metabolite retinal (*2*). An outline of the structure and functioning of bacterio-rhodopsin is given in *Chapter 15*.

Recently, the involvement of an intact carotenoid in a similar system has been demonstrated. The eubacterium *Salinibacter ruber* employs a light-driven proton-pumping mechanism to generate energy [8]. The light-harvesting pigment, xanthorhodopsin, is again a retinal-opsin protein similar to bacteriorhodopsin, but the carotenoid salinixanthin (*3*) [9] is also present in a 1:1 ratio with retinal, and acts as a light-harvesting antenna, passing energy on to retinal with a quantum efficiency of about 40%.

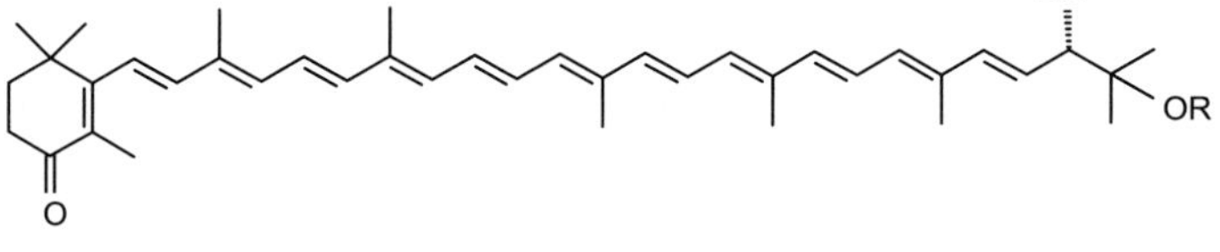

(*3*) salinixanthin [R = 6-O-(13-methyltetradecanoyl)-β-D-glucopyranosyl]

D. Protection against Oxidative Damage

1. Photoprotection

High on the list of functions attributed to carotenoids is 'photoprotection', *i.e.* protection against damage mediated by light. Its fundamental importance to life has already been mentioned (Section C.1). Because harvesting light energy in photosynthesis is essential to support life on Earth, light is usually thought of as beneficial. Light energy, however, especially in combination with oxygen, can be very harmful, causing damage to cells and tissues *via* singlet oxygen ($^1O_2$) and oxidizing free radicals. Living organisms have evolved defences to prevent or minimize this damage. Carotenoids are a major part of this defence and can be effective in a number of ways. They can act as a filter, preventing harmful wavelengths of light from penetrating into susceptible tissues. The 'secondary carotenoids' that can accumulate in very high concentrations in some green algae under stress conditions [5], including high intensity solar radiation, are very effective filters of harmful radiation. Commercial exploitation of this for the production of β-carotene (**3**) by *Dunaliella salina* and astaxanthin (**406**) by *Haematococcus pluvialis* is described in *Volume 5, Chapter 5*.

2. Singlet oxygen quenching

a) Photosensitized by (bacterio)chlorophyll

Carotenoids can act as quenchers to prevent the formation and damaging affects of $^1O_2$ (Fig. 1). Cells of carotenoidless mutants of photosynthetic organisms, *e.g. Rhodobacter sphaeroides* R26, are killed when illuminated in the presence of oxygen [10]. Bacteriochlorophyll acts a a sensitizer (SENS), absorbs light, and generates the singlet excited state ^{1}SENS* which may undergo conversion into the longer lived, lower-energy triplet state ^{3}SENS* (Eqn. 1). This in turn can undergo energy transfer to oxygen to form the highly reactive singlet oxygen $^1O_2$ (Eqn. 2). If unchecked, singlet oxygen can cause extensive damage to cellular components, either directly or *via* oxidizing free radicals. The triplet energy level of carotenoids is comparatively low (*Chapter 9*), so that carotenoids with more than seven conjugated double bonds can quench, *i.e.* accept triplet energy from ^{3}SENS* (Eqn. 3), or from $^1O_2$ (Eqn. 4), thereby preventing or minimizing damage [11]. The triplet state carotenoid (^{3}CAR*) is of such low energy that it cannot generate $^1O_2$ and dissipates its excitation energy harmlessly to its surroundings (Eqn. 5). Protection against the production and damaging effects of $^1O_2$ in the chloroplasts of higher plants, mentioned in Section C.1, occurs by a similar mechanism in which chlorophyll acts as the sensitizer. Some herbicides act by inhibiting carotenoid biosynthesis so that the plants do not have this protection and suffer photodynamic killing [12]. This role of carotenoids in photosynthesis is described in *Chapter 14*.

$$SENS \xrightarrow{h\nu} {}^1SENS^* \longrightarrow {}^3SENS^* \qquad (1)$$

$${}^3SENS^* + {}^3O_2 \longrightarrow SENS + {}^1O_2^* \qquad (2)$$

$${}^3SENS^* + CAR \longrightarrow SENS + {}^3CAR^* \qquad (3)$$

$${}^1O_2^* + CAR \longrightarrow {}^3O_2 + {}^3CAR^* \qquad (4)$$

$${}^3CAR^* \longrightarrow CAR + (heat) \qquad (5)$$

Fig. 1. Equations showing the formation of triplet state sensitizer ${}^3SENS^*$, the formation of singlet oxygen ${}^1O_2^*$ and the quenching of ${}^3SENS^*$ and ${}^1O_2^*$ by carotenoid CAR. The same mechanism applies to the examples described in the text (Sections 2 a-c) when the sensitizer SENS is (bacterio)chlorophyll, methylene blue, or protoporphyrin IX, respectively.

b) Exogenous photosensitizers

Similar photoprotection is also important in non-photosynthetic organisms if an endogenous or exogenous sensitizer is present. In non-photosynthetic bacteria, exogenous photo-sensitizers such as methylene blue (SENS in Fig. 1) can form triplet states and sensitize the formation of ${}^1O_2$. Endogenous carotenoids protect against this but, in carotenoidless mutants of normally carotenogenic bacteria, the cells are killed by light and oxygen [13]. In many species of bacteria and moulds, light induces the biosynthesis of carotenoids to protect the cells against photodynamic damage (*Volume 3, Chapter 2*).

c) Human erythropoietic protoporphyria

In humans, use is made of the ability of β-carotene (**3**) to quench the photochemical formation of ${}^1O_2$ in patients suffering from the photosensitivity condition erythropoietic protoporphyria, in which defective haem metabolism leads to the accumulation of free protoporphyrin IX (SENS in Fig. 1). This is an effective photosensitizer of ${}^1O_2$ formation, leading to damage to skin and tissues. β-Carotene, given as a high-dose supplement, is deposited in the skin and acts to quench the photosensitization, thereby preventing ${}^1O_2$ formation, preventing damage and alleviating the symptoms [14]. Carotenoids may also help to reduce discomfort and cell damage due the inflammation (erythema) of sunburn, in which ${}^1O_2$ may be involved. Protective roles of carotenoids in skin are discussed in *Volume 5, Chapter 16*.

d) Singlet oxygen produced non-photochemically

Singlet oxygen can be generated in ways other than photosensitization. In the body's defences against infection, neutrophils in the blood attack bacteria by an 'oxidative burst' in which ${}^1O_2$ is generated non-photochemically. Recent work has shown that a carotenoidless mutant of the human pathogen *Staphylococcus aureus* is much more susceptible to oxidative killing and, in particular, that its survival rate against attack by neutrophils is greatly reduced [15], compared with the wild-type, which contains the C_{30} carotenoid staphyloxanthin (*4*) [16]. The virulence of the wild-type *S. aureus* is attributed, at least in part, to the ability of staphyloxanthin to quench $^1O_2$ and possibly also protect against other oxidizing free radicals.

(*4*) staphyloxanthin [R = 6-O-(12-methyltetradecanoyl)-glucosyl]

3. Protection against oxidizing free radicals (antioxidant action)

The electron-rich conjugated polyene system renders carotenoids very susceptible to oxidation and to free-radical reactions. Oxidizing conditions and oxidative free radicals can be generated in many ways in cells and tissues, and living organisms need efficient defences to protect against this. In the presence of oxidizing free radicals, unsaturated lipids are destroyed by chain reactions involving peroxy radicals. A scheme of the reactions involved is given in *Chapter 7, Fig. 2*. It was reported in 1984 that, in solution, β-carotene can react with peroxy radicals to form a resonance-stabilized carbon-centred radical, and that, at relatively low oxygen concentration, this process would consume peroxy radicals so that the carotenoid could act as a chain-breaking antioxidant [17]. Since then, the possibility that carotenoids could be protective antioxidants in biological systems has been a major area of interest in carotenoid research. Many experiments have confirmed that, under defined conditions in model systems, carotenoids can be effective antioxidants. It is also clear that, under different conditions, especially in the presence of relatively high oxygen concentrations, carotenoids can have a pro-oxidant action.

A wide-ranging quantitative assessment of antioxidant properties of natural colourants and phytochemicals included natural carotenoids and synthetic carotenoid analogues and derivatives [18]. The factors assayed were singlet oxygen quenching, peroxide formation and oxygen pressure dependence. The efficient $^1O_2$ quenching by carotenoids was confirmed and some synthetic models and analogues were prepared that proved to be better quenchers than the natural carotenoids. Astaxanthin (**406**) and actinioerythrol (**552**) were classified as excellent antioxidants that quench very efficiently both excited triplet states, thereby preventing $^1O_2$ formation, and ground-state radicals. β-Carotene (**3**) and lycopene (**31**) were good antioxidants that could strongly inhibit peroxide formation but suffered considerable degradation, whereas shorter chromophore carotenoids, *e.g.* ζ-carotene (**38**) had only moderate effects.

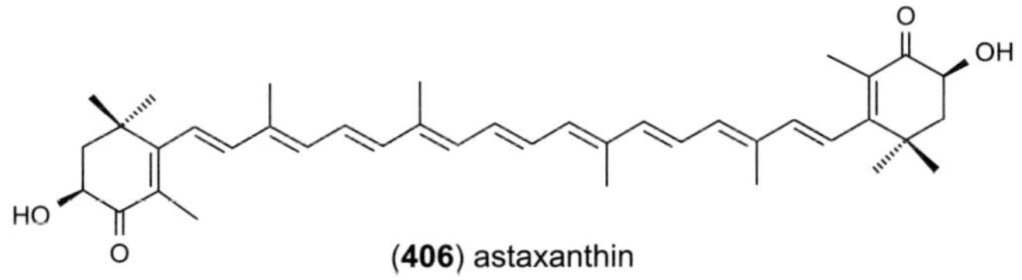

(406) astaxanthin

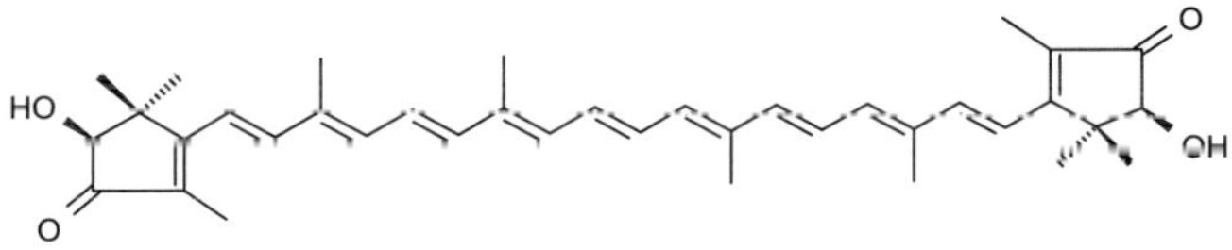

(552) actinioerythrol

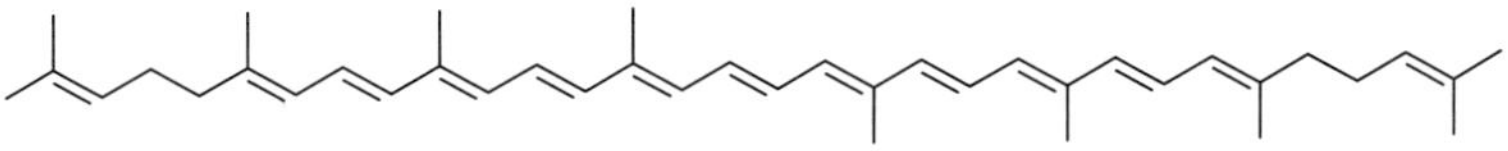

(31) lycopene

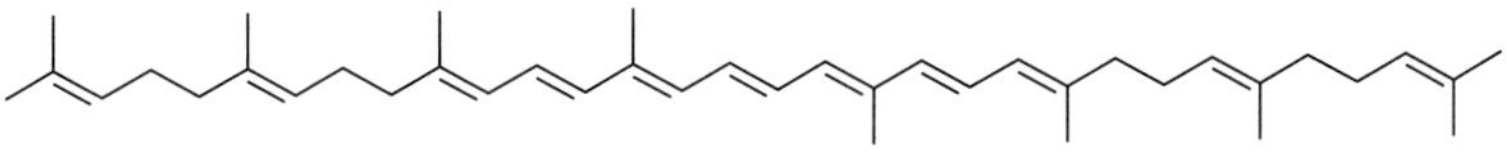

(38) ζ-carotene

The ability of carotenoids to be effective antioxidants in solution is beyond dispute. It would be surprising if living organisms did not make full use of this ability. It is difficult, however, to find controlled experiments that demonstrate unequivocally a radical quenching antioxidant action of carotenoids *in vivo*. Even so, the idea that carotenoids may be protective antioxidants in all kinds of organisms and tissues has been widely accepted and they are often included, with vitamin E (tocopherol) and others, in the list of lipophilic natural antioxidants. Carotenoid peroxides can be quenched and the carotenoid restored by anti-oxidant vitamins, vitamin C in the aqueous and vitamin E in the lipid phase [19]. Protection of carotenoids against oxidation may be a key requirement for carotenoids to be effective in any biological action.

Cautious critical evaluation of the evidence in each case is recommended. The basic chemistry of carotenoid radicals and the free radical reactions on which the antioxidant behaviour of carotenoids is based is given in *Chapter 7*. The topic is discussed in detail, and the significance of carotenoids as antioxidants (and pro-oxidants) in relation to human health and disease is assessed in *Volume 5, Chapter 12*.

E. Photoreceptors

1. Blue-light receptors

Light is a signal that initiates various kinds of responses in living organisms. This requires a sensitive photoreceptor to detect the light signal and a biochemical mechanism to convert the signal into a response. Different photoreceptors are activated by different light wavelengths. A variety of responses are known that are activated by blue light in the wavelength range 400-500 nm, the region where carotenoids absorb maximally [20]. Action spectra determined for some of these responses are reminiscent of the absorption spectrum of β-carotene (**3**), but riboflavin, free or in flavoproteins, has a very similar spectrum. There has been a long debate about whether β-carotene could be the blue-light receptor, but it now seems that the usual biological blue-light receptor is a flavoprotein. [*N.B.* The blue light receptor is often called 'cryptochrome' but should not be confused with the carotenoid cryptochrome, the 5,8:5',8'-diepoxide of β-cryptoxanthin].

2. Phototropism and phototaxis

a) Phototropism

Amongst the blue-light-triggered responses are the similar, in principle, phenomena photo-tropism and phototaxis. Phototropism is the light-induced growth or curvature of plants and fungi towards (positive) or away from (negative) the incident light. The phototropic bending of oat (*Avena sativa*) coleoptiles has been studied for more than a hundred years. Another well known example is the curvature of hyphae of the mould *Phycomyces blakesleeanus* towards light. In spite of arguments in support of β-carotene, it now seems clear that the photoreceptor is a flavin or flavoprotein [21]. Even so, in *Avena*, carotenoids may have a role as screening pigments, increasing irradiance differences and enhancing sensitivity.

b) Phototaxis

Phototaxis describes the direct motion or change of motion of an organism (usually unicellular), or of cells or organelles within an organism, in response to light. Again it may be positive or negative. There is more evidence for a role of carotenoids in phototaxis in unicellular algae, though not as the photoreceptor pigments. Phototaxis of *Euglena gracilis* was recognized over a hundred years ago. In constant uniform illumination there is no photoresponse, but photostimulation occurs when the light intensity varies above or below a certain threshold level. The primary photoreceptor almost certainly consists of an array of flavin molecules in the paraflagellar body at the base of the flagellum that drives the motion.

Close to the paraflagellar body is another area, the eyespot or stigma, which has a high concentration of carotenoid, mainly ketocarotenoids, *e.g.* echinenone (**283**) [22]. It is believed that the carotenoids in the stigma screen the primary photoreceptor, and that the *Euglena* cell responds to light by moving so that the screening is minimized [23]. Other algae and diatoms may have similar mechanisms, but use different photoreceptors. The green alga *Chlamydomonas reinhardtii*, for example, uses a rhodopsin-type photoreceptor which forms a patch in the plasma membrane over a carotenoid-rich eyespot in the adjoining chloroplast membrane.

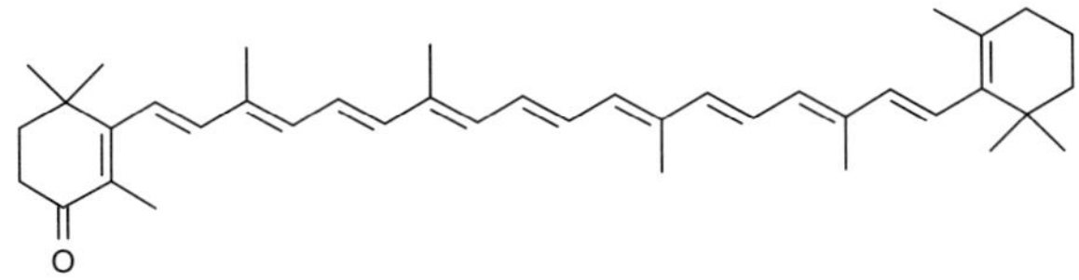

(**283**) echinenone

There is an example of a phototactic response in higher plants that is mediated by carotenoids. Stomatal response to blue light has been implicated in modulating the opening of stomata at dawn, when the incident light has a high blue component. Zeaxanthin (**119**) in guard cell chloroplasts has been identified as the blue-light photoreceptor, and also appears to mediate interactions between light and CO_2 in guard cells [25].

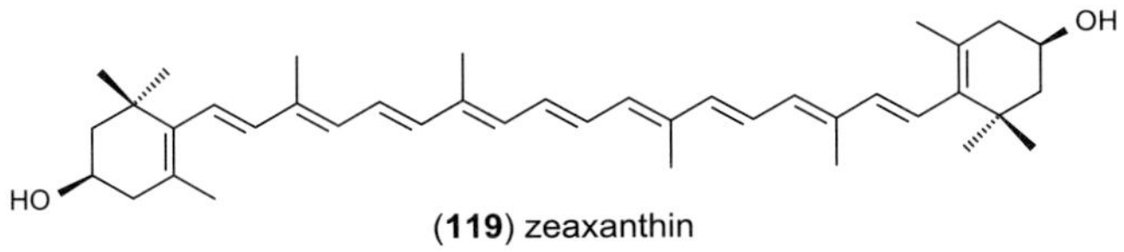

(**119**) zeaxanthin

F. Vision

1. Visual pigments

The ability to see requires photoreceptors. In the eyes of all animals, the photoreceptors consist of the carotenoid metabolite retinal (*2*), or a closely related derivative, complexed with an opsin protein. The structure and functioning of the most extensively studied of the visual pigments, mammalian rhodopsin, are described in *Chapter 15*. In addition to this, however, intact carotenoids are involved in some important ways in the visual process.

2. Macular pigments

The fovea, the most sensitive region of the retina of the eye of humans and other primates, contains a small yellow area, the macula lutea, which contains a very high concentration of the xanthophylls zeaxanthin, mainly as the *meso* (3*R*,3'*S*) isomer (**120**) and lutein (**133**) [26]. The carotenoid affords protection against the the high intensity light that is focused onto the foveal region, and is an important defence against age-related macular degeneration, a major cause of impaired vision and blindness in the elderly. Whether the carotenoid acts as a light filter, removing harmful high-energy blue light, or has a protective role against oxidative damage is still under discussion. The role of lutein and zeaxanthin in the macula is treated in detail in *Volume 5, Chapter 15*.

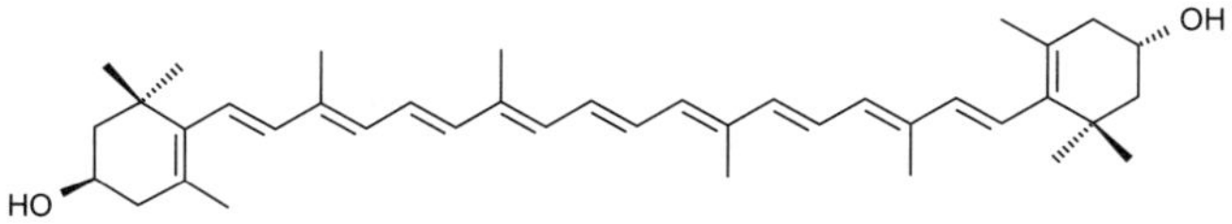

(**120**) (3*R*,3'*S*)-zeaxanthin

3. Retinal filters: oil droplets

In some birds, in addition to different photoreceptors sensitive to blue, red and green light, the retinal cells also have coloured oil droplets (red, yellow, greenish-yellow) through which light has to pass to reach the photoreceptors [27]. In the retina of the turkey (*Meleagris gallopavo*), some twelve carotenoids have been detected; the three main components, astaxanthin (mainly 3*S*,3'*S*, **406**), galloxanthin [(3*R*)-10'-apo-β-carotene-3,10'-diol, **495.2**] and (6*S*,6'*S*)-ε,ε-carotene (**20.1**) comprise 65% of the total [28]. None of the droplets contains just a single pure carotenoid, but three types of droplet can be identified, in each of which one of the three carotenoids predominates, giving different filter cut-off wavelengths over a range of 100 nm. This provides a flexible and sensitive means of restricting short wavelengths and enhancing the sensitivity and perception of colour. A similar situation is found in some reptiles, such as the sea turtle *Chelonia* [29].

The pigmentation of fish corneas by carotenoids appears not to have any significance for colour vision but has been assigned a role as a protective blue-light filter [30].

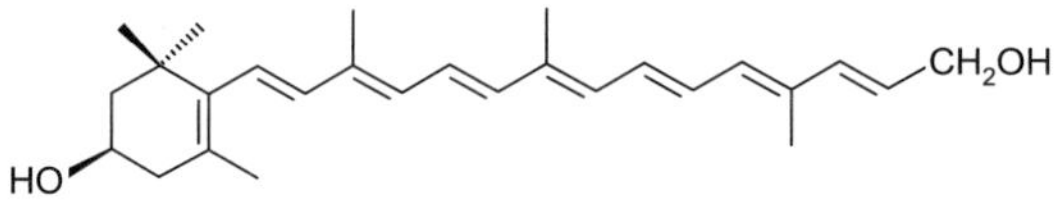

(**495.2**) galloxanthin

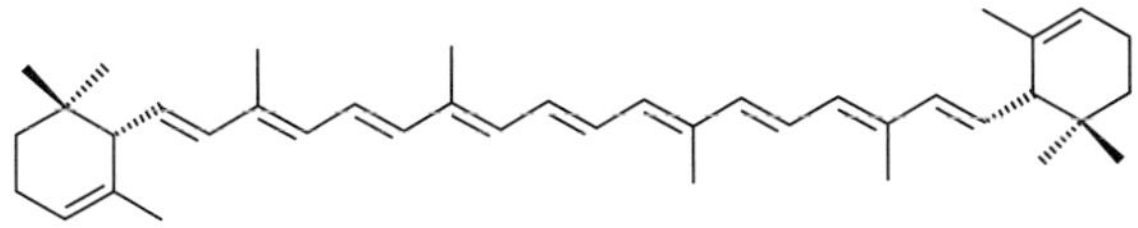

(20.1) (6*S*,6'*S*)-ε-carotene

G. Colour

In Nature, colouration is not a trivial role. Colour provides some of the most important ecological signals and is a major influence on behaviour and interactions between species and between individuals of the same species (*Chapter 11*). But to see and identify colours needs a very sophisticated visual system, with different photoreceptors that respond to different wavelength ranges. In the absence of such colour vision, colour has no meaning. It is probable, therefore, that colour and colour vision evolved together. In the plant kingdom, with very few exceptions, colour is due to pigments. Pigments also provide many important colours in animals, but some animal colours ('structural colours') are produced in a different way, by physical phenomena arising from structural features within the tissues [2,3].

1. Pigmentary colours

Substances that have the property of absorbing light in one part of the visible range of the electromagnetic spectrum (*ca.* 380-750 nm) reflect or transmit the remaining part. The colour seen therefore depends on the wavelengths of maximum absorption. Carotenoids have a very wide natural distribution in plants, animals and microorganisms and have an extremely important place among the various groups of natural pigments, together with the poly-phenolic flavanoids, tetrapyrroles such as chlorophyll, and the inert polymeric melanins. Most pigment classes are of ancient origin and had other important biological functions before colour became a significant factor in life.

2. Structural colours

In the animal kingdom, some of the most remarkable colours owe their origin not to the presence of light-absorbing pigments but to optical phenomena such as light scattering, interference or diffraction by microscopic structures present in the tissues [31].

Very small particles, smaller in diameter than the wavelength of red and yellow light, will reflect or scatter more of the short-wavelength (blue) than of the long-wavelength (red) components of white light. This process, known as Rayleigh or Tyndall scattering, produces the colour Tyndall blue. Most non-iridescent blue colours in animals are Tyndall blues. In blue feathers of birds such as the blue tit (*Parus ceruleus*), the keratin of the feather barbs oontaino light scattering particles in the form of microscopic air-filled lamellae. Green colour in feathers is often due to a superimposition of Tyndall blue and a yellow pigment, commonly carotenoid. Blue colours in marine invertebrate animals, however, are typically due to carotenoproteins (see *Chapter 6*).

When the scattering particles are larger than the wavelength of light, there is no wavelength discrimination and the scattered light is white. This Mie scattering from solid particles or lipid droplets gives rise to natural white colours, illustrated by, for example, milk, white butterflies and white flowers. When the reflecting particles are crystalline, *e.g.* guanine, a metallic sheen is produced. A good example is the flashing silver appearance common in fish. Overlaying this with an orange or yellow pigment, often carotenoid, gives a metallic gold effect as in goldfish (*Carassius auratus*).

The striking iridescent colours frequently encountered in the animal kingdom, particularly in birds, insects and fishes, are also structural and usually are produced by interference phenomena, or occasionally by diffraction, due to thin laminar structures in the tissues. Peacock blue is a typical and spectacular example.

3. Colour in plants

a) Plant pigments

The plant kingdom is predominantly green because of the presence of chlorophylls. The colour of carotenoids, which occur universally in the chloroplasts of green leaves, is normally masked by the chlorophylls. Yellow carotenoid colours may be seen in etiolated (dark-grown) seedlings and in species with decorative foliage and they are also revealed in autumn leaves of some trees, as the green leaves senesce and chlorophyll is degraded [5]; the red of autumn leaves is due to anthocyanins. A transient red colour often seen in young leaves is usually due to anthocyanins, but high concentrations of ketocarotenoids may give a red colour to the young leaves of cycads and some conifers [32,33]. The carotenoids accumulate in oil droplets but their functional role is unknown.

Against this green background, the contrasting vivid colours of flowers or bracts and fruits catch the eye and attract insects for pollination and birds for seed dispersal, and so are vital for the survival and propagation of the species. Flowers and fruit come in all colours. Most pink, red, purple and blue shades are due to anthocyanins (water-soluble phenolic compounds) or, in some families, betalains (betacyanins). Phenolic compounds (chalkones) or betalains (betaxanthins) are also responsible for some yellow and orange colours, but these colours are most commonly due to carotenoids, as are reds in some fruit, familiar examples being tomato (*Lycopersicon esculentum*) and red pepper (*Capsicum annuum*). Combinations of lipid-soluble carotenoids and water-soluble anthocyanins can give some unusual colours, such as brown and maroon as in wallflowers (*Cheiranthus cheiri*).

In flowers, carotenoids are found in all anatomical parts: sepals, pollen, anthers, stamens and petals. In fruit, they may be in the skin, flesh, juice and sometimes seeds [21]. Carotenoids are not widespread in roots but there are some notable examples, *e.g.* carrot (*Daucus carota*) and sweet potato (*Ipomea batatas*). The functional significance of carotenoids in seeds and roots, which normally are not exposed to light, is not known.

b) Chromoplasts

In flowers, fruits and roots, the carotenoids are synthesized and located in chromoplasts, organelles related to and often derived from chloroplasts. There is great diversity in the morphology of chromoplasts [34]. The most conspicuous process during the early part of chloroplast-chromoplast transformation is the degradation of the photosynthetic apparatus. Simultaneous with this, an active synthesis of carotenoids begins and special carotenoid-bearing structures start to develop. These are structurally diverse, and several types are often present in the same organelle. Some carotenoid-containing structures may be transient and disappear again in senescent chromoplasts.

The most common carotenoid-containing structures encountered in the ripe or senescent chromoplasts are plastoglobules, spherical lipid droplets 100 - 1000 nm or more in diameter, that lie singly or in groups in the chromoplast stroma. In some cases, the concentration of carotenoids inside the plastoglobules or connected with them can increase so much that they crystallize. The best known examples are the large crystals of β-carotene (**3**) in carrot roots and in narcissus flowers, and of lycopene (**31**) in tomato fruit [35].

In tubulous chromoplasts of some flower petals (*e.g. Chelidonium majus* and fruit (*e.g. Capsicum annuum*) [36], chiral carotenoid superstructures, typically of xanthophyll esters, are found. Structural aspects of the natural assemblies and aggregates of carotenoids in chromoplasts of flowers and fruit are discussed in *Chapter 5*.

The different internal structures of chromoplasts appear to have developed to accommodate the high concentration of carotenoids in ways that allow the carotenoid colour to be displayed most effectively. The main function of chromoplasts is colour but products of carotenoid breakdown, enzymic or non-enzymic, may make a significant contribution to perfume and aroma, and in attracting, repelling or confusing insects (see *Chapter 15*)

4. Carotenoids in moulds, yeasts and bacteria

We see many of these microorganisms in culture as coloured, frequently by carotenoids [21]. Indeed some important examples are cultured for commercial production of carotenoids (*Volume 5, Chapter 5*). The natural colour is probably fortuitous, however; the carotenoids are not there for colouration but for other purposes such as photoprotection, strengthening of membranes *etc.*

5. Colour in animals

a) Colouration by carotenoids

Animals use colour for many purposes, such as advertizing and warning. Ecological aspects of these signals are described in detail in *Chapter 11*. Here the discussion is limited to how colours are produced and displayed, and the relative importance of carotenoids in this. Colouration by carotenoids is widespread, especially in birds, fish, insects and marine invertebrate animals [37]. Carotenoids are responsible for most, though not all, yellow, orange and red colours; phaeomelanins provide the orange-red of hair and some feathers, and pterins the yellow of some reptiles, amphibians and insects [2,38-41]. In feathers, the carotenoids are associated with keratin, the structural protein of feather barbules.

Animals do not biosynthesize the carotenoids they accumulate; they must obtain them from the diet, though the dietary carotenoids can be modified structurally (*Volume 3, Chapter 7*). Animals in captivity, *e.g.* aviary birds, aquarium ornamental fish, and fish and crustaceans in aquaculture, must be supplied with carotenoids in the feed in order to maintain or enhance colour and brightness (see *Chapters 12* and *13*).

Surprisingly, besides the typical yellow, orange and red colour range, carotenoids are also responsible for some purple, green and blue colours. These colours are due to carotenoprotein complexes, which are specific, stoichiometric combinations of carotenoid and protein. The binding is normally not covalent but the carotenoid is bound in such a way that its light-absorption properties are drastically altered, leading to substantial bathochromic shifts and purple-blue colours. Structural features and the mechanism of carotenoid-protein interactions are described in *Chapter 6*. Carotenoproteins are widely distributed in many phyla of invertebrate animals, especially ones from a marine environment. It has been suggested that the carotenoproteins may serve to camouflage the animals in the prevailing undersea light conditions, serve as general photoreceptors, or give protection against possible harmful effects of light.

b) Animal pigment cells: chromatophores

In the integuments of invertebrate and poikilothermic ('cold-blooded') vertebrate animals, pigments are frequently located in special cells known as 'chromatophores' (sometimes 'chromatocytes') [38,42]. Pigment cells containing melanin are known as melanophores or melanocytes. They are usually black, though some, which contain phaeomelanin, may be brown, yellow or orange-red. Most yellow, orange and red integumental cells are xantho-phores (yellow) or erythrophores (red). The predominant pigments in these, present in lipid droplets or vesicles, are carotenoids or, in many insects and poikilothermic vertebrates such as frogs, pterins. Pterins, which are synthesized in the developing xanthophores and erythrophores, are often the first yellow pigments to be seen, and the carotenoids, which must be obtained from the diet, do not appear until later.

Iridophores, are not strictly pigment-containing cells, but they contain organelle structures that are oriented so that they reflect light efficiently. The white or colourless purines (usually guanine) that they contain may be arranged in stacks of crystalline platelets which reflect white light and give rise to metallic sheens.

c) Colour changes

In the dermis of some vertebrates, the various chromatophore types, usually not in equal numbers, are associated into a dermal chromatophore unit that can bring about rapid colour changes [38,42]. Colour changes occur rapidly by variations in the contributions made by the different cell types in the functional unit, usually by dispersal or aggregation of pigment particles or granules within the chromatophores, including the yellow-orange carotenoids in the xanthophores or erythrophores, so that they make a greater or lesser contribution to the colour, for example to allow the animal to match the colour of its background. Slower changes during long-term adaptation to background require the accumulation of increased amounts of pigment. The main regulatory factor is light, and the process is under hormonal control. In crustaceans, separate pigment-dispersing hormones and pigment-concentrating hormones have been recognized for black, white and red-orange pigments [43]. All these hormones are produced by the eyestalks where the receptor pigments that detect background colour changes are located.

H. Carotenoids in Membranes

1. Membrane structure

The basic structure of a natural membrane is a bilayer of polar lipids, *e.g.* phosphatidylcholine (Fig. 2a). The acyl chains of the lipids constitute the fluid hydrophobic core of the membrane and the polar head groups on the outer surfaces of the membrane interact with the aqueous medium. Living organisms use a variety of polar lipids with different hydrocarbon chain lengths and levels of unsaturation, and different polar head groups. Natural membranes are not just simple bilayers, however. They contain other lipid components such as cholesterol, and proteins, which may be integral or peripheral. Carotenoids are also common components, though often in low concentration.

The acyl lipid chains are undergoing different kinds of molecular motion: the very fast (nanoseconds) *gauche-trans* isomerization of alkyl chains and the slower rotational and lateral diffusion of the whole lipid molecule. At different temperatures, phospholipid membranes can exist in different phases, *i.e.* in a tightly ordered gel or solid phase below the temperature of main phase transitions, or as a liquid-crystalline phase above this temperature. Details of the localization and orientation of carotenoids in liposomes and natural membranes and the influence of carotenoids on membrane structure and properties are described in a comprehensive review [44]. The main features and conclusions are summarized here.

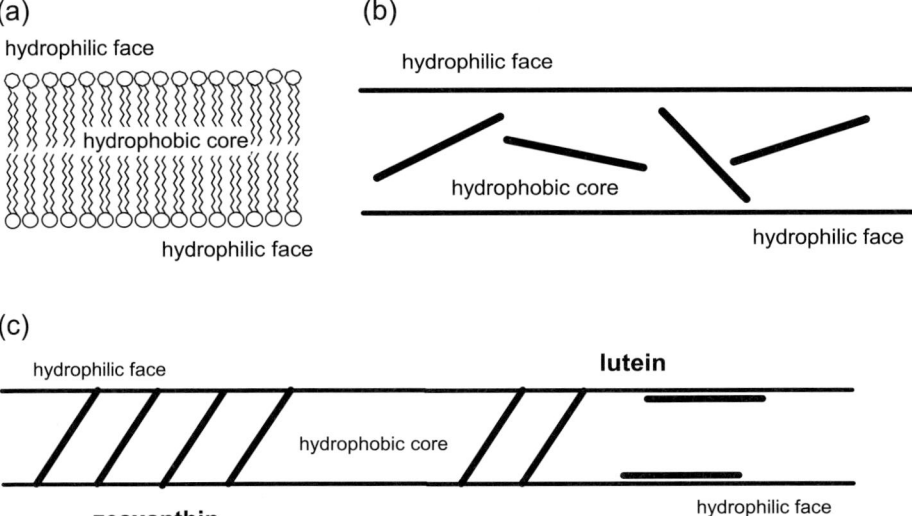

Fig. 2. (a) Diagrammatic representation of a phospholipid bilayer. (b) Localization of hydrocarbons such as β-carotene (**3**) and lycopene (**31**) in various orientations in the hydrophobic core of the bilayer. (c) Orientation (left) of zeaxanthin (**119**), spanning the membrane bilayer, and (right) of lutein (**133**), with some molecules spanning the bilayer, others lying horizontally near the hydrophilic surfaces.

2. Localization and orientation of carotenoids

Most studies of the structure and dynamics of membranes and interactions of the various components have used not natural membranes but model membrane bilayers (liposomes) that are prepared from saturated or unsaturated phospholipid molecules which spontaneously form vesicles when dispersed in aqueous media. Carotenoids and other components can be incorporated at chosen concentrations during preparation of the liposomes. The rod-like structure, the presence of polar end groups, and the molecular dimensions of a typical carotenoid, which match the thickness of the bilayer, are directly responsible for the localization and orientation of carotenoid molecules within the membrane and for effects on the membrane properties [44-46]. Carotenes such as β-carotene (**3**) or lycopene (**31**) are entirely lipophilic and remain within the hydrocarbon inner part of the bilayer. Polarized light spectroscopy studies, especially linear dichroism, have indicated an average orientation of these carotenoids with their long axis roughly perpendicular to the lipid acyl chains (Fig. 2b).

Polar substituents in a carotenoid, especially hydroxy groups, cannot readily penetrate into the hydrophobic core of the bilaycr, so they associate with the polar head-group region of the membrane. The dihydroxycarotenoid zeaxanthin (119) spans across the bilayer with one polar end group associated with each polar surface (Fig. 2c, left).

Minor differences in structure can lead to significant differences in the behaviour of very similar carotenoids. Zeaxanthin (119) and lutein (133) differ only in the position of one double bond in one of the rings and in the stereochemistry at C(3') but, whereas zeaxanthin adopts only an orientation spanning the membrane, two pools of lutein are seen, one, like zeaxanthin, spanning the membrane but the other oriented horizontally with respect to the membrane (Fig. 2c, right). The possible significance of this in relation to the functioning of these xanthophylls in the macula of the eye is discussed in *Volume 5, Chapter 15*.

The degree of aggregation of carotenoid molecules in membranes (*Chapter 5*) and the effects of carotenoids on membrane structure and dynamics are strongly influenced by the physical state of the membrane lipid phase.

3. Effects of carotenoids on membrane structure and properties

If molecules bearing spin labels are introduced into model membrane structures, the fluidity of different parts of the membrane, in the presence or absence of carotenoid can be probed by EPR spectroscopy.

Natural abundance ^{1}H-, ^{13}C-, and ^{31}P-NMR spectroscopy allows features of molecular dynamics to be observed directly without the need to apply probes which could possibly affect the system.

The EPR and NMR studies give a consistent and complementary picture of the effects of carotenoids on the structure and dynamics of the lipid membranes [44]. Studies by other methods, such as differential scanning calorimetry, support this picture. The main findings are as follows.

i) The polar carotenoids lutein (133) and zeaxanthin (119) restrict the molecular motion of lipids and increase the rigidity of the membrane in its fluid state. This is consistent with the 'molecular rivet' role of xanthophylls. β-Carotene (3), which is not anchored across the membrane by polar end groups, does not have this effect.

ii) β-Carotene increases the motional freedom of lipids in the headgroup region, in contrast to zeaxanthin which decreases motion in this region as it does in the hydrophobic core.

iii) β-Carotene and, to a lesser extent, zeaxanthin increase the penetration of small molecules and ions into the polar zone of the membrane. The hydrophobic barrier is lower in the headgroup region but greater in the membrane interior in carotenoid-containing membranes than in carotenoidless controls.

iv) Penetration of molecular oxygen into the lipid bilayer is limited when polar carotenoids are present.

There is, however, little direct evidence that the effects of carotenoids seen in studies with model membranes have any real physiological significance *in vivo*.

4. Reinforcement of bacterial membranes

In mammals, cholesterol plays a crucial role in strengthening and maintaining the membrane structure and in optimizing the properties and dynamics of the membrane. A similar role has been identified for carotenoids in the membranes of bacteria, which do not biosynthesize or accumulate cholesterol. Support for this comes from studies with *Acholeplasma laidlawii*, the membranes of which normally contain cholesterol, obtained from its host [47]. When grown in the absence of cholesterol, the organism biosynthesizes polar carotenoids which take the place of cholesterol in the membrane. Whereas the size of cholesterol is such that it fits into just one leaflet of the bilayer and spans only half the membrane, the carotenoid molecule is of suitable length to span across the membrane, with polar substituents in the end groups associated with the polar outer faces of the bilayer, and the hydrocarbon polyene chain in the hydrophobic lipid core. Evolution has ensured that the carotenoids in different bacteria are of a suitable length to match the thickness of the membrane. In most cases, normal C_{40} carotenoids will do this, but some species with thinner or thicker membranes biosynthesize shorter (C_{30} diapocarotenoids) or longer (C_{45} and C_{50}) carotenoids, respectively. Glycosylation of carotenoid end groups strengthens the association with the polar head groups of the bilayers. As an example, the acyclic C_{50} xanthophyll bacterioruberin (**456**) which is characteristic of *Halobacterium* species (Archaea) is a good fit for the thicker diphytanyl-lipid membranes of these organisms. Reconstitution experiments with various combinations of lipids and carotenoids have confirmed that the incorporation of carotenoids into model membranes is best when the length of the carotenoid is close to the thickness of the bilayer. It is concluded, therefore, that the carotenoid acts as a 'rivet' mechanically reinforcing and strengthening the bilayer [48,49].

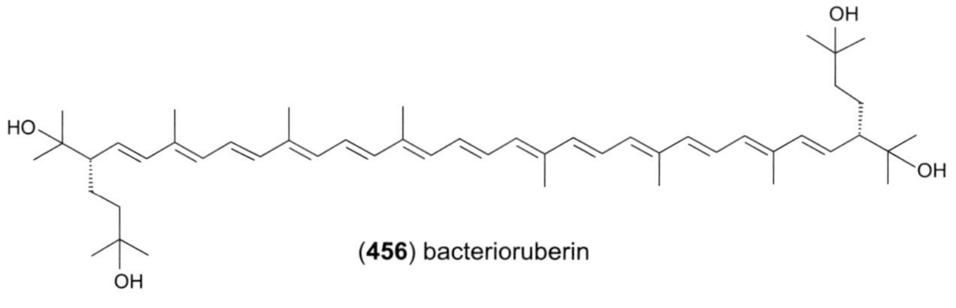

(**456**) bacterioruberin

5. Protection of membranes against oxidizing free radicals

Peroxidation of unsaturated acyl chains causes disruption of membranes. The ability of carotenoids to protect liposome membranes against oxidative damage caused by free radicals generated *via* azo-initiators has been studied [50]. The hydrocarbons β-carotene (**3**) and lycopene (**31**), which are located in the inner hydrophobic part of the bilayer, were only effective against free radicals generated in this inner core. In contrast, zeaxanthin (**119**), which spans the membrane with its hydroxy groups penetrating into the polar outer zone, was able also to protect against damage by radicals generated in the aqueous phase. Any protective effect of carotenoids against membrane damage *in vivo* would therefore depend on which carotenoid is present, its positioning and orientation in the bilayer, and where the oxidizing free radicals are generated. As yet, however, studies have been restricted to model liposome systems.

6. Membrane-associated processes

Many essential molecular processes in cells are associated with the membrane, *e.g.* receptors and signal transduction systems which involve movement of proteins in the membrane. Modification of the structure and dynamics of a membrane, *e.g.* by carotenoids, therefore my have an effect on such membrane-associated processes. Various molecular and cellular processes in human and mammalian cells *in vitro* are affected when the cells are cultured in the presence of carotenoids (*Volume 5, Chapter 11*). This leads to the hypothesis that the activity of the cellular regulatory systems *in vivo* could be influenced by the presence and concentration of particular carotenoids in the natural membranes, through their effect on membrane fluidity or by stabilizing and protecting the membrane and the system against disruption by oxidizing radicals.

7. Zeaxanthin in thylakoid membranes

There is increasing evidence that, in plants suffering stress, and during the operation of the xanthophyll cycle (*Chapter 14*), zeaxanthin (**119**) appears and may accumulate in the thylakoid membrane, leading to decreased fluidity, increased stability and increased resistance to oxidation [38,51]. The physiological relevance of this is not yet clear, however.

I. Fertility and Reproduction

Carotenoids, sometimes as carotenolipoproteins (lipovitellins) *e.g.* ovoverdin in eggs of lobsters and other crustaceans, are frequently concentrated in eggs and reproductive tissues of vertebrate and invertebrate animals [37]. It is assumed that the carotenoid in egg yolk provides a ready source of provitamin A or intact carotenoids for the developing embryos or newly hatched young. The importance of this in fish, and of the relationship between carotenoids and reproductive success and health and survival of the young, is treated in *Chapter 12*. Aspects of carotenoids in sexual signalling and display and in relation to indicating fecundity, especially in birds, are discussed in *Chapter 11*.

Mammals do not normally use carotenoids for colouration, but carotenoids are found in reproductive tissues of many mammals, including humans. The presence, distribution and possible significance of carotenoids in human male and female reproductive tissues is outlined in *Volume 5 Chapter 7*. The most striking example of the accumulation of carotenoids in mammalian reproductive tissues is the high concentration of β-carotene in the *corpus luteum* in the ovaries of cattle and other animals. It was once thought that the carotene must be biosynthesized within the tissue [52], though it is now known that this is not the case, and the carotene is of dietary origin. This accumulation surely cannot be fortuitous, though the role that the carotene plays has not been established.

Various studies have reported that β-carotene may have a positive effect on fertility in farm animals, especially cattle, horses and pigs [28]. There is as yet no explanation for such effects, at the physiological, cellular or molecular level.

J. Miscellaneous Proposed Functions

The carotenoid functions and actions described above are those that have been studied most extensively and for which some understanding is therefore developing. Many other imaginative suggestions have been made over the years, *e.g.* of roles for carotenoids in oxygen storage and transport in invertebrate animals, and applications of this for pollution treatment [53]. The reader is recommended to read the many fascinating observations and ideas summarized in older books and reviews [37-41,54,55], and see if any of these observations are interesting enough to merit reinvestigation by modern methods.

K. Breakdown Products and Metabolites

It has long been known that some carotenoid metabolites, notably vitamin A (retinol) and its derivatives, have essential functions, and that other metabolites or breakdown products make an important contribution to natural perfumes and aromas. There is now increasing interest in the possibility that other products of enzymic or non-enzymic oxidative breakdown of carotenoids may have biological activity relevant to health. The functions of such products are discussed in *Chapter 15*.

L. Carotenoids in Human Health and Nutrition

As well as their role as provitamin A, and their importance for eye health, protecting against macular degeneration, dietary carotenoids are associated with reduced risks of some cancers and other serious conditions, stimulation of the immune system and benefits for skin health. An entire volume, *Volume 5*, has been allocated to this important and topical area of carotenoid research.

References

[1]   J. A. Olson, *J. Nutr.*, **119**, 94 (1989).
[2]   G. Britton, *The Biochemistry of Natural Pigments*, Cambridge University Press, Cambridge (1983).
[3]   H.-D. Martin, *Chimia,* **49**, 45 (1995).
[4]   G. Britton, *FASEB J.*, **9**, 1551 (1995).
[5]   A. J. Young, in *Carotenoids in Photosynthesis* (ed. A.J. Young and G. Britton), p. 16, Chapman and Hall, London (1993).
[6]   G. Guglielmi, J. Lavaud, B. Rousseau, A.-L. Etienne, J. Houmard and A. V. Ruban, *FEBS J.*, **272**, 4339 (2005).
[7]   R. A. Mathies, S. W. Lin, J. B. Ames and W. T. Pollard, *Ann. Rev. Biophys. Biophys. Chem.*, **20**, 491 (1991).
[8]   S. P. Balashov, E. S. Imasheva, V. A. Boichenko, J. Anton, J. M. Wang and J. K. Lanyi, *Science*, **309**, 2061 (2005).
[9]   B. F. Lutnaes, A. Oren and S. Liaaen-Jensen, *J. Nat. Prod.*, **65**, 1340 (2002).
[10] M. Griffiths, W. R. Sistrom, G. Cohen-Bazire and R. Y. Stanier, *Nature*, **176**, 1211 (1955).
[11] C. S. Foote, in *Free Radicals and Biological Systems* (ed. W. A. Pryor), p. 85, Academic Press, New York (1976).
[12] P. M. Bramley, in *Carotenoids in Photosynthesis* (ed. A. J. Young and G. Britton), p. 127, Chapman and Hall, London (1993).
[13] M. M. Mathews-Roth and N. I. Krinsky, *Photochem. Photobiol.*, **11**, 419 (1970).
[14] M. M. Mathews-Roth, *Ann. N. Y. Acad. Sci.*, **691**, 127 (1993).
[15] G. Y. Liu, A. Essex, J. T. Buchanan, V. Datta, H. M. Hoffman, J. F. Bastian, J. Fierer and V. Nizet, *J. Exp. Med.*, **202**, 209 (2005).
[16] A. Pelz, K. P. Wieland, K. Putzbach, P. Hentschel, K. Albert and F. Götz, *J. Biol. Chem.*, **280**, 32493 (2005).
[17] G. W. Burton and K. U. Ingold, *Science*, **224**, 569 (1984).
[18] S. Beutner, B. Bloedorn, S. Frixel, I. H. Blanco, T. Hoffmann, H.-D. Martin, B. Mayer, P. Noack, C. Ruck, M. Schmidt, I Schülke, S. Sell, H. Ernst, S. Haremza, G Seybold, H. Sies, W. Stahl and R. Walsh, *J. Sci. Food Agric.*, **81**, 559 (2001).
[19] F. Bohm, R. Edge, E. J. Land and T. G. Truscott, *J. Am. Chem. Soc.*, **119**, 621 (1997).
[20] H. Senger (ed.), *Blue Light Responses: Phenomena and Occurrence in Plants and Microorganisms, Vols. 1 and 2*, CRC Press, Boca Raton (1987).
[21] T. W. Goodwin, *The Biochemistry of the Carotenoids, 2nd. Edn., Vol.1: Plants*, Chapman and Hall, London (1980).
[22] H. Stransky and A. Hager, *Arch. Mikrobiol.*, **72**, 84 (1970).
[23] E. Bünning and G. Schneiderhörn, *Arch. Mikrobiol.*, **24**, 80 (1956).

[24]   R. D. Smyth, J. Saranak and K. W. Foster, *Prog. Phycol. Res.*, **6**, 255 (1988).

[25]   E. Zeiger and J. Zhu, *J. Exp. Bot.*, **49**, 433 (1998).

[26]   R. A. Bone, J. T. Landrum, G. W. Hime, A. Cains and J. Zamor, *Invest. Ophthalmol. Vis. Sci.*, **34**, 2033 (1993).

[27]   B. H. Davies, *Pure Appl. Chem.*, **51**, 623 (1979).

[28]   B. H. Davies, *Pure Appl. Chem.*, **63**, 131 (1991).

[29]   B. H. Davies, A. Akers, R. J. Lewis-Jones, S. Pollard and A. J. Shufflebotham, *Vision Res.*, **24**, 1701 (1984).

[30]   W. R. A. Muntz, in *Handbook of Sensory Physiology, Vol. VII/1: Photochemistry of Vision* (ed. H. J. A. Dartnell), p. 530, Springer, Berlin (1972).

[31]   K. Nassau, *The Physics and Chemistry of Color*, Wiley, New York (1983).

[32]   B. Czeczuga, *Biochem. Syst. Ecol.*, **14**, 203 (1986).

[33]   F. Cardini, M. Giananneschi, A. Selva and M. Chelli, *Phytochemistry*, **26**, 2029 (1987).

[34]   P. Sitte, H. Falk and B. Liedvogel, in *Pigments in Plants* (ed. F.-C. Czygan), p. 117, Gustav Fischer, Stuttgart (1980).

[35]   G. Britton, L. Gambelli, P. Dunphy, P. Pudney and M. Gidley, *Proc. 2nd. Int. Congr. Pigments in Food, Lisbon* (ed. J. A. Empis), p. 151 (2002).

[36]   S. Nechifor, C. Socaciu, F. Zsila and G. Britton, *Proc. 2nd. Int. Congr. Pigments in Food, Lisbon* (ed. J. A. Empis), p. 155 (2002).

[37]   T. W. Goodwin, *The Biochemistry of the Carotenoids, 2nd. Edn., Vol. 2: Animals*, Chapman and Hall, London (1984).

[38]   A. E. Needham, *The Significance of Zoochromes*, Springer, Berlin (1974).

[39]   D. L. Fox, *Animal Biochromes and Structural Colors, 2nd. Edn.*, University of California Press, Berkeley (1976).

[40]   D. L. Fox, *Biochromy: Natural Coloration of Living Things*, University of California Press, Berkeley (1979).

[41]   H. M. Fox and G. Vevers, *The Nature of Animal Colours*, Sidgwick and Jackson, London (1960).

[42]   J. T. Bagnara and M. E. Hadley, *Chromatophores and Color Change: The Comparative Physiology of Animal Pigmentation*, Prentice-Hall, Englewood Cliffs, NJ (1973).

[43]   M. Fingermann, in *Endocrinology of Selected Endocrine Types* (ed. H. Laufer and R. G. H. Downer), p. 357, A. R. Liss, New York (1988).

[44]   W. I. Gruszecki, in *The Photochemistry of Carotenoids* (ed. H. A. Frank, A. J. Young, G. Britton and R. J. Cogdell), p. 363, Kluwer, Dordrecht (1999).

[45]   A. Sujak, W. Okulski and W. I. Gruszecki, *Biochim. Biophys. Acta*, **1509**, 255 (2000).

[46]   W. Okulski, A. Sujak and W. I. Gruszecki, *Biochim. Biophys. Acta*, **1509**, 216 (2000).

[47]   S. Rottem amd O. Markowitz, *J. Bacteriol.*, **140**, 944 (1979).

[48]   M. Rohmer, P. Bouvier and G. Ourisson, *Proc. Natl. Acad. Sci. U.S.A.*, **76**, 847 (1979).

[49]   G. Ourisson and Y. Nakatani, in *Carotenoids: Chemistry and Biology* (ed. N. I. Krinsky, M. M. Mathews-Roth and R. F. Taylor), p. 237, Plenum Press, New York (1989).

[50]   A. A. Woodall, G. Britton and M. J. Jackson, *Biochem. Soc. Trans.*, **22**, 133S (1994).

[51]   W. I. Gruszecki and K. Strzalka, *Biochim. Biophys. Acta*, **1060**, 310 (1991).

[52]   B. M. Austern and A. M. Gawienowski, *Lipids*, **4**, 229 (1969).

[53]   V. N. Karnaukhov, *The Biological Functions of Carotenoids*, Nauka, Moscow (1988).

[54]   T. W. Goodwin, *The Comparative Biochemistry of the Carotenoids*, Chapman and Hall, London (1952).

[55]   N. I. Krinsky, in *Carotenoids* (ed. O. Isler), p. 669, Birkhäuser, Basel (1971).

Carotenoids
Volume 4: Natural Functions
© 2008 Birkhäuser Verlag Basel

Chapter 11

Signal Functions of Carotenoid Colouration

Jonathan D. Blount and Kevin J. McGraw

A. Introduction

The importance of carotenoids for natural colouration, in relation to other classes of pigments and structural colours, has been outlined in *Chapter 10*. But colour only has significance if it is perceived, identified and interpreted by other organisms (animals). In other words, colour is a means of communication, a signal. Now, in this Chapter, this new direction for carotenoid research, behavioural ecology, is highlighted. Various hypotheses that have been proposed to explain the signal functions of colour, and particularly of carotenoids, in plants and animals are discussed and the empirical evidence to support these hypotheses is presented.

1. Biological signals and their reliability

A biological signal can be defined as any morphological, behavioural, or chemical feature that has evolved because it alters or affects the behaviour of other individuals. A signal may therefore convey information, which is of relevance to the receiver, about the current state of the signaller or its future behaviour [1]. The information contained in signals need not always be correct, but it must be correct often enough for selection to predispose the receiver to respond to it [1]. Otherwise, at some point the signal will be ignored because it has lost its meaning. This raises the question of what maintains the reliability, or 'honesty' of signals? The 'handicap principle' suggests that signals are honest because they are costly to produce [2]. For example, the tail of a peacock (*Pavo cristatus*) might be costly to grow in terms of resource utilization, or it might be costly to bear because it impedes efficient flight and therefore makes the bird more vulnerable to predation. So individuals with longer tails should be higher quality individuals, because relatively poor quality individuals should not be able to afford to pay such high costs of resource allocation or predation risk [2]. The handicap principle has been evoked to explain the evolution of a broad range of signal types, some of which are based on colouration.

In certain instances, however, signals need not be costly to be reliable. If the interests of the signaller and receiver coincide, then signals can be honest. For example, the bright red, warningly coloured toxic insect and its potential predator both benefit if the insect is not attacked. Secondly, certain signals are honest because it is impossible to lie, namely indices of quality such as body size [1].

2. Carotenoids as signals

For a resource to be subject to allocation trade-offs, as envisaged in the handicap principle, it must (i) be needed for more than one body function, and (ii) be in limited supply. Carotenoids fulfil these criteria, because they have a variety of other functions as well as being pigments responsible for colouration. In addition, the physiological supply of carotenoids may be limiting because, at least in animals, these compounds cannot be produced *de novo* and must be obtained in the diet [3]. Indeed carotenoid colouration abounds in Nature in many different signalling contexts. In animals, carotenoid-based signals include those designed for communication both between species (*e.g.* warning colouration) and within species (*e.g.* epigamic colouration – colours associated with courtship). In contrast, the signal functions of carotenoids in plants have been relatively little studied.

B. Carotenoid Signals in Plants

1. Fruits and flowers

Reproductive structures of plants, *i.e.* fruits and flowers, use colour and odour to attract consumers who then serve to disperse the seeds or pollen. Compared to animal signals, plant colour signals are generally simpler in expression, because they lack a behavioural component and are displayed against a relatively unchanging background [4]. This simplicity makes plants good models for studying signal evolution. Little is known, however, about evolutionary changes to optimize the attractiveness of flower colour to animals, or evolutionary changes in animals' visual systems or behavioural responses to plant colouration [5]. Moreover, little is known of the signal functions of carotenoids, as a specific class of pigments, in fruits and flowers. Most yellow and orange, and some red reproductive structures in plants are coloured by carotenoids, whereas pink, purple, blue-black and often red fruits and flowers use anthocyanins [6]. The one study relating bird colour vision (as modelled mathematically) to natural fruit colour and carotenoid content showed that carotenoid content is poorly predicted by fruit colouration and that birds cannot discriminate carotenoid contents of fruits based on their colour [7] .

Carotenoids in many pollens may serve to improve pollen detectability [6], but there is no empirical evidence to support this. In fruits and flowers, carotenoids may perform several functions independent of signalling, such as photoprotection (*Chapter 10*), and they play essential roles in photosynthesis (*Chapter 14*).

Choices made by consumers between different fruits or flowers are based on several factors, including innate preference, past experience of relative rewards, familiarity or ease of handling, and sensory adaptations and constraints that may make them more or less able to detect particular colours or odours. One hypothesis for the diversity of flower and fruit colouration proposes that particular classes of consumers are associated with particular colour signals, because pollination or seed dispersal is most efficient when done by specific animal taxa [5]. For example, pipevine swallowtail butterflies (*Battus philenor*) innately prefer yellow colour, and to a lesser extent blue and purple [8], whereas Indian red admirals (*Vanessa indica*) prefer yellow and blue [9]. However, learning can play an important role in determining colour preferences. It has classically been considered that red flowers are innately attractive to hummingbirds and simultaneously invisible to bees but, in fact, hummingbirds' colour preferences are learned and based on nectar rewards [10], and this typically leads to preferences for red flowers (at least in N. America) because these provide the greatest nectar rewards in those regions where hummingbirds live [11]. Bees do visit red flowers [12], but take longer to locate red than blue flowers because red contrasts relatively poorly against the background as perceived by the bees. Red floral colour may therefore be a strategy to reduce detectability by bees [5]. Because peak visual sensitivities are species-specific, there is some scope for directing signals at particular taxa, or avoiding others.

Fruit and flower colour is expressed maximally when a fruit is ripe or a flower fully open, *i.e.* when seed viability or nectar availability is highest [4]. But do such signals provide an honest indicator of the nutritional reward to consumers, *e.g.* availability of nectar, or energy? Few studies have addressed this question. In the sky lupin (*Lupinus nanus*), the petal banner spot changes from white to red once pollination has taken place, and red colour increases the attractiveness of the floral display, so optimizing the foraging and pollination efficiency of consumers [13]. Nutritional components, including lipid, protein, carbohydrate and water content, were found not to correlate with colouration indices in red or black fruits, however, but did so in fruits of other colours such as blue and green [14]. Conspicuous colours including red and black may, therefore, signal fruit detectability whereas less conspicuous colours (*e.g.* blue or green) may signal fruit nutritive reward. Similarly, fruit nutritional components, including mass, energy and sugar concentration, did not correlate with colour scored qualitatively by human observers [15].

The possibility that the pigments responsible for fruit colour may themselves be a valuable nutritional reward to consumers was not considered, however. Indeed, the brightly coloured seeds of *Margaritaria* spp. have been suggested to be a dishonest signal because they offer no nutritional reward to the consumer [4], but this overlooks the possible nutritional value of carotenoids or other pigments [16]. Do foragers seek fruits and flowers richer in carotenoids? Carotenoids are a potentially limiting resource for animals (see Section **D**). What is the potential cost to fruits and plants of allocating carotenoids to signal production as opposed to other somatic or reproductive functions, and is this balanced by benefits of more efficient seed or pollen dispersal?

An alternative hypothesis, that such colour displays may serve to render invertebrate herbivores (*i.e.* parasites) more conspicuous to predators and therefore lead them to avoid foraging on such plants [17], has not been tested experimentally.

2. Leaves

Why do leaves in many deciduous tree species become yellow or red in autumn? The conventional explanation is that such colour (which is due to the presence of carotenoids and anthocyanins) is a non-adaptive consequence of leaf senescence; as chlorophyll is degraded, the underlying carotenoids are exposed and large amounts of red anthocyanins may be synthesized. This spectacular autumn leaf colour may perform a signalling role, having arisen because of co-evolution between trees and insects [18,19]. Red and yellow leaf colour may make the defensive commitment of the plant clear to insects (*e.g.* aphids) that infest the plant in autumn and exploit it as a host for the winter [18]. Well-defended plants may thus reduce their parasite load, and the parasites benefit by infesting a more exploitable host. Because such allocation of carotenoids and anthocyanins to leaves may be costly to plants, autumn leaf colouration is suggested to be an honest (handicap) signal of plant defensive capacity [18].

Those plant species that suffer greater exploitation by insects should therefore invest more resources in defence and signalling of such defence [18,19]. A comparative study has indeed shown a positive association between the degree of autumn colouration and the diversity of monophagous aphids across 262 species of deciduous trees [18]. The most effectively defended individuals of a species should produce the most intense colour and be the most likely to be avoided by parasitic insects [18,19]. Indeed, it has been shown that individuals of mountain birch (*Betula pubescens*) that are in better condition produce more intense autumn colouration [20,21] and suffer less insect damage the following season [20], and *Prunus padus* trees with a higher percentage of red-yellow leaves were colonized by fewer aphids (*Rhopalosiphum padi*) [22]. All these studies are correlational, however, and an experimental approach is required to demonstrate cause-effect relationships. It would be valuable to test directly whether aphids are averse to the colour yellow [22], or attracted to it [19]. And, of course, alternative explanations for autumn leaf colour (*e.g.* photoprotection) may not be mutually exclusive [23].

C. Carotenoid Signals in Animals

1. Signalling to other species (heterospecifics)

a) Species recognition

Amongst bird species, for example, there is tremendous variation in the colour of plumage and bare parts (*e.g.* skin). The traditional explanation is that such colour variation has evolved to enable different species to recognize (and avoid) each other and thus reduce the risk of hybridization [24,25]. The morphological or ecological differences between closely related species should, therefore, increase (diverge) in sympatric species (those that occur in the same area), a phenomenon called 'character displacement'. This prediction has been supported by comparisons among sympatric *Ficedula* species of flycatchers in Europe [26], which display melanin colouration, and among *Agelaius* species of blackbirds in North America [27].

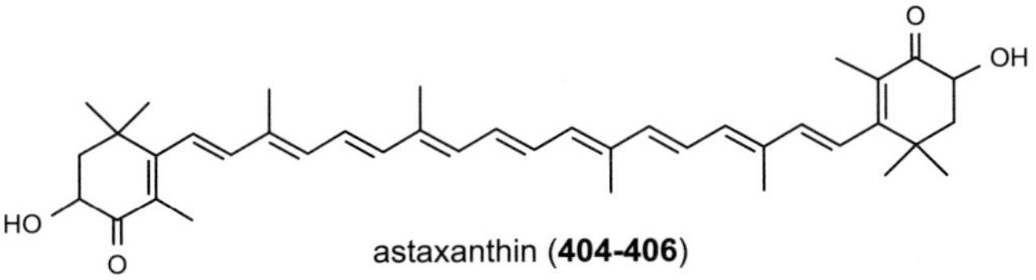

astaxanthin (**404-406**)

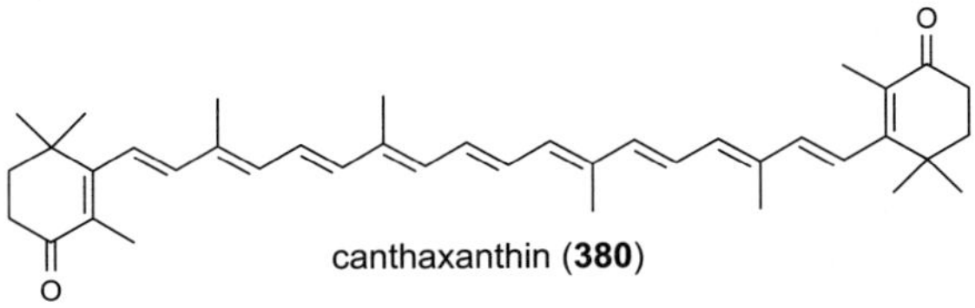

canthaxanthin (**380**)

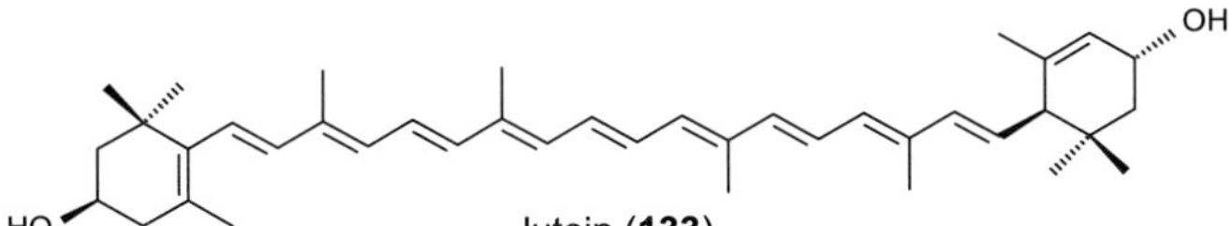

lutein (**133**)

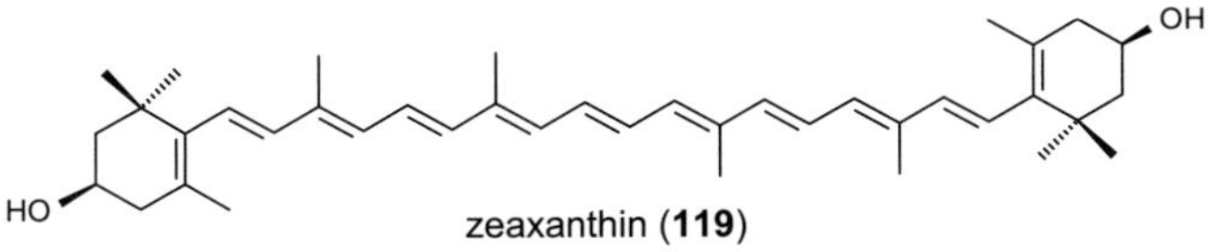

zeaxanthin (**119**)

Tricoloured blackbirds (*A. tricolour*) and red-winged blackbirds (*A. phoeniceus*) have striking red and yellow, or red and white, epaulette feather colour, respectively [27]. In *A. phoeniceus*, red epaulettes have been shown to contain astaxanthin (**404-6**), canthaxanthin (**380**), lutein (**133**) and zeaxanthin (**119**) [28]. Carotenoids are presumably also responsible for the red colour in *A. tricolour*, though this has not been verified. Apart from a few classic examples, however, such character displacement does not appear to be a widespread phenomenon. A recent comparative analysis concluded that, in general, sympatric species are not more divergent in respect of plumage colour than would be expected by chance [29]. Instead, between-species differences in plumage colour were associated with the habitats (particularly the lighting conditions) in which the birds live [29]. Differences in signalling conditions may have led plumage colour to be adapted to provide maximum chromatic contrast against those backgrounds and in those light environments that are relevant to individual species [2,30].

Between-species differences in colour may also have evolved to enable individuals of the same species to recognize each other, rather than to avoid members of other species [31]. Importantly, signal traits under selection for species recognition may have evolved for a different purpose; for example, the strikingly coloured epaulettes in *Agelaius* blackbirds function in social status signalling.

Due to practical and technical constraints, data used in large-scale comparative analyses relate to feather colour (being obtained from museum specimens) [29] to the neglect of bare parts such as legs, wattles and gapes, which could, in theory, have evolved to facilitate species isolation. This could be an important caveat because carotenoid colouration in birds is more common in bare parts than in plumage [32]. There is also a dearth of information about how species differences in colouration may relate to sympatry *etc.* in taxa other than birds.

b) Warning (aposematic) colouration and mimicry

An important evolutionary mechanism thought to be responsible for conspicuous colour and pattern in animals is aposematism, *i.e.* where appearance advertises unprofitability (unpalatability, toxicity, or ability to evade capture) to predators. Carotenoids are responsible for some such signals in a wide diversity of organisms including insects, amphibians, reptiles and birds [33-35] but, typically, relatively few species have been studied in detail. Seven-spot ladybird beetles (*Coccinella septempunctata*) have bright red elytra (wing cases) coloured by several carotenoids including torulene (**11**) and β-carotene (**3**) [36]. When threatened, they discharge from leg joints fluid (reflex blood) which contains an alkaloid (coccineline) that is distasteful and toxic to some bird predators [37]. Asian ladybird beetles (*Harmonia axyridis*) with more extensive red carotenoid colouration in their elytra are endowed with more alkaloid (harmonine) [38].

torulene (**11**)

β-carotene (**3**)

Several aposematic butterfly and moth species have been analysed for carotenoid profiles, for example small ermine moth (*Yponomeuta mahalebellus*), magpie moth (*Abraxas grossulariata*) and cinnabar moth (*Tyria jacobaeae*) [39], swallowtail butterflies (*Pachliopta aristolochiae, Battus philenor* and *B. polydamas*) [40], and narrow-bordered five-spot burnet moth (*Zygaena lonicerae*) [41]. It is impossible, however, to know whether carotenoids play any role in colour signalling, because the carotenoids were extracted from whole carcasses. Carotenoids contribute to aposematic colouration in some Lepidopteran species [42]. For example, kite swallowtail butterflies (*Graphium* spp.) circulate lutein (**133**) in their wing veins, apparently giving yellow-green or emerald colouration [43]. The accumulation is selective; with *Graphium* species, for example, lutein is the only one of the eight carotenoids found in the leaves of its food plant (*Cinnamomum camphora*) that appears in the wings [43]. Caterpillars of the large white butterfly (*Pieris brassicae*) reared on a carotenoid-free diet lose their ability to develop background-contrasting aposematic colouration. Addition of lutein to the caterpillars' diet [44] restores the aposematic colouration even though this is due to melanin and bile pigments as well as carotenoids in the cuticle and epidermis, suggesting that dietary lutein plays an important role in mediating the endocrine response which gives rise to melanization and bile pigment accumulation [44].

Among vertebrates, poison frogs (Dendrobatidae) produce some of the most toxic alkaloid poisons known, and many species have bright aposematic colouration [45,46], at least some of which is due to carotenoids [42]. In birds, chemical defence, presumably against predators and/or parasites, has been described only in five species of the genus *Pitohui* and one species of a separate genus (*Ifrita kowaldi*), all of which occur in New Guinea. These species possess in their feathers batrachotoxin, the same potent neurotoxic alkaloid used by some poison frogs of the genus *Phyllobates* [47]. Pitohuis have bright orange feather colouration, but it is not known whether this is carotenoid-based or phaeomelanic.

Mimicry (copying the colour or pattern of a more dangerous, perhaps toxic, species) is a form of colour signalling related to aposematism, and may also confer defence against predation, but there are no confirmed reports of the use of carotenoids for this purpose.

c) Crypsis

Animals that are camouflaged in their environment, so that detection of them is challenging, are described as being cryptic. It may seem unlikely that brightly coloured carotenoids should confer crypsis to animals. However, experiments have shown that long-jawed goby

(*Gillichthys mirabilis*), greenfish (*Girella nigricans*), and Pacific killifish (*Fundulus parvipinnis*), when kept in tanks against different coloured backgrounds (*e.g.* white, yellow, black) for considerable periods of time (up to 2 months), gradually changed their skin colour towards that of the background [42]. Analysis showed no significant differences in carotenoid content between individuals that had been exposed to control (white) and coloured backgrounds; the colour changes were due only to expansion or contraction of the pigment masses within the xanthophores [42] (see *Chapter 10*). These, however, were studies of captive animals against artificially coloured backgrounds, and the conclusion that colour changes are adaptive and confer crypsis is speculation. It is not known whether any prey species in the wild reduce their carotenoid colour signals when at risk from predators. Copepods may reduce their carotenoid reserves when exposed to predators [48], but a signal function has not been determined for carotenoid colouration in these taxa. It has been speculated that tunaxanthin (**149**) in the skin of Trinidadian guppies (*Poecilia reticulata*) may play a role in crypsis [49].

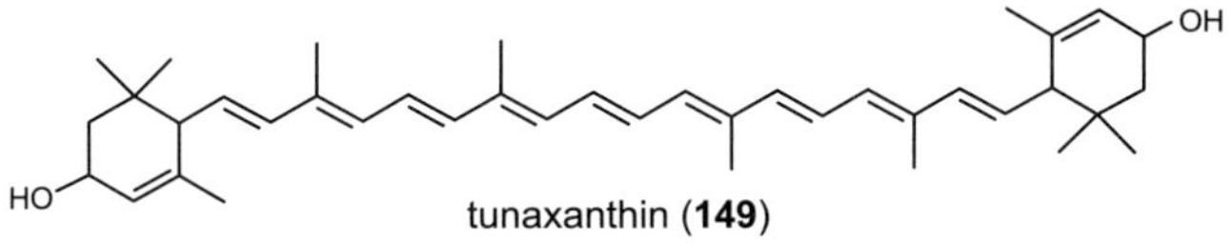

tunaxanthin (**149**)

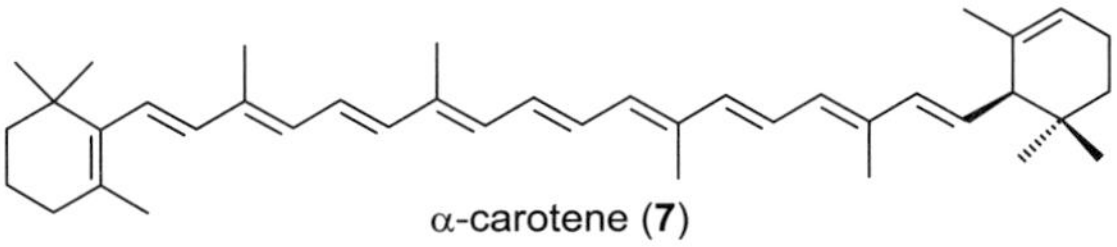

α-carotene (**7**)

In some insects, cryptic colouration is produced by a combination of carotenoids with other pigments. For example, the stick insect *Dixippus morosus* uses α-carotene (**7**) and β-carotene (**3**) together with an undetermined 'blue compound' to produce green colouration [42]. There are examples of species where carotenoids facilitate crypsis, rather than being directly responsible for it. In certain deep sea animals (fish, cephalopods, and euphausiid and decapod shrimps) that use bioluminescent signals, the bioluminescent cells form part of a complex organ (photophore) which incorporates different pigments that either reflect or absorb light, and is designed to rotate to modify the direction, intensity, spectral or angular distribution of the emitted light [50]. Carotenoids are located in the sheath cells of the photophores and are presumed to prevent lateral leakage of light; astaxanthin (**406**) and its esters are responsible for this function in the decapod shrimps *Oplophorus* and *Systellaspis*. The photophores are most abundant on the animals' ventral surface, where luminescence is directed downwards to match closely the characteristics of the light penetrating from the sea surface. This obscures the silhouette that would otherwise be visible from below [50,51].

Carotenoid-protein complexes (*Chapter 6*) can produce colours such as blue, green and grey in marine animals, including certain amphipods and crustaceans. For example, American lobsters (*Homarus americanus*) are predominantly almost black due to a carotenoid-protein complex [42]. Such carotenoid-protein complexes could confer crypsis, though this has not been investigated.

2. Signalling to members of the same species (conspecifics)

a) Sexual signalling

Most animals carefully select the individuals with which they mate and, in many instances, pursue mates of the highest quality, in order to maximize the number and quality of offspring they produce [52]. Colours (often carotenoid-based) are among the many traits, including songs, dances, pheromones, horns/antlers/spurs, and body size/symmetry, on which animals base their mating decisions [53]. Females are typically the more selective sex when it comes to mating, due to their larger investment in gametes (eggs), and thus males have developed traits like bright colours to attract female mates [54]. This sexual dichromatism (colour differences between the sexes) is often (but not always) a first indication that sexual selection by female mate choice is driving extravagant displays in male animals.

Although the first evidence for a sexual-signalling role of carotenoid-based colours in male animals came from studies in the early 1900s (*e.g.* in three-spined sticklebacks *Gasterosteus aculeatus* [55]), it has been only within the past few decades that evolutionary theories of mate choice have been widely accepted, mathematically modelled, tested, and ultimately validated. Pioneering work [56] on carotenoid pigmentation in the orange skin of Trinidadian guppies (*Poecilia reticulata*) stimulated research on sexual selection for carotenoid pigmentation in fishes and birds [57]. Direct mate selection, independent of other processes such as intrasexual competition, is a difficult behaviour to observe in the field, but some examples, namely two fish species, one lizard species, and seven bird species, where careful experimental studies have shown that females choose mates based on variation in the extent or quality of male colours that are confirmed to be carotenoid-derived, are listed in Table 1. These studies are of different kinds (*e.g.* field *v.* laboratory, correlational *v.* experimental), but their results are consistent with the notion that brighter or more extensive colours are sexually favoured. There are other studies on mate choice for bright colouration in these taxa [reviewed in 58,59] but, in these cases, the chemical basis for the colour has not been identified. Evidence is lacking for a sexual signalling role of carotenoid colours in invertebrates, amphibians, and mammals, which are less commonly sexually dichromatic. In contrast, in some birds (*e.g.* *Euplectes* widowbirds [60] and common yellowthroat *Geothlypis trichas* [61]), carotenoid-containing colours in males are not always sexually attractive and may serve other purposes.

Table 1. Animal species for which the mate-choice function of bright carotenoid-based colours in males has been tested. Only those studies where careful experimental work has been conducted are included.

Vertebrate class	Species name	Common name	References
Osteichthyes	*Poecilia reticulata*	Guppy	[56,62-65]
	Gasterosteus aculeatus	Three-spined stickleback	[55,66]
Reptilia	*Sauromalus obesus*[a]	Chuckwalla	[60]
Aves	*Carpodacus mexicanus*	House finch	[67,68]
	Carduelis tristis	American goldfinch	[69]
	Carduelis spinus	Eurasian siskin	[70]
	Malurus melanocephalus	Red-backed fairy wren	[71]
	Ploceus cucullatus	Village weaverbird	[72]
	Emberiza citrinella	Yellowhammer	[73,74]
	Taeniopygia guttata	Zebra finch	[75,76]

[a]Yellow/orange skin also derives some colour from pterin pigments.

Some female animals also display large, rich regions of carotenoid-based colour, in a few species matching or even surpassing those of males in colour and size. These female colours may be non-functional and simply genetically correlated with expression in males, especially when females display only vestiges of the male trait [77]. In several evolutionary lineages, however, female colour expression has become more elaborate [78,79], independent of any changes in male colour, which suggests an adaptive value. There is now good evidence that carotenoid-based colours in females of one fish species and three bird species catch the attention of males in mating contexts [80] (Table 2). For several other species, studies show assortative mating based on carotenoid-based colour (*i.e.* where more carotenoid-rich males tend to pair with more carotenoid-rich females, which has been interpreted by some as evidence for male mate choice), but such studies cannot rule out condition-dependent female mate choice or the role of intrasexual competition in pair-bond formation.

There is clear adaptive value to male mate choice in species where males invest heavily in offspring care, if carotenoid colouration is linked to the reproductive quality of females. Males do not always select the most colourful females, however (Table 2), and sometimes do not use colour at all in mate choice (*e.g.* red-winged blackbirds, *Agelaius phoeniceus*) [81].

Table 2. Animal species in which female carotenoid-based colours are used by males in mate selection. Only species where direct mate choice was tested and confirmed by experiment are included.

Vertebrate class	Species name	Common name	References
Osteichthyes	*Gobiusculus flavescens*	Two-spotted goby	[82]
Aves	*Carpodacus mexicanus*	House finch[a]	[83]
	Taeniopygia guttata	Zebra finch[b]	[75]
	Petronia petronia	Rock petronia	[84]

[a] Only captive males exhibit this preference. In the wild, males pair with older females, who actually tend to be less colourful than young females.
[b] Males prefer intermediately coloured females.

b) Social status signalling

i) Intra-sexual competition. Animals of the same sex often compete for limited or valuable space, food or mates and use signals that communicate their intent and ability to fight; this ensures that not all fights escalate to the point of injury or death [1]. Male birds that live in non-breeding foraging flocks frequently encounter new, unfamiliar rivals and could benefit from using signals that reveal their aggressive nature [85]. Carotenoid-containing colours are among the traits that serve as signals of social status in both fishes and birds [86]. The few examples where males with vibrant or larger colour patches have been shown to be dominant to males with smaller, less intense colours are listed in Table 3. This is not a universal role, however. In other species (*e.g.* house finch [87]; northern cardinal *Cardinalis cardinalis* [88]), there is no consistent link between male colour and aggression.

Females of many colourful species are also aggressive but, in the only example tested (the house finch), no relationship was found between plumage colouration and dominance [89].

Table 3. Animal species in which male carotenoid-based colours have been shown experimentally to serve as status signals.

Vertebrate class	Species name	Common name	References
Chondrichthyes	*Cichlasoma meeki*	Firemouth cichlid	[90]
Aves	*Euplectes ardens*	Red-collared widowbird[a]	[91,92]
	Euplectes axillaris	Red-shouldered widowbird[a]	[93,94]
	Agelaius phoeniceus	Red-winged blackbird[a]	[95,96]
	Xanthocephalus xanthocephalus	Yellow-headed blackbird	[97]

[a] These species also deposit melanin pigments into their colourful feather patches.

ii) Individual recognition. Signalling an individual's identity is common in animals that exhibit discrete, genetically determined morphotypes and that live in crowded environments, where otherwise it would be difficult to recognize neighbours and strangers [98]. Carotenoid-based plumage is thought to serve this very function in the red-billed quelea (*Quelea quelea*) [99], which is perhaps the most abundant bird on Earth and is found in densities of up to 60,000 birds per hectare.

c) Parent - offspring signalling

One of the classic examples of a behavioural releaser, *i.e.* a simple cue by one animal that activates an instinctive response in another, is the carotenoid-based red beak spot of adult herring gulls (*Larus argentatus*), which induces a chick to open its mouth to receive food [100]. Young birds also occasionally bear brilliant colours that serve to reveal to parents their nutritional needs or their worth as offspring [101]. For example, the red mouth flush of begging nestling common canaries (*Serinus canaria*) directly reflects hunger [102]. In

American coot (*Fulicula americana*) chicks, orange head plumes attract parental attention; offspring lacking them receive less food and grow more slowly [103]. Ephemeral mouth colours in young birds are largely thought to be caused by increased flow of blood (red haemoglobin), whereas orange colour in the feathers can be produced by carotenoid, phaeomelanin, or pterin pigments [104]. In one example, however, dietary supplementation with lutein (**133**) led to reddening of the mouth of nestling barn swallows (*Hirundo rustica*) [105].

d) Other conspecific signal functions

Sexually variable traits can serve other conspecific-signalling functions [98] as (i) signals of genetic relatedness or compatibility between prospective mates, (ii) signals of reproductive strategy (*e.g.* sneaker *versus* territorial males), (iii) signals of presence (*e.g.* promoting group cohesion), and (iv) amplifiers of other signals, to enhance detection and discrimination. To date, there have been no studies of whether carotenoid-based colour in animals functions in any of these ways.

Morphological traits such as colours may not serve only a single function, but may be used for multiple purposes, *e.g.* signalling to both males and females, or be presented in conjunction with other signals that reveal similar or different information [106]. For example, male widowbirds use carotenoid-containing colour to signal social status to other males, but use tail length to attract female mates [107]. No carotenoid-based ornament, however, has yet been found to have dual utility in epigamic and intrasexual contexts (*e.g.* unlike melanic plumage colouration in common yellowthroats [61]). It is also possible that carotenoid intake may broadly influence the expression of many ornament types and serve as a 'missing link' underlying redundant signal expression in animals.

D. Information Content of Carotenoid Signals

Animal signals, including carotenoid-based colours, have the potential to reveal information about nutritional state, disease status/health, genetic quality, aggressiveness, and fertility, among other characteristics [1].

1. Nutritional state

Clearly, because carotenoids are derived from the diet, their acquisition from food is critical for developing carotenoid-based colouration. Numerous studies of captive birds and fishes have demonstrated that dietary deprivation of carotenoids reduces, and supplementation enhances, carotenoid-based colour expression [53,108,109]. Typically, only colour intensity is affected and not patch size, which is thought to be under stronger genetic control. Various xanthophylls and carotenes can brighten the integument in colourful species [109], though there are some exceptions; lycopene (**31**), violaxanthin (**259**), and β-carotene (**3**) have no effect in common canaries [110], nor does β-carotene in American goldfinches (*Carduelis tristis*) [109] or canthaxanthin (**380**) in carmine bee-eaters (*Merops nubicus*) [111]. In some species it is clear that the ability to take up certain carotenoids from the diet is lacking (*e.g.* β-carotene in green iguanas (*Iguana iguana*) [112]; lutein (**133**), zeaxanthin (**119**) and lycopene (**31**) in American flamingos (*Phoenicopterus ruber*) [113]), though effects on colouration have not been explored. In other species, notably the house finch, selectivity for colouration has been identified; ingestion of common dietary xanthophylls (lutein and zeaxanthin) creates a yellow plumage, whereas β-cryptoxanthin (**55**) is metabolized to ketocarotenoids such as 3′-hydroxyechinenone (**295**) to make the feathers red [114,115].

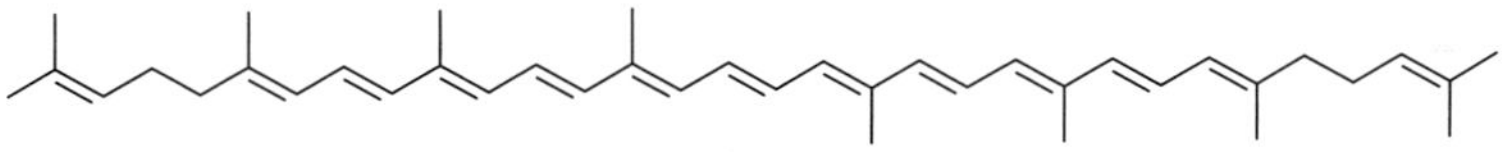

lycopene (**31**)

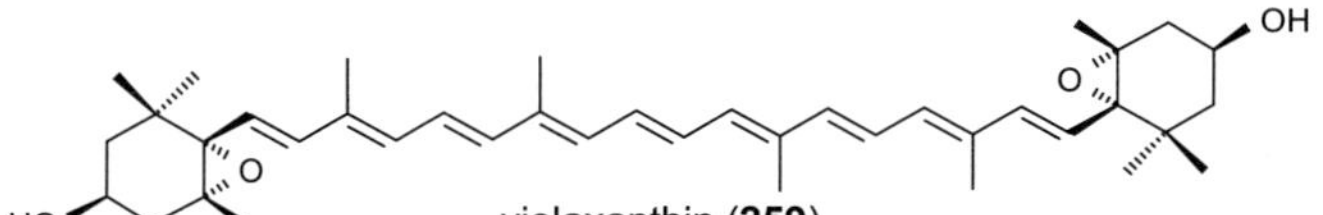

violaxanthin (**259**)

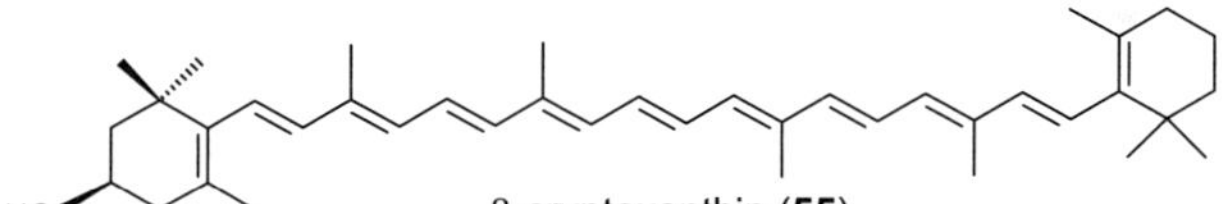

β-cryptoxanthin (**55**)

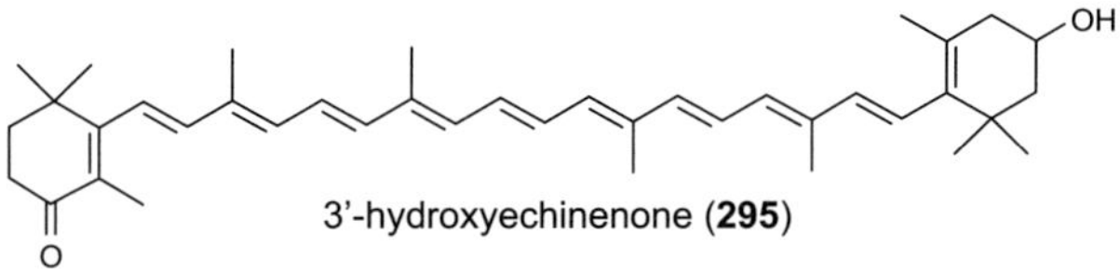

3′-hydroxyechinenone (**295**)

But what nutritional factors are responsible for the colour variation seen in wild animals? Does natural intake of carotenoids predict colour intensity and allow individuals to signal nutritional proficiency with their colour signals? This appears to be the case in two species that have been studied. Male guppies [116] and house finches [117] that had higher concentrations of carotenoids in their gut contents exhibited deeper colours. Other studies have accumulated consistent, though indirect, evidence that more colourful birds are superior foragers, *e.g.* in Eurasian siskins (*Carduelis spinus*) [118]. Generally, however, little is known about what carotenoids animals are ingesting in their food at the time that colour is being developed.

There remains the question of whether general 'nutritional state' or total food intake influences carotenoid-based colour intensity. The first study of this showed that male house finches that grew feathers at a faster rate were more colourful [119]. By manipulating food access, while controlling carotenoid intake, it was demonstrated that total caloric intake influences plumage redness in male house finches [114]. This relationship was also recently confirmed for female house finches [89] and for male American goldfinches (*Carduelis tristis*) [120]. It is not known what controls this relationship. In American goldfinches, the reduction in carotenoid levels was detected only in blood, so it was inferred that carotenoid assimilation from food is sensitive to general nutrition [120]. Cholesterol plays an important role in carotenoid accumulation and colouration in male zebra finches [121].

Recent studies of captive birds have revealed that the quality of nutrition received early in life affects carotenoid assimilation at adulthood. Thus, poor early nutrition results in reduced carotenoid colouration in adult ring-necked pheasants (*Phasianus colchicus*) [122], and in reduced blood carotenoid levels in adult zebra finches [123]. Despite the reduced blood carotenoid levels, zebra finches did not have reduced carotenoid pigmentation (bill redness) and, in mate choice trials, were no less attractive to females than were control males [123]. Possibly, individuals of relatively short-lived species such as the zebra finch that have received poor early nutrition may maximize sexual attractiveness to obtain a mate in the short term, even if this may have adverse effects on health and potential lifespan [123].

2. Parasite load

Parasites and pathogens exert a broad range of effects on animal behaviour and physiology, so it is not surprising that carotenoid-based colours are sensitive to disease status. Initial studies supporting this link were done with poultry; 'pale-bird syndrome', *i.e.* the fading of yellow tissues when chickens are infected with coccidian parasites, has long been documented [124]. The earliest empirical studies on the relationship between disease and colouration in wild animals showed that, in two species of fish (three-spined sticklebacks and guppies), parasitized males were less colourful and hence less preferred as mates by females [66,125].

The effects of different parasites (*e.g.* haematozoans, pox virus, coccidians, and ectoparasites like fleas and ticks) on carotenoid pigmentation in birds have also been studied. Coccidians have the strongest negative influence (*e.g.* in male and female house finches [89,126], American goldfinches [127], and greenfinches, *Carduelis chloris* [128]), perhaps because they inhabit the gut lining and directly disrupt nutrient absorption and lipoprotein production [129], and drain internal carotenoid stores that may have a role in fighting infections [130]. Mycoplasmal conjunctivitis (caused by a bacterium) and avian pox (a virus) also reduce carotenoid colouration in male house finches [131]. Macroparasites like mites and fleas act similarly in a wide range of birds [132,133]. Haematozoans have less clear, consistent effects on avian carotenoid pigmentation [108].

3. Immune defence

Brighter carotenoid colouration may signal an individual's greater capacity to defend against parasites and diseases, and hence the greater ability to provide direct benefits such as parental care [134]. The beneficial effects of carotenoids on the immune system are discussed elsewhere (*Volume 5, Chapter 17*). Individuals with a lower burden of parasites or diseases may have more carotenoid available, so that their superior health is advertised to prospective mates [134]. This hypothesis is supported by numerous correlational studies [135], but there has been limited experimental work. Dietary carotenoid supplementation of captive male zebra finches confirmed that carotenoid availability can be limiting for sexual attractiveness [76], and for cell-mediated and humoral immune responses following experimentally applied challenge [76,136]. In European blackbirds (*Turdus merula*), inoculation with a novel antigen, sheep red blood cells, resulted in diminished bill colouration, suggesting that humoral immune activation depleted the body pool of carotenoids [137]. Similar findings have been reported for zebra finches [136] and mallards [138].

Are animals that are more colourful more resistant to infections or able to clear them more quickly? In two species of birds, more colourful individuals exhibited stronger responses to disease challenges. Greenfinches that displayed larger yellow patches in tail feathers were better able to resist and clear infection by Sindbis virus [139]. Redder male house finches cleared mycoplasmal conjunctivitis infection faster [140]. Vivid yellow leg colour is strongly correlated with the ability of American kestrels (*Falco sparverius*) to counter blood-parasite infections [141]. However, dietary carotenoid supplementation of captive male American goldfinches had no effect on immune responses or disease resistance [142].

4. Antioxidant activity

It has been suggested that the expression of signals, including carotenoid colouration, may be sensitive to an individual's oxidant-antioxidant balance, and ability to resist oxidative stress [143]. Very few experimental studies have been performed, but one, with zebra finches, has shown that carotenoid-supplemented individuals have greater resistance to oxidative damage, supporting the hypothesis that more intense carotenoid colouration may advertise an individual's superior capacity for antioxidant protection [144] (see *Volume 5, Chapter 12*).

5. Fecundity

a) Egg production

Carotenoids are an important constituent of eggs, at least in taxa that produce yolky follicles [145]. If insufficient carotenoid is available, this may constrain a female's ability to lay, or the quality of her eggs. Studies of non-domesticated bird species in the wild and in captivity have shown that dietary supplementation with carotenoids can result in a range of beneficial effects including increased production of clutches [146], increased yolk carotenoid concentrations [147-149], reduced yolk susceptibility to lipid peroxidation [149,150], and increased offspring survival [149]. Similar effects have been observed in fish [151]. Females of many bird and fish species have carotenoid colouration, but it is not clear whether this may signal fecundity to males. In the two-spotted goby, a small marine fish, males prefer to mate with those females displaying the reddest abdomen [82], which reveals the pigmentation of the eggs visible through the abdominal wall [152]. There is little evidence, however, of a correlation between female colour (or egg carotenoid levels) and egg quality in this species [153]. In birds, house finch females with the most intense carotenoid plumage display are preferred as mates, but no association between female colouration and reproductive success was found [83]. In wild blue tits (*Parus [Cyanistes] caeruleus*) there was no relationship between the carotenoid-based plumage colouration of adult females and the degree of carotenoid deposition in eggs [147]. In contrast, both male and female captive zebra finches that were fed lutein and zeaxanthin developed a more colourful (carotenoid-based) beak which, in the female, directly correlated with the concentration of carotenoids delivered to the eggs [149]. However, female birds may invest more in offspring production (*e.g.* in egg carotenoids) when paired to a more attractive male [154].

b) Sperm production

It has been suggested that carotenoid colouration in males may be an indicator of fertility, *i.e.* the capacity to produce high quality sperm [155,156]. There is no evidence in the literature that carotenoids are themselves present in sperm or seminal plasma in any species bearing carotenoid colouration. However, if carotenoid colouration is a general indicator of overall health, this may include sperm quality.

Work with Trinidadian guppies has shown that carotenoid colouration predicts sperm reserves [157] and fertilizing success [158]. In mallards (*Anas platyrhynchos*), a link has been reported between bill colouration, higher blood levels of carotenoids and faster swimming sperm [159]. Other studies, however, have found no correlation or even negative correlations between carotenoid colouration and sperm quality parameters [156].

c) Offspring rearing

Studies of a range of bird species have suggested that individuals with greater carotenoid colouration work harder at providing food for mates and nestlings [*e.g.* 8,160-162]. For example, in a cross-fostering experiment with blue tits, chick growth was related to the plumage yellowness of the foster father but not to that of the genetic parents [161].

6. Genetic quality

By mating with an elaborately ornamented male, a female may obtain 'good genes', *i.e.* genes that aid in the survival and reproduction of her progeny. For this advantage to be conferred by intense carotenoid-coloured males, colouration must be heritable and genetically coupled to some measure of offspring quality. In one species (the zebra finch), quantitative genetic studies show that carotenoid-based beak colour in males is highly heritable [163]. However, this work was performed before it was known that maternal sources of yolk carotenoids could confound studies of colour development in offspring. Future studies should be aimed at determining whether carotenoid colouration and the fitness advantages it confers are linked to candidate genes known to have fitness consequences in wild animals [164].

7. Photoprotection

Although it is visible to many animal taxa and provides important information during mate choice and foraging [165], UV radiation has the potential to cause serious damage to biomolecules. Most bird species moult once per year, at which time carotenoids are deposited into growing feathers. Once feathers are fully grown, they are metabolically inert, so the carotenoids contained in them are non-retrievable. Individuals in better health at the time of moult may be able to allocate greater amounts of carotenoids to plumage display. Oxidation caused by exposure to UV radiation could explain the seasonal fading of feather carotenoids observed in green jays (*Cyanocorax yncas longirostris*) [166], great tits (*Parus major*) [167] and house finches [168]. If so, the rate of bleaching of feather carotenoids could be inversely related to the initial concentration of carotenoids deposited into feathers at the time of moult. Feather colouration may, therefore, become progressively more reliable as an indicator of individual condition over the course of a season, especially at the time that mate choice decisions are made, which is typically several weeks or months post-moult. Bare parts (bill, legs, *etc.*) are metabolically active, and could constitute a dynamic signal of defence against UV exposure.

E. Measurement of Carotenoids and Colour in Ecological Studies

1. Carotenoid analysis and colour measurement

To provide the most information, accurate, reliable, detailed analysis of carotenoids is essential. Thin-layer chromatography (TLC) is useful for screening, qualitative comparison and isolation, but the method of choice for routine analysis is HPLC. All analyses should be supported by the rigorous identification of the compounds involved. The principles and practice of HPLC are described in *Volume 1A*, and the identification of carotenoids in

Volumes 1A and *1B*. Recent procedures, guidelines and recommendations for HPLC and HPLC-MS are given in *Volume 5, Chapter 2*.

HPLC is used routinely to analyse plants (leaves, fruit, flowers, roots *etc*). With animals, however, the technique can only be used to analyse museum and collected specimens, pathological samples, blood and feathers. It cannot be used in the field. For information about living animals and populations in the field, non-invasive methods for estimating carotenoids have to be used.

In most ecological studies, carotenoid colouration is estimated qualitatively (*e.g.* by categorizing individuals into 'brighter' and 'duller' groups), semi-quantitatively by colour chips, or quantitatively by digital analysis of photographs or by reflectance spectrophotometry [14,15,102-104]. Such non-invasive approaches do not give information about the carotenoids and carotenoid profile [169]. Actual carotenoid profiles are known for relatively few species (*e.g.* for less than 1% of the 8000 or more species of birds), and this detailed analysis needs to be extended to a much wider range of animals, as a basis for ecological studies. Once the relationship between colouration and carotenoid content has been determined for a particular species, non-invasive colour measurements may then provide a good indication of actual carotenoid levels in that species [169].

2. Sampling range

Most studies of the signal functions of carotenoid colouration have focused on vertebrates, especially birds and fishes. Carotenoid colouration in signalling contexts has been remarkably little studied in invertebrates, despite there being a considerable amount of detailed information on the carotenoid profiles of these taxa [3]. Insects are notably under-represented in the literature on carotenoid signalling, though this group accounts for over half of the 1.4 million species of animals that have so far been described [170]. There is, therefore, a need for more studies of the signal functions of carotenoid colouration in taxa other than birds and fishes.

3. Temporal changes in colouration

Studies of the information content of carotenoid colouration need to take into account the potential for colour change. Bare parts such as skin may change colouration over relatively short timescales, *e.g.* within a breeding season in male three-spined sticklebacks [171]. Even feathers are not exempt from colour change, however. In birds, repeated measurements of carotenoid colouration of plumage in the same individuals over successive years can show remarkable variation [161]. Certain taxa may exhibit rapid and ephemeral colour changes as stress responses, for example to capture or crowding, or in response to changes in physical condition, motivation, or social status [172].

F. Applied Value of Carotenoid Colouration

1. Maintaining colouration and health in captive animals

The aesthetic value of carotenoid colour to humans cannot be measured scientifically, but its financial value is clear to those involved in managing animals in the pet trade, in zoos, and for commercial production. Pet owners and zoo visitors want to own or see attractively coloured, healthy animals. Consumers want to purchase vibrant, fresh-looking meat and eggs. Carotenoid-coloured species and products have long been a part of the pet trade (*e.g.* canaries), zoo exhibits (*e.g.* flamingos), and produce markets (*e.g.* salmon steaks, chicken meat and eggs), and decades of scientific research have been put into the need to maintain bright colour in these and other domesticated animals [173-175]. Just as with the wild animals described above, providing a carotenoid-rich diet to captive animals and housing them under sufficiently healthy conditions is necessary to achieve optimal pigmentation.

Managers seeking pigmentation strategies for domestic animals can learn much from evolutionary ecology. First and foremost, it should be determined whether or not the focal trait or tissue in the captive species is carotenoid-based; not all red, orange, and yellow colours are carotenoid-based (see [104] for a review). Parrots, for example, do not use carotenoids to develop such colours, but instead use a novel class of non-isoprenoid linear polyenal pigments (psittacofulvins [176,177]) whose production is not as sensitive to diet as are carotenoid colours.

2. Conservation implications of carotenoid signals

Animals use their carotenoid colours to communicate with one another about their quality and viability, so there is every reason to believe that humans can similarly use these features in 'indicator species' to monitor attributes of population and environmental health *via* the spatial and temporal variability that carotenoid colours exhibit [178,179]. Sexually selected traits generally have this utility for being sensitive to, and hence revealing, recent environmental perturbations and, for example, have been used to monitor effects of the Chernobyl nuclear disaster in the 1980s [180]. To date, little empirical work has been done to assess the conservation value of carotenoid colours, but several related observations emphasize their potential in this respect. For example, the average carotenoid colouration in a population of northern cardinals was found to decrease after a harsh winter that presumably killed many fruit crops [181]. Moreover, a successfully re-introduced population of the endangered hihi (*Notiomystis cincta*) from New Zealand was found to have the highest carotenoid levels among all populations studied, suggesting that carotenoids were somehow more available from the environment (or less demanded by the immune system) in these birds [182].

Conservation plans are far too often adopted *after* significant mortality events or environmental changes, perhaps because of the lack of easily collectable data that may point to potential ecological or population declines. Because carotenoids are so closely linked to the current environment and to an animal's current health state, they are good candidates for background monitoring of the health and viability of animal populations and habitats. Samples that can be collected relatively non-invasively, such as eggs or moulted feathers, may be especially useful in this context.

G. Conclusions

The study of carotenoid colouration has grown apace over the past decade, as ecologists increasingly embrace the tools of chemical analysis to glean insights into the information content of carotenoid colouration. Investigations of epigamic traits in particular have been fruitful in the development of theory, and improving our understanding of the information content of carotenoid colouration. Yet few studies have included rigorous carotenoid analysis, and in every area covered by this Chapter, unanswered questions remain. For example, can carotenoid availability mediate crypsis? Can carotenoid colouration in female animals signal their fecundity to prospective mates? Does carotenoid-based gape colouration signal condition and neediness in chicks? Studies of taxa other than birds and fishes, in particular insects, offer much potential for furthering our understanding of carotenoid colouration. There is also a need for studies that bridge traditional ecological research boundaries. For example do foragers target more carotenoid-rich flowers and fruits, and does greater carotenoid allocation to signal expression ensure enhanced reproductive success in plants? Are there costs to plants of allocating greater amounts of carotenoids to colour signals as opposed to alternative somatic functions? Carotenoid colouration therefore promises to remain a rich source of inspiration for new ideas and discovery, especially when this leads to collaboration amongst behavioural ecologists, botanists, biochemists and chemists.

References

[1]   J. Maynard-Smith and D. Harper, *Animal Signals*, Oxford University Press, Oxford (2003).
[2]   A. Zahavi and A. Zahavi, *The Handicap Principle*, Oxford University Press, New York (1997).
[3]   T. W. Goodwin, *The Biochemistry of the Carotenoids, Vol. II, Animals*, Chapman and Hall, London (1984).
[4]   H. M. Schaefer, V. Schaefer and D. J. Levey, *Trends Ecol. Evol.*, **19**, 577 (2004).
[5]   L. Chittka, J. Spaethe, A. Schmidt and A. Hickelsberger, in *Cognitive Ecology of Pollination: Animal Behavior and Floral Evolution* (ed. L. Chittka and J. D. Thomson), p. 106, Cambridge University Press, Cambridge (2001).
[6]   J. B. Harborne, *Nat. Prod. Rep.*, **18**, 361 (2001).
[7]   H. M. Schaeffer, K. McGraw and C. Catoni, *Funct. Ecol.*, **22**, 303 (2008).
[8]   M .R. Weiss, *Anim. Behav.*, **53**, 1043 (1997).
[9]   H. Omura and K. Honda, *Oecologia*, **142**, 588 (2005).
[10]   E. Melendez-Ackerman, D. R. Campbell and N. M. Waser, *Ecol.*, **78**, 2532 (1997).
[11]   S. D. Healy and T. A. Hurly, in *Cognitive Ecology of Pollination: Animal Behavior and Floral Evolution* (ed. L. Chittka and J. D. Thomson), p. 127, Cambridge University Press, Cambridge (2001).

[12] L. Chittka and N. M. Waser, *Israel J. Plant Sci.*, **45**, 169 (1997).
[13] M. R. Weiss, *Nature*, **354**, 227 (1991).
[14] H. M. Schaefer and V. Schmidt, *Proc. Roy. Soc. Lond. B (Suppl.)*, **271**, S370 (2004).
[15] N. T. Wheelwright and C. H. Janson, *American Naturalist*, **126**, 777 (1985).
[16] K. J. McGraw, *Anim. Behav.*, **69**, 757 (2005).
[17] S. Lev-Yadun, A. Dafni, M. A. Flaishman, M. Inbar, I. Izhaki, G. Katzir and G. Ne'eman, *Bioessays*, **26**, 1126 (2004).
[18] W. D. Hamilton and S. P. Brown, *Proc. Roy. Soc. Lond. B*, **268**, 1489 (2001).
[19] M. Archetti and S. P. Brown, *Proc. Roy. Soc. Lond.. B*, **271**, 1219 (2004).
[20] S. B. Hagen, I. Folstad and S. W. Jacobsen, *Ecol. Lett.*, **6**, 807 (2003).
[21] S. B. Hagen, S. Debeausse, N. G. Yoccoz and I. Folstad, *Proc. Roy. Soc. Lond. B (Supp.)*, **271**, S184 (2004).
[22] M. Archetti and S. R. Leather, *Oikos*, **110**, 339 (2005).
[23] H. M. Schaefer and D. M. Wilkinson, *Trends Ecol. Evol.*, **19**, 616 (2004).
[24] A. R. Wallace, *Darwinism: An Exposition of the Theory of Natural Selection with Some of its Applications*, Macmillan Publishers, London (1889).
[25] D. Lack, *Ecological Isolation in Birds*, William Clowes and Sons Ltd., London (1971).
[26] R. V. Alatalo, L. Gustafsson and A. Lundberg, *Proc. Roy. Soc. Lond. B*, **256**, 113 (1994).
[27] J. W. Hardy and R. W. Dickerman, *Living Bird*, **4**, 107 (1965).
[28] K. J. McGraw, K. Wakamatsu, A. B. Clark and K. Yasukawa, *J. Avian Biol.*, **35**, 543 (2004).
[29] M. K. McNaught and I. P. F. Owens, *J. Evol. Biol.*, **15**, 505 (2002).
[30] J. A. Endler, *Ecol. Monographs*, **63**, 1 (1993).
[31] H. E. H. Paterson, *Evolution and the Recognition Concept of Species*, Johns Hopkins University Press, Baltimore (1993).
[32] V. A. Olson and I. P. F. Owens, *J. Evol. Biol.*, **18**, 1534 (2005).
[33] E. B. Poulton, *The Colours of Animals: Their Meaning and Use Especially Considered in the Case of Insects*, Keegan Paul, Trench, Trübner (1890).
[34] R. R. Baker and G. A. Parker, *Phil. Trans. Roy. Soc. Lond. B*, **287**, 63 (1979).
[35] G. D. Ruxton, T. N. Sherratt and M. P. Speed, *Avoiding Attack: The Evolutionary Ecology of Crypsis, Warning Signals and Mimicry*, Oxford University Press, Oxford (2004).
[36] G. Britton, W. J. S. Lockley, G. A. Harriman and T. W. Goodwin, *Nature*, **266**, 49 (1977).
[37] N. M. Marples, P. M. Brakefield and R. J. Cowie, *Ecol. Entomol.*, **14**, 79 (1989).
[38] A. L. Bezzerides, K. J. McGraw, R. S. Parker and J. Husseini, *Behav. Ecol. Sociobiol.*, **61**, 1401 (2007).
[39] J. Feltwell and M. Rothschild, *J. Zool.*, **174**, 441 (1974).
[40] M. Rothschild and R. Mummery, *Biol. J. Linn. Soc.*, **28**, 359 (1986).
[41] T. Bornefeld and F.-C. Czygan, *Z. Naturforsch*, **30**, 298 (1975).
[42] D. L. Fox, *Animal Biochromes and Structural Colours*, Cambridge University Press, Cambridge (1953).
[43] M. Rothschild and R. Mummery, *Biol. J. Linn. Soc.*, **24**, 1 (1985).
[44] M. Rothschild, B. Gardiner, G. Valadon and R. Mummery, *Nature*, **254**, 592 (1975).
[45] K. Summers and M. E. Clough, *Proc. Natl. Acad. Sci. USA*, **98**, 6227 (2001).
[46] H. B. Cott, *Adaptive Coloration in Animals*, Oxford University Press, Oxford and New York (1940).
[47] J. P Dumbacher, T. F. Spande and J. W. Daly, *Proc. Natl. Acad. Sci. USA*, **97**, 12970 (2000).
[48] I. T. Van der Veen, *J. Evol. Biol.*, **18**, 992 (2005).
[49] J. Hudon, G. F. Grether and D. F. Millie, *Physiol. Biochem. Zool.*, **76**, 776 (2003).
[50] P. J. Herring, *J. Optics A – Pure Appl. Optics*, **2**, R29 (2000).
[51] M. S. Nowel, P. M. J. Shelton and P. J. Herring, *Biol. Bull.*, **195**, 290 (1998).
[52] M. Andersson, *Sexual Selection,* Princeton University Press, Princeton, N. J. (1994).
[53] G. E. Hill, *Proc. Int. Ornithol. Congr.*, **22**, 1654 (1999).
[54] A. V. Badyaev and G. E. Hill, *Biol. J. Linn. Soc.*, **69**, 153 (2000).
[55] J. J. Pelkwijk and N. Tinbergen, *L. Zeitschr. Tierpsychol.*, **1**, 193 (1937).
[56] J. A. Endler, *Environ. Biol. Fishes,* **9**, 173 (1983).
[57] G. E. Hill, in *Bird Coloration. Volume II. Function and Evolution* (ed. G. E. Hill and K .J. McGraw), p. 137, Harvard University Press, Cambridge, Massachusetts (2006).
[58] M. A. Kwiatkowski and B. K. Sullivan, *Evolution*, **56**, 2039 (2002).
[59] C. Macías-Garcia and E. Ramirez, *Nature*, **434**, 501 (2005).
[60] S. R. Pryke, S. Andersson and M. J. Lawes, *Evolution,* **55**, 1452 (2001).
[61] S. A. Tarof., P. O. Dunn and L. A. Whittingham, *Proc. Roy. Soc. Lond. B,* **272**, 1121 (2005).
[62] A. Kodric-Brown, *Behav. Ecol. Sociobiol.*, **17**, 199 (1985).
[63] A. E. Houde, *Evolution,* **41**, 1 (1987).
[64] G. F. Grether, *Evolution,* **54**, 1712 (2000).

[65]   K. Karino and S. Shinjo, *Ichthyolog. Res.*, **51**, 316 (2004).
[66]   M. Milinski and T. C. M. Bakker, *Nature*, **344**, 330 (1990).
[67]   G. E. Hill, *Anim. Behav.*, **40**, 563 (1990).
[68]   G. E. Hill, *Nature*, **350**, 337 (1991).
[69]   K. Johnson, R. Dalton and N. Burley, *Behav. Ecol.*, **4**, 138 (1993).
[70]   J. C. Senar, J. Domenech and M. Camerino, *Behav. Ecol. Sociobiol.*, **57**, 465 (2005).
[71]   J. Karubian, *Evolution*, **56**, 1673 (2002).
[72]   E. C. Collias, N. E. Collias, C. H. Jacobs, F. McAlary and J. T. Fujimoto, *Condor*, **81**, 91 (1979).
[73]   J. Sundberg, *Behav. Ecol. Sociobiol.*, **37**, 275 (1995).
[74]   J. Sundberg and A. Dixon, *Anim. Behav.*, **52**, 113 (1996).
[75]   N. Burley and C .B. Coopersmith, *Ethology*, **76**, 133 (1987).
[76]   J. D. Blount, N. B. Metcalfe, T. R. Birkhead and P. F. Surai, *Science*, **300**, 125 (2003).
[77]   T. Amundsen, *Trends Ecol. Evol.*, **15**, 149 (2000).
[78]   R. E. Irwin, *American Naturalist*, **144**, 890 (1994).
[79]   K. J. Burns, *Evolution*, **52**, 1219 (1998).
[80]   T. Amundsen and H. Parn, in *Bird Coloration, Volume II. Function and Evolution* (ed. G. E. Hill and K. J. McGraw), p. 280, Harvard University Press, Cambridge, Massachusetts (2006).
[81]   K. E. Muma and P .J. Weatherhead, *Behav. Ecol. Sociobiol.*, **25**, 23 (1989).
[82]   T. Amundsen and E. Forsgren, *Proc. Natl. Acad. Sci. USA*, **98**, 13155 (2001).
[83]   G. E. Hill, *Evolution*, **47**, 1515 (1993).
[84]   M. Griggio, F. Valera, A. Casas and A. Pilastro, *Anim. Behav.*, **69**, 1243 (2005).
[85]   S. A. Rohwer, *Evolution* **29**, 593 (1975).
[86]   J. C. Senar, in *Bird Coloration, Volume II. Function and Evolution* (ed. G. E. Hill and K. J. McGraw), p. 87, Harvard University Press, Cambridge, MA (2006).
[87]   K. J. McGraw and G. E. Hill, *Behav. Ecol.,* **11**, 520 (2000).
[88]   L. L. Wolfenbarger, *Condor,* **101**, 655 (1999).
[89]   G. E. Hill, *A Red Bird in a Brown Bag: The Function and Evolution of Colorful Plumage in the House Finch*, Oxford University Press, Oxford, U.K. (2002).
[90]   M. R. Evans and K. Norris, *Behav. Ecol.*, **7**, 1 (1996).
[91]   S. R. Pryke, M. J. Lawes and S. Andersson, *Anim. Behav.*, **62**, 695 (2001).
[92]   S. R. Pryke, S. Andersson, M. J. Lawes and S. E. Piper, *Behav. Ecol.*, **13**, 662 (2002).
[93]   S. R. Pryke and S. Andersson, *Anim. Behav.*, **66**, 217 (2003).
[94]   S. R. Pryke and S. Andersson, *Behav. Ecol. Sociobiol.*, **53**, 393 (2003).
[95]   C. G. Eckert and P J. Weatherhead, *Behav. Ecol. Sociobiol.*, **20**, 143 (1987).
[96]   E. Roskaft and S. Rohwer, *Anim. Behav.*, **35**, 1070 (1987).
[97]   S. A. Rohwer and E. Roskaft, *Behav. Ecol. Sociobiol.*, **25**, 39 (1989).
[98]   J. Dale, in *Bird Coloration. Volume II. Function and Evolution* (ed. G. E. Hill and K. J. McGraw), p. 36, Harvard University Press, Cambridge, Massachusetts (2006).
[99]   J. Dale, *Proc. Roy. Soc. Lond. Ser. B,* **267**, 2143 (2000).
[100]   N. Tinbergen and A. C. Perdeck, *Behaviour*, **3**, 1 (1950).
[101]   R. M. Kilner, in *Bird Coloration. Volume II. Function and Evolution* (ed. G. E. Hill and K. J. McGraw), p. 201, Harvard University Press, Cambridge, Massachusetts (2006).
[102]   R. Kilner, *Proc. Roy. Soc. Lond. B,* **264**, 963 (1997).
[103]   B. E. Lyon, J. M. Eadie and L. D. Hamilton, *Nature,* **371**, 240 (1994).
[104]   K. J. McGraw, in *Bird Coloration. Volume I. Mechanisms and Measurements* (ed. G. E. Hill and K. J. McGraw), p, 354. Harvard University Press, Cambridge, Massachusetts (2006).
[105]   N. Saino, P. Ninni, S. Calza, R. Martinelli, F. De Bernardi and A. P. Møller, *Proc. Roy. Soc. Lond. B,* **267**, 57 (2000).
[106]   R. A. Johnstone, *Phil. Trans. Roy. Soc. Lond. B,* **351**, 329 (1996).
[107]   S. Andersson, S. R. Pryke, M. J. Lawes, J. Örnborg and M. Andersson, *American Naturalist,* **160**, 683 (2002).
[108]   G. E. Hill, in *Bird Coloration. Volume 1. Mechanisms and Measurements* (ed. G. E. Hill and K. J. McGraw), p. 507, Harvard University Press, Cambridge, Massachusetts (2006).
[109]   K. J. McGraw, in *Bird Coloration. Volume I. Mechanisms and Measurements* (ed. G. E. Hill and K. J. McGraw), p. 175, Harvard University Press, Cambridge, Massachusetts (2006).
[110]   H. Brockmann and O. Völker, *Hoppe-Seyl. Z.*, **224**, 193 (1934).
[111]   E. S. Dierenfeld and C. D. Sheppard, *Zoo Biol.*, **15**, 183 (1996).
[112]   J. Raila, A. Schuhmacher, J. Gropp and F. J. Schweigert, *Comp. Biochem. Physiol. A*, **132**, 513 (2002).
[113]   D. L. Fox and J. W. McBeth, *Comp. Biochem. Physiol.*, **34**, 707 (1970).
[114]   G. E. Hill, *J. Avian Biol.,* **31**, 559 (2000).

[115]   C. Y. Inouye, G. E. Hill, R. Stradi and R. Montgomerie, *Auk,* **118**, 900 (2001).
[116]   G. F. Grether, J. Hudon and D. F. Millie, *Proc. Roy. Soc. Lond. B*, **266**, 1 (1999).
[117]   G. E. Hill, C.Y. Inouye and R. Montgomerie, *Proc. Roy. Soc. Lond. B,* **269**, 1119 (2002).
[118]   J. C. Senar and D. Escobar, *Avian Sci., 2*, 19 (2002).
[119]   G. E. Hill and R. Montgomerie, *Proc. Roy. Soc. Lond. B*, **258**, 47 (1994).
[120]   K. J. McGraw, G. E. Hill and R. S. Parker, *Anim. Behav.*, **69**, 653 (2005).
[121]   K. J. McGraw and R. S. Parker, *Physiol. Behav.*, **87**, 103 (2006).
[122]   T. Ohlsson, H. G. Smith, L. Råberg and D. Hasselquist, *Proc. Roy. Soc. Lond. B*, **269**, 21 (2002).
[123]   J. D. Blount, N. B. Metcalfe, K. E. Arnold, P. F. Surai, G. L. Devevey and P. Monaghan, *Proc. Roy. Soc. Lond. B,* **270**, 1691 (2003).
[124]   J. K. Bletner, R. P. Mitchell and R. L. Tugwell, *Poult. Sci.*, **45**, 689 (1966).
[125]   A. E. Houde and A. J. Torio, *Behav. Ecol.*, **3**, 346 (1992).
[126]   W. R. Brawner III, G. E. Hill and C. A. Sunderman, *Auk*, **117**, 952 (2000).
[127]   K. J. McGraw and G. E. Hill, *Proc. Roy. Soc. Lond. B,* **267**, 1525 (2000).
[128]   P. Horak, L. Saks, U. Karu, I. Ots, P. F. Surai and K. J. McGraw, *J. Anim. Ecol.*, **73**, 935 (2004).
[129]   P. C. Allen, *Comp. Biochem. Physiol. B*, **87**, 313 (1987).
[130]   J. K. Tyczkowski, P. B. Hamilton and M. D. Ruff, *Poult. Sci.*, **70**, 2074 (1991).
[131]   G. E. Hill, K. L. Farmer and M. L. Beck, *J. Exp. Biol.*, **207**, 2095 (2004).
[132]   D. G. C. Harper, *Anim. Behav.*, **58**, 553 (1999).
[133]   J. Figuerola, J. Domenech and J. C. Senar, *Anim. Behav.*, **65**, 551 (2003).
[134]   G. A. Lozano, *Oikos, 70*, 309 (1994).
[135]   A. P. Møller, C. Biard, J. D. Blount, D. C. Houston, P. Ninni, N. Saino and P. F. Surai, *Avian Poultry Biol. Rev., 11*, 137 (2000).
[136]   K. J. McGraw and D. R. Ardia, *American Naturalist*, **162**, 704 (2003).
[137]   B. Faivre, A. Grégoire, M. Préault, F. Cézilly and G. Sorci, *Science, 300*, 103 (2003).
[138]   A. Peters, K. Delhey, A. G. Denk and B. Kempenaers, *American Naturalist, 164*, 51 (2004).
[139]   K. Lindstrom and J. Lundstrom, *Behav. Ecol. Sociobiol.*, **48**, 44 (2000).
[140]   G. E. Hill and K. L. Farmer, *Naturwissenschaften*, **92**, 30 (2005).
[141]   R. D. Dawson and G. R. Bortolotti, *Naturwissenschaften*, **93**, 597 (2006).
[142]   K. J. Navara and G. E. Hill, *Behav. Ecol.*, **14**, 909 (2003).
[143]   T. von Schantz, S. Bensch, M. Grahn, D. Hasselquist and H. Wittzell, *Proc. Roy. Soc. Lond. B,* **266**, 1 (1999).
[144]   C. Alonso-Alvarez, S. Bertrand, G. Devevey, M. Gaillard, J. Prost, B. Faivre and G. Sorci, *American Naturalist, 164*, 651 (2004).
[145]   J. D. Blount, D. C. Houston and A. P. Møller, *Trends Ecol. Evol,*. **15**, 47 (2000).
[146]   J. D. Blount, D. C. Houston, P. F. Surai and A. P. Møller, *Proc Roy. Soc. Lond. B*, **271**, S79 (2004).
[147]   C. Biard, P. F. Surai and A. P. Møller, *Oecologia*, **144**, 32 (2005).
[148]   J. D. Blount, P. F. Surai, R. G. Nager, D. C. Houston, A. P. Møller, M. L. Trewby and M. W. Kennedy, *Proc Roy. Soc. Lond. B*, **269**, 29 (2002).
[149]   K. J. McGraw, E. Adkins-Regan and R. S. Parker, *Naturwissenschaften*, **92**, 375 (2005).
[150]   J. D. Blount, P. F. Surai, D. C. Houston and A. P. Møller, *Funct. Ecol.*, **16**, 445 (2002).
[151]   R. Vassallo-Agius, H. Imaizumi, T. Watanabe, T. Yamazaki, S. Satoh and V. Kiron, *Fish. Sci.*, **67**, 260 (2001).
[152]   P .A. Svensson, E. Forsgren, T. Amundsen and H. Nilsson Sköld, *J. Exp. Biol.*, **208**, 4391 (2005).
[153]   P. A. Svensson, C. Pélabon, J. D. Blount, P. F. Surai and T. Amundsen, *Funct. Ecol.*, **20**, 689 (2006).
[154]   N. Saino, V. Bertacche, R. P. Ferrari, R. Martinelli, A. P. Møller and R. Stradi, *Proc Roy. Soc. Lond. B*, **269**, 1729 (2002).
[155]   B. C. Sheldon, *Proc. Roy. Soc. Lond. B*, **257**, 25 (1994).
[155]   J. D. Blount, A. P. Møller and D. C. Houston, *Ecol. Lett.*, **4**, 393 (2001).
[157]   T. E. Pitcher and J. P. Evans, *Can. J. Zool.*, **79**, 1891 (2001).
[158]   J. P. Evans, L. Zane, S. Francescato and A. Pilastro, *Nature*, **421**, 360 (2003).
[159]   A. Peters, A. G. Denk, K. Delhey and B. Kempenaers, *J. Evol. Biol.*, **17**, 1111 (2004).
[160]   S. U. Linville, R. Breitwisch and A. J. Schilling, *Anim. Behav.*, **55**, 119 (1998).
[161]   J.C. Senar, J. Figuerola and J. Pascual, *Proc. Roy. Soc. Lond. B,* **269**, 257 (2002).
[162]   M. Preault, O. Chastel, F. Cezilly and B. Faivre, *Behav. Ecol. Sociobiol.*, **58**, 497 (2005).
[163]   D. K. Price and N. T. Burley, *Heredity*, **71**, 405 (1993).
[164]   R, Buchholz, M. D. Jones-Dukes, S. Hecht and A. M. Findley, *J. Anim. Breed. Genet., 121*, 176 (2004).

[165] I. C. Cuthill, J. C. Partridge and A. T. D. Bennett, in *Animal Signals: Signalling and Signal Design in Animal Communication* (ed. Y. Espmark, T. Amundsen and G. Rosenqvist), p. 61, Tapir Academic Press, Trondheim (2000).
[166] N. K. Johnson and R. E. Jones, *Wil. Bull.*, **105**, 389 (1993).
[167] J. Figuerola and J. C. Senar, *Ibis*, **147**, 797 (2005).
[168] K. J. McGraw and G. E. Hill, *Can. J. Zool.*, **82**, 734 (2004).
[169] L. Saks, K. McGraw and P Hõrak, *Funct. Ecol.*, **17**, 555 (2003).
[170] E. O. Wilson, *The Diversity of Life*. Penguin Books Ltd., London (2001).
[171] U. Candolin, *Proc Roy. Soc. Lond. B*, **267**, 2425 (2000).
[172] A. Kodric-Brown, *Amer. Zool.*, **38**, 70 (1998).
[173] D. L. Fox, J. W. McBeth and G. Mackinney, *Comp. Biochem. Physiol.*, **36**, 253 (1970).
[174] W. L. Marusich and J. C. Bauernfeind, in *Carotenoids as Colorants and Vitamin A Precursors* (ed. J. C. Bauernfeind), p. 320, New York, Academic Press (1981).
[175] T. R. Birkhead, *A Brand-New Bird*, Basic Books, New York (2003).
[176] R. Stradi, E. Pini and G. Celentano, *Comp. Biochem. Physiol. B*, **130**, 57 (2001).
[177] K. J. McGraw and M.C. Nogare, *Biol. Lett.*, **1**, 38 (2005).
[178] G. E. Hill, *Bioscience*, **45**, 25 (1995).
[179] C. Isaksson, J. Örnborg, E. Stephensen and S. Andersson, *EcoHealth*, **2**, 138 (2005).
       [180] A. P. Møller and T. A. Mousseau, *Evolution*, **55**, 2097 (2001).
[181] S. U. Linville and R. Breitwisch, *Auk*, **114**, 796 (1997).
[182] J. G. Ewen, P. Surai, R. Stradi, A. P. Moller, B. Vittorio, R. Griffiths and D. P. Armstrong, *Animal Conservation*, **9**, 229 (2006).

Carotenoids
Volume 4: Natural Functions
© 2008 Birkhäuser Verlag Basel

Chapter 12

Carotenoids in Aquaculture: Fish and Crustaceans

Bjørn Bjerkeng

A. Introduction

This Chapter deals with selected topics on the use of carotenoids for colouration in aquaculture and incudes examples from ecological studies which support our understanding of functions and actions of carotenoids and colouration in fishes and crustaceans. Animal colours may be physical or structural in origin [1], *e.g.* Tyndall blues and iridescent diffraction colours, or they may be due to pigments, including carotenoids (*Chapter 10*).

Many marine and freshwater animals, including fish and crustaceans, owe their bright colouration to carotenoids. In captivity, such animals require a diet supplemented with carotenoids to obtain a colour that is typical for the species and to meet other requirements.

As discussed in *Chapter 11*, natural animal colouration conveys information to other individuals *via,* for instance, carotenoid-based sexual signals that influence mate choice [2] or warning colours that deter predators [3] and thus play an important role in certain ecological interactions. For humans, colours in food elicit in the consumer psychological and physiological expectations based on experience, tradition and customs, and are linked to anticipated quality [4]. The colour of a food item is a cue used to form judgements about desirability.

Carotenoids have been used extensively as additives to provide food colours, either by applying them to foods directly, or by supplying them indirectly in the diets (feed) of animals that are used for food [5]. The conspicuous colouration of most seafood is due to carotenoids

[6]. Most of the applications to farmed species involve indirect colouration. External colouration is important to the ornamental fish/animal hobbyist and the farming industry that supplies them. Carotenoids should be included in the diet of many ornamental species to avoid the dull colouration that would otherwise be acquired by many animals kept in captivity.

Capture fisheries are not expanding, but food fish production still grew at an annual rate of 3.1% in the period between 1987 and 1997, driven by aquaculture, which showed a global increase of 11.2% [7]. In 1997, production of high-value finfish (certain salmonid and sparid fish species) represented only 5% of the total aquaculture production, but generated 39% of the export revenue; 76% of the production was in developed countries. In contrast, developing countries accounted for 98% of the production of crustaceans. Flesh colour is among the most important quality parameters for salmonid fishes [8], and pigment feeding is regarded as the most important management practice for successful marketing of farmed Atlantic salmon *(Salmo salar)* [9]. The difficulty of obtaining natural colouration in captivity has been a bottleneck for successful commercialization of other species such as red seabream *(Chrysophrys major)*, gilthead seabream *(Sparus aurata)*, and red porgy *(Pagrus pagrus)*.

Astaxanthin (**404-406**) and canthaxanthin (**380**), either alone or in combination, are the carotenoids most commonly used for pigmentation in farming of aquatic animals.

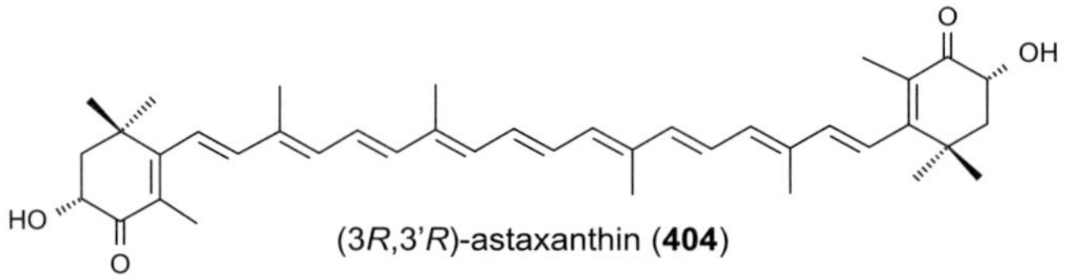

(3R,3'R)-astaxanthin (**404**)

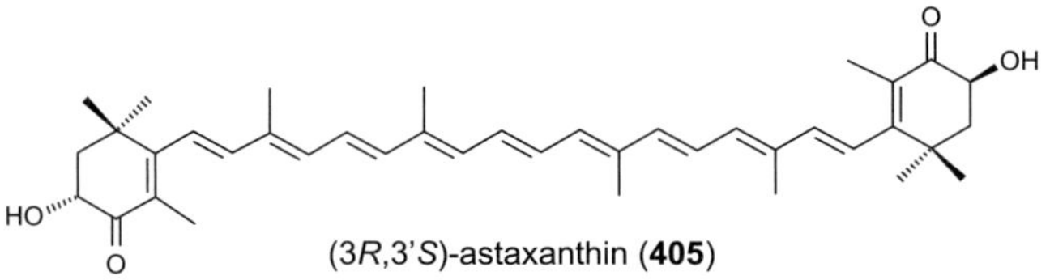

(3R,3'S)-astaxanthin (**405**)

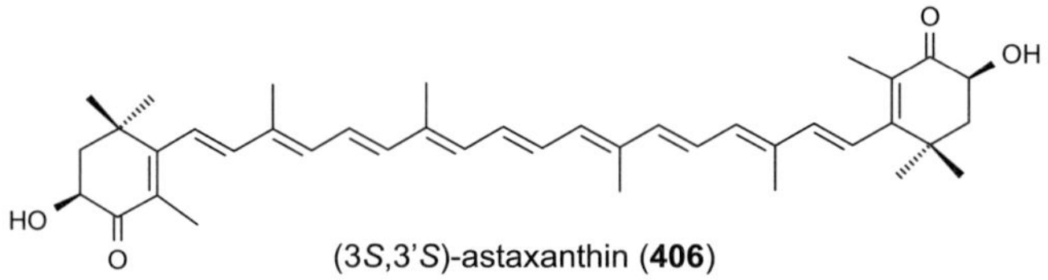

(3S,3'S)-astaxanthin (**406**)

canthaxanthin (**380**)

This Chapter covers developments in research on carotenoid pigmentation in fish and crustaceans in general and highlights collected experiments on aquacultured species during the past decade. The main focus is on muscle of salmonid fishes and integumentary pigmentation of certain other fish species and crustaceans. Comprehensive reviews have dealt with carotenoid distribution and biochemistry in animals [10], and marine animal carotenoids [11,12]. Earlier literature on carotenoid pigmentation in aquaculture is treated in previous reviews [6,13-19]. The absorption and metabolism of carotenoids in fishes and crustaceans was treated in *Volume 3, Chapter 7* [20]; only more recent developments are reviewed here.

B. Market Issues

Aquaculture is developing, expanding, intensifying and diversifying in most regions of the world, and represented 33.7% of the total world fisheries in 2005 [21]. Due to the lack of growth in capture fisheries since the late 1980s, aquaculture has to meet the growing global demand for aquatic food. According to FAO, global aquaculture production has increased at an average annual growth rate of 8.8% over the past 50 years to 59.4 million tonnes (including plants) with a total value of US$ 70.3 billions by 2004. Of this, China produced almost 70% of the production volume and the rest of Asia and the Pacific region 22% [21]. Of the global production of cultured penaeid shrimp (2.5 million tonnes), oysters (4.3 million tonnes), cyprinid fish (18.3 million tonnes) and plants (13.9 million tonnes), 87% or more is produced in Asia and the Pacific region, whereas about 56% of the farmed salmonid fishes are produced in the northern part of Western Europe. In 2004, the global production of salmonid fishes was about 2 million tonnes, and of crustaceans about 3.7 million tonnes [21]. For Norway, in 2004 the second largest fish exporter after China, sales of salmonid fishes were about 0.65 million tonnes in 2005 [22], but Norwegian fish production from aquaculture is estimated to reach 5 million tonnes by 2020, with Atlantic salmon representing 2 million tonnes [23]. Asia provides more than 50% of the global supply of ornamental freshwater and marine fish, which had an estimated wholesale value of US$ 900 million and a retail value of US$ 3 billion in 2000 [24]. The number of species traded globally is into the thousands, and USA is considered the largest market for ornamental fish [25]. Approximately 90% of the freshwater ornamental fish that are traded are cultured, whereas the marine ornamental species predominantly are collected from the wild [26]. The ornamental animal and plant trade is expanding, as illustrated by a recent survey of fish species in the upper Paraná River flood plain in Brazil, where more than 83% of the 101 captured taxa were considered ornamental [27]. Capture of marine ornamental fish, especially, raises important issues related to mortality, conservation and trade regulation [28].

Only a small fraction of the total volume of aquacultured species, reared intensively or semi-intensively, require diets supplemented with carotenoids to obtain acceptable colouration. The world market for carotenoids is discussed in *Volume 5, Chapter 4*. Astaxanthin is the major carotenoid used to supplement diets, and it would require about 130 tonnes of astaxanthin to feed the global salmonids produced by aquaculture a diet containing 50 mg astaxanthin per kg. About 90% of the astaxanthin is produced by chemical synthesis by two major companies and the current price is in the range from US$ 1500 to US$ 2000 per kg. In comparison, algal products containing astaxanthin esters that are intended for the human market sell for about US$ 5400 and upwards per kg astaxanthin. The recently developed polar water-soluble astaxanthin derivatives such as the disuccinates and diphosphates have interesting properties and may find a future application in aquaculture [29]. Since most of the shrimp and prawn production is extensive or semi-intensive, only a relatively small amount of feed supplemented with carotenoids is used. It is required in intensive production systems, however.

Nutrient requirements for ornamental fish are poorly known [30], and only a few papers are found on the supplementation of various carotenoid sources to such species. Among recent investigations, it was found that inclusion of 8% *Spirulina* was required to obtain optimum colouration of red swordtail (*Xiphophorus helleri*) [31], and that formulated synthetic astaxanthin was a better source than the microalga *Chlorella vulgaris* for skin pigmentation of goldfish (*Carassius auratus*) in terms of total carotenoid content, though different hues may be obtained [32,33]. The global production of tilapia (a common name used for about 70 species in the family Cichlidae) is about 1.8 million tonnes [21]. The red skin colour of the tilapia *Oreochromis niloticus* shows single gene inheritance [34]; marketing has benefited from the introduction and mass selection of red coloured strains [35].

C. Importance of Pigmentation

1. Colouration - ecological and evolutionary aspects

Animal colour patterns may serve roles in regulation of temperature, intraspecific communication and evasion of predators. The signalling functions of carotenoids in relation to ecology and evolution are treated in detail in *Chapter 11*. Carotenoid-based colours often have signal effects that have evolved because they elicit responses in the recipients, such as in the decisions of mate choice in guppies [2]. Whereas the coral fish *Hypoplectrus* exhibits extraordinary gene-based colour polymorphism that may drive speciation [36], other species, such as salmonid fishes, exhibit sexually dimorphic colour patterns. Many prey species have evolved to match their surroundings by adopting colour patterns that represent a more or less random sample of their background [37]. Crustaceans have cryptic colouration. An example is the blue-black of the carapace of lobster *Homarus gammarus* due to astaxanthin-based carotenoproteins (*Chapter 6*).

Sexual selection often relies on carotenoid-based signals. Colourful ornaments are often displayed by males, but occasionally also by females (as with two-spotted gobies [38]) and, as a determinant for sexual selection, play an important role in reproduction. The Pacific salmon (*Oncorhynchus nerka*) displays the most extreme (red) nuptial colour among the salmonid fishes. In a study with various abstract colour models it was shown that males have a strong preference, which apparently is innate, and spawn with models with a red hue [39]. Carotenoid-based ornaments may have evolved because they are limiting due to the scarcity of the carotenoids [40], which may be required for other functions such as provitamin A, immune function or as antioxidants. Thus, in the salmonid fish Arctic charr or char (*Salvelinus alpinus*) the intensity of red integumental colouration was negatively related to lymphocyte counts [41] and may therefore signal immunocompetence. In three-spined stickleback (*Gasterosteus aculeatus*), in which the male is the parental caretaker following spawning, a high astaxanthin intake was associated with increased reproductive investment and a longer lifespan [42]. A lower suceptibility to oxidative stress may explain the increased longevity. As indicated above, a male preference for redder females in two-spotted gobies may have evolved because improved phototaxis in the offspring improves foraging capability [43].

Although the information is not required for studies in evolution *per se*, it would be useful for the carotenoid biochemist to establish the physiological basis for carotenoid signals and why they are working. Controlled feeding trials with carotenoids would facilitate such studies. Poor and unpredictable zygote quality hamper the development of several aquacultured species, notably marine species. It has been suggested that introduction of breeding systems in which broodfish have the opportunity to choose their mates voluntarily (based on, for instance, external carotenoid-based colouration) may improve the offspring quality in terms of survival and growth due to improved genetic compatibility of the parents [44].

2. Salmon muscle colour

Consumers have a preference for pink-coloured salmonid fish products, and associate redder salmon with a fresher fish that has better flavour and higher quality [45]. Consumers are willing to pay significantly more for fillets of Atlantic salmon that are normal or above normal in redness [46,47]. The pleasure of eating salmon is related positively to the red colour which is interpreted as an indicator of quality and superior flavour [48-50]. Older scientific studies apparently supported this notion by indicating a favourable relationship between muscle colouration and flavour attributes in salmonid fishes; more recently, astaxanthin concentration of raw fillets was found to correlate significantly with more intense smoke odour and less off-odour of the smoked fish [51].

The relationship between colouration and flavour may be rationalized in a number of ways. First, colouration may interact with various sensory perceptions of properties of food commodities or beverages such as flavour intensity [52]. The correspondence between colour and taste, *e.g.* sweetness, may be learned rather than innate, however [53], so that it may be consumer expectations about salmon flavour that are related to colouration. Second, carotenoids may themselves be flavour-active, may be precursors of flavour-active degradation products [54], or may interact with chemical reactions in which flavour-active compounds are produced.

The putative effects of astaxanthin concentration and redness on the intensity of salmon flavour perception have been addressed by tests with patés of salmonid fish [55], including presentation of the samples to the assessor panel under red light to mask the colour differences. Astaxanthin or redness was found to have, at the most, only a minor influence on salmon flavour, which implied that colourants can be applied directly to such products. Astaxanthin utilization can be increased from about 10% to 100% in such instances. Sexual maturation in salmonid fishes is associated with a considerable drop in muscle and whole-body concentrations of astaxanthin and canthaxanthin [56], and induces changes in chemical composition of the muscle that are related to inferior watery and tough texture, and less pronounced odour and flavour [57]. Fallacies are frequently encountered in sensory science due, among other things, to the lesser acuity of the lower senses (taste, smell) compared to vision [58], and previous reports regarding relationships between colour and taste of salmonid fishes could be attributed to factors that not were controlled for, *e.g.* the use of shrimp extracts as astaxanthin source. Traditional opinions regarding the relationship between colour and flavour may instead reflect not the carotenoids themselves but other dietary components that are able to affect flavour, or the sexual maturation status, and may have contributed to current consumer expectations and preferences.

3. Embryonic development and larval growth

Carotenoids in fish eggs and fry and the role of xanthophylls as vitamin A precursors in fish were treated thoroughly in *Volume 3, Chapter 7* [20] and elsewhere [59]. Carotenoids accumulate in the reproductive organs of a wide range of organisms but, in an early assessment, they were found apparently to be unnecessary for normal embryonic development [60]. Until recently, firm knowledge on this topic was hampered by the lack of controlled experiments in which carotenoid supply was the only variable [61].

Fertilization rate has often been used as an indicator of egg or embryo quality, but this may not be valid [62]. In Atlantic salmon, supplementation of semi-purified casein/gelatine-based diets with astaxanthin led to improved growth performance in first feeding fry (*ca.* 0.2 g) [63,64], juveniles (*ca.* 1.8 g) [65] and parr (*ca.* 16 g) [66]. In contrast, no effect of feeding diets supplemented with canthaxanthin to rainbow trout broodstock, before spawning, was found on subsequent growth of fry [67], indicating that any role for this carotenoid in reproduction would be restricted either to long-term sub-clinical effects or to fish exposed to poor fish culture. More recently, in rainbow trout broodstock, maternal dietary supplementation with astaxanthin was found to improve the fertilization rate and percentage of eyed and hatched eggs, and to reduce mortality of eyed eggs. Astaxanthin also has a positive paternal effect on fertilization rate [68].

A pioneering study [69] suggested that all the egg carotenoids of brown trout (*Salmo trutta*) were transferred to the embryo and fry, in which astaxanthin was found predominantly as esters in the integument. Later studies with wild Atlantic salmon, however, showed that about 30% of the egg astaxanthin disappeared during development to fry [70]. The appearance of relatively large amounts of astaxanthin esters at the hatching stage indicates that it is diverted to the integument [71]. In rainbow trout, the loss of egg carotenoids is about 60% when the fish reach the start feeding stage [72]. Since these fishes are not eating at this stage, the carotenoids must have been transformed into colourless metabolites; xanthophylls such as canthaxanthin and astaxanthin serve as vitamin A precursors in salmonid fishes [20]. Astaxanthin is probably not an important vitamin A source at the egg stage whereas it may serve as a precursor at the fry stage when a functional liver has been developed [73]. Irreversible transformation into vitamin A and retinoic acids and subsequent excretion *via* catabolic pathways may be responsible for at least part of the observed loss. In juvenile hybrid tilapia (*Oreochromis niloticus* x *O. aureus*), β-carotene (**3**) may fulfil the vitamin A requirements [74].

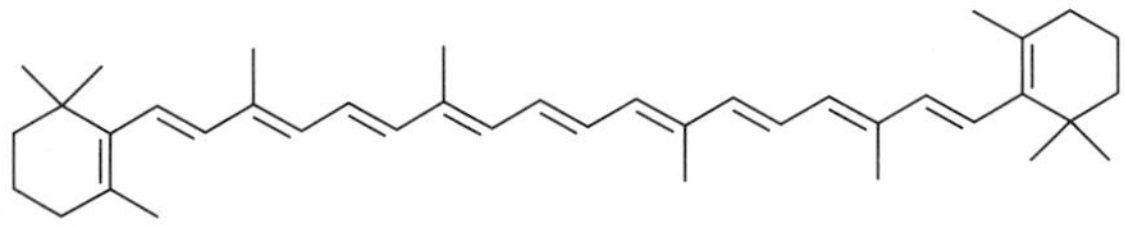

β-carotene (**3**)

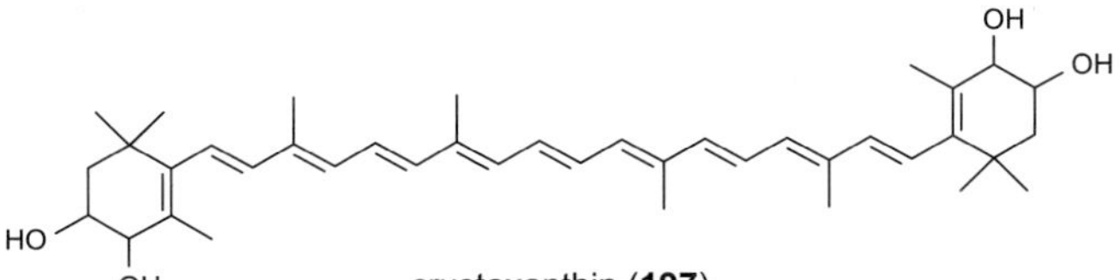

crustaxanthin (**197**)

In Arctic charr (*Salvelinus alpinus*) fed a diet supplemented with astaxanthin, the observed ratio between the (3,4,3',4'-di-*cis*), (3,4-*trans*-3',4'-*cis*), and (3,4,3',4'-di-*trans*) isomers of crustaxanthin (**197**) in the ovaries was approximately 2.6:3.1:1 (18-21% of total carotenoids) [75]. This suggests a relatively strongly stereoselective enzymic reduction of astaxanthin to crustaxanthin in favour of the sterically hindered (3,4-*cis*)-glycolic forms in the Arctic charr. Thus, the role of crustaxanthin in the retinoid metabolism of fish with a reductive carotenoid metabolism requires investigation. No metabolites were detected in the carp (*Cyprinus carpio*) after administration of radioactive crustaxanthin [76], but this species has an oxidative carotenoid metabolism [20]. Also, crustaxanthin is not present in Atlantic salmon [71]. The role of the integumental carotenoids remains to be determined firmly. In addition to serving as a reserve of provitamin A, they may contribute to camouflage, signalling, photoprotection and immunocompetence.

At present there is no knowledge about the biochemical basis of the effects of carotenoids during embryogenesis and later stages in salmonid fishes. However, a gene (*bco1*) from zebrafish (*Danio rerio*) encoding the enzyme responsible for vitamin A formation, β-carotene 15,15′-oxygenase, has been cloned [77] (*Chapter 16*). Targeted gene knockdown resulted in severe malformations of eyes, craniofacial skeleton and pectoral fins during embryonic development. Furthermore, retinoic acid formation, dependent on local formation of retinal *de novo* from provitamin A, appeared to be essential. This firmly establishes a crucial role of carotenoids in the early development of fish. However, the substrate requirements and putative roles of carotene-cleavage enzymes in formation of retinoids in developing embryos of salmonid fishes and other farmed species require investigation. The contribution of astaxanthin to the retinoid pool in eggs and fry of salmonid fishes remains to be determined. Direct effects of intact carotenoids, or metabolites other than vitamin A, should also be considered. Some studies have found carotenoids to have a positive effect on egg production and offspring growth or survival whereas several other studies have failed to detect such effects. This could be due to the choice of parameters. Female two-spotted gobies (*Gobiusculus flavescens*), fed diets supplemented with 135 mg astaxanthin per kg diet, developed a stronger nuptial coloration, were more likely to spawn and produced larvae that had a higher phototactic response than unsupplemented fish [43]. Phototaxis (see *Chapter 10*) is crucial for survival of the larvae. Moreover, commonly reported parameters for reproductive success, such as fertilization and hatching rates, were unaffected.

Broodstock diet formulation is essential in aquaculture for mass producing offspring with good quality [43,78]. Supplementation with 30 mg astaxanthin per kg soft-dry pellets for 5 months before spawning increased total egg production, egg quality and number of normal larvae of yellowtail (*Seriola quinqueradiata*) [79]. Supplementation with paprika powder as an ample source of carotenoids appears to improve larval survival more than supplementation with astaxanthin [80]. It should be noted that, in this species, astaxanthin comprises only about 1% of the total carotenoids of the eggs after the broodstock are fed a diet with astaxanthin, while zeaxanthin (**119**) and lutein (**133**) represent about 90% of the total carotenoids [80]. In striped jack (*Pseudocaranx dentex*), supplementation with 10 mg astaxanthin per kg dry pellets increased the number of eggs produced three-fold [81].

Relatively little is known about the role of dietary carotenoids in crustaceans except for their effect on colouration [82]. Retinoids were not detected in the eggs of the prawn *Penaeus semisulcatus* [83], and a role of carotenoids to provide retinoids is therefore plausible. Supplementation of broodstock diets for the prawn *P. monodon* with astaxanthin improves ovarian development and spawning, which may suggest a role in reproduction [84]. Dietary astaxanthin supplementation (30 - 120 mg/kg) apparently does not affect growth in the spiny lobster (*Panulirus ornatus*) (weight 18 g) [85].

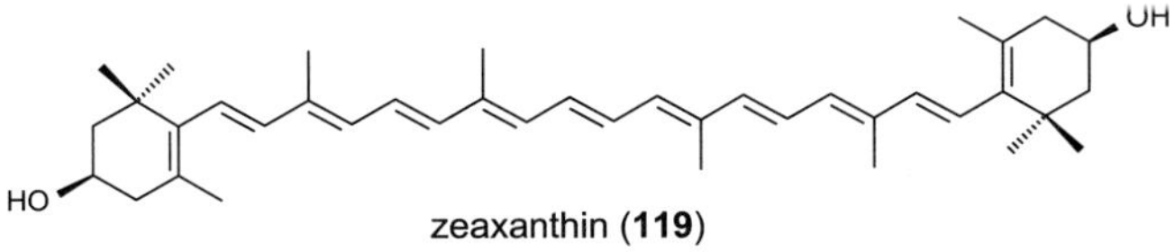

zeaxanthin (**119**)

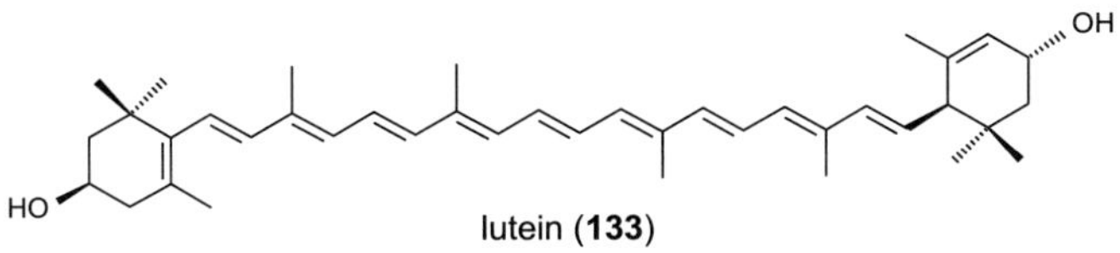

lutein (**133**)

Nutritionally enriched brine shrimps (*Artemia* spp.) are used as live feed for larval start feeding of marine fish species. Some *Z* isomers of canthaxanthin (**380**) accumulate in the ovaries, eggs and encysted embryos, but not in growing animals [20]. The possible presence of *E/Z* isomers of carotenoids in reproductive tissues of other species of crustaceans or fish has largely been ignored, but a similar geometrical isomer composition of astaxanthin was found in the diet and eggs (*Z* isomers 16% of total astaxanthin) of rainbow trout [86], which indicates that the *Z* isomers are slightly enriched in eggs compared to plasma (in which *Z* isomers comprise 7-8% of total astaxanthin) of fish fed a similar astaxanthin source [87]. The subsequent metabolic fate of these *Z* isomers is not known, but one speculation is that they may serve as precursors for the local formation of (9*Z*)- and (13*Z*)-retinoic acids which are involved in the early events of cell differentiation [88].

It is well recognized that the ratio of optical (*R/S*) isomers of astaxanthin (**404-406**) in the eggs of salmonid fishes is similar to that of the diet [89]. Nothing is presently known about the metabolism of these isomers in the early stages of development.

D. Sources of Dietary Carotenoids

The most important carotenoid sources used in aquaculture are synthetic formulated astaxanthin (**404-406**) and canthaxanthin (**380**) and, as natural astaxanthin sources, the red yeast *Xanthophyllomyces rhodorhous* (formerly *Phaffia rhodozyma*) (3*R*,3'*R*, **404**) and the alga *Haematococcus pluvialis* (3*S*,3'*S*, **406**). Other carotenoid sources with lesser commercial importance are crustaceans and their by-products for astaxanthin, and microorganisms, mostly algae, producing various carotenoids. Algae are important for direct consumption in shellfish and shrimp aquaculture and are used indirectly as food for live prey fed to fish larvae [90]. Flowers of *Adonis* spp. biosynthesize (3*S*,3'*S*)-astaxanthin (**406**) esters, but there is currently no commercial production of carotenoids from these or other plant sources. The industrial synthesis and formulation of astaxanthin, canthaxanthin and other xanthophylls is covered in *Volume 2, Chapter 3 Part VII*, and natural production by microbial and plant biotechnology in *Volume 5, Chapters 5* and *6*.

Current products containing astaxanthin or canthaxanthin are formulated to contain about 10% carotenoid. To avoid carotenoid loss during extrusion and drying of the feed pellets it is most common to use a cold-water dispersible product that can be applied to the pellets post-extrusion. The corn-starch covered beadlets (size about 400 µm) consist of a matrix, typically lignosulphonate, in which droplets of astaxanthin in antioxidants are embedded. The optical isomer ratio of synthetic astaxanthin is 1:2:1 for the (3*R*,3'*R*)-, (3*R*,3'*S*)- and (3*S*,3'*S*)-isomers, (**404, 405, 406**, respectively), and it contains about 20% *Z*-isomers. Aggregates or crystallites of carotenoids [91] (*Chapter 5*) may form before the fish are fed and this may influence the gastrointestinal absorption rate.

Whereas astaxanthin is more efficiently accumulated in the muscle of rainbow trout than is canthaxanthin [15], the opposite appears to be true for Atlantic salmon [92-94]. When the dietary carotenoid concentration exceeds 30 mg/kg, more canthaxanthin than astaxanthin is taken up into the blood of Atlantic salmon when the inclusion levels are similar. When the diet contains both carotenoids, there is a reduction in the uptake of both carotenoids, the effect being more prominent for astaxanthin [94]. In salmonid fishes, astaxanthin dipalmitate is utilized more poorly than is unesterified astaxanthin [95]. This appears to be true also for natural astaxanthin esters from *H. pluvialis* [96,97]. For skin pigmentation of the sparid fish, red porgy (*Pagrus pagrus*), *H. pluvialis* was utilized more efficiently than a commercially formulated astaxanthin [98].

Mechanical or enzymic disruption of the rigid cell wall of the yeast *X. rhodorhous* is essential for the efficient emptying of the cell content and the utilization of the astaxanthin (the 3*R*,3'*R* isomer, **404**). Although feed production losses of astaxanthin are higher with increasing degree of cell disruption [99], this is more than outweighed by the increased bioavailability of the astaxanthin for muscle pigmentation of rainbow trout [100]. A higher apparent digestibility coefficient (ADC) of astaxanthin (65-70%) is found for Atlantic salmon fed diets supplemented with a modern product of *X. rhodorhous* than for salmon fed the diet supplemented with a formulated synthetic astaxanthin (40%) [101]. This reflected the higher proportion of dietary astaxanthin utilized for muscle pigmentation (6.3%), which was 86% higher than for salmon fed the synthetic formulated astaxanthin. The large difference in utilization warrants further studies into the biochemical basis for intestinal absorption of carotenoids in fishes and the effects of the matrix within which the carotenoids are found. It is important to consider effects of both species and source with respect to carotenoid utilization.

E. Carotenoid Utilization

1. Uptake from the gastrointestinal tract

Carotenoids are poorly utilized by fish, and the retention of astaxanthin in muscle of Atlantic salmon is usually less than 12% [102-105]. One reason for this is the relatively poor absorption of carotenoids from the gut. In general, the factors that influence the bioavailability of carotenoids in humans (*Volume 5, Chapter 7*) are also relevant for fish. In addition, temperature and other environmental factors have to be taken into account for poikilothermic animals.

Determination of apparent digestibility coefficients (ADC) is a classical method for estimating nutrient absorption. In fish, the ADC is often determined by an indirect method that relates concentration of a nutrient in diet and faeces to that of yttrium oxide (Y_2O_3) or another inert digestibility marker supplemented to the diet [106]. Inhibition by phloridzin indicates that absorption of astaxanthin by chinook salmon (*Oncorhynchus tshawytscha*) is an active process [107]. In rainbow trout, however, passive absorption of astaxanthin is indicated by the linear response in blood plasma astaxanthin over a dietary concentration range 12.5-200 mg/kg [108]. Recent evidence for rainbow trout [109,110] also indicates a facilitated uptake from the gastrointestinal tract. Geometrical isomers of astaxanthin had different uptake rates from the intestine (in decreasing order all-E >13Z >9Z), whereas the optical (R,S) isomers were taken up to the same extent. The gastrointestinal absorption of astaxanthin into blood is a slow process, as indicated by the relatively high values for T_{max} (time to achieve maximum blood concentration) of about 18 to 30 hours [111-113]. Several factors such as the composition of the diet, the dietary carotenoid content, carotenoid species and molecular linkage (esterification) may influence carotenoid digestibility [14,15,17], as do water temperature and ration size (see below). The ADCs of 4-ketocarotenoids reported in the earlier literature vary widely [14,15]; figures ranging from 4-97% are reported.

It has been common practice to lyophilize faeces samples for digestibility measurements, and high digestibility estimates (>80%) and variation have been reported for freeze-dried faeces samples [114-116]. In part, this variation may be ascribed to the use of different techniques for faeces collection and sample treatment. Breakdown of carotenoids in the faeces during extraction and analysis may explain exceptionally high ADC estimates whereas incomplete extraction may explain unrealistically low values. Low water activity renders carotenoids susceptible to oxidation and degradation. Typical ADC-values for astaxanthin and canthaxanthin range between 35 and 55%, when analyses are performed on frozen non-lyophilized faeces [109,117]. Different values may be found, however, depending on factors such as dose, carotenoid source and type, and diet composition. Thus, a higher dietary lipid level (up to at least 40%) appears to have a positive effect on carotenoid uptake and deposition [118].

Astaxanthin ADC is affected negatively by ration size and, therefore, indirectly by growth [119]. Astaxanthin ADC was 1.5 times higher when Atlantic salmon were fed a low (40%) as opposed to a full ration (100%) but, due to the low feed intake, the total amount of digested astaxanthin was only about 50% of that in fish fed a full ration. Digestibility measurements of astaxanthin, conducted in a commercial sea farm in northern Norway during autumn when the growth rate was very high, indicated that the apparent digestibility of astaxanthin was as low as 14.5% when the fish were fed a full ration but 38% at half the ration [120]. The ADC of astaxanthin is also influenced by temperature, and was about 11% higher in Atlantic salmon kept at 8°C than in ones kept at 12°C [121]. Similarly, Arctic charr reared at 8°C were significantly redder than fish kept at 12°C [122,123].

Recent studies with transgenic mice and the fruit fly *Drosophila,* and with human Caco2 cells in culture have revealed the importance of intestinal receptors such as SR-BI for the facilitated uptake of carotenoids (*Volume 5, Chapter 7*). It is considered likely that the situation in fish and crustaceans is similar to this, and that the gastrointestinal uptake, uptake in liver, and ultimately uptake and deposition of dietary carotenoids in the muscle and integumental cells, are governed by receptors and transport proteins. This is an important area for future study.

2. Distribution in muscle and integument

The different types of pigment cells in the skin of poikilothermic vertebrates are referred to as chromatophores, notably xanthophores (yellow) and erythrophores (red) (*Chapter 10*). These have the ability to translocate intracellular pigment organelles, under nervous and endocrine control, thus enabling fish to change colour more rapidly than other vertebrates. The biology of invertebrate and vertebrate integuments has been reviewed [124,125]. The primary pigmentary organelles of xanthophores and erythrophores are carotenoid droplets. Crustaceans have four types of chromatophores, including erythrophores and xanthophores, which are loaded with pigment granules. The red-pigment-concentrating hormone was the first neuropeptide hormone to be characterized and was isolated in 1972 from the shrimp *Pandalus borealis* [126]. Marine invertebrate animals contain carotenoproteins which are carotenoids bound stoichiometrically to proteins (*Chapter 6*).

Carotenoids are associated with the protein fraction of the muscle of salmonid fishes, and astaxanthin and canthaxanthin may be combined with the actomyosin complex by non-specific hydrophobic bonds [127-129]. Recently, it was found that α-actinin is the only myofibrillar protein that correlates significantly with astaxanthin binding [130,131]. α-Actinin is a component of the myofibrillar sarcomeric Z-disk which forms the borders of individual sarcomeres of the myofibrils and serves to crosslink opposing thin filaments that interdigitate at the Z-line. Combination studies *in vitro* show that astaxanthin combines, in a close to stoichiometric 1:1 relationship, not only with α-actinin isolated from Atlantic salmon but also with that from the normally unpigmented Atlantic halibut (*Hippoglossus hippoglossus*) [131]. Based on the astaxanthin:actomyosin ratio in muscle of coho salmon

(*Oncorhynchus kisutch*) [128] a theoretical saturation level of nearly 100 mg astaxanthin/kg flesh was indicated for salmonid fishes [132], which is comparable to the highest actual levels (59 mg astaxanthin/kg muscle) reported for sockeye salmon (*O. nerka*) [133]. This is considerably higher than the levels found in large Atlantic salmon where concentrations around 10 mg/kg are found after astaxanthin is fed in the diet [134,135].

The amount of carotenoids that ultimately reach the target tissue(s) can serve as a convenient measure of carotenoid bioavailability. A commonly used measure of carotenoid bioavailability in salmonid fishes is muscle retention (amount of carotenoids deposited in the muscle as a percentage of the ingested amount). The amount of dietary astaxanthin that is utilized for flesh pigmentation rarely exceeds 15% in Atlantic salmon [14] and 18% in rainbow trout [15]. The digestibility and retention of carotenoids are negatively correlated with the dietary carotenoid concentration. The amount retained also depends on fish species, size and growth rate. The muscle retention of dietary astaxanthin, fed at 66 mg/kg diet, was only 3.9% in Atlantic salmon growing from about 0.14 to 0.74 kg [117], but was 30-42% in rainbow trout fed a dietary concentration of about 35 mg astaxanthin/kg diet [109]. This may be explained by extensive metabolic transformation. The muscle tissue carries 93-95% of the total body burden of astaxanthin in Atlantic salmon and rainbow trout [116]. If the percentage of absorbed intact astaxanthin that is excreted *via* the gills and/or urine is close to zero, the results [109] indicate that about 55-67% of the absorbed astaxanthin undergoes metabolic transformation in rainbow trout. Similarly, in Atlantic salmon given various astaxanthin sources at a dietary level of 37-50 mg/kg, the retention of digested astaxanthin ranged from 22.2 to 33.1%. This indicates that about 67% of the astaxanthin that was absorbed by the salmon was transformed metabolically or excreted (not with faeces). Studies are needed to identify these metabolites and their fate.

3. Alternative administration

The bioavailability, determined as the area under the time-concentration curve after a single low dose of intraperitoneally injected astaxanthin (*ca.* 0.5 mg/kg body weight), in Atlantic salmon was about 12-fold higher than for an orally administered dose [113]. A similar difference was obtained after intra-arterial compared to oral administration of [6,7,6',7'-^{14}C$_4$]-astaxanthin in rainbow trout [136].

idoxanthin (**349**)

The dose-response relationships for astaxanthin E/Z isomers and the metabolite idoxanthin (349) in plasma, muscle, liver, kidney and skin, in fish species that usually have white flesh (Atlantic cod, *Gadus morhua*) or red flesh (Atlantic salmon) were compared following intraperitoneal injection of 100 mg astaxanthin [137]. Astaxanthin concentrations increased linearly in a dose-dependent manner in plasma and muscle of both species, and were highly correlated. Extreme astaxanthin concentrations up to 90 and 50 mg/litre in plasma and 30 and 1 mg/kg in muscle were detected in Atlantic salmon (*ca.* 0.5 kg) and cod, respectively, after 4 weeks. Rapid astaxanthin uptake was also found in rainbow trout [138]. The capacity for muscle binding of astaxanthin is therefore much higher than can be achieved by regular feeding. The linear dose-response showed that muscle astaxanthin-binding capacity had not reached saturation, so there is a higher potential for astaxanthin incorporation. Accumulation of astaxanthin E/Z isomers in the various tissues was selective in favour of the Z-isomers. Higher astaxanthin levels in plasma, muscle, skin, kidney and liver of Atlantic salmon and Atlantic cod may thus be obtained by intraperitoneal injection than by regular feeding. Differences in uptake mechanisms for cellular incorporation in the muscle may explain the differences in astaxanthin uptake in different fish species. Uptake of carotenoids is easier from the intraperitoneum than the gastrointestinal tract.

F. Conclusion

Although some progress has been made, little is yet known about the genes and proteins involved in carotenoid pigmentation, such as those involved in absorption, transport, uptake, metabolic transformation and their spatial and temporal expression [139]. Determination of the factors that govern the metabolic turnover of carotenoids and the physiological effects of carotenoids and their metabolites in different life-stages of the various species is very important for the aquaculture industry.

References

[1]   D. L. Fox, *Animal Biochromes and Structural Colours*, 2nd ed., Univ. Calif. Press, Berkeley (1976).

[2]   A. E. Houde, *Sex, Colour, and Mate Choice in Guppies*, Princeton Univ. Press, Princeton NJ (1997).

[3]   M. Edmunds, *Malacologia*, **32**, 241 (1991).

[4]   B. Lyman, *A Psychology of Food*, Van Nostrand Reinhold, New York (1989).

[5]   H. Kläui and J. C. Bauernfeid, in *Carotenoids as Colorants and Vitamin A Precursors*, (ed. J. C. Bauernfeind), p. 47. Academic Press, New York (1981).

[6]   F. Shahidi, Metusalach and J. A. Brown, *Crit. Rev. Food Sci.*, **38**, 1 (1998).

[7]   C. L. Delgado, N. Wada, M. W. Rosegrant, S. Meijer and M. Ahmed, *Fish to 2020: Supply and Demand in Changing Global Markets*, *(WorldFish Center Tech. Rep. 62)*, Int. Food Policy Res. Inst., Washington DC and WorldFish Center, Penang, Malaysia (2003).

[8]   S. Sigurgisladottir, O. Torrissen, Ø. Lie, M. Thomassen and H. Hafsteinsson, *Rev. Fish. Sci.*, **5**, 223
       (1997).

[9]   N. H. Moe, *Proc. Nutr. Soc. New Zealand*, **15**, 16 (1990).

[10]   T. W. Goodwin, *The Biochemistry of the Carotenoids, Vol. 2, Animals*, 2nd ed., Chapman and Hall,
       London (1984).

[11]   T. Matsuno, *Fish. Sci.*, **67**, 771 (2001).

[12]   T. Matsuno and S. Hirao, in *Marine Biogenic Lipids, Fats, and Oils, Vol. 1*, (ed. R. G. Ackman), p. 251,
       CRC Press, Boca Raton (1989).

[13]   K. L. Simpson, T. Katayama and C. O. Chichester, in *Carotenoids as Colorants and Vitamin A
       Precursors*, (ed. J. C. Bauernfeind), p. 463, Academic Press, New York (1981).

[14]   O. J. Torrissen, R. W. Hardy and K. D. Shearer, *CRC Crit. Rev. Aquat. Sci.*, **1**, 209 (1989).

[15]   T. Storebakken and H. K. No, *Aquaculture*, **100**, 209 (1992).

[16]   S. P. Meyers, *Pure Appl. Chem.*, **66**, 1069 (1994).

[17]   B. Bjerkeng, in *Avances en Nutrición Acuícola V*, (ed. L. E. Cruz-Suárez, D. Rique-Marie, M. Tapia-
       Salazar, M. A. Olvera-Novoa and R. Civera-Cerecedo), p. 71, Univ. Autónoma Nuevo León, Nuevo
       León, México (2000).

[18]   D. C. Nickell and J. R. C. Springate, in *Farmed Fish Quality*, (ed. S. C. Kestin and P. D. Wariss), p. 58,
       Fishing New Books, Oxford (2001).

[19]   N. F. Haard, in *Advances in Seafood Biochemistry – Composition and Quality* (ed. G. J. Flick and R. E.
       Martin), p. 305, Technomic Publ., Lancaster, PE (1992).

[20]   K. Schiedt, in *Carotenoids, Vol. 3 Biosynthesis and Metabolism*, (ed. G. Britton, S. Liaaen-Jensen and H.
       Pfander), p. 285, Birkhäuser, Basel (1998).

[21]   FAO Fisheries and Aquaculture Department, *The State of World Fisheries and Aquaculture 2006*, FAO,
       Rome (2007).

[22]   The Directorate of Fisheries, *Key Figures from the Aquaculture Industry 2005*, The Directorate of
       Fisheries, Bergen (2006).

[23]   The Research Council of Norway, Aquaculture 2020, *Transcending the Barriers – as long as…A
       Foresight Analysis*, The Research Council of Norway, Oslo (2005).

[24]   FAO Fisheries Department, *State of the World Aquaculture 2006*, FAO Fish. Tech. Paper. No. 500, FAO,
       Rome (2006).

[25]   F. A. Chapman, S. A. Fitz-Coy, E. M. Thunberg and C. M. Adams, *J. World Aquacult. Soc.*, **28**, 1 (1997).

[26]   M. Tlusty, *Aquaculture*, **205**, 203 (2002).

[27]   F. M. Pelicice and A. A. Agostinho, *Fish. Res.*, **72**, 109 (2005).

[28]   C. Monteiro-Neto, F. E. D. A. Cunha, M. C. Nottingham, M. E. Araújo, I. L. Rosa and G. M. L. Barros,
       Biodiv. Conserv., **12**, 1287 (2003).

[29]   B. J. Foss, G. Nadolski and S. F. Lockwood, *Mini-Rev. Med. Chem.*, **6**, 953 (2006).

[30]   J. Sales and G. P. J. Janssens, *Aquat. Living Resourc.*, **16**, 533 (2003).

[31]   R. James, K. Sampath, R. Thangarathinam and I. Vasudevan, *Israeli J. Aquacult.*, **58**, 97 (2006).

[32]   L. Gouveia, P. Rema, O. Pereira and J. Empis, *Aquacult. Nutr.*, **9**, 123 (2003).

[33]   L. Gouveia and P. Rema, *Aquacult. Nutr.*, **11**, (2005).

[34]   K. C. Majumdar, K. Nasaruddin and K. Ravinder, *Aquacult. Res.*, **28**, 581 (1997).

[35]   M. Garduño-Lugo, G. Muñoz-Córdova and M. A. Olvera-Novoa, *Aquacult. Res.*, **35**, 340 (2004).

[36]   O. Puebla, E. Bermingham, F. Guichard and E. Whiteman, *Proc. Roy. Soc. B*, **274**, 1265 (2007).

[37]   J. A. Endler, *Evol. Biol.*, **11**, 319 (1978).

[38]   T. Amundsen and E. Forsgren, *Proc. Nat. Acad. Sci. USA*, **98**, 13155 (2001).

[39]   C. J. Foote, G. S. Brown and C. W. Hawryshyn, *Anim. Behav.*, **67**, 69 (2004).

[40]   G. F. Grether, J. Hudon and D. F. Millie, *Proc. Roy. Soc. B*, **266**, 1317 (1999).

[41]   F. Skarstein and I. Folstad, *Oikos*, **76**, 359 (1996).

[42]   T. W. Pike, J. D. Blount, B. Bjerkeng, J. Lindström and N. B. Metcalfe, *Proc. Roy. Soc. B*, **274**, 1591 (2007).

[43]   P. A. Svensson, J. D. Blount, E. Forsgren, B. Bjerkeng and T. Amundsen, unpublished results.

[44]   J. T. Nordeide, *Aquacult. Res.*, **38**, 1 (2007).

[45]   S. Anderson, *Proc. 10th Bienn. Conf. Int. Inst. Fish. Econ. Trade*, Oregon State Univ., Corvallis   (2000).

[46]   G. Steine, F. Alfnes and M. B. Rørå, *Mar. Resourc. Econ.*, **20**, 211 (2005).

[47]   F. Alfnes, A. G. Guttormsen, G. Steine and K. Kolstad, *Am. J. Agric. Econ.*, **88**, 1050 (2006).

[48]   G. Sylvia, M. T. Morrissey, T. Graham and S. Garcia, *J. Aquat. Food Prod. Technol.*, **4**, 51 (1995).

[49]   G. Sylvia, M. T. Morrissey, T. Graham, T. and S. Garcia, *J. Food Prod. Market.*, **3**, 49 (1996).

[50]   E. E. Prince, *Trans. Am. Fish. Soc.*, **46**, 50 (1916).

[51]   O. Einen and G. Skrede, *Aquacult. Nutr.*, **4**, 99 (1998).

[52]   F. M. Clydesdale, *Crit. Rev. Food Sci. Nutr.*, **33**, 83 (1993).

[53]   P. Reardon and E. W. Bushnell, *Infant Behav. Developm.*, **11**, 245 (1988).

[54]   P. Winterhalter, in *Biotechnology for Improved Foods and Flavors*, (ed. G. R. Takeoka), p. 295. Am. Chem. Soc., Washington (1996).

[55]   M. Østerlie, B. Bjerkeng, H. Karlsen and H. M. Storrø, *J. Aquat. Food Prod. Technol.*, **10**, 65 (2001).

[56]   B. Bjerkeng, T. Storebakken and S. Liaaen-Jensen, *Aquaculture*, **108**, 333 (1992).

[57]   A. Aksnes, B. Gjerde and S. O. Roald, *Aquaculture*, **53**, 7 (1986).

[58]   E. P. Köster, *Food Qual. Pref.*, **14**, 359 (2003).

[59]   V. P. Palace and J. Werner, *Sci. Mar.*, **70S2**, 41, (2006).

[60]   T. W. Goodwin, *Biol. Rev. Cambr.*, **25**, 391 (1950).

[61]   O. J. Torrissen, in *The Current Status of Fish Nutrition in Aquaculture*, (ed. M. Takeda and T. Watanabe), p. 387. Tokyo University of Fisheries, Tokyo (1990).

[62]   R. Christiansen and O. J. Torrissen, *Aquaculture*, **153**, 51 (1997).

[63]   R. Christiansen, Ø. Lie and O. J. Torrissen, *Aquacult. Fish Management*, **25**, 903 (1994).

[64]   R. Christiansen, O. Lie and O. J. Torrissen, *Aquacult. Nutr.*, **1**, 189 (1995.)

[65]   R. Christiansen, O. Lie and O. J. Torrissen, *Aquacult. Nutr.*, **2**, 55 (1996).

[66]   R. Christiansen, J. Glette, Ø. Lie, O. J. Torrissen and R. Waagbø, *J. Fish Diseases*, **18**, 317 (1995).

[67]   G. Choubert, J.-M. Blanc and H. Poisson, *Aquacult. Nutr.*, **4**, 249 (1998).

[68]   M. R. Ahmadi A. A. Bazyar S. Safi, T. Ytrestøyl and B. Bjerkeng, *J. Appl. Ichthyol.*, **22**, 388 (2006).

[69]   D. M. Steven, *J. Exp. Biol.*, **26**, 295 (1949).

[70]   J. C. A. Craik and S. M. Harvey, *J. Fish Biol.*, **29**, 549 (1986).

[71]   A. Pettersson and Å. Lignell, *Ambio*, **28**, 43 (1999).

[72]   A. A. Bazyar Lakeh, M. R. Ahmadi, S. Safi, T. Ytrestøyl and B. Bjerkeng, unpublished results.

[73]   R. Ørnsrud, A. Wargelius, Ø. Sæle, K. Pittman and R. Waagbø, *J. Fish Biol.*, **64**, 399 (2004).

[74]   C.-J. Hu, S.-M. Chen, C.-H. Pan and C.-H. Huang, *Aquaculture*, **253**, 602 (2006).

[75]   B. Bjerkeng, B. Hatlen and M. Jobling, *Comp. Biochem. Physiol.*, **125B**, 395 (2000).

[76]   J. Boonjawat and J. A. Olson, *Comp. Biochem. Physiol.*, **50B**, 363 (1975).

[77]   J. M. Lampert, J. Holzschuh, S. Hessel, W. Driever, K. Vogt and J. von Lintig, *Development*, **130**, 2173 (2003).

[78]   M. S. Izquierdo, H. Fernández-Palacios and A. G. J. Tacon, *Aquaculture*, **197**, 25 (2001).

[79]   T. Watanabe and R. Vassallo-Agius, *Aquaculture*, **227**, 35 (2003).

[80]   V. Verakunpiriya, K. Mushiake, K. Kawano and T. Watanabe, *Fish. Sci.*, **63**, 816 (1997).

[81]   R. V. Agius, T. Watanabe, S. Satoh, V. Kiron, H. Imaizumi, T. Yamazaki and K. Kawano, *Aquacult. Res.*, **32S**, 263 (2001).

[82]   R. Vassallo-Agius, H. Imaizumi, T. Watanabe, T. Yamazaki, S. Satoh and V. Kiron, *Fish. Sci.*, **67**, 260 (2001).

[83]   M. A. Liñán-Cabello, J. Paniagua and P. M. Hopkins, *Aquacult. Nutr.*, **8**, 299 (2002).

[84] W. Dall, *Mar. Biol.*, **124**, 209 (1995).

[85] M. P. Pangantihon-Kühlmann, O. Millamena and Y. Chern, *Aquat. Living Resourc.*, **11**, 403 (1998).

[86] M. C. Barclay, S. J. Irvin, K. C. Williams and D. M. Smith, *Aquacult. Nutr.*, **12**, 117 (2006).

[87] M. Østerlie, B. Bjerkeng and S. Liaaen-Jensen, *J. Nutr.*, **129**, 391 (1999).

[88] P. McCaffrey and U. C. Dräger, *Cytokine Growth Factor Rev.*, **11**, 233 (2000).

[89] B. Bjerkeng, *Prog. Fish-Cult.*, **59**, 129 (1997).

[90] P. Spolaore, C. Joannis-Cassan and A. Isambert, *J. Biosci. Bioeng.*, **101**, 87 (2006).

[91] D. Horn and J. Rieger, *Angew. Chem. Int. Ed.*, **40**, 4330 (2001).

[92] L. G. Buttle, V.O. Crampton and P. D. Williams, *Aquacult. Res.*, **32**, 103 (2001).

[93] R. T. M. Baker, A.-M. Pfeiffer, F.-J. Schöner and L. Smith Lømmon, *Anim. Feed Sci. Technol.*, **99**, 97 (2002).

[94] A. Kiessling, R.-E. Olsen and L. Buttle, *Aquacult. Nutr.*, **9**, 253 (2003).

[95] T. Storebakken, P. Foss, K. Schiedt, E. Austreng, S. Liaaen-Jensen and U. Manz, *Aquaculture*, **65**, 279 (1987).

[96] D. A. White, G. I. Page, J. Swaile, A. J. Moody and S. J. Davies, *Aquacult. Res.*, **33**, 343 (2002).

[97] D. A. White, A. J. Moody, R. D. Serwata, J. Bowen, C. Soutar, A. J. Young and S. J. Davies, *Aquacult. Nutr.*, **9**, 247 (2003).

[98] N. Tejera, J. R. Cejas, C. Rodríguez, B. Bjerkeng, S. Jerez, A. Bolaños and A. Lorenzo, A., *Aquaculture*, **270**, 218 (2007).

[99] T. Storebakken, M. Sørensen, B. Bjerkeng, J. Harris, P. Monahan and S. Hiu, *Aquaculture*, **231**, 489 (2004).

[100] T. Storebakken, M. Sørensen, B. Bjerkeng and S. Hiu, *Aquaculture*, **236**, 391 (2004).

[101] B. Bjerkeng, M. Peisker, K. von Schwartzenberg, T. Ytrestøyl and T. Åsgård, *Aquaculture*, **269**, 476 (2007)

[102] E. Wathne, B. Bjerkeng, T. Storebakken, V. Vassvik and A. B. Odland, *Aquaculture* **159**, 217 (1998)

[103] B. Bjerkeng, B. Hatlen and E. Wathne, *Aquaculture*, **180**, 307 (1999).

[104] B. Bjerkeng, B. Hatlen and E. Wathne, *Aquaculture*, **189**, 389 (2000).

[105] B. Bjerkeng, K. Hamre, B. Hatlen and E. Wathne, *Aquacult. Res.*, **30**, 637 (1999).

[106] E. Austreng, T. Storebakken, M. S. Thomassen, S. Refstie and Y. Thomassen, *Aquaculture*, **188**, 65 (2000).

[107] J. N. Bird and G. P. Savage, *Proc. Nutr. Soc. New Zealand*, **15**, 45 (1990).

[108] G. Choubert, J.-C. G. Milicua and R. Gomez, *Comp. Biochem. Physiol.*, **108A**, 1001 (1994).

[109] B. Bjerkeng, M. Følling, S. Lagocki, T. Storebakken, J. J. Olli and N. Alsted, *Aquaculture*, **157**, 63 (1997).

[110] M. Østerlie, B. Bjerkeng and S. Liaaen-Jensen, *J. Nutr.*, **129**, 391 (1999).

[111] I. Gobantes, G. Choubert, M. Laurentie, J-C. G. Milicua and R. Gomez, *J. Agric. Food Chem.*, **45**, 454 (1997).

[112] G. H. Aas, B. Bjerkeng, T. Storebakken and B. Ruyter, *Fish Physiol. Biochem.*, **21**, 325 (1999).

[113] J. B. Maltby, L. J. Albright, C. J. Kennedy and D. A. Higgs, *Aquacult. Res.*, **34**, 829 (2003).

[114] P. Foss, T. Storebakken, E. Austreng and S. Liaaen-Jensen, *Aquaculture*, **65**, 293 (1987).

[115] G. Choubert and T. Storebakken, *Ann. Zootech.*, **45**, 445 (1996).

[116] G. I. Page and S. J. Davies, *Comp. Biochem. Physiol.*, **143A**, 125 (2006).

[117] B. Bjerkeng and G. M. Berge, *Comp. Biochem. Physiol.*, **127B**, 423 (2000).

[118] B. Bjerkeng, S. Refstie, K. T. Fjalestad, T. Storebakken, M. Rødbotten and A. Roem, *Aquaculture*, **157**, 297 (1997).

[119] T. Ytrestøyl, G. Struksnæs, K.-A. Rørvik, W. Koppe and B. Bjerkeng, *Aquaculture*, **261**, 215 (2006).

[120] K.-A- Rørvik, E. Lundberg, F. A. Jakobsen, A. A. Jakobsen, T. Ytrestøyl and B. Bjerkeng, unpublished results.

[121] T. Ytrestøyl, G. Struksnæs, W. Koppe and B. Bjerkeng, *Comp. Biochem. Physiol.*, **142B**, 445 (2005).

[122] R. E. Olsen and A. Mortensen, *Aquacult. Res.*, **28**, 51 (1997).

[123] R. Ginés, T. Valdimarsdottir, K. Sveinsdottir and H. Thorarensen, *Food Qual. Pref.*, **15**, 177 (2004).

[124] J. Bereiter-Hahn, A. G. Matoltsy and K. S. Richards, *Biology of the Integument, Vol. 1, Invertebrates*, Springer, Berlin (1984).

[125] J. Bereiter-Hahn, A.G. Matoltsy and K.S. Richards, *Biology of the Integument, Vol. 2, Vertebrates*, Springer, Berlin (1986).

[126] A. Huberman, *Aquaculture*, **191**, 191 (2000).

[127] H. Henmi, T. Iwata, M. Hata and M. Hata, *Tohoku J. Agric. Res.*, **37**, 101 (1987).

[128] H. Henmi, M. Hata and M. Hata, *Bull. Jap. Soc. Sci. Fish.*, **55**, 1583 (1989).

[129] H. Henmi, M. Hata and M. Hata, *Bull. Jap. Soc. Sci. Fish.*, **56**, 1821 (1990).

[130] S. J. Matthews, N. W. Ross, S. P. Lall and T. A. Gill, *Comp. Biochem. Physiol.*, **144B**, 206 (2006).

[131] M. R. Saha, N. W. Ross, R. E. Olsen and S. P. Lall, *Comp. Biochem. Physiol.*, **144B**, 488 (2006).

[132] B. Bjerkeng, T. Storebakken and S. Liaaen-Jensen, *Aquaculture*, **108**, 333 (1992).

[133] S. A. Turujman, W. G. Wamer, R. R. Wie and R. H. Albert, *J. Assoc. Off. Am. Chem. Int.*, **80**, 622 (1997).

[134] R. Christiansen, G. Struksnæs, R. Estermann and O. J. Torrissen, *Aquacult. Res.*, **26**, 311 (1995).

[135] O. I. Forsberg and A .G. Guttormsen, *Aquaculture*, **253**, 415 (2006).

[136] G. Choubert, J.-P. Cravedi and M. Laurentie, *Aquacult. Res.*, **36**, 1526 (2005).

[137] T. Ytrestøyl and B. Bjerkeng, *Aquaculture*, **263**, 179 (2007).

[138] T. Ytrestøyl and B. Bjerkeng, *Comp. Biochem. Physiol.*, **147B**, 250 (2007).

[139] J. Hudon, *Biotech. Adv.*, **12**, 49 (1994).

Carotenoids
Volume 4: Natural Functions
© 2008 Birkhäuser Verlag Basel

Chapter 13

Xanthophylls in Poultry Feeding

Dietmar E. Breithaupt

A. Introduction

Since most consumers associate an intense colour of food with healthy animals and high food quality, xanthophylls are widely used as feed additives to generate products that meet consumers' demands. An important large-scale application is in poultry farming, where xanthophylls are added to feed to give the golden colour of egg yolk that is so much appreciated. Now, with numerous new applications in human food, in the pharmaceutical industry, and in cosmetic products, there is an increasing demand for xanthophylls on the international market (*Volume 5, Chapter 4*).

For most people living in developed countries, good nutrition is assured, so consumers are no longer interested simply in the nutritional value of their food but also in the content of natural substances that improve the aesthetic appeal of the food (colour, flavour) or may exhibit special effects on health [1]. There is increasing public interest in the xanthophylls that are used to fortify chicken feed, especially lutein (**133**) and zeaxanthin (**119**), in relation to human eye health and age-related macular degeneration [2-4] (*Volume 5, Chapter 15*). This has led to a large and successful market for lutein in the past decade. Further investigations revealed that lutein has additional health benefits for the chickens themselves, for example an influence on the immune response in laying hens [5]. Dietary lutein significantly boosted the secondary antibody response to vaccination against infectious bronchitis virus. It was concluded likely that, in addition to producing 'designer eggs', lutein supplementation benefits the health of the chicken by increasing the efficacy of vaccination.

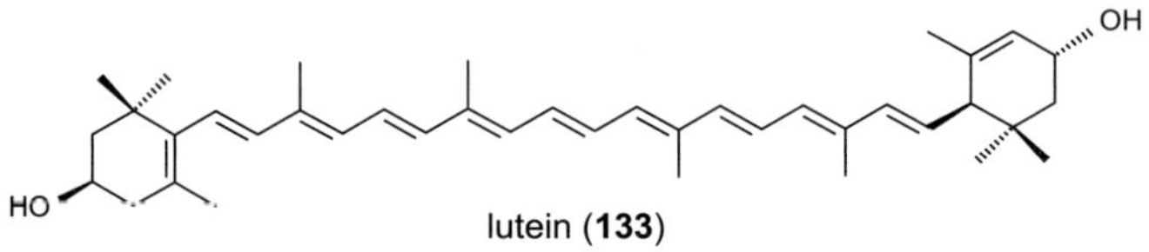

lutein (**133**)

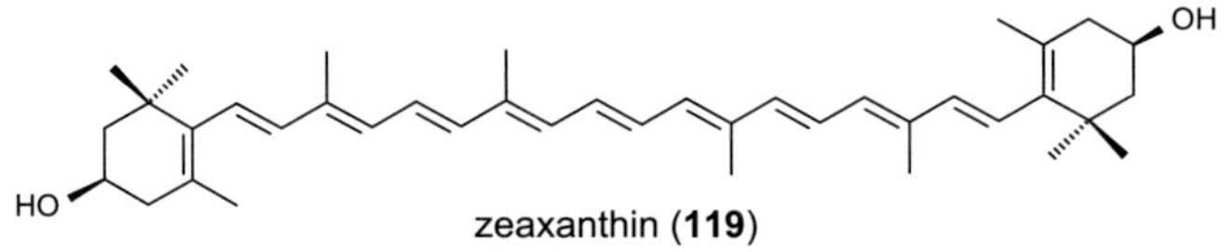

zeaxanthin (**119**)

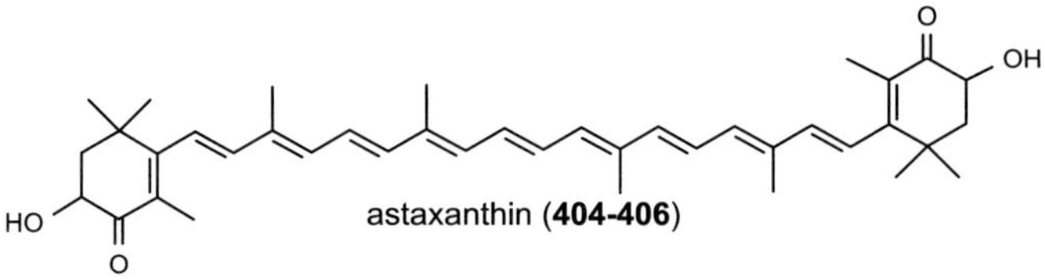

astaxanthin (**404-406**)

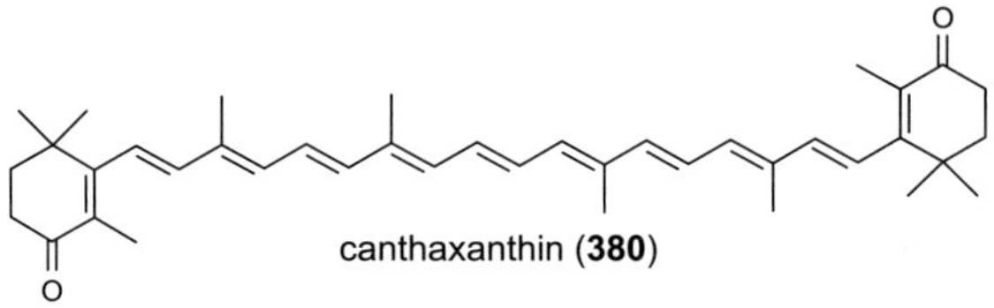

canthaxanthin (**380**)

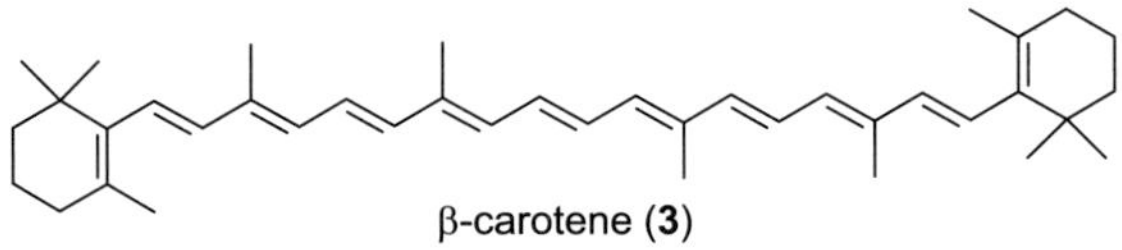

β-carotene (**3**)

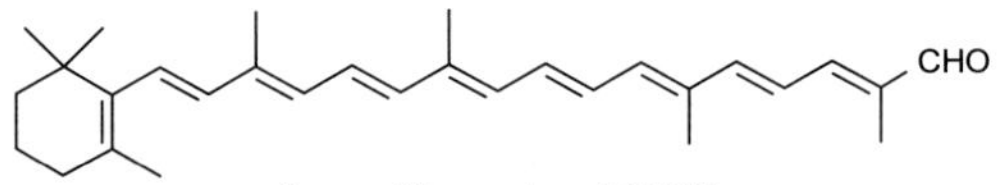

β-apo-8'-carotenal (**482**)

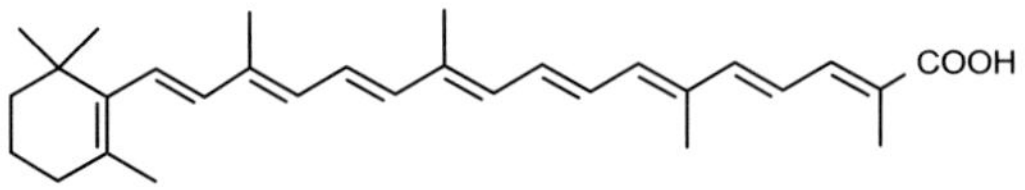

β-apo-8'-caroten-8'-oic acid (**486**)

These findings are in agreement with other results [6] which showed that deficiency in carotenoids, either *in ovo* or post-hatch, increased parameters of systemic inflammation in the chickens.

Although many different xanthophylls with interesting biological properties are found in Nature, only a few are of industrial importance. The current worldwide commercial market for carotenoids was estimated to be US$ 887 million for 2004 (*Volume 5, Chapter 4*). The largest field of commercial application of carotenoids is animal feed, mainly because of the importance of astaxanthin (**404-406**) and canthaxanthin (**380**) in aquaculture (*Chapter 12*). In the food market, β-carotene (**3**) is still the carotenoid with highest economic importance.

Xanthophylls such as β-apo-8'-carotenal [8'-apo-β-caroten-8'-al (**482**)] and β-apo-8'-carotenoic acid [8'-apo-β-caroten-8'-oic acid (**486**)] ethyl ester, which are produced by chemical synthesis, play a minor role on the international market [7]. Lutein, which is obtained from natural sources and not produced commercially by synthesis due to the complexity of the process, is in third place in the global market of xanthophylls [7]. Other xanthophylls, such as β-cryptoxanthin (**55**) and zeaxanthin (**119**) have no significant market importance individually as poultry-feed additives, though they are minor constituents in fruits such as red pepper and, consequently, are present in their oleoresins.

B. Legal Situation in the European Union (EU)

Legislation concerning the use of xanthophylls in animal feed is different around the world and is strictly regulated in colour directives, leading to lists of feed additives. Within the EU, the fundamental 'Regulation (EC) No 1831/2003 of the European Parliament and of the Council' [8] determines the use of additives in animal nutrition and sets out rules for the authorization, marketing, and labelling of feed additives. Colourants belong to the group of 'sensory additives', covering substances that improve or change the organoleptic properties of the feed or the visual characteristics of food derived from animals that consumed the respective feed. Referring to colourants, the category 'sensory additives' (annex 1, 2.a) includes three subgroups: (i) substances that add or restore colour in feeding stuff; (ii) substances which, when fed to animals, add colours to food of animal origin; (iii) substances which favourably affect the colour of ornamental fish or birds. Based on this annex, the 'Community Register of Feed Additives pursuant to Regulation EC No 1831/2003' [9] lists eight xanthophylls that may be added to poultry feed (with references to Community legal acts). The maximum content of a xanthophyll permitted in the feed, and further information, are given separately for each country in the 'List of the authorised additives in feedingstuffs published in application of Article 9t (b) of Council Directive 70/524/EEC concerning additives in feedingstuff' [10]. Table 1 provides a summary of the situation in the EU. There is no time limit for the period of authorization.

C. Analysis

Table 1 summarizes the chemical data needed to identify the xanthophylls by spectrophotometric methods (Vis maxima) and by LC-(APCI)MS measurements (main ions) that are used to facilitate peak assignment. The Vis maxima are of particular importance for identification of unknown compounds in HPLC analysis. Some characteristic Vis spectra of relevant xanthophylls are presented in Fig. 1. An extensive coverage of UV/Vis spectroscopy of carotenoids is given in *Volume 1B, Chapter 2*, and UV/Vis data for all carotenoids are given in the *Carotenoids Handbook*. Lutein (**133**), zeaxanthin (**119**) and β-cryptoxanthin (**55**) show similar Vis spectra though lutein absorbs at 5nm shorter wavelength. The ketocarotenoids canthaxanthin, citranaxanthin (**466**) and capsanthin (**335**) show only one round maximum, in ethanol, at approximately 474 nm. The bathochromic shift of roughly 28 nm compared to lutein is responsible for the orange to red appearance of these xanthophylls.

Table 1. Spectroscopic and LC-(APCI)MS data used for identification of xanthophylls. The numbering of the xanthophylls 1 - 8 corresponds to the numbering of peaks in Fig. 1. The 'E-numbers' of the xanthophylls are also given.

Xanthophyll / EC No	Vis maxima [nm] [a]; ε [L/(mol x cm)] (solvent[b], wavelength) [c]	main ion (m/z; 100 % intensity)
1. Lutein (**119**) / E 161b	420, 446, 472 145100 (E, 445 nm)	551.4 $[M+H-H_2O]^+$
2. Capsanthin (**335**) / E 160c	474 121000 (T, 483 nm)	585.4 $[M+H]^+$
3. Zeaxanthin (**119**) / E 161h	426, 452, 478 144500 (E, 450 nm)	569.4 $[M+H]^+$
4. 8'-Apo-β-caroten-8'-al (**482**) / E 160e	460 110000 (P, 457 nm)	417.3 $[M+H]^+$
5. Canthaxanthin (**380**) / E 161g	476 107300 (P, 463 nm)	565.4 $[M+H]^+$
6. 8'-Apo-β-caroten-8'-oic acid (**486**) ethyl ester / E 160f	446 87700 (P, 430 nm)	461.3 $[M+H]^+$
7. Citranaxanthin (**466**) / E 161i	474 98000 (P, 463 nm)	457.3 $[M+H]^+$
8. β-Cryptoxanthin (**55**) / E 161c	426, 452, 478 131000 (P, 452 nm)	553.4 $[M+H]^+$

[a]Values determined in the HPLC eluent (*tert.*-butyl methyl ether / methanol / water) according to [11].

[b]Solvents: E, ethanol; T, toluene; P, light petroleum

[c]According to [12].

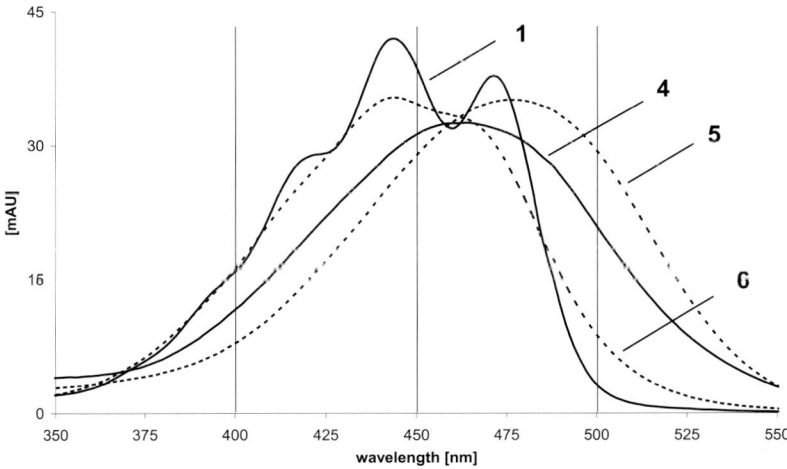

Fig. 1. Characteristic Vis spectra of xanthophylls used in poultry feeding, obtained by a photo-diode array detector (PDAD) during an HPLC run of standard compounds (illustrated in Fig. 2) [11]. The assignment of spectra is as follows: 1: lutein (**133**); 4: 8'-apo-β-caroten-8'-al (**482**); 5: canthaxanthin (**380**); 6: 8'-apo-β-caroten-8'-oic acid (**486**) ethyl ester.

The different Vis spectral characteristics determine the selection of the best wavelength to use for detection of xanthophylls in HPLC analysis (Fig. 2). The wavelength usually applied for detection is 450 nm (Fig. 2, trace B), which permits an acceptable sensitivity in determination of each xanthophyll. However, if the red xanthophylls canthaxanthin, citranaxanthin, and capsanthin are of main interest, a longer wavelength, up to 500 nm, should be chosen to enhance the limit of detection (trace C). Applying 425 nm (trace A) results in a loss of sensitivity for all xanthophylls. For xanthophyll identification, LC-(APCI)MS is used. Since the interface is usually operated in the positive mode, only positively charged ions are detected. The data set for identification of the listed xanthophylls is also given in Table 1. Apart from lutein, all the xanthophylls listed give intense quasimolecular ions ($[M+H]^+$). To scan for lutein, the high-intensity ion at m/z 551.4, formed by loss of water ($[M+H-H_2O]^+$), is used rather than the quasimolecular ion (m/z 569.4), which usually is present at only low abundance.

If the total xanthophyll uptake is of interest, it must be borne in mind that, in addition to being accumulated without modification, lutein (and other xanthophylls) may be metabolized, leading to a range of structural modifications and to the formation of breakdown products with different chemical and spectroscopic properties (*Volume 3, Chapter 7*). Thus, a complex mixture of apo-carotenoids (*e.g.* aldehydes) may be found in the egg yolk, the subcutaneous fat, the skin, the liver, the integuments, and, in ornamental birds, in the feathers.

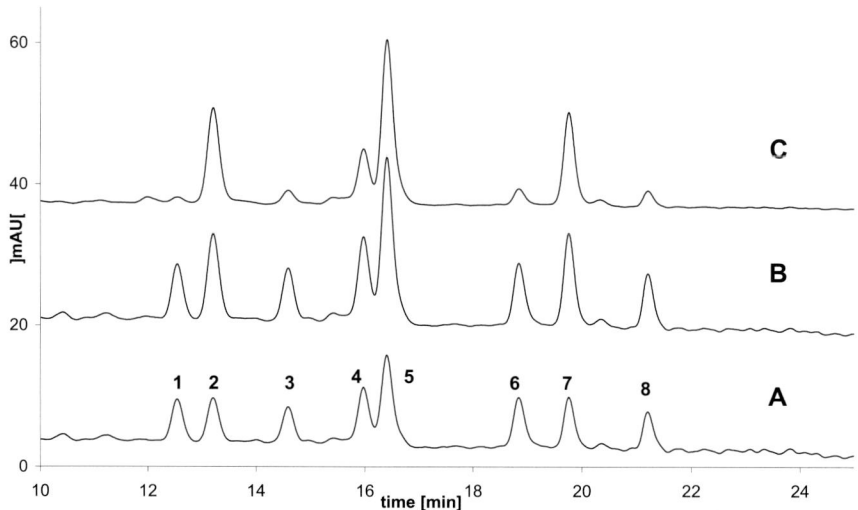

Fig. 2. Representative HPLC chromatograms (extended sections) of the free forms of xanthophylls used in poultry feeding, detected by a photodiode array detector (PDAD) at three different wavelengths (trace **A**: 425 nm, trace **B**: 450 nm, trace **C**: 500 nm). The HPLC conditions are in accordance with [11], individual concentrations are in the range 7 - 14 µmol/L. The peak assignment is as follows: 1: lutein (**133**); 2: capsanthin (**335**); 3: zeaxanthin (**119**); 4: 8′-apo-β-caroten-8′-al (**482**); 5: canthaxanthin (**380**); 6: 8′-apo-β-caroten-8′-oic acid (**486**) ethyl ester; 7: citranaxanthin (**466**); 8: β-cryptoxanthin (**55**).

β-cryptoxanthin (**55**)

citranaxanthin (**466**)

capsanthin (**335**)

D. Application of Xanthophylls in Poultry Farming

It is a characteristic of chickens that xanthophylls are accumulated whilst carotenes do not contribute to the pigmentation [13]. Consequently, xanthophylls have gained economic interest for colouring chicken skin and especially egg yolk. It is well known that the colour intensity as well as the colour hue (yellow to red) can be controlled by the concentration and type of dietary xanthophylls used [14,15], and that yolk pigmentation is affected by factors such as the birds' physiology and health, diet, feed production, and product characteristics [2]. An important application is for laying hens, and the eight xanthophylls listed in Table 1 can be present in commercial egg yolks. The use of astaxanthin would result in a pink yolk, currently unwanted by most consumers within the EU.

The maximum concentration allowed in the feed was set to 80 mg/kg for all xanthophylls except canthaxanthin, for which a maximum amount of only 8 mg/kg is allowed for laying hens and 25 mg/kg for other poultry [10], to avoid the risk of crystalline canthaxanthin being deposited in the eyes of the consumer [16,17]. If mixtures of canthaxanthin and other xanthophylls are used, the total carotenoid content must not exceed 80 mg/kg feed. If eggs are produced in accordance with guidelines of organic farming, the use of synthetic compounds and of colour additives obtained by organic solvent extraction from plants is prohibited. Xanthophylls found in 'organic' egg yolks are due to application of xanthophyll-rich plants, e.g. alfalfa (*Medicago sativa*) or corn (*Zea mays*) in the feed.

It is generally accepted that it is easier to generate a yellow – orange hue than a red hue in egg yolks. Lutein is the traditional colouring principle used for an intense yellow hue. It is the predominant xanthophyll in green leafy plants (*e.g.* spinach, alfalfa) and it is also present in high concentration, as acyl esters, in petals of marigold flowers (*Tagetes erecta*). To obtain the lutein (which is accompanied in *T. erecta* by roughly 5% zeaxanthin [18]), the petals are fermented and dried, then extracted with organic solvents, which are subsequently removed by distillation to leave the oleoresin. The fatty acid constituents of the lutein diesters in marigolds are mainly the saturated lauric (C12:0), myristic (C14:0), palmitic (C16:0), and stearic (C18:0) acids [19,20]. The lutein esters in the extract may be hydrolysed before formulation of the final feed product. Thus, lutein can be found on the market in either its native esterified or its free form. The Vis maximum is not influenced by esterification [21]. Latin-American countries and China play important roles in the global market for *T. erecta*.

Usually, lutein is not applied as the only xanthophyll in poultry feed because the egg yolk may take on a greenish hue, resulting in a grey tone of processed food such as noodles. To avoid this, orange – red xanthophylls are added after an initial phase of feeding yellow xanthophylls. Efforts have been made to select varieties of *T. erecta* that possess an elevated zeaxanthin concentration, and would impart an orange hue to the yolk. Lutein from *T. erecta* has to be applied in a 6-fold higher concentration than 8'-apo-β-caroten-8'-al to attain the same pigmentation effect [M. Grashorn, personal communication].

For the second feeding step, several xanthophylls may be used to obtain a red hue. The most prominent are canthaxanthin and dried red pepper, which contains mainly capsanthin, accompanied by minor components, including capsorubin (**413**), zeaxanthin and β-carotene.

capsorubin (**413**)

In the ripe red fruit, the capsanthin is mostly esterified with various fatty acids [22,23]. Red pepper, therefore, provides a natural mixture, depending on the variety used for production and on the ripeness of the fruit. Only natural cultivars can be used for feed production; varieties obtained by genetic manipulation techniques are not allowed. Usually this feed additive is obtained by milling dried red peppers since capsanthin is not produced by chemical synthesis. A drawback of natural red pepper products is that the concentration of the xanthophyll in the feed has to be roughly 3.5-fold higher than that used for synthetic canthaxanthin. The widespread acceptability of synthetic red xanthophylls, especially canthaxanthin, may be due not only to lower application concentrations but also to the lower costs of synthetic compounds. Canthaxanthin is currently the preferred red xanthophyll in poultry farming because its pigmentation efficacy is higher than that of citranaxanthin. Canthaxanthin occurs naturally, but it is produced in bulk by chemical synthesis, and is a target for microbial production (*Volume 5, Chapter 5*).

In the egg yolk, xanthophylls are typically found in their free form, regardless of whether free or acylated xanthophylls were fed, proving that laying hens efficiently hydrolyse xanthophyll acyl esters prior to deposition in the yolk. It was suggested, therefore, that prior saponification might improve the bioavailability of xanthophylls, thereby possibly enhancing the efficiency of deposition [24]. This topic remains controversial. One study [25] has found an enhanced bioavailability of free red pepper xanthophylls, and another [24] reported that the more lipophilic lutein esters are utilized 'better' than crystalline lutein. Other studies, however, have found no significant difference in the colours of yolks whether the birds were fed natural or saponified red pepper oleoresin [15], and comparable concentrations of lutein and capsanthin were found in chicken plasma, whether free or esterified xanthophylls were fed [26]. The discussion continues about whether free or esterified xanthophylls have higher bioavailability, not only with respect to animal feed, but also in view of human dietary supplements that contain free or esterified forms of lutein [27] (*Volume 5, Chapter 4*).

The formulation used has a strong influence on the absorption and deposition of added xanthophylls in poultry feed. A complex product examined [28] contained additionally vegetable oil, a surfactant, a chelating agent, an antioxidant, alkali and a solvent. Eggs produced by application of this formulation contained more than 0.8 mg carotenoid per 100 g of the edible portion, proving that eggs may serve as an important source of xanthophylls in the human diet.

E. Conclusion

Discoloured or inadequately coloured food suggests an unhealthy animal and consequently an inferior consumer product. Thus, there is an increasing demand for xanthophyll feed additives in poultry farming. The intensive use of synthetic xanthophylls is mainly due to the lower amounts needed and the lower costs, the guaranteed quality, and the enhanced stability of formulated products, offering enhanced pigmenting efficiency. On the other hand, however, there is a growing market for organic food, in the production of which the use of synthetic additives or of unnatural solvents in the preparation of natural extracts is not acceptable. This market is driven mainly by the consumers' demand and may intensify the search for plants or microbial products that can provide high colouring power.

References

[1]   M. Grashorn, *Pol. J. Food Nutr. Sci.*, **14**, 15 (2005).

[2]   R. Baker and C. Günther, *Trends Food Sci. Technol.*, **15**, 484 (2004).

[3]   J. T. Landrum and R. A. Bone, *Arch. Biochem. Biophys.*, **385**, 28 (2001).

[4]   B. Olmedilla, F. Granado, I. Blanco, M. Vaquero and C. Cajigal, *J. Sci. Food Agric.*, **81**, 904 (2001).

[5]   G. Y. Bedecarrats and S. Leeson, *J. Appl. Poultry Res.*, **15**, 183 (2006).

[6]   E. A. Koutsos, J. C. G. Lopez and K. C. Klasing, *J. Nutr.*, **136**, 1027 (2006).

[7]   M. Rajan, http://www.bccresearch.com/editors/RGA-110R.html (2005).

[8]   Regulation (EC) No 1831/2003 of the European Parliament and the Council of 22 September 2003 on additives for use in animal nutrition. *Official Journal of the European Union*, L268/29, 18.10.2003.

[9]   Community Register of Feed Additives pursuant to Regulation (EC) No 1831/2003 (http://europa.eu.int/comm/food/food/animalnutrition/feedadditives/registeradditives_en.htm).

[10] List of the authorised additives in feedingstuffs published in application of Article 9t(b) of Council Directive 70/524/EEC concerning additives in feedingstuffs. *Official Journal of the European Union*, 2004/C 50/01.

[11] J. Schlatterer and D. E. Breithaupt, *J. Agric. Food Chem.*, **54**, 2267 (2005).

[12] H.-P. Köst, (Ed.), *CRC Handbook of Chromatography, Plant Pigments, Vol. I*, CRC Press, Boca Raton, USA (1988).

[13] H. Hencken, *Poultry Sci.*, **71**, 711 (1992).

[14] B. H. Chen and S. H. Yang, *Food Chem.*, **44**, 61 (1992).

[15] S. M. Lai, J. I. Gray and C. J. Flegal, *J. Sci. Food Agric.*, **72**, 166 (1996).

[16] G. B. Arden and F. M. Barker, *J. Toxicol., Cutaneous Ocular Toxicol.*, **10**, 115 (1991).

[17] R. T. M. Baker, *Trends Food Sci. Technol.*, **12**, 240 (2001).

[18] W. L. Hadden, R. H. Watkins, L. W. Levy, E. Regalado, D. M. Rivadeneira, R. B. van Breemen and S. J. Schwartz, *J. Agric. Food Chem.*, **47**, 4189 (1999).

[19] R. Piccaglia, M. Marotti and S. Grandi, *Indust. Crops Prod.,* **8**, 45 (1998).

[20] D. E. Breithaupt, U. Wirt and A. Bamedi, *J. Agric. Food Chem.,* **50**, 66 (2002).

[21] G. K. Gregory, T. S. Chen and T. Philip, *J. Food Sci.,* **51**, 1093 (1986).

[22] D. E. Breithaupt and W. Schwack, *Eur. Food Res. Technol.,* **211**, 52 (2000).

[23] Y. Goda, T. Nakanishi, S. Sakamoto, K. Sato, T. Maitani and T. Yamada, *J. Food Hygien. Soc. Japan,* **37**, 20 (1996).

[24] T. Philip, C. W. Weber and J. W. Berry, *J. Food Sci.,* **41**, 23 (1976).

[25] P. B. Hamilton, F. I. Tirado and F. Garcia-Hernandez, *Poultry Sci.,* **69**, 462 (1990).

[26] D. E. Breithaupt, P. Weller and M. A. Grashorn, *Poultry Sci.,* **82**, 395 (2003).

[27] P. E. Bowen, S. M. Herbst-Espinosa, E. A. Hussain and M. Stacewicz-Sapuntzakis, *J. Nutr.,* **132**, 3668 (2002).

[28] L. S. Kaw, G. S. Keng, A. Xavier, I. Jesuadimai and H. M. Tan, U.S. Pat. Appl. Publ. USXXCO US 2006171995 A1 20060803 (2006).

Carotenoids
Volume 4: Natural Functions
© 2008 Birkhäuser Verlag Basel

Chapter 14

Carotenoids in Photosynthesis

Alison Telfer, Andrew Pascal and Andrew Gall

A. Introduction

1. General comments

Carotenoids are the secret ingredient in photosynthesis; masked by the green of chlorophyll, they are only revealed in their true glory during senescence, when chlorophyll is degraded to display the glowing colours of autumn. Yet the presence of these orange and yellow pigments is absolutely essential for oxygenic photosynthesis. This Chapter will explain the importance of carotenoids to oxygenic organisms and also their roles in anoxygenic photosynthetic bacteria, where their presence is often more obvious but in other ways may be less crucial.

2. Photosynthesis

Photosynthesis is a two-step process, consisting of 'light reactions' and 'dark reactions'. During the 'light reactions', solar energy is converted into chemical energy (ATP) and reducing equivalents [NAD(P)H], which in turn catalyse the 'dark reactions' in which CO_2 is reduced to form carbohydrate, $(CH_2O)_n$.

Light reaction:
$$2H_2A + 4h\nu \ = \ 4H^+ + 4e + 2A \qquad \text{[NAD(P)H and ATP formed]}$$

Dark reaction:
$$CO_2 + 4H^+ + 4e \ = \ (CH_2O) + H_2O \qquad \text{[NAD(P)H and ATP consumed]}$$

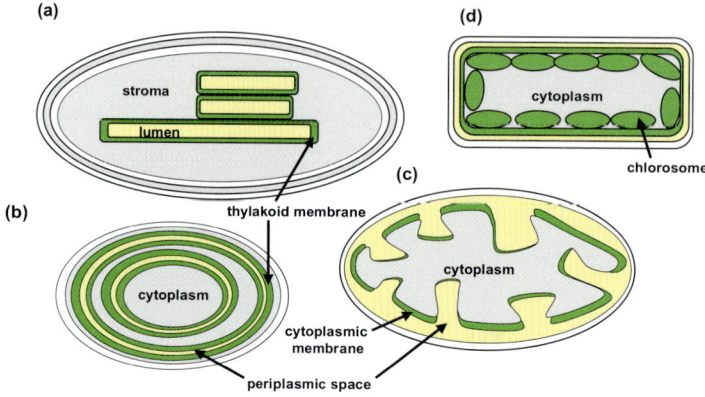

Fig. 1. The different architectures involved in photosynthesis: (a) chloroplast, (b) cyanobacterium, (c) purple photosynthetic bacterium, (d) chlorosome-containing green sulphur bacterium.

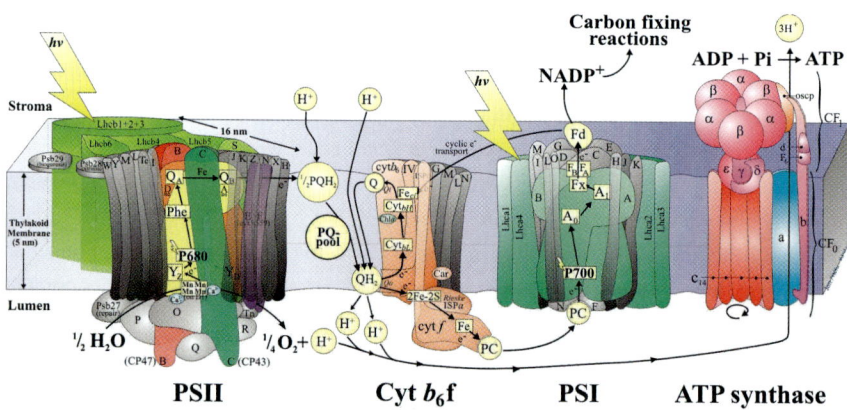

Fig. 2. A cartoon of the photosynthetic membrane in a typical chloroplast-containing plant cell. The major integral membrane proteins involved in the light reactions of photosynthesis are depicted and the black arrows indicate the electron and proton pathways. The electron transport system pumps protons across the membrane, from the lumenal side to the stromal side, and the resultant transmembrane proton motive force is used by the ATP-synthase to generate ATP. Courtesy of J. Nield (Imperial College London).

The light reaction takes place in pigment-protein complexes which, in eukaryotic organisms, span the inner chloroplast membranes, the thylakoids. The dark reaction occurs in the stroma (Fig. 1a). In prokaryotic organisms, the complexes are located in the cytoplasmic membrane and the dark reaction occurs in the cytoplasm (Fig. 1b-d).

In general, there are two major types of chlorophyll-protein complexes, namely (i) light-harvesting antenna (LH) complexes, which bind many chlorophyll molecules that carry out energy transfer both within and between different types of complex and (ii) reaction centre (RC) complexes, which catalyse the primary photochemical reactions and electron transfer. In oxygenic organisms (higher plants, algae and cyanobacteria), which use water as the electron source, the overall reaction occurs *via* two sequential light-driven chlorophyll-protein complexes, the photosystems PSII and PSI, from water (*i.e.* $H_2A = H_2O$ and oxygen is the by-product, 2A) to NADPH *via* the cytochrome b_6f complex (Fig. 2).

In anoxygenic organisms, the electron donor may be organic acids, H_2S or other sulphur compounds. Photosynthesis in these organisms occurs under anaerobic conditions, and the photosynthetic apparatus is simpler in that there is only one photosystem which, in some cases, is similar to PSI (Type 1) and reduces $NADP^+$ directly, in other cases similar to PSII (Type 2) which reduces $NADP^+$ indirectly, by reverse electron transfer [1] (Fig. 3).

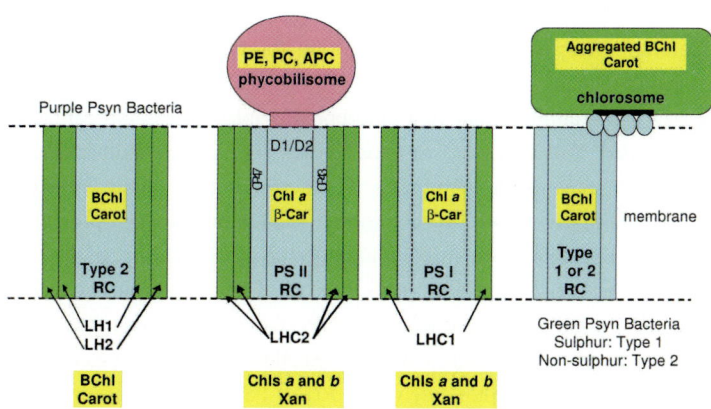

Fig. 3. A cartoon of the different photosystems and antenna complexes found in photosynthetic organisms. The reaction centre (RC) complexes are in blue, the carotenoid-containing outer antenna complexes are in green. The membrane-associated phycobilisome is coloured purple. The yellow boxes highlight the major pigment composition in each of the pigment-protein complexes. Key: (B)Chl, (Bacterio)chlorophyll; β-Car, β-carotene; Carot, carotenoid; PE, phycoerythrin (in red algae only); PC, phycocyanin; APC, allophycocyanin; Xan, xanthophyll.

Photosynthetic organisms are very diverse and have a wide variety of pigment-protein complexes, which contain chlorophylls (or bacteriochlorophyll) and accessory pigments: carotenoids and phycobilins. Most of these pigment-protein complexes serve as an antenna, *i.e.* they harvest light energy. Upon absorption of light by a pigment molecule, excitation energy is transferred throughout the pigment bed and ultimately to specialized (bacterio) chlorophyll molecule(s) in the RCs. The RCs carry out the initial photochemical reaction:

charge separation to form a radical pair, $D^{•+}A^{•−}$ (D, donor; A, acceptor). There are then further electron transfer steps to secondary acceptor molecules which stabilize the charge-separated state. This decreases the likelihood of the back reaction between the radical pair and thus prevents the loss of stored energy as heat.

In both oxygenic and anoxygenic photosynthesis, the pigment-protein complexes are arranged such that, during light-driven electron transfer, protons are transported across the photosynthetic membrane into the lumen or periplasmic space (Figs. 1 and 2), thereby building up an electrochemical potential gradient which drives ATP formation *via* the ATP synthase (Fig. 2). The dark reactions, in which CO_2 is converted into carbohydrate, utilize energy stored in ATP, along with the reducing equivalents of NAD(P)H.

3. Role of carotenoids

Carotenoids are involved in photosynthesis in a number of ways (Fig. 4) which are described and discussed in subsequent sections. A major role is to harvest light energy and pass this energy on to chlorophyll by singlet-singlet excitation transfer. In an aerobic environment, however, carotenoids also have a role in photoprotection; they can prevent formation of singlet oxygen ($^1O_2$), by rapidly quenching chlorophyll triplet states, and can scavenge $^1O_2$ directly, if any is formed. This very reactive toxic oxygen species would otherwise oxidize the pigments, proteins and lipids of the membrane, thus destroying the photosynthetic apparatus [2]. Indeed, under natural conditions, all chlorophyll-containing pigment-protein complexes bind carotenoids at almost always one carotenoid molecule per 3-4 chlorophylls.

There is also another very important protection mechanism present in most oxygenic organisms, namely non-photochemical quenching (NPQ), in which carotenoids quench excess excitation energy, releasing it harmlessly as heat (Section E.3). This lessens the extent of over-reduction at the electron acceptor side of the RC of PSII, so that the formation of other toxic oxygen species (in particular superoxide and hydroxyl radicals) is minimized.

Carotenoids within the PSII RC have been shown to be oxidized if the lifetime of highly oxidizing forms of chlorophyll is increased beyond a few microseconds. The importance of this electron transfer is discussed in Section **F**. Recently carotenoids and chlorophylls have been discovered in the cytochrome b_6f complex, though there is only one molecule of each bound to the protein (Section **E**.5). These pigment molecules are probably not involved in light harvesting or directly in the light-driven electron-transfer reactions. It has been suggested that they may have a light-signalling role during assembly and degradation of the complexes (Section **G**). Carotenoids often are required to stabilize the structure of complexes; these sometimes cannot be assembled unless critical carotenoids are present [3]. Finally, how carotenoids act as indicators of a membrane potential gradient across the photosynthetic membrane is discussed in Section **H**.

The principles and recent advances in carotenoid photochemistry are presented in *Chapter 9*. Now, in this Chapter, the carotenoids involved in photosynthesis, their roles and the various types bound to protein complexes will be reviewed. Their functions will be related to the many recently published X-ray structures of the pigment-protein complexes involved in photosynthetic electron transfer, in which the positions of many carotenoids have been identified. For more background on this subject the reader is directed to references [4-7].

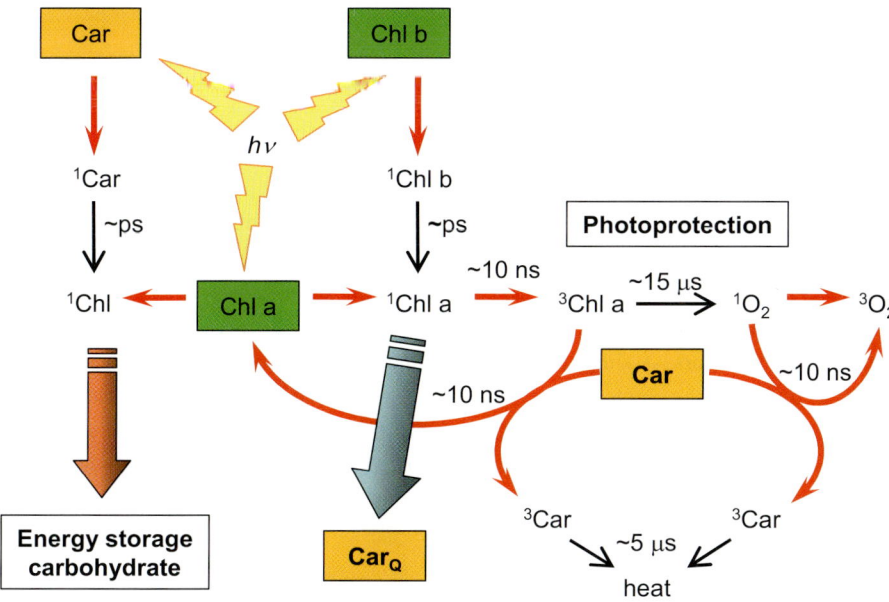

Fig. 4. A model of the energy dissipation pathways of the different excited states of carotenoids and chlorophylls following photon excitation. Car$_Q$: carotenoid non-photochemical quenching mechanism, NPQ.

B. Photosynthetic Pigment-Protein Complexes

The light reactions of the different photosynthetic organisms are driven by a variety of pigment-binding complexes. Figure 3 illustrates schematically how they are organized. In some organisms, the RCs (both Type 1 and 2) are surrounded by membrane-intrinsic LHCs. In other organisms, extrinsic complexes associated with the surface of the membrane, *e.g.* chlorosomes and phycobilisomes, may replace the LHCs or be present in addition to them. The X-ray crystal structures of most of the major antenna and RC complexes have been elucidated and in most of them the carotenoids bound to the complexes have been resolved.

Although oxygenic organisms dominate the earth, it is often convenient to describe the purple bacterial complexes and their carotenoids first. Historically, it was found to be easier to study the structure and function of bacterial photosynthetic systems because (i) the photosynthetic apparatus is less complex than that of plants and algae, (ii) the organisms are easy and quick to culture, (iii) the proteins are relatively easy to isolate and purify in large quantities, and (iv) the bacteria are amenable to genetic manipulation. As a result, much of the detailed photochemistry of carotenoids has been studied in the purple bacterial system.

C. Reaction Centres: Structure and Function

1. Type 2 reaction centres

Purple bacteria, green non-sulphur bacteria and PSII all have Type 2, quinone-reducing RCs. The bacteriochlorophyll b-containing pBRC (purple bacterial reaction centre) from *Blastochloris* (*Blc.*) *viridis* (formerly known as *Rhodopseudomonas viridis*) was the first membrane protein to have its structure elucidated [8]; the authors received the Nobel Prize in Chemistry for this work. A simplified version of the X-ray structures of purple bacterial and PSII RCs, including the carotenoids, is shown in Fig. 5. In contrast to the bacterial RCs, the PSII RC-core protein complex crystallizes as a dimer, which is also the natural state in the membrane. There are some examples of dimeric RC-LH1 complexes observed in the photosynthetic membrane from purple bacteria but these are an exception.

a) General features

In Type 2 RCs, charge separation results in formation of oxidized (bacterio)chlorophyll, [P^+], and reduced (bacterio)phaeophytin, [(B)Phaeo$^-$]. Subsequently, the electron passes to a quinone molecule (Q_A) to form a semiquinone and then this electron is transferred directly to a second quinone (Q_B). This is a two-electron, two-proton gate, *i.e.* it receives two electrons sequentially upon two charge-separation events, it is protonated and the quinol then leaves the binding site, allowing a new quinone molecule to bind and receive the next pair of electrons. The protein scaffold of the RC that encapsulates the pigments consists of two entwined polypeptides (L and M in the pBRC and D1 and D2 in PSII) each with five membrane-spanning helices. These two polypeptides bear pairs of electron transfer cofactors in a pseudo-two-fold symmetry. There are other peptides present, such as the H-subunit of the pBRC, that do not interact with pigments; within the framework of this Chapter, they can be ignored. Charge separation and electron transfer steps are only on one side, *i.e.* the 'active' branch, to Q_A and then the electron moves across to the second bound quinone, Q_B, on the 'inactive' branch (Fig. 5; the arrows indicate the active electron transfer branch). The two RCs differ greatly in their electron donors. PSII must be very strongly oxidizing (>+1.1 V) in order to split water, whereas the purple bacterial RC only has to generate ~+500 mV.

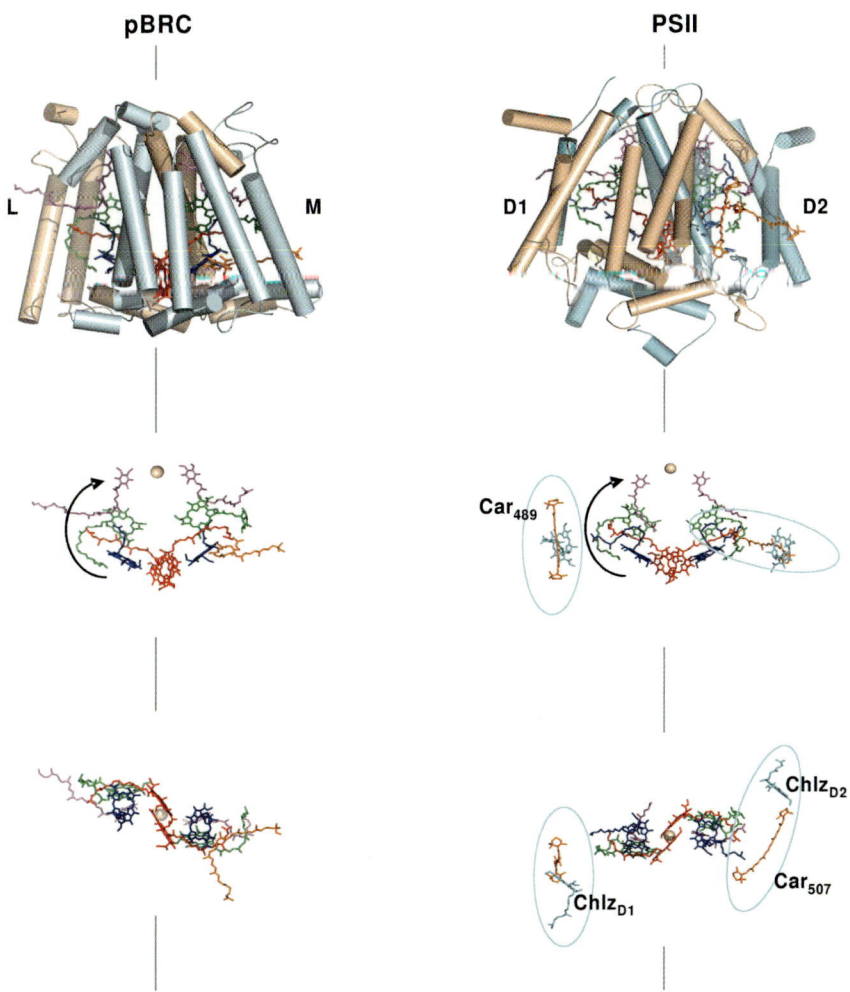

Fig. 5. The structure of the purple bacterial (pBRC) and PSII reaction centre complexes. (Top) Side view showing the transmembrane α-helices that enclose the pigments. (Middle) The same side view depicting only the cofactors. (Bottom) Looking down from the cytoplasmic (pBRC) and lumenal (PSII) membrane surface. PSII contains two additional carotenoid (Car) and chlorophyll (Chl) molecules (located in the ovals) whereas the pBRC has only one Car. The solid line represents the pseudo C_2 axis of symmetry. The primary donor (P), accessory (B)Chls (B), (B)Phaeophytins (Ph), quinones (Q_A and Q_B) and carotenoids are coloured red, blue, green, magenta and orange, respectively. The spheres represent the central Fe^{2+} atoms. The additional Chls (Chlz$_{D1}$ and Chlz$_{D2}$) in PSII are coloured cyan. The arrows indicate the active electron transfer branch. The pBRC and PSII structures are from *Rhodobacter sphaeroides* and *Thermosynechococcus elongatus*, respectively (Protein Data Bank accession numbers 1PSS and 2AXT). The figure was produced by use of PyMOL [9].

The bacteriochlorophyll primary donor in purple bacteria, P870, oxidizes a cytochrome whereas, in PSII, the chlorophyll primary donor P680 oxidizes Tyr$_{161}$, of the D1 protein, and this in turn oxidizes a metal complex composed of 4 Mn atoms and one Ca atom After four oxidizing equivalents have been accumulated on the 'Mn cluster', water is oxidized and oxygen is released as a by-product.

The actual electron transfer reactions occur from specialized (bacterio)chlorophyll *a* molecules to a (bacterio)phaeophytin acceptor (Fig. 5). In PSII it is thought that the nature of the binding sites of these specialized molecules on the protein induces the very high potential when they are oxidized, much higher than that of the normal potential of chlorophyll *a* (~0.7 V). There is still discussion about what is the first electron transfer step; in low temperature experiments it is almost certainly from the accessory chlorophyll on the active branch (Chl$_{D1}$ or Chl$_L$, also known as B$_L$), but the positive charge is stabilized mainly on P$_{D1}$/P$_L$ though it is shared to a certain degree with P$_{D2}$/P$_M$. The electron passes first to the (bacterio)phaeophytin, on the active branch, before moving to Q$_A$ and then to Q$_B$.

b) Location of carotenoids

Both the purple bacterial and PSII RCs bind carotenoids, but in different places, as they have very different roles. [Note that the structure and carotenoid content of green bacterial RCs (both Types 1 and 2) are unknown]. In the pBRC there is a single carotenoid molecule (but the kind of carotenoid varies from species to species). In the known structures [10], it is in the 15-*cis* configuration and is bound within van der Waals distance of the accessory bacteriochlorophyll on the 'inactive M-branch' (Fig. 5). A detailed discussion on the structure and spectroscopy of the pBRC can be found in two reviews [10,11]. Purple bacterial RC complexes can be isolated and purified from mutant purple photosynthetic bacteria that contain no carotenoids. Indeed the X-ray crystal structure of the RC from the carotenoidless strain R26.1 of *Rhodobacter (Rba.) sphaeroides* (formerly known as *Rhodopseudomonas sphaeroides*) is the 'same' as that which contains a carotenoid molecule, except that there is a void that contains detergent molecules where the carotenoid is usually located.

Recently, crystal structures have been obtained after different carotenoid molecules have been introduced into the R26.1 RC complex [12,13]. These studies have suggested that the carotenoid molecules are incorporated unidirectionally, since a specific amino acid residue acts as a 'gate-keeper' which then permits a second amino acid to form hydrogen bonds with the carotenoid and lock it in place. This physical mechanism is optimized for a 15-*cis* carotenoid molecule. When the pBRC is used as a model system, resonance Raman spectroscopy, together with normal-coordinate analysis, of the RC with 15-*cis* carotenoid bound, and related studies *in vitro*, have revealed twisting and considerable changes in bond order of the conjugated polyene backbone upon generation of carotenoid triplets. These changes are proposed to enhance the rate of relaxation to the ground state and the dissipation of triplet energy [14-17]. Recent carotenoid reconstitution experiments have started to yield a more quantitative view of the photoprotective function of (15-*cis*)-carotenoids containing different numbers of conjugated double bonds [12]. No such quenching has been observed in *Blc. viridis* as the triplet state energy of bacteriochlorophyll *b* is lower than that of the corresponding bacteriochlorophyll *a* in *Rba. sphaeroides*, so that triplet energy transfer from the primary donor is, in effect, blocked.

The PSII RC, on the other hand, binds two carotenoid molecules, which, in all structures known so far, are both β-carotenes (**3**). Their location has been hotly debated, though it is clear from a number of spectroscopic observations that one is more or less horizontal to the plane of the membrane (Car$_{507}$) and the other perpendicular to this plane (Car$_{489}$) (Fig. 5) [18]. The latest 3.0 Å X-ray structure places Car$_{507}$ on the D2 protein and Car$_{489}$ on the D1 protein [19] although it was argued earlier [20], on the basis of spectroscopic evidence and an earlier structure in which only the D2-side carotenoid was resolved [21], that they might both be on the D2-side. (*N.B.* the designation of these two carotenoid molecules is based on their light absorption properties and not on their relative position in the X-ray model). In this latest structure, both carotenoids, at their closest points, are still at least 16 Å from the nearest of the four central chlorophyll molecules. The PSII RC has two additional chlorophylls compared to the pBRC; these are located out towards the periphery of the protein. They are known as Chlz$_{D1}$ and Chlz$_{D2}$ and, in the latest structure [19], the two carotenoids are within van der Waals distance of them (Fig. 5).

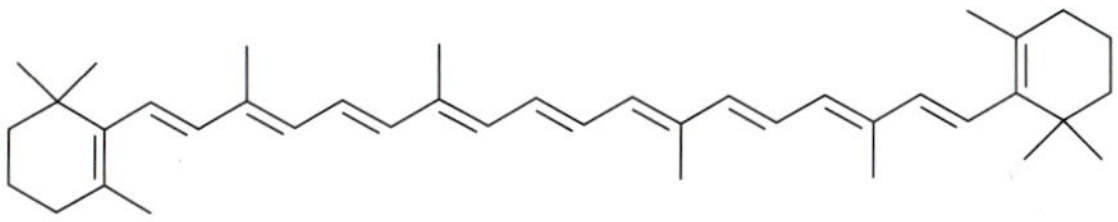

β-carotene (**3**)

2. Type 1 reaction centres

Green sulphur bacteria, heliobacteria and PSI all have Type 1 RCs. These are characterized by having as their main electron acceptors very low-potential iron-sulphur proteins that can reduce NAD(P)$^{+}$. The X-ray crystal structure of the trimeric PSI RC isolated from the cyanobacterium *Thermosynechococcus* (*T.*) *elongatus* has been resolved [22]; there are 90 chlorophyll and 22 carotenoid molecules per monomer. Unfortunately in the X-ray structure of higher plant PSI, which is always a monomeric complex binding an extra antenna (LHCI), the carotenoids have not yet all been resolved [23,24]. Nor have any of the green bacterial or the *Heliobacterium* RC structures been determined.

The structure of one of the monomers in the trimeric *T. elongatus* structure [22] is shown in Fig. 6. Each of the main polypeptides, PsaA and PsaB, has eleven membrane-spanning helices, five of which are remarkably similar to those of D1 and D2 in the PSII RC. They bear all the electron transfer cofactors, forming a curved basket shape around them, and can thus be considered as the RC region. The remaining six helices of each polypeptide are closely related to the inner antenna complexes (CP47 and CP43) seen in the PSII core (see Figs. 2 and 3) and only have an antenna function (see Section **D**.2).

The electron transfer cofactors bound to the central, RC, part of this large complex, are organized similarly to those in Type 2 RCs. There is a pseudo-two-fold symmetry in Type 1 RCs, just as there is in PSII. However, in Type 1 RCs, both electron transport chains are active, with the electrons converging (when they leave the phylloquinone acceptors at the end of each branch) into a single chain of low-potential iron-sulphur clusters, F_X, F_A and F_B. These reduce a soluble iron-sulphur protein, ferredoxin, which in turn reduces $NADP^+$ (see Fig. 2). There are additional regions at the N-terminal domain of both PsaA and PsaB and a number of other small membrane-intrinsic polypeptides which also bind chlorophyll and carotenoid (see Section **D**).

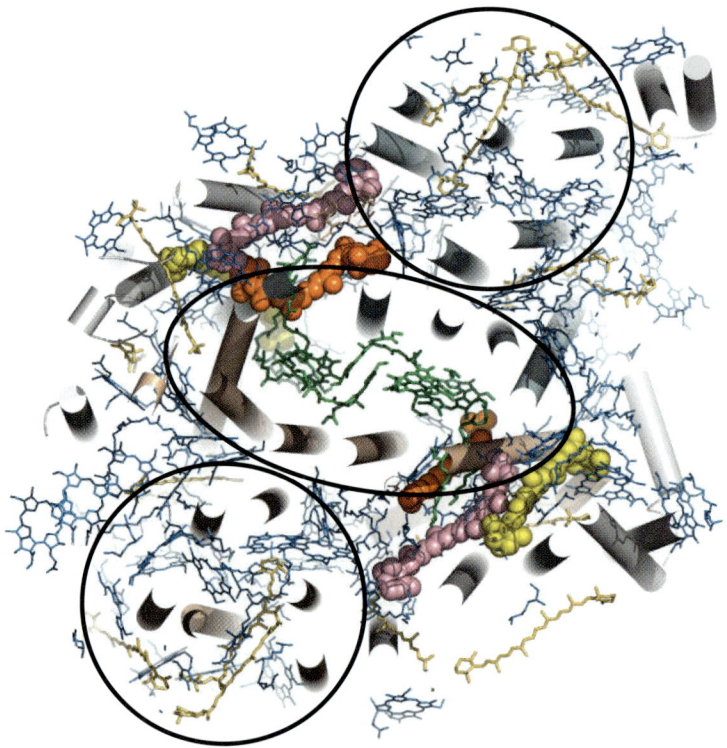

Fig. 6. The structure of the PSI reaction centre complex from *Thermosynechococcus elongatus*. View of a monomer along the membrane normal from the stromal side. The position of the RC is highlighted by the oval. The upper and lower circles indicate the relative locations of the core antennae. The central RC pigments are coloured green. Carotenoid molecules in close proximity are represented as space filling models. Protein Data Bank accession number 1JBO. The figure was produced by use of PyMOL [9].

A model [22] shows all the carotenoid molecules resolved as β-carotene (3) as *per* the chemical analysis; sixteen are all-*trans*, and five have one or two *cis* double bonds. None is in very close contact with any of the RC's chlorophyll cofactors. Almost all are in the outer antenna regions of the protein complex, where they function both as antenna and in photoprotection as chlorophyll triplet quenchers (see Sections **D** and **E**). Figure 6 shows the six closest carotenoids, two of which approach to within about 6-7 Å of the chlorophyll acceptor molecules (A_0) on each side, and one within *ca* 9 Å of the accessory chlorophyll on one branch. Even in the very recent structure of the higher plant PSI [24], where five carotenoids have been modelled, the nearest approach of a carotenoid to P700 is 10 Å. None of the carotenoids are within van der Waals distance of the electron transfer cofactors, so they would be poor at quenching any triplet states formed within the RC (see Section **E**).

D. Light Harvesting: Antenna Structure and Function

1. General aspects

Solar energy spans a large spectral range, approximately from 300 to 2000 nm, with the peak at around 500 nm. This represents an extraordinary potential resource for any organism that can exploit it [4]. As effector molecules in photosynthetic bio-engines, (bacterio)chlorophylls are ideally suited light-energy transducers. They convert this energy, through their excited states, into chemical energy (initially as an electrochemical gradient, as outlined above). They exhibit strong absorption transitions in two parts of the solar spectrum, the so-called Soret (300-400 and 350-460 nm) for bacteriochlorophyll and chlorophyll, respectively and Qy (640-750 and 700-960 nm) regions, plus a weaker absorption band called Qx (~590, 630 nm). Carotenoids absorb light energy in the blue and green region of the visible spectrum (350-550 nm), thus extending the range of solar radiation outside that absorbed by (bacterio) chlorophylls. The direct result of this is an increase in the efficiency of photosynthetic energy conversion, because the absorption cross-section of the antenna pigment bed is increased. To fulfil this function, the carotenoids must be located at tightly-controlled distances and orientations relative to the chlorophyll molecules, in order to ensure efficient transfer of the absorbed energy. Thus the antenna structure of the membrane-intrinsic complexes is highly conserved, and the membrane-intrinsic antenna proteins can conveniently be grouped into only three 'superfamilies', namely the core-type, LHC-type and LH-type proteins.

Antenna proteins use a variety of different carotenoid species, which extend the absorption range within the blue-green region of the spectrum. This can be further augmented by a tuning of the absorption properties of specific carotenoids *via* interactions with their protein-binding pocket. While only small shifts in absorption maxima can be achieved through such mechanisms, they can potentially be of great importance, for example when they extend the absorption cross-section at the edge of the spectrum.

Purple bacteria live in the anaerobic layers of ponds, lakes, estuaries and streams [25]. In this ecological niche, the spectrum of the solar energy that reaches these organisms has been filtered by oxygenic phototrophs (algae, cyanobacteria and plants) that are located above them. The purple bacteria, therefore, must harvest wavelengths above 760 nm. They achieve this by having a photosynthetic apparatus that is based on bacteriochlorophyll (Bchl) where the Qy absorption transitions range, as stated above, from 700 nm to beyond 900 nm [26] (Fig. 7). The carotenoids in these organisms harvest blue/green light, up to 550 nm. This is some 10-40 nm further into the red than the majority of the carotenoids in oxygenic phototrophs. Thus photosynthetic bacteria not only use light far into the infrared to drive photosynthesis but also, by using specific different carotenoids, take advantage of part of the visible spectrum that is less efficiently absorbed by oxygenic organisms.

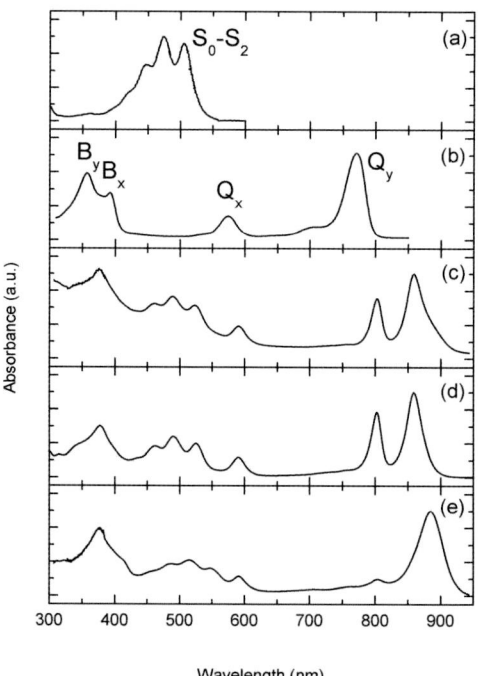

Fig. 7. Room-temperature absorption spectra of free (in solvent) and non-covalently bound bacteriochlorophyll (Bchl *a*) and carotenoid pigments from *Rbl. acidophilus* 10050. (a) The ground state (S_0-S_2) absorption peaks of solvent-extracted rhodopin glucoside (**95**) in hexane. (b) Monomeric Bchl *a* in 7:2 (v/v) acetone:methanol showing the Soret (B_x and B_y), Q_x (590 nm) and Q_y (772 nm) electronic transitions. (c) Absorption spectrum of isolated membranes. The absorption spectra of the carotenoids and Q_y-transitions of the bacteriochlorophylls (and bacteriophaeophytins) are both red-shifted when the pigments are bound to their apoprotein scaffold. (d) Absorption spectrum of the detergent-purified LH2 complex in which rhodopin glucoside is the principal carotenoid (see Table 1). (e) Absorption spectrum of the detergent-purified RC-LH1 'core' complex.

2. Antenna complexes

The three membrane-intrinsic antenna complex 'superfamilies' (LH-type, LHC-type and core-type), each with somewhat different protein and pigment organizations, promote efficient light absorption and energy transfer to the RCs. The LH family is found in green non-sulphur bacteria (of which *Chloroflexus aurantiacus* is an example) and all purple bacteria. These are organisms with Type 2 RCs.

The LH-type complexes are absent from green sulphur bacteria and heliobacteria, which have Type 1 RCs. However, these organisms have extrinsic light harvesting complexes, chlorosomes, associated with the surface of the cytoplasmic membrane (Figs. 1 and 3), and containing chlorophyll *a* and carotenoids. Chlorosomes are also present in the green non-sulphur bacteria in addition to LH-type complexes (Fig. 3). The membrane-intrinsic antennae of oxygenic organisms can be classified into two types; the 'outer' or peripheral LHC-type complexes transfer excitation energy *via* the 'inner' or core-type complexes to either the PSII or PSI RCs. Higher plants and most algae have both the LHC-type and core-type antenna complexes (Fig. 3).

In some prokaryotes but also in many algae there are also extrinsic antenna complexes which add to the complexity of the light-harvesting process. The cyanobacteria and red algae have extrinsically bound phycobilisomes [which contain phycobilins (linear tetrapyrroles), especially phycocyanin and phycoerythrin, but do not bind carotenoids], in addition to the intrinsic core-type antenna (which binds chlorophylls and carotenoids). Under Fe-rich conditions they have no LHC-type complexes. On the other hand, the prochlorophytes, the chlorophyll *d*-containing cyanobacterium *Acaryochloris* (*A.*) *marina* and, under low Fe-stress, some other cyanobacteria have core-type intrinsic antennae *e.g.* Pcb or IsiA (which are also known as CP43′) proteins. Other extrinsic complexes include chlorosomes and the special carotenoid-rich extrinsic antenna complex, the peridin-chlorophyll-protein (PCP), that is found in some dinoflagellate algae. The wealth of different kinds of antennae discovered so far, especially in algae and photosynthetic prokaryotes, suggests that there may be new forms of light-harvesting system still to be discovered.

a) LH complexes

The LH family is found in bacteria with Type 2 RCs, *i.e.* purple bacteria. They have two types of integral-membrane light-harvesting antenna complexes called LH2 and LH1; the latter forms a ring around the RC whilst the former consists of individual rings in the plane of the membrane. The X-ray crystal structures of these bacterial proteins have been elucidated [27-30]. These two pigment-protein complexes are also sometimes known as the 'peripheral' and 'core' light-harvesting complexes, respectively, due to their relative location in the photosynthetic apparatus with respect to the pBRC. Some bacterial species [*e.g.* *Rhodospirillum* (*Rsp.*) *rubrum*] do not synthesize LH2 complexes but all species assemble the

RC-LH1 'core' complex [26]. In the near IR wavelengths, LH2 has two strong bacteriochlorophyll absorption bands at *ca.* 800 and 850 nm (Fig. 7d) and is also referred to as the B800-850 complex, whereas the RC-LH1 complex has its main bacteriochlorophyll absorption band at *ca.* 875 nm (Fig. 7e) and is referred to as the B875 complex [31]. The carotenoid molecules play a vital role in the fine-tuning of these bacteriochlorophyll absorption bands.

The LH-type proteins are constructed from a basic modular system. This consists of bacteriochlorophyll and carotenoid molecules non-covalently bound to two, low-molecular weight, hydrophobic apoproteins, called α and β. Each apoprotein spans the photosynthetic membrane once. The α/β-apoprotein dimers oligomerize to produce the intact native annular structures (Figs. 8 and 9). The α/β-dimers are small compared to the other types of antenna complexes and have only two transmembrane helices, one (all-*trans*)-carotenoid molecule, which spans the membrane diagonally, and either two (in LH1) or three (in LH2) bacteriochlorophyll molecules (Fig. 8). The primary sequence of the α and β apoproteins determines whether an LH1 or LH2 complex is assembled [26].

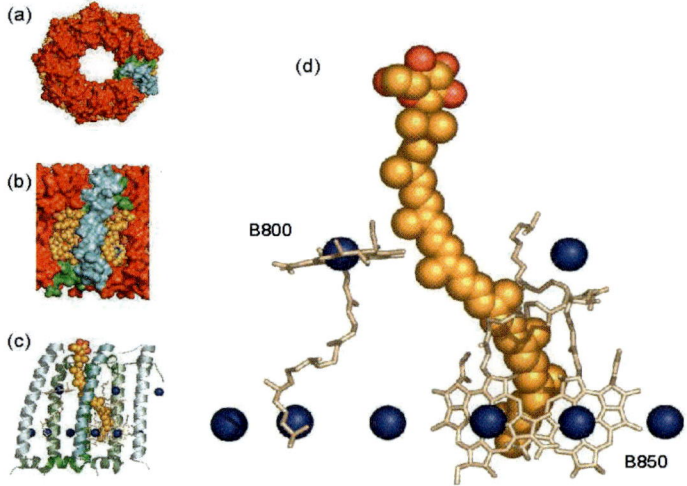

Fig. 8. Structure of the LH2 complex from *Rbl. acidophilus* strain 10050. (a) Viewed from the periplasmic side of the photosynthetic membrane. The nonameric structure is shown in red except for one α/β-dimer for which the inner α-apoprotein and outer β-apoprotein are coloured green and cyan, respectively. The bacteriochlorophyll (Bchl) *a* and carotenoid (Car) pigments (yellow) are sandwiched between the neighbouring apoproteins. (b) A view of three α/β-dimers perpendicular to the direction of the transmembrane α-helices viewed from the outside of the antenna complex. (c) As (b) but the apoproteins are displayed in ribbon format that allows the carotenoid to be seen along its long axis. The three α-apoproteins are coloured in different shades of green, the three β-apoproteins are coloured in different shades of cyan. The Mg^{2+} ions of the Bchl molecules from the three α/β-dimers are depicted as blue spheres. The B800-Bchl from the left-hand dimer and the B850-Bchl molecules from the central α/β-dimer are shown in wireframe. (d) An expanded view of (c) but with the apoproteins removed. Redrawn from [32].

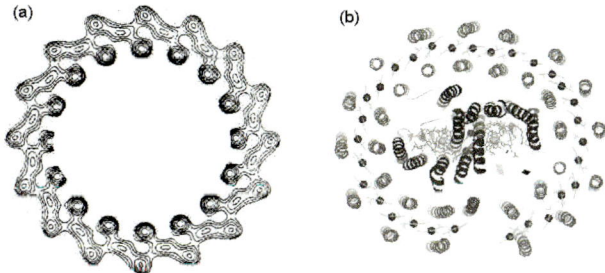

Fig. 9. The annular organization of the LH1 complex. (a) An 8.5 Å resolution projection map of the reconstituted LH1 complex from *Rsp. rubrum*, comprising 16-α/β-dimers [33]. (b) The structure of RC-LH1 complex from *Rps. palustris* at 4.8Å resolution [30], looking down from the surface of the cytoplasm where the complex has been sectioned at the level of the ring of bacteriochlorophylls, which are shown in wireframe. Compared to (a), one α/β-dimer has been 'substituted' by the peptide W. The figure was produced by use of PyMOL [9].

LH2 forms a ring of either eight or nine α/β-dimers (depending on the species) whilst LH1 is made up of sixteen α/β-dimers (this number may also vary). The LH1 encloses the RC. The green non-sulphur bacteria (with type 2 RCs) have, in addition to chlorosomes, an intrinsic antenna complex similar to LH1. The detailed structure has not been determined, however.

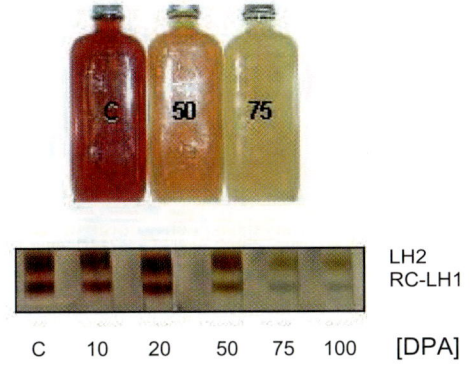

Fig. 10. The effect of increasing diphenylamine (DPA) concentration (zero, C), in cultures of *Rps. palustris*, and the resulting distribution of coloured carotenoids in the RC-LH1 and LH2 antenna complexes after separation by sucrose density gradient centrifugation. Redrawn from [34]. The carotenoid compositions of the complexes are given in Table 1.

When the LH2 complexes and the RC-LH1 complexes are solubilized with detergent, they have different apparent densities and are thus easily separated by centrifugation on sucrose gradients. Because the human eye is not able to see NIR light corresponding to the long wavelength absorption bands of bacteriochlorophyll, the colours of the bands in a sucrose gradient (*i.e.* the LH complexes) are due mainly to the strong absorption by the carotenoid molecules. Furthermore, different colours of the LH2 and RC-LH1 can easily be observed, *i.e.*

this is a direct and simple way to observe the preferential uptake of different carotenoid molecules by the two antenna complexes of the same photosynthetic apparatus (Fig. 10).

i) Ultrastructure of LH2. The first high-resolution crystal structure of any LH2 complex was obtained from *Rhodoblastus (Rbl.) acidophilus* strain 10050 (formerly *Rhodopseudomonas acidophila*). This LH2 complex is composed of nine α/β-dimers [29] (Fig. 8a). The inner and outer walls of the ring are formed by the nine α-apoprotein helices and the nine β-apoprotein helices, respectively. The LH2 complex is sealed on either side by the N- and C-termini of the apoproteins that fold over and interact with one another. There is one monomeric bacterio-chlorophyll per α/β-dimer. The various intermolecular interactions give rise to the 800 nm absorption band (Fig. 7d) and thus these bacteriochlorophylls are termed the B800 molecules. Towards the N-terminal side is located a second ring of 18 bacteriochlorophyll molecules where the bacteriochlorin rings are much closer to each other (Fig. 8c). The various molecular interactions give rise to the 850 nm absorption band (Fig. 7d) and thus these bacterio-chlorophylls are termed the B850 molecules. There are two B850 molecules per α/β-dimer.

rhodopin glucoside (**95**)

The LH2 complex from *Rbl. acidophilus* also contains nine molecules of the carotenoid (all-*trans*)-rhodopin glucoside (**95**). It was recently postulated that a second, di-*cis*, rhodopin glucoside molecule was present in a less than 1:1 stoichiometric ratio with the α/β-dimer [35]. However, based on rigorous pigment analysis, spectroscopic measurements and the most recent *meso* LH2 crystal structure, it has been confirmed that there is indeed only one, all-*trans*, carotenoid per α/β-dimer [36-38]. The 'second carotenoid' molecule is actually the result of electron density attributable to partial detergent occupancy [37]. The (all-*trans*)-rhodopin glucoside in LH2 has 11 conjugated double bonds and, when viewed down its long axis, forms a twist that resembles a half helix (Fig. 8c-d). The carotenoid runs along nearly the whole length of the transmembrane α-helices, passing in proximity to the edge of the bacteriochlorin ring of a B800 molecule (closest contact 3.4Å). It then passes into the next α/β-dimer and runs over the face of the bacteriochlorin ring of the α-protein-bound B850 bacteriochlorophyll. Thus this carotenoid has an important structural role in LH2 as it interacts with the B800 and B850 molecules of neighbouring α/β-dimers. This important structural role is often overlooked but it is widely known that nearly all species of purple photosynthetic bacteria require the presence of carotenoids for the natural assembly of LH2 complexes (Fig. 10) [34,39-43].

ii) Ultrastructure of LH1. The LH1 complex has a modular structure similar to that of the LH2 complex. The non-covalently bound carotenoid molecules are all-*trans* but this class of LH-antenna lacks the ring of monomeric bacteriochlorophyll molecules. The ring of dimeric bacteriochlorophyll *a* molecules absorb further into the near IR wavelengths up to and sometimes beyond 900 nm. From reconstitution experiments of LH1 in the absence of carotenoid molecules, a two-dimensional map of the carotenoid-less LH1 complex from *Rsp. rubrum* was derived [33]. In this antenna there are 16 α/β-dimers per ring (Fig. 9a). RC-LH1 'core' complexes were first visualized in EM pictures of membranes from *Blc. viridis* [44]. It then took 24 years before the 4.8Å resolution X-ray structure of an RC-LH1 complex was solved where the 16th α/β-dimer is replaced by the protein W [30] (Fig. 9b). Unfortunately, at this rather low resolution the structure of the bacteriochlorophylls, carotenoids and amino acid side chains cannot be resolved. The exact number of α/β-dimers in the ring may vary from species to species and is still hotly debated. For a more detailed discussion on this topic the reader should consult a recent review [32].

b) LHC-type complexes: chlorophyll *a/b* proteins

The 'outer' or peripheral LHC antenna protein complexes are found only in plants and some algae. LHCII and LHCI complexes are normally associated with PSII and PSI, respectively, and are products of the *lhcb* (LHCII) and *lhca* (LHCI) gene families.

i) Ultrastructure of LHCII. The crystal structure of the main LHCII complex in plants and green algae [45,46] shown in Fig. 11 reveals a trimeric organization, which is its oligomer- ization state in the membrane and under most isolation conditions. Each monomer has three membrane-spanning helices and binds eight chlorophyll *a*, six chlorophyll *b* and four xanthophylls (two luteins (**133**), in sites L1 and L2, one neoxanthin (**234**) in site N1, and one xanthophyll cycle carotenoid in site V1; see Section **E.4**).

lutein (**233**)

neoxanthin (**234**)

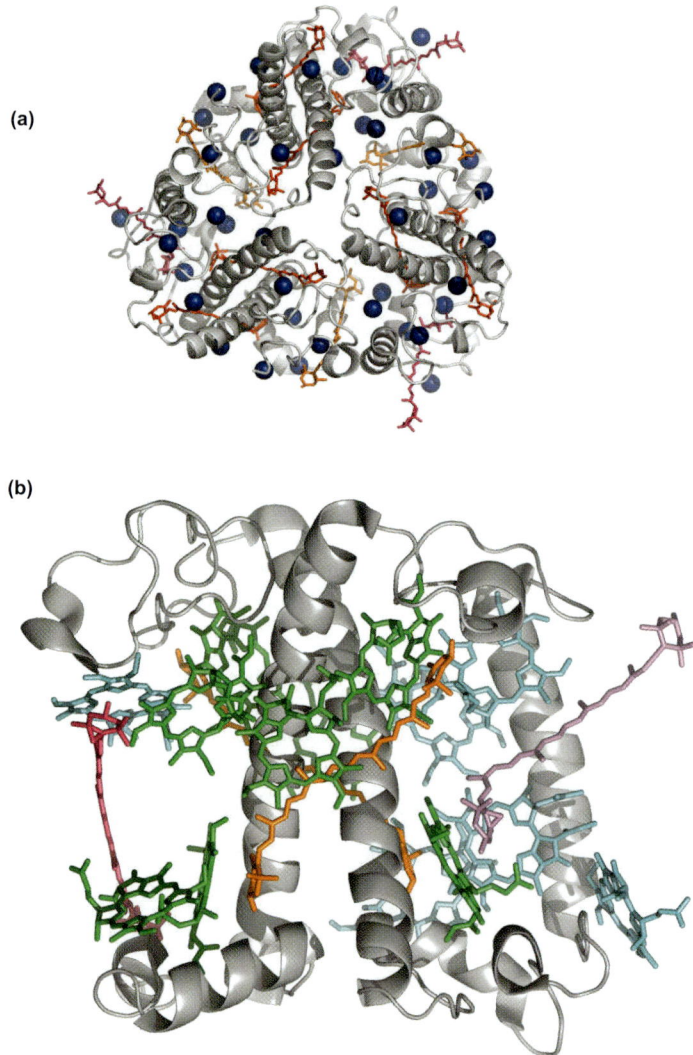

Fig. 11. Structure of the LHCII complex from spinach (*Spinacia oleracea*). (a) View of the trimer as seen along the membrane normal from the stromal side. Lutein (**133**), neoxanthin (**234**) and xanthophyll cycle carotenoids are shown as orange, purple and magenta, respectively. For clarity only the central magnesium atoms of the chlorophyll (Chl) molecules have been depicted, as solid blue spheres. (b) A monomer of LHCII viewed in parallel with the membrane plane. The carotenoids are coloured as in (a) and the Chl *a* and Chl *b* molecules are coloured green and blue, respectively. Protein Data Bank accession number 1RWT. The figure was produced by use of PyMOL [9].

The most striking feature is the cross-brace arrangement of the two central helices and the (all-*trans*)-luteins (**133**), which are both at an angle of *ca.* 60° to the membrane normal. These carotenoid molecules are essential for correct reconstitution of LHCII *in vitro* [47]. Refolding experiments have demonstrated that the L1 site is highly lutein-specific whereas the L2 site can also bind violaxanthin (**259**); the N1 site is highly specific for neoxanthin [48], which is in the 9-*cis* configuration and so has a hook-shaped structure, and is located in a chlorophyll *b* rich region near the third helix. Its polyene backbone is also at an angle of *ca.* 60° to the membrane normal. In dark-adapted samples, the fourth site (V1) contains violaxanthin. This carotenoid constitutes part of the xanthophyll cycle, and can be converted into zeaxanthin (**119**) *via* antheraxanthin (**231**) in high-light conditions by the violaxanthin de-epoxidase enzyme (Scheme 1) as discussed in Section **E.4**. This site is often not fully occupied [49].

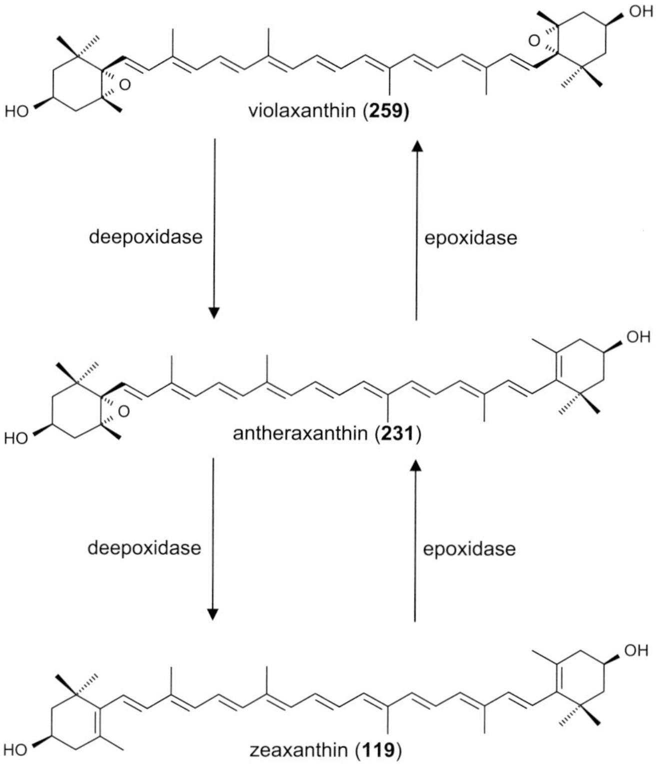

Scheme 1

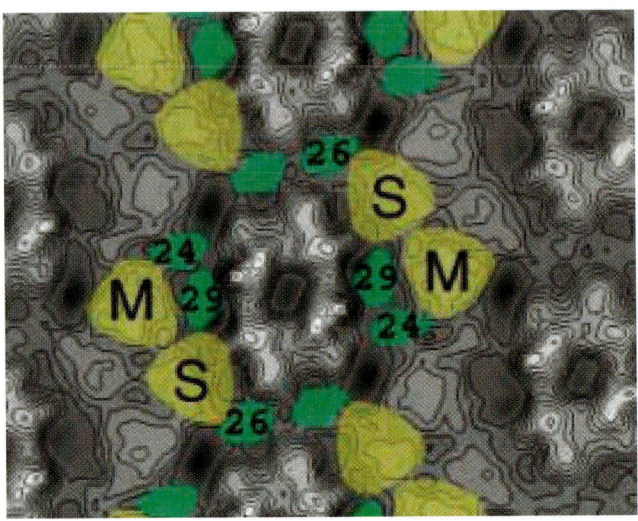

Fig. 12. The spatial organization of the *Arabidopsis thaliana* PSII supercomplex: RC, Core antenna, CP24, CP26 and CP29, and LHCII trimers in a native membrane array. The central PSII core is surrounded by the major S (strongly bound) and M (medium bound) LHCII trimers (yellow) and minor monomeric antenna complexes CP24, CP26 and CP29 (green). Redrawn from [52].

ii) Other LHC proteins ('minor complexes'). Associated with PSII there are other LHC proteins which are often termed 'minor complexes', as they are present in lower amounts than the major LHCII (they are monomeric and usually stoichiometric to PSII). Also called CP24, CP26 and CP29, they exhibit the characteristic three membrane-spanning helices and bind both chlorophylls (*a* and *b*) and xanthophylls. Their pigment-binding stoichiometry is lower than that of the main LHCII (and varies between them) and they have a higher chlorophyll *a/b* ratio, but the majority of binding sites are common. In particular, they do not contain the V1 site; violaxanthin can bind to one or more of the other three sites and can thus be present to a higher level relative to both chlorophyll and overall carotenoid, but it appears to be less amenable to the de-epoxidase enzyme [49]. Indeed, refolding experiments suggest that the carotenoid complement is more plastic than that of LHCII. CP26 can, for example, be induced to bind only violaxanthin [50]. The principal role of the minor LHCIIs is to connect the bulk LHCII with the core antenna. They have not yet been crystallized, but they can be modelled into the PSII superstructure that has been observed by electron microscopy based on the structure of the main complex (Fig. 12) [51,52].

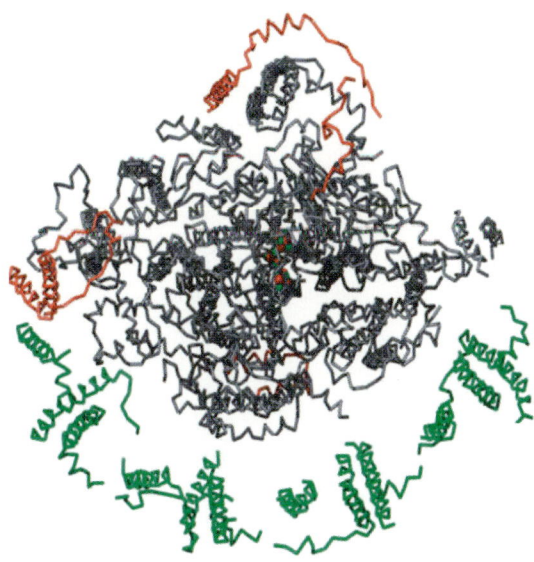

Fig. 13. A structural model of PSI showing the location the LHCI monomers as viewed from the stromal side of the membrane. The PSI core complex is coloured red and grey, the four associated LHCI monomers green. Redrawn from [24].

iii) Ultrastructure of LHCI. LHCI complexes also have three membrane-spanning helices (Fig. 13). Four LHCI monomers (Lhca1-4) associate with one side of the monomeric PSI core complex [24]. This is very different from the situation in the cyanobacteria which have no LHCI and in which PSI is always a trimer. The structure of these complexes is otherwise very similar to that of LHCII with two of the three transmembrane helices forming a cross brace, suggesting that there will be a similar arrangement of carotenoids to that seen in LHCII. Because of the difficulty in separating the individual LHCI proteins from each other, most of the data on these proteins are based on reconstitution experiments. However, partly because of these problems, reports of their carotenoid composition are somewhat variable [53,54].

iv) Unusual algal LHC complexes. Outside the Chlorophyta, some algae also possess more distantly-related LHC-type antenna proteins [55]. Red algae contain LHCI-like complexes, thought to be associated exclusively with PSI, but which bind only chlorophyll *a* and zeaxanthin (**119**). In brown algae and diatoms, chlorophyll *c* rather than chlorophyll *b* is the second chlorophyll species present, while the major carotenoid is fucoxanthin (**369**). There is a surprisingly high carotenoid to chlorophyll ratio in these fucoxanthin-chlorophyll *a/c*-proteins (FCP); in some cases carotenoid ≥ chlorophyll is reported [56,57]. Presumably the high carotenoid level is useful in the very low light habitat of these organisms. Brown algae and diatoms also appear to be the only algae outside the Chlorophyta to exhibit a light-regulated xanthophyll cycle. In this the monoepoxide diadinoxanthin (**230**) is converted into diatoxanthin (**118**) in high light (Scheme 2). The function of the cycle in photoprotection nevertheless appears to be similar or identical to that in green organisms (see Section **E.4**).

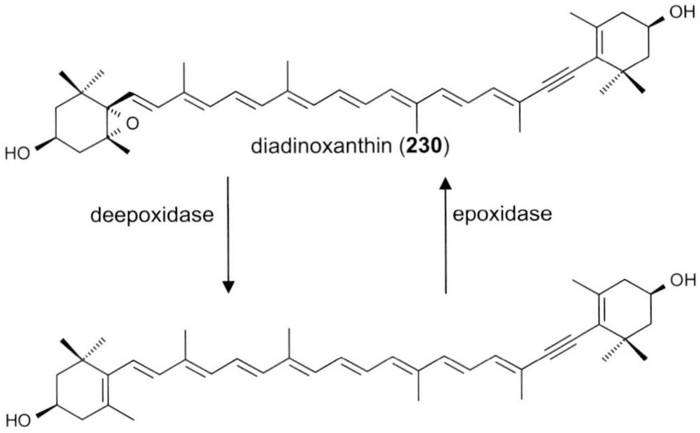

diadinoxanthin (**230**)

deepoxidase   epoxidase

diatoxanthin (**118**)

Scheme 2

Finally, dinoflagellates contain FCP-like antenna proteins, but these bind only chlorophyll *a* and the carotenoid peridinin (**558**). This variation in pigment binding between algal species and green plants is a reflection of the variability in polypeptide sequences as well as the pigments they synthesize, and has evolved, in the main, to allow the different classes of algae to adapt to their individual ecological niches. For the sake of clarity, however, these differences will not be discussed further as the basic chlorophyll and carotenoid binding is similar, so their carotenoids can perform essentially the same functions.

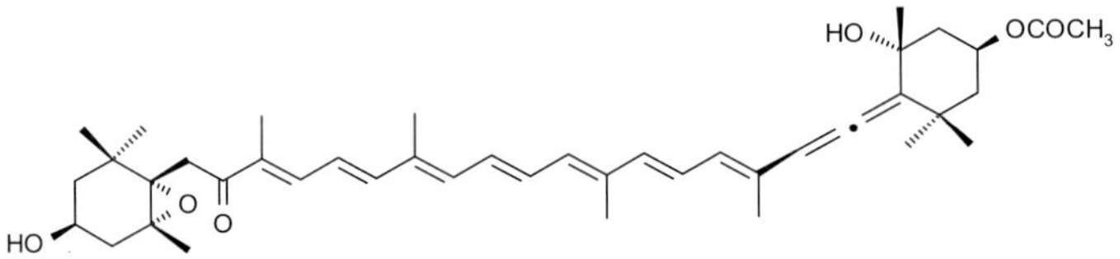

fucoxanthin (**369**)

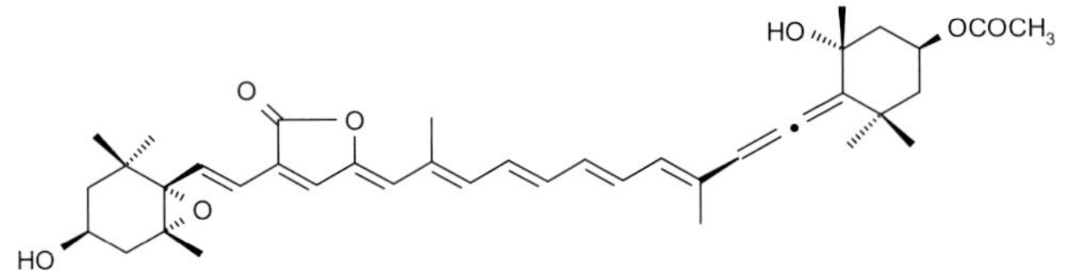

peridinin (**558**)

c) Core-type complexes: chlorophyll *a* proteins

Antenna complexes of this group characteristically have six transmembrane helices (arranged as three pairs) forming a ring structure. This is true of CP47 and CP43, which are the inner antennae associated with the PSII RC (Fig. 14) and of the amino-terminal domains of PsaA and PsaB in PSI (Fig. 6, circles). Another consistent feature is that these complexes (or domains), which are found in higher plants and green algae, normally only bind chlorophyll *a* and β-carotene. They act as intermediary antenna complexes and funnel light captured by LHC complexes (higher plants and algae) or extrinsic light harvesting complexes (prokaryotes and red algae) to the RCs.

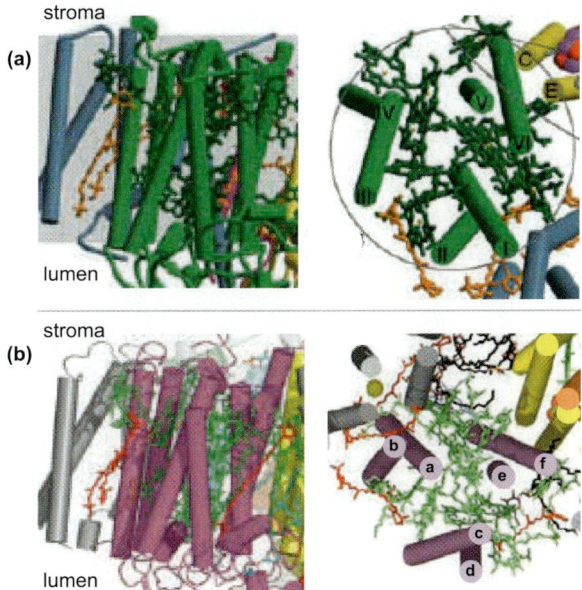

Fig. 14. The structure of the PSII core-type complex CP43 from *Thermosynechococcus elongatus* (a) as described in [21] (peptides coloured green) and (b) as described in [19] (peptides coloured purple). On the left: side view along membrane plane. On the right: view from the lumenal (a) and stromal (b) surface of the membrane. Reproduced from [21] and [19], with permission.

This is a simplistic view, however, and is only true for PSII. The core antenna system in PSI is more complicated. As Fig. 6 shows, in addition to the 'six-ring' structure there are more chlorophylls and carotenoids interacting partially with PsaA and PsaB but also with a number of small polypeptides (with one to three transmembrane helices). This increases the antenna size of PSI (>90 chlorophyll *a* and 22 carotenoids/RC) considerably relative to that of PSII (35-36 chlorophyll *a* and 11 carotenoids/RC). Another difference is that, in PSII, all the carotenoids are in the all-*trans* configuration whereas, in PSI, of the 22 carotenoids present in the structure, five have either one or two *cis* double bonds. All the carotenoids are within van der Waals distance of a large number of chlorophyll molecules, ensuring good energy transfer between the two pigment types and efficient light harvesting and photoprotection (Fig. 4).

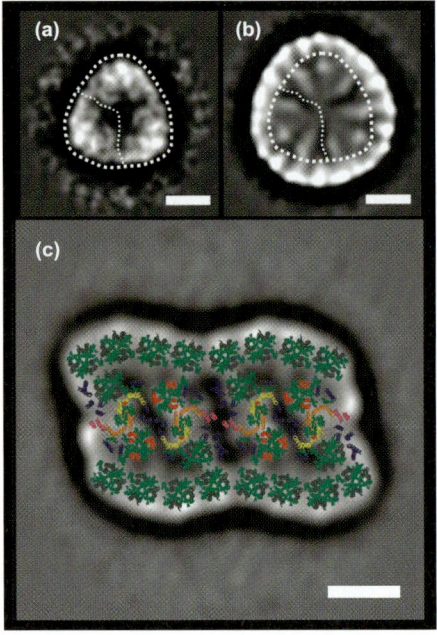

Fig. 15. Electron micrographs of PSI and PSII supercomplexes surrounded by rings of Pcb antenna (isolated from *Acaryochloris marina*). (a) An averaged EM image showing PSI trimers lacking Pcb proteins from Fe-sufficient cells. The likely three-fold symmetry axis has been marked. (b) PSI trimers surrounded by an 18 Pcb-subunit ring arrangement when cells are grown in Fe-depleted medium. (c) Insertion of the X-ray structures of PSII and CP43 into an EM image of the Pcb-PSII supercomplex, viewed from its lumenal surface. The chlorophylls of the PSII monomer and Pcb are shown in light green. The location of the transmembrane helices of D1 (yellow), D2 (orange), CP47 (red), CP43 and Pcb proteins (dark green), cytochrome b_{559} (pink) and low molecular weight subunits (blue) are also shown. Bars = 10 nm. Redrawn from [61,62].

Proteins constituting a more distant branch of the core-type antenna family are present in various prokaryotes. These are found constitutively in prochlorophytes and *Acaryochloris marina* (Pcb proteins) and facultatively in other cyanobacteria, under low Fe-stress (IsiA proteins, also known as CP43') [58]. Despite the fact that, in prochlorophytes, these complexes also bind some chlorophyll *b*, it has been shown through gene sequence analysis that they are close in primary structure to CP43 and not to the chlorophyll *a/b* LHC complexes of higher plants and algae. Chlorophyll *d* replaces almost all of the chlorophyll *a* in the Pcbs of *A. marina* [59]. Structures at atomic resolution are not available, but single particle analysis has demonstrated the formation of supercomplexes with rings of IsiA/Pcb complexes around the PSI trimer (in various species grown in low Fe-conditions only) and rows of Pcb proteins along the sides of dimers (*Prochloron*) or double dimers (*A. marina*) of the PSII core complexes (in cells grown under normal Fe levels) (Fig. 15) [60-62 and references therein].

In addition to the different chlorophyll content, there are also differences in the carotenoids in these complexes. This is particularly true of *A. marina* which contains zeaxanthin (**119**) and α-carotene (**7**) almost exclusively, with only very low levels of other carotenoids. It is thought that the Pcb proteins of this organism probably bind both zeaxanthin and α-carotene - so that there is xanthophyll present in the Pcbs of this organism, in contrast to the other 'normal' core complexes.

α-carotene (**7**)

d) Chlorosomes

Chlorosomes are the major light-harvesting complexes of green photosynthetic bacteria [63]. There are two types of green bacteria that contain them: the green sulphur bacteria (with Type 1 reaction centres) and the green filamentous (non-sulphur) bacteria (with Type 2 reaction centres) (see Fig. 3). The former species are strict anaerobes whilst the latter are able to grow aerobically and chemotrophically in the dark or phototrophically under anaerobic conditions. In contrast to other carotenoproteins involved in photosynthesis there is comparatively little information on the detailed structure and function of chlorosomes. These light-harvesting complexes exhibit major structural differences compared with other types of antenna. For example, (i) they contain bacteriochlorophyll *c, d* or *e* that self-assembles into large aggregates which are not bound to an apoprotein scaffold and (ii) the carotenoid/ bacteriochlorophyll ratio is not fixed. This system appears to be an adaptation of the light-harvesting apparatus to maximize light absorption in the extremely low light environment these bacteria inhabit [4,63].

The green sulphur bacteria with Type 1 RCs have no membrane-intrinsic antenna complex; they simply rely on chlorosomes. In the absence of a high-resolution structure, Fig. 16 shows a schematic representation of a typical chlorosome. It is like a large bag of bacteriochlorophyll and carotenoid molecules held together by a monolayer membrane that is enriched in glycolipids but also contains the Csm family of proteins. The baseplate (with bacteriochlorophyll) and a number of soluble Fenner-Matthews-Olson (FMO) proteins connect the chlorosome to the RC complex in the case of the green sulphur bacterium, *Chlorobium tepidum*. The FMO protein was the first chlorophyll-protein structure solved. It is a small trimeric structure binding seven bacteriochlorophyll molecules per monomer but no carotenoid.

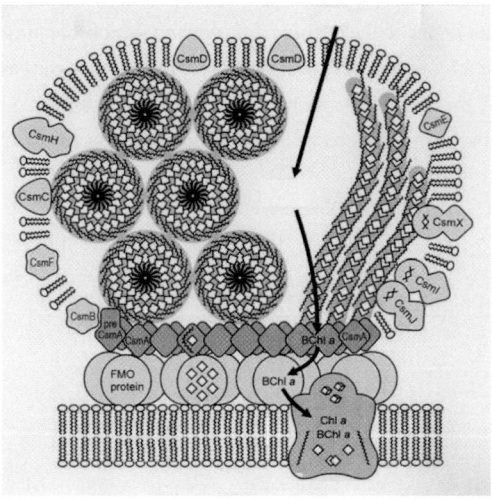

Fig. 16. A schematic representation of a typical chlorosome. The chlorosome is anchored to the membrane surface *via* the carotenoid-less FMO protein. The black arrows represent the excitation energy transfer pathway. The opposing 'rod' (left) and 'lamellar' (right) structural models are depicted. Redrawn from [64].

Originally, the bacteriochlorophyll aggregates were believed to form rods [65] and this structural model is displayed on the left in the chlorosome model (Fig. 16). More recently a lamellar model has been proposed as shown on the right in Fig 16 [66,67]. In this model, the carotenoids occupy the hydrophobic space between the chlorin planes and interact with the esterifying alcohols [66]. Indeed, recent EM and X-ray diffraction experiments have confirmed the spatial location of the carotenoids [66]. This was achieved by correlating the observed spacing between the Bchl aggregates as a function of carotenoid content.

In the green non-sulphur bacteria (Type 2 RCs), the chlorosome probably has a similar structure but there is no FMO protein and it transfers energy indirectly to the RC *via* an intrinsic LH1 type antenna complex. Both types of green bacteria live at extremely low light intensities and thus require this very large antenna system.

e) The peridinin-chlorophyll-protein complex (PCP) in dinoflagellate algae

Dinoflagellates are unusual organisms that have, in addition to their LHC-type antenna, an extrinsic light-harvesting complex which binds only chlorophyll *a* and the carotenoid peridinin (**558**). Uniquely, this complex has much more carotenoid than chlorophyll bound to the protein (4:1). The protein is a soluble trimeric complex which associates with the photosynthetic membrane. The monomer is a folded double domain, with each domain binding four peridinin molecules and one chlorophyll *a*, *i.e.* the trimer has 24 carotenoids and 6 chlorophylls (Fig. 17) [68].

Energy transfer is very rapid from carotenoid to chlorophyll as they are within van der Waals distance of each other, while the chlorophyll to chlorophyll distance is 17 Å. Energy is then transferred from the PCP to a membrane-intrinsic chlorophyll *a/c* antenna and subsequently to the RCs.

(a)   **(b)**

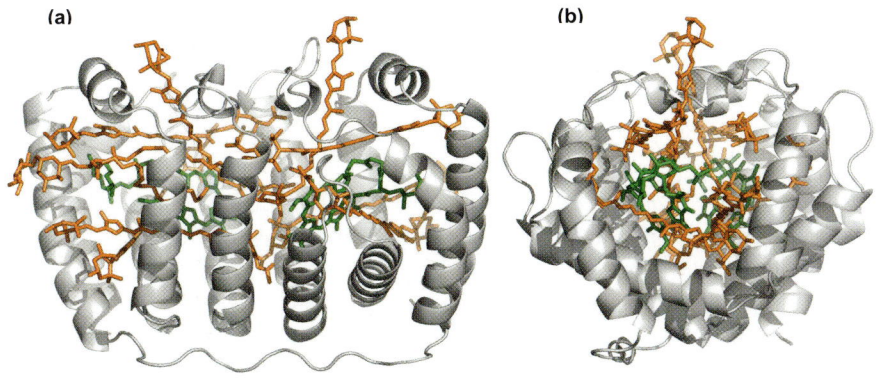

Fig. 17. The structure of the peridinin-chlorophyll-protein. (a) View perpendicular to the long axis. (b) Side view from the C-terminal region with the last two α-helices removed to reveal the spatial arrangement of the carotenoid (orange) and chlorophyll *a* (green) molecules. Protein Data Bank accession number PPR1. The figure was produced by use of PyMOL [9].

f) Carotenes or xanthophylls?

Oxygenic organisms bind both xanthophylls and carotenes but, as discussed above, there is a distinct segregation into different complexes. As a general rule, the outer antennae bind xanthophylls (almost) exclusively, whereas the inner core-type antenna binds carotenes. Cyanobacteria and red algae have phycobilisomes and no membrane-intrinsic peripheral LHC antenna complexes. They only have the RC and 'inner' or 'core' antenna proteins of PSII and PSI (which bind only chlorophyll *a* and not chlorophylls *b* or *c*). They also contain carotene, but xanthophylls are normally absent except in some odd cases such as when cyanobacteria accumulate zeaxanthin (**119**) under stress conditions. The precise location of the xanthophylls is unknown, however, and it is possible that they are deposited between the thylakoid membranes in oily droplets, as seen in the green alga *Dunaliella viridis* when it accumulates high levels of β-carotene (**3**) under stress [69].

Also, zeaxanthin is present in the prochlorophytes and the unusual, chlorophyll *d* dominated cyanobacterium, *A. marina*. The zeaxanthin is thought to be associated with the intrinsic antenna proteins known as Pcbs. In *A. marina* as well as the prochlorophytes, (*Prochlorococcus* spp), there is α-carotene (**7**) in place of β-carotene (**3**). Purified PSII core preparations [70], PSII RCs [19] and PSI preparations [71] all have only α-carotene *i.e.* there is no xanthophyll present in CP43, CP47 or RC complexes and it is likely that the xanthophyll binds to the Pcbs.

No explanation is yet available for the selective location of carotenes in the RC and inner antenna and of xanthophylls in the outer antenna complexes.

E. Photoprotection

1. Over-excitation of photosystems and production of toxic oxygen species

It is well known that high light intensities can cause problems during photosynthesis. Plants can be damaged under high light conditions so that certain chlorophyll-protein complexes need to be repaired; this phenomenon is known as photoinhibition [72-74].

Light absorption increases with light intensity and, as the energy and electron transfer steps become rate limiting, the probability of the excited singlet state of chlorophyll in the antenna undergoing intersystem crossing to form the chlorophyll triplet state will increase (Fig. 4), as will the accumulation of reduced electron acceptor molecules. Neither chlorophyll triplets nor highly reducing species cause any problem under anaerobic conditions. However, in oxygenic organisms, which live aerobically and actually produce oxygen at a high rate in PSII, the probability of chlorophyll triplets being quenched by triplet ground-state oxygen, thereby yielding $^1O_2$, will increase (Fig. 4). Also, as the intensity increases, reduced quinones and iron-sulphur proteins (in PSII and PSI, respectively) will accumulate and thus will have the chance of reacting with oxygen to form other toxic oxygen species, notably superoxide and the extremely reactive hydroxyl radical. Plants and other photosynthetic organisms would thus not survive in an aerobic environment if there were not efficient protection mechanisms in place to deal with these challenges.

In the antenna, the problem of chlorophyll triplets is dealt with by carotenoids, which are always present in chlorophyll-protein complexes, and are bound within van der Waals distance of chlorophyll molecules. Their main photoprotective function is to quench chlorophyll triplets before they react with oxygen to form $^1O_2$, but they can also scavenge directly any singlet oxygen that is produced (Fig. 4). Carotenoid triplets decay harmlessly, emitting the absorbed energy as heat.

There is also an additional quenching mechanism, non-photochemical quenching (NPQ) which involves certain specific xanthophylls (Section E.4). If there is over-reduction on the acceptor side of the two photosystems, production of toxic oxygen species is mostly dealt with by the action of scavengers of radical oxygen species, particularly superoxide dismutase, ascorbate and glutathione peroxidases. This does not involve carotenoids and is not covered further here. It appears, however, that the action of the xanthophyll cycle also plays a role in protection of the lipid bilayer from oxidative damage [75], through a mechanism that is unknown but may occur at specific sites on the periphery of LHC proteins [76].

2. The valve reaction: chlorophyll triplet quenching by carotenoids

The increase in the yield of carotenoid triplets as a function of light intensity once photosynthetic electron transport is saturated was originally called the 'valve' reaction [77,78]. At high light intensity, the possibility of forming the chlorophyll triplet state increases. In chlorophyll-protein complexes, carotenoids are normally bound within van der Waals contact (*i.e.* <3.6 Å) with at least some of the chlorophylls. The close proximity not only facilitates singlet-singlet energy transfer during the antenna function of carotenoids but also triplet-triplet transfer, in which chlorophyll triplets generate carotenoid triplets. This occurs on a ns timescale (Fig. 4); carotenoid triplets are detected easily by their absorption at around 510 nm. All the carotenoids present in photosynthetic complexes have nine or more conjugated double bonds and their triplet excited state energy levels are therefore below that of singlet oxygen. Hence, the carotenoid triplets cannot generate singlet oxygen and they decay harmlessly, in *ca.*10 μs, releasing heat [79].

The normal photoprotection mechanism in photosynthetic pigment complexes, as already introduced above, is for carotenoid to prevent formation of $^1O_2$ in the first place (Fig. 4). In the LHC-type antenna complexes, found in oxygenic organisms, where both chlorophylls *a* and *b* are present, the main triplet state formed is that of chlorophyll *a*, as excitation energy transfer from chlorophyll *a* to chlorophyll *b* is much faster (~ps) than triplet formation (~10 ns [80]). Thus it is mainly chlorophyll *a* triplets which need to be quenched by carotenoid.

As shown in the previous sections, there is always at least one carotenoid per three or four (bacterio)chlorophylls in the photosynthetic pigment-protein complexes. The pigments in LH complexes may be ordered regularly or they may be arranged apparently randomly, as in LHCII and core antenna complexes. In all antenna complexes, however, the carotenoids are always bound within van der Waals distance of a number of the (bacterio)chlorophylls. These in turn are bound closely enough for rapid electron transfer within the complex, and hence the carotenoids are ideally organized for quenching any (bacterio)chlorophyll triplets that are formed by intersystem crossing.

The RCs of PSI and PSII are different from the antenna complexes in that there are no carotenoids bound in van der Waals contact with the chlorophyll electron transfer cofactors. The consequence of this is discussed in detail in Section **F**. On isolation, the PSII RC loses its secondary electron acceptors so, when the primary radical pair is formed, it recombines, with a 30% probability of forming a triplet state. This chlorophyll triplet cannot be quenched by carotenoid because of distance so, under aerobic conditions, $^1O_2$ is formed. This then rapidly bleaches chlorophyll, oxidizes the protein and inactivates the complex [81]. The rate of formation and decay of this toxic species has been measured by its luminescence at 1270 nm. The $t_{1/2}$ for $^1O_2$ formation is around 15 μs which is ~3 orders of magnitude slower than the quenching of the chlorophyll triplet state by carotenoid [82]. Thus quenching by carotenoids very effectively prevents the formation of significant amounts of $^1O_2$ and protects against the inevitable damage that would ensue if this toxic species were formed (Fig. 4). The necessity for this protection is obviously restricted to organisms in an aerobic environment. Presumably bacterial photosynthetic complexes bind carotenoids for light harvesting and structure stabilization. However, as not all of these organisms are strict anaerobes, the carotenoids present can again be considered photoprotective in aerobic conditions.

3. Singlet oxygen quenching

Despite the efficiency of triplet-triplet transfer from chlorophyll to carotenoid, some unquenched chlorophyll triplets could remain, especially under high excitation pressure. These triplets, in an aerobic environment, will form $^1O_2$ which, because of its extreme toxicity even at very low levels, would cause damage to pigments, protein and the lipids of the thylakoid membrane. Although there have been claims that there is 100% energy transfer from chlorophyll to carotenoid [79], it has been shown [83] that only 92% of the triplets formed on chlorophyll *a* in isolated LHCII were transferred to carotene *i.e.* there is a possibility of significant $^1O_2$ formation.

There is, however, a second line of defence by carotenoids, as they can also quench $^1O_2$ directly. This will further reduce the level of $^1O_2$ available to cause oxidative damage to the photosynthetic apparatus. Recombinant studies of LHCII with modified carotenoid composition have suggested that the two lutein (**133**) sites are more involved in chlorophyll triplet quenching whilst neoxanthin (**234**) has a more important role in $^1O_2$ scavenging [48]. On balance, carotenoids are extremely effective in preventing oxidative damage, as demonstrated by the fact that, when carotenoid synthesis is prevented, plants become exceedingly light-sensitive; young seedlings grown in the dark bleach and die as soon as they are exposed to light [84]. The $^1O_2$ is so dangerous, however, that even very low levels cause gradual oxidative damage, which is exacerbated at high light intensities. When this occurs in the PSII RC it probably contributes to the susceptibility of plants to photoinhibition [85].

In RCs there is rapid energy transfer from the cofactor chlorophylls to the primary electron donor which undergoes charge separation. If the primary radical pair ($D^{\bullet+}A^{\bullet-}$) is not stabilized rapidly, by forward electron transfer to secondary acceptors, there are three possibilities: (i) a back reaction to the excited singlet state, (ii) conversion to the triplet state *via* a radical pair mechanism or (iii) direct decay to the ground state, by a non-radiative mechanism,. The triplet state of the primary electron donor would react with oxygen, to form $^1O_2$, unless there is carotenoid bound close enough to pre-empt this. This situation occurs in purple bacterial RCs. Here the redox potential of the oxidized primary donor is not high enough to extract electrons from carotenoid. The single carotenoid present, in the pBRC, is very near the accessory chlorophyll on the inactive electron transfer branch and so can quench the chlorophyll triplet state (see Fig. 4). In PSI, *in vivo*, the chance of the back reaction occurring and consequently populating the triplet state ^{3}P700 is extremely unlikely. This is because if, for some reason, CO_2 fixation and cyclic electron transfer are inhibited, the very low redox potential electron acceptors will react rapidly with oxygen to form other toxic species such as superoxide [78].

In the PSII RC, however, the primary electron donor, P680, is highly oxidizing (>1.1 V), in order to be able to split water (0.8 V). If a carotenoid molecule were bound close enough for triplet-triplet transfer (Fig. 4) it could also be oxidized by P680$^+$, when present, thus short-circuiting the normal function of PSII [86]. As described in Section **C**.1, the carotenoids in the PSII RC are at least 16 Å away from the electron-transfer chlorophyll cofactors, thus preventing this short circuit, but also making quenching of ^{3}P680 very inefficient.

What are the chances of triplet formation in PSII? It has been argued that, even at normal light intensities, in addition to high excitation pressure, there is the chance of the reaction centre triplet state (^{3}P680) being formed [87,88]. As it is impossible for carotenoid to quench the triplet directly, $^1O_2$ will inevitably be formed. Consequently the presence of the carotenoids in the RC can protect against damage to chlorophyll by directly quenching $^1O_2$ [89]. This was demonstrated by observing that both the luminescence due to $^1O_2$ (at 1270 nm) and the degree of photodamage to chlorophyll were proportionately increased as the carotenoid level of isolated PSII RCs was lowered. It was concluded, therefore, that the two β-carotenes play a role in quenching $^1O_2$. This is a diffusion-dependent reaction so, although the carotenes provide a significant degree of protection against oxidative damage to chlorophyll and the protein moiety, this protection is incomplete [89]. The inevitable damage is thought to be a component in the factors leading to the rapid turnover of the D1 protein during photoinhibition [74].

The question arises of why there are normally only carotenes in 'inner' core antenna and reaction centres and only xanthophylls in 'outer' antenna, in oxygenic organisms with intrinsic LHCs. It is well established that a carotenoid must have at least nine conjugated double bonds to quench singlet oxygen and that the greater the chromophore length, the more effective the carotenoid is at quenching. β-Carotene (**3**) is slightly more efficient than the xanthophylls lutein (**133**) and violaxanthin (**259**) [90] but the reason for the specific distribution of carotenes and xanthophylls between the photosynthetic pigment protein complexes remains unclear.

4. Non-photochemical quenching and the xanthophyll cycle

A result of the high oxidizing potential of $P680^+$, and of the nearby presence of molecular oxygen in PSII, is the need to regulate excitation pressure within the PSII antenna system. In PSI, when the light intensity increases above the level that can be used in photochemistry, excess excitation can be safely quenched by simply allowing $P700^+$ to persist. This is, however, not possible in the case of PSII, as problems are associated with long-lived forms of $P680^+$ because they are so strongly oxidizing (>1.0 V). Thus, the system has to use more subtle mechanisms to regulate excess excitation energy. Carotenoids appear to carry out a number of roles in this allosteric mechanism, certainly in its control and also possibly by being (or being part of) the quenching entity. Traditionally, the mechanism has been monitored by chlorophyll fluorescence measurements, where quenching of the fluorescence yield may be due to energy use through photochemistry or through non-photochemical, dissipatory phenomena. The major form of non-photochemical quenching, or NPQ, is directly correlated with the ΔpH across the thylakoid membrane, and involves formation of an energy trap in the antenna pigment bed. For a review of NPQ measurements, see [91].

One of the principal features of NPQ is its association with the xanthophyll cycle (Scheme 1). Under high light, violaxanthin (259) is converted into zeaxanthin (119), by removal of the epoxide group on each β-ring by the enzyme violaxanthin de-epoxidase [92]. This enzyme is activated by a low lumenal pH (indicative of a high photochemical rate). The reverse reaction is catalysed constitutively at a slower rate by zeaxanthin epoxidase. NPQ also appears to involve the PSII protein PsbS [93], a more distant relative of the LHC family, thought to be associated with the antenna. Energy dissipation is associated with organizational changes within the antenna system, which may induce conformational changes of these proteins [94]. When isolated from the photosynthetic membrane, the main LHCII complex can be manipulated to undergo conformational changes that affect its pigment structure, including the formation of efficient energy quenchers [95]. The conversion of violaxanthin into zeaxanthin appears to potentiate these changes, making the host LHC proteins more sensitive to the ΔpH [94].

The quenching centres themselves have not yet been definitely identified, but a number of possibilities exist. Most hypotheses describe either a chlorophyll-chlorophyll dimer or a chlorophyll-carotenoid heterodimer, or even a combination of the two. In the past, zeaxanthin has been implicated directly as being (part of) the quenching entity [75], possibly through its binding to PsbS [97,98] but, as the presence of zeaxanthin is not required for quenching either *in vitro* or *in vivo*, this clearly cannot be the case all the time. An alternative suggestion is that one of the observed chlorophyll dimers in LHCII is important, possibly through interaction with L1 in the terminal emitter domain [95,98]). In each case, quenching by carotenoid could occur through a lowered S_1 energy of the carotenoid involved [99], or possibly by formation of a Car^+ cation [97]. Significant advances have been made in recent years, and it is expected that the remaining questions should be answered soon.

Diatoms also exhibit an NPQ mechanism which involves its own xanthophyll cycle (Scheme 2), involving diadinoxanthin (**230**) and diatoxanthin (**118**), bound to FCP proteins. While quenching in these organisms can be much stronger than in higher plants, it nevertheless appears to involve a similar mechanism [99].

5. Cytochrome b₆f complex

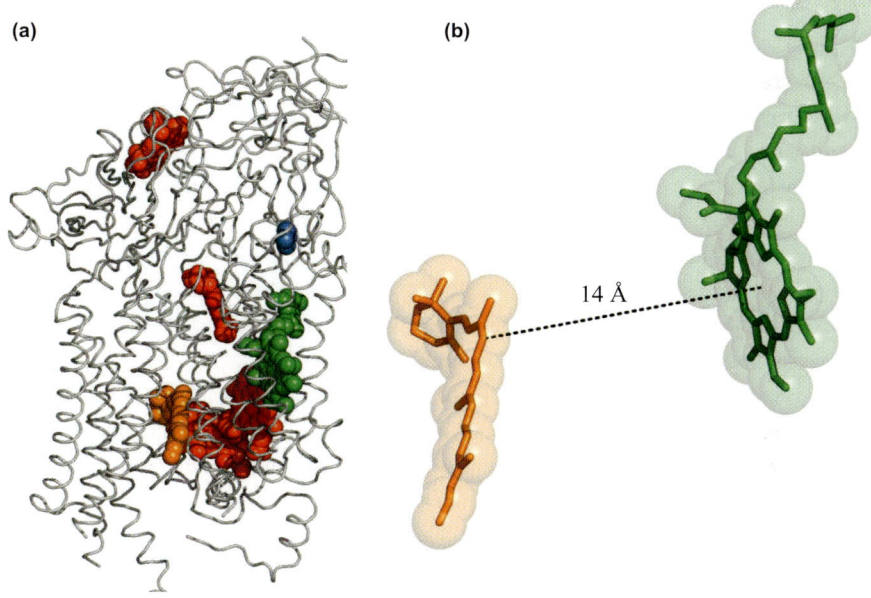

Fig. 18.   The location of the chlorophyll and carotenoid molecules in the cytochrome b_6f complex. (a) The structure of cytochrome b_6f as seen perpendicular to the membrane plane. The location of the carotenoid (orange), chlorophyll (green), haems, (red) and Fe_2S_2 cluster (blue) are marked. (b) The closest distance between the carotenoid (orange) and chlorophyll (green) molecules is 14 Å. Protein Data Bank accession number 1Q90. The figure was produced by use of PyMOL [9].

It was discovered recently that the cytochrome b_6f complex has both a chlorophyll *a* and a β-carotene molecule bound to it [100-102] (Fig. 18). Before the crystal structure of this complex was determined, it was expected that this carotene would be positioned within van der Waals distance of the chlorophyll, so that any chlorophyll triplet that was formed would be quenched before it could interact with ground state oxygen to form $^1O_2$. It was, therefore, a surprise when it was found that the nearest approach of the carotene to the chlorophyll is 14 Å, so not only can there be no triplet-triplet transfer but even $^1O_2$ quenching will not be particularly efficient as it will be diffusion dependent and the possibility of oxidation of amino acids by $^1O_2$ will compete [101]. It is probable that the chlorophyll is quenched by some unknown mechanism although it is also possible [100] that there may be additional carotene molecules which have not been determined in the structure and which interact directly with the chlorophyll. The chlorophyll fluorescence is quenched so its singlet lifetime might be too short to allow formation of the triplet state, and $^1O_2$ would not be formed [103].

F. Electron Donation: Carotenoid Oxidation in PSII

Oxidation of carotenoid by chlorophyll does occur during photosynthesis, but only in one specific case, in the PSII RC, because of the very large oxidizing potential of the $P680^+/P680$ redox couple (>1.1 V) which is so much higher than that of normal Chl^+/Chl (~0.7 V) and of Car^+/Car (~1.0 V) [86,104]. Therefore, in order to extract electrons from water, via the redox active tyrosine, Y_z, it is necessary for oxygenic organisms to raise the redox potential of the special chlorophylls involved in PSII charge separation above that of water (~0.8 V). A consequence of this high potential, however, is that if, under any circumstances, re-reduction of $P680^+$ from the water-oxidizing complex is slowed or halted, this highly oxidizing species will remove electrons from the surrounding environment, i.e. from other pigments including the two β-carotenes bound to the D1 and D2 complex (Fig. 5).

As discussed in Section E.3, two carotenoid molecules are present in the PSII reaction centre but they apparently serve to scavenge $^1O_2$ rather than to quench chlorophyll triplets directly (Fig. 4). It has been proposed that this occurs because, due to rapid energy transfer from the cofactor chlorophylls to the primary electron donor, the only triplet formed within the PSII RC is that of P680 itself. The mechanism of chlorophyll triplet formation is normally via intersystem crossing, but in the PSII RC there is an alternative mechanism. If the primary radical pair $(D^{\bullet+}A^{\bullet-})$ is not stabilized rapidly by forward electron transfer to secondary acceptors, there is the possibility of a back reaction and the chance of the radical-pair excited singlet being converted into the radical-pair triplet state. This triplet would also react with oxygen to form $^1O_2$ unless there is carotenoid bound close enough to pre-empt this by quenching the chlorophyll triplet before ground state oxygen is able to. This situation occurs in purple bacterial RCs, where the redox potential of the oxidized primary donor is not high enough to extract electrons from carotenoid. The single carotenoid molecule present in the pBRC is very near the accessory chlorophyll on the inactive electron transfer branch and so can quench the chlorophyll triplet state (Fig. 4). In PSII, however, the primary electron donor, P680, is highly oxidizing (>1.1 V), in order to be able to split water (0.8 V), so if a carotenoid molecule is bound close enough for triplet-triplet transfer (Fig. 4) it could also be oxidized by $P680^+$, thus short-circuiting the normal function of PSII [86].

It has been argued that, even at normal light intensities, but especially at high excitation pressure, there is an increased likelihood of the reaction centre triplet state ($^3P680$) being formed [88]. As discussed above, in the PSII RC it is impossible for carotenoid to quench the triplet directly, so $^1O_2$ will inevitably be formed. It has been shown, however, that the presence of the carotenoids in the RC does protect against damage to chlorophyll [89], by quenching $^1O_2$. This was demonstrated by observing that both photodamage and the luminescence due to $^1O_2$ increased when the carotenoid level of isolated PSII RCs was lowered. The conclusion is that the two β-carotenes play a role in quenching $^1O_2$ directly in a diffusion-dependent reaction. This provides significant but incomplete protection against oxidative damage to chlorophyll and the protein complex [89]. It is the eventual oxidation of the D1 polypeptide which probably leads to the rapid turnover of this protein [74].

If electron transfer from the water-splitting complex is slowed, as occurs during photoinhibition, the lifetime of P680$^+$ is extended and the two carotenoids closest to the accessory chlorophylls on D1 and D2, at 20 Å and 13 Å from the chlorophyll chlorin rings, respectively [19], can be oxidized. For isolated RCs or PSII core complexes, both absorption and resonance Raman spectroscopic data have been obtained for oxidation of both of the carotenoids in the RC [105,106]. This oxidation has been reported to be in the ms time range, consistent with a distance >18 Å from the high-potential chlorophylls of the RC [20,107]. Oxidation of Car$_{D1}$ is expected to be much slower than that of Car$_{D2}$, because of the difference in proximity to the central chlorophyll cofactors, but clear evidence for this is still required.

The oxidized carotenoid on the D2-side can be re-reduced by Cyt b$_{559}$, if the latter is initially in the pre-reduced form, so rapidly that the oxidation of Car$_{D2}$ cannot be detected. It is unclear how the carotenoid on the D1 side is re-reduced or if this occurs at a physiologically useful rate. It has been suggested that, because the Car$^+$ cation is very unstable, then if Car$_{D1}$ is oxidized *in vivo*, it might be destroyed and hence lead to instability of the D1 polypeptide [20]. This could account, at least partially, for the very rapid turnover of the D1 polypeptide, compared to the rest of the PSII RC complex and indeed all other photosynthetic protein complexes, that is seen during photoinhibition [20]. Thus Car$_{D1}$ may have a signalling role in the turnover of the D1 protein. *In vivo,* however, if the P680$^+$ lifetime is prolonged, the most likely carotenoid to be oxidized will be Car$_{D2}$, which is much closer to the RC chlorophylls. The electron donation by carotenoid is seen as an inevitable consequence of the high redox potential of P680$^+$ and the subsequent re-reduction by Cyt b$_{559}$ is required to 'repair' the oxidized carotenoid, so that its important role in quenching $^1O_2$ is not impaired.

G. Assembly and Reconstitution of Complexes

1. Bacteria

It is possible to study the assembly and stability of bacterial light-harvesting complexes *in vivo* or *in vitro*. The former approach uses site-directed mutagenesis or the availability of naturally occurring strains of the same species. This approach relies on the natural machinery within the living cell that provides all the required pigments and lipids. Alternatively, it is possible to reconstitute some LH complexes from their individual components (peptides, bacteriochlorophylls and sometimes lipids, as well as carotenoids) in order to follow the assembly mechanisms *in vitro*. These two approaches have been reviewed in detail [32,108]. Only the more salient findings pertaining to carotenoids will be discussed here.

It has been well documented that the absence of carotenoid molecules inhibits the assembly of the LH2 complex but not necessarily that of the RC-LH1 complex in photosynthetic purple non-sulphur bacteria [34,39-43,109]. This has been interpreted as indicating that the stability of LH2 requires the presence of the carotenoid molecule whereas the RC-LH1 complex can be successfully assembled in the absence of coloured carotenoids. What happens under carotenoid-limiting conditions? How does limiting the pool of coloured carotenoids influence the development of the light-harvesting complexes?

In purple bacteria it is now considered that there are five main carotenoid biosynthetic pathways called the spirilloxanthin (**166**)/spheroidene (**97**), okenone (**317**), isorenieratene (**24**), γ-carotene (**12**)/β-carotene (**3**), and diapocarotene (*e.g.* **515**) pathways, respectively [110]. Apart from the diapocarotenoids (C_{30}) carotenoids are usually C_{40} molecules composed of eight isoprene units.

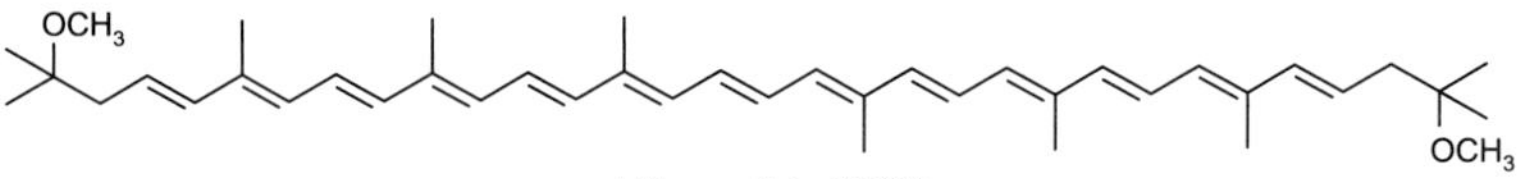

spirilloxanthin (**166**)

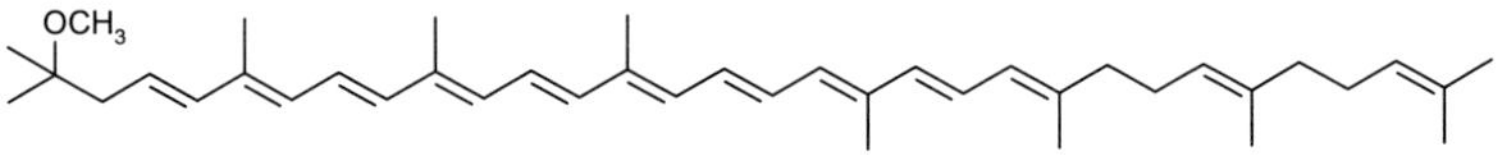

spheroidene (**97**)

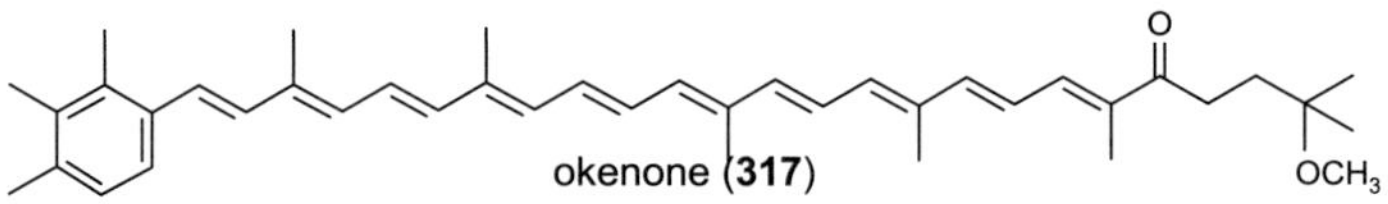

okenone (**317**)

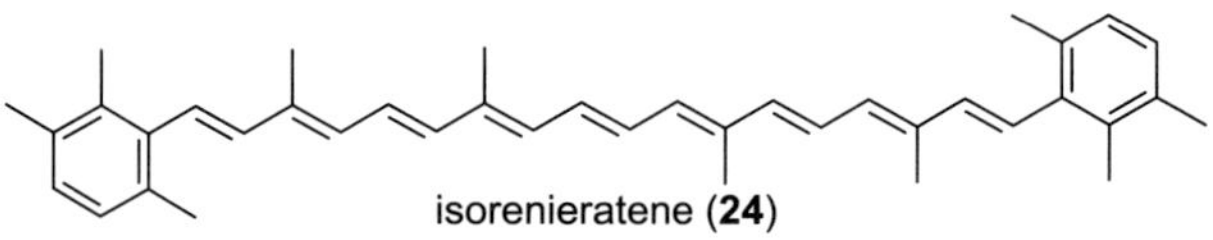

isorenieratene (**24**)

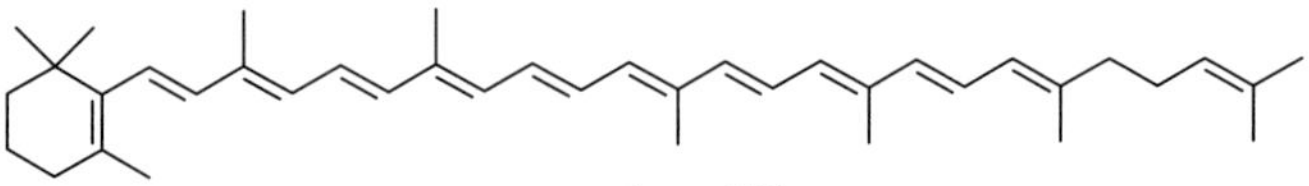

γ-carotene (**12**)

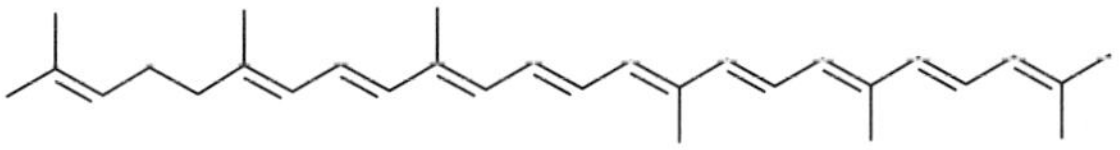

4,4'-diaponeurosporene (**515**)

The pathway to all C_{40} carotenoids starts with phytoene (**44**) as the precursor. The diapocarotenoids are biosynthesized by an analogous C_{30} pathway. Phytoene desaturase then catalyses a three step or four-step desaturation process that produces lycopene (**31**) or neurosporene (**34**), respectively (Scheme 3).

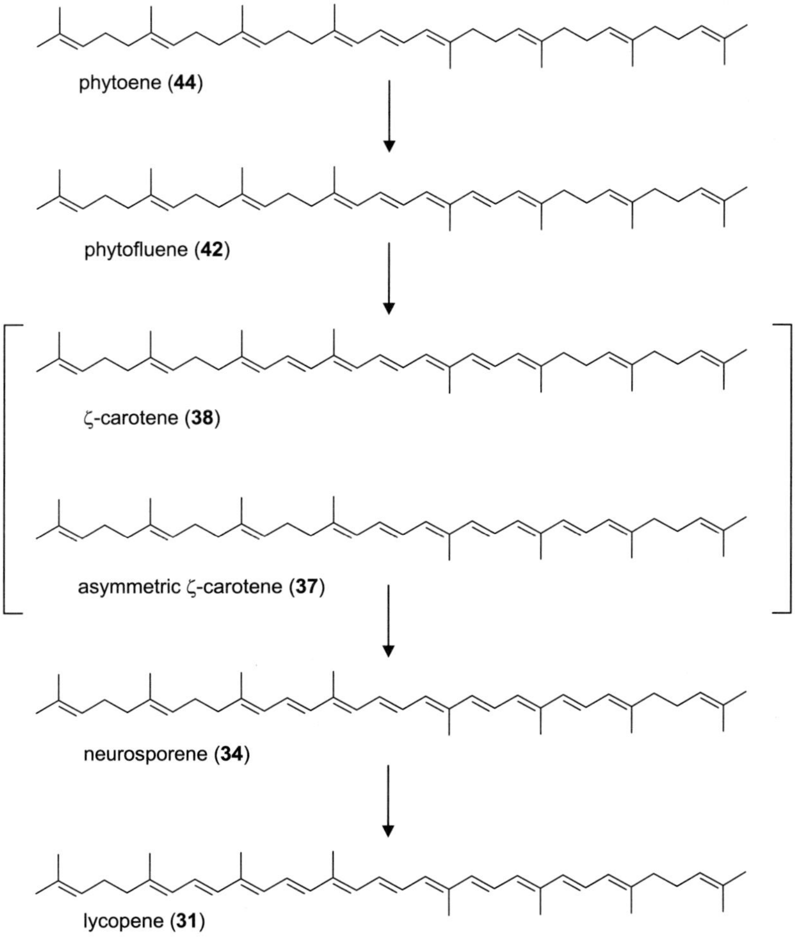

phytoene (**44**)

phytofluene (**42**)

ζ-carotene (**38**)

asymmetric ζ-carotene (**37**)

neurosporene (**34**)

lycopene (**31**)

Scheme 3

This can then be followed by desaturation, saturation, cyclization and aromatization steps, and/or the introduction of oxygen functions and glucosyl groups to give the large diversity of carotenoid molecules found in anoxygenic photosynthetic bacteria (for details see *Volume 3, Chapter 2*). Phytoene desaturase enzyme activity can be interrupted by the addition of diphenylamine (DPA) [39,111], causing a build up of phytoene and related precursors such as phytofluene (**42**), ζ-carotene (**38**) and asymmetrical ζ-carotene (**37**) (Table 1).

Table 1 Composition (mol% of total carotenoids) of coloured carotenoids found in the LH antenna complexes from *Rps. palustris* (upper table) and *Rbl. acidophilus* (lower table) ranked in order of their biosynthesis [36].

Rps. palustris complex	LH2			RC-LH1		
DPA concentration (μM)	0	50	75	0	50	75
Lycopene (**31**)	22	15(a)	22(a)	2	26(a)	n/d
Rhodopin (**93**)	63	77(b)	59	4	33	n/d
Rhodovibrin (**167**)	2	3	3	7	20	n/d
Anhydrorhodovibrin (**91**)	12	2	5	12	5	n/d
Spirilloxanthin (**166**)	1	3	11	75	16	n/d
Rbl. acidophilus complex	LH2			RC-LH1		
DPA concentration (μM)	0	50	75	0	50	75
Lycopene (**31**)	3	9(a)	12(a)	1	6(a)	17(a)
Rhodopin (**91**)	17	62	67(b)	1	16	23(b)
Rhodopin glucoside (**95**)	78	10	15	5	6	11
Other glucosides	1	0	0	0	0	0
Rhodovibrin (**167**)	0	6	4	0	18	18
Anhydrorhodovibrin (**91**)	1	6	3	9	11	11(c)
Spirilloxanthin (**166**)	<1	6	0	84	44	19

(a) Includes the biosynthetic intermediates neurosporene (**34**), asymmetric ζ-carotene (**37**), phytofluene (**42**) and phytoene (**44**).
(b) Includes compounds tentatively identified by HPLC as hydroxy derivatives of shorter chromophore carotenoids, especially asymmetric ζ-carotene and neurosporene
(c) Includes a carotenoid tentatively identified as a hydroxymethoxy derivative of asymmetrc ζ-carotene
n/d Insufficient carotenoid to determine

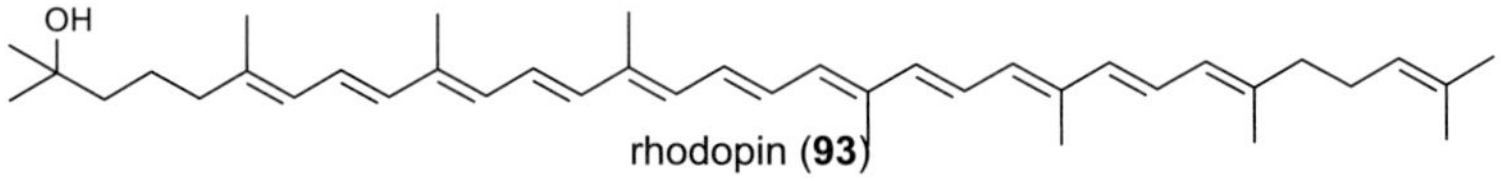

rhodopin (**93**)

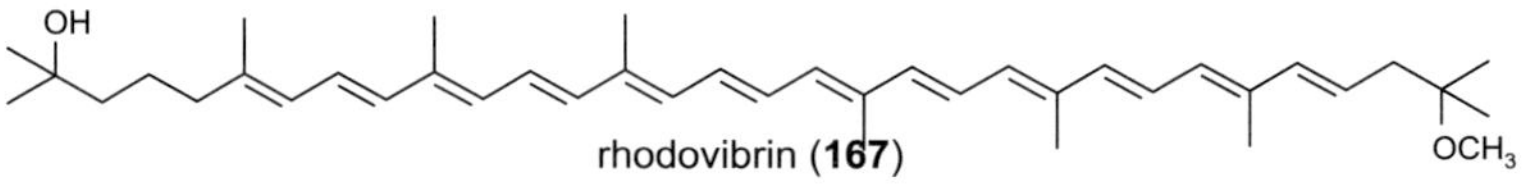

rhodovibrin (**167**)

From the species so far studied, most purple photosynthetic bacteria appear to use the spirilloxanthin biogenesis pathway that is further sub-divided into the 'normal spirilloxanthin', 'unusual spirilloxanthin', 'spheroidene' and 'carotenal' variants. Examples of commonly studied bacteria that use these different spirilloxanthin sub-pathways are *Blc. viridis, Rps. palustris, Rba. sphaeroides* and *Rbl. acidophilus,* respectively. In the case of *Rbl. acidophilus* 10050 the major carotenoid present is rhodopin glucoside (**95**), as observed in the X-ray crystal structure [29], while rhodopin (**93**) and rhodovibrin (**167**) are favoured in *Rps. palustris* [34]. It is informative to follow what happens during cell growth of two typical purple bacteria, *Rbl. acidophilus* and *Rps. palustris,* under carotenoid-limiting conditions.

The hierarchy of sequestration of carotenoids into the two types of antenna complex favours the preferential incorporation of longer chromophore carotenoids into LH2 rather than into the RC-LH1 (Fig. 10 and Table 1). At higher DPA levels, the LH2 becomes a distinctive brown colour while the RC-LH1 is blue-green, somewhat reminiscent of antennae of carotenoid-less mutants such as *Rba. sphaeroides* R26 and *Rsp. rubrum* G9+. At even higher levels of DPA the bacteria only assemble carotenoid-less 'core' complexes, but they are very unhealthy and eventually the culture dies. At these extreme concentrations the inhibitor DPA can also disrupt the biosynthesis of other isoprenoid compounds, including ubiquinols.

Although the altered light-absorption properties are due to an increased mixing of carotenoids (Table 1) that have reduced numbers of conjugated C=C double bonds, the overall process of carotenoid to bacteriochlorophyll energy transfer remains remarkably similar to that in control LH2 complexes [34]. This suggests that, although the carotenoid molecules are vital for the assembly of LH2 complexes *in vivo,* then, as long as there are polyene molecules of the right length present, the overall native architecture of the peptide scaffold is maintained; there are thus only minor effects on the light absorption properties of the B800 and B850 bacteriochlorophyll molecules and the structure of their binding pockets [34,41-43,112,113]. It is interesting to note that photosynthetic purple sulphur bacteria are able to assemble a 'native' carotenoid-less LH2 complex in the membrane when the culture is treated with DPA [40]; the features which allow these cells to assemble intact LH2 complexes are unknown, however.

It is possible to reconstitute LH complexes *in vitro*. The 8.5Å resolution projection map of the 2D-crystal of the carotenoid-less LH1 complex from *Rsp. rubrum* is an excellent example of this [33] (Fig. 9a). The inclusion of carotenoids in the mixture reconstituted *in vitro* shifts the equilibrium in favour of the intact LH1 complex but, to date, no LH1 truly like that *in vivo* has been achieved [114,115], because the reconstitution protocol has not been optimized.

2. Higher plants

Reconstitution *in vitro* has also proved to be a most valuable tool in the study of the role of carotenoids in LHCII and the other light-harvesting complexes, although no such study has been carried out with the inner core antenna complexes or RCs.

By use of different ratios of carotenoids, it has been shown that the two lutein (**133**) sites constituting the cross-brace in the centre of the LHCII complex (Fig. 11) are more or less essential for protein folding and stability. These sites can accommodate most of the other LHC xanthophylls, with a greater or lesser binding efficiency, except for neoxanthin (**234**) [48,116]. Competition experiments show these lutein sites to have >200-fold preference for lutein over neoxanthin whilst the neoxanthin site has only a 25-fold preference for neoxanthin over lutein [116]. The site specificity for the two central sites is thus: lutein (**133**) > zeaxanthin (**119**) > violaxanthin (**259**) > neoxanthin, (**234**).

The third site most strongly binds neoxanthin and does not bind lutein or zeaxanthin. The specificity for lutein over zeaxanthin was less marked, with two sites having a five-fold and a third only a three-fold preference for lutein. It was concluded that this flexibility is required for these complexes to function not only as light-harvesting antennae, but also as quenching centres under high excitation pressure. It should be noted that, so far, no structures of refolded proteins have been determined and, therefore, the absolute positioning and identification of particular carotenoids has had to be deduced indirectly.

H. Carotenoid Bandshift: a Monitor of the Membrane Potential

Electric fields cause a shift in the light-absorption properties of pigments as the field changes the energy levels of the pigment. This is known as a Stark effect. It was in the 1950s that effects of an electric field on the carotenoids in photosynthetic membranes were first discovered [117] in thylakoids and in isolated purple bacterial photosynthetic membranes [118]. The absorption change at 515 nm (or 518 nm) is induced by the membrane potential which develops across the 'coupled' membranes during electron transport. The fast change is due to charge separation in PSI and PSII (50% per photosystem) [119] and a slower rise is due to the electrogenic component linked to cytochrome reduction in the bc1 or b_6f complex during the so-called Q-cycle. The flash-induced absorption change decay rate is increased by uncouplers, such as the ionophore gramicidin, and the slow rise can be inhibited specifically by inhibitors of the cytochrome complexes (*e.g.* antimycin or myxothiazol). This fast 515 nm change has been used to measure the membrane potential [120] and, much more recently, has revealed the details of a modified Q-cycle in the bc1 complex of *Rba. sphaeroides* [121].

Recently, the Stark effect has been used as a tool to examine energy transfer between chlorophylls *a* and *b* and the xanthophylls in native and reconstituted LHCII trimers [122].

I. Evolutionary Considerations

In our current aerobic environment on Earth we are accustomed to accepting that carotenoids are required to provide protection against the destructive effect of $^1O_2$, which is formed as a consequence of chlorophyll triplet formation. Indeed, all chlorophyll-containing pigment-protein complexes contain also carotenoid. However, in those photosynthetic bacteria that inhabit environments which are strictly anaerobic, there is no problem if chlorophyll triplets are formed, so why, in these organisms that arose in the early anaerobic atmosphere 3.5 billion years ago, have carotenoids evolved to be there within van der Waals distance of chlorophyll? The closeness of the carotenoid to chlorophyll suggests that their light-harvesting role was very important and it is possible that as an aerobic environment developed their potential as photo-protectants was exploited to the full.

References

[1]   W. Nitschke and A. W. Rutherford, *Trends Biochem. Sci.*, **16**, 241 (1991).

[2]   D. Siefermann-Harms, *Physiol. Plantarum.*, **69**, 561 (1987).

[3]   Y. Wang, L. Mao and X. Hu, *Biophys. J.*, **86**, 3097 (2004).

[4]   R. E. Blankenship (ed.), *Molecular Mechanisms of Photosynthesis*, Marston Books Services, Oxford (2002).

[5]   R. J. Cogdell, *Pure Appl. Chem.*, **57**, 723 (1985).

[6]   H. A. Frank, A. J. Young, G. Britton and R. J. Cogdell (eds.), *The Photochemistry of Carotenoids*, Kluwer Academic Publisher, Dordrecht (1999).

[7]   A. J. Young and G. Britton (eds.), *Carotenoids in Photosynthesis*, Chapman and Hall, London (1993).

[8]   J. Deisenhofer, O. Epp, K. Miki, R. Huber and H. Michel, *Nature*, **318**, 618 (1985).

[9]   W. L. Delano (ed.), *The Pymol User's Manual*, Delano Scientific, Palo Alto (2002).

[10]   C. R. D. Lancaster, U. Ermler and H. Michel, in *Anoxygenic Photosynthetic Bacteria* (ed. R. E. Blankenship, M. T. Madigan and C. E. Bauer), p. 503, Kluwer Academic Publishers, Dordrecht (1995).

[11]   A. J. Hoff and J. Deisenhofer, *Physics Rep.*, **287**, 1 (1997).

[12]   H. Hashimoto, R. Fujii, K. Yanagi, T. Kusumoto, A. T. Gardiner, R. J. Cogdell, A. W. Roszak, N. W. Issacs, Z. Pendon, D. Niedzwiedski and H. A. Frank, *Pure Appl. Chem.*, **78**, 1505 (2006).

[13]   A. W. Roszak, K. McKendrick, A. T. Gardiner, I. A. Mitchell, N. W. Isaacs, R. J. Cogdell, H. Hashimoto and H. A. Frank, *Structure*, **12**, 765 (2004).

[14]   Y. Koyama and R. Fujii, in *The Photochemistry of Carotenoids* (ed. H. A. Frank, A. J. Young, G. Britton and R. J. Cogdell), p. 161, Kluwer Academic Publisher, Dordrecht (1999).

[15]   M. Lutz, L. Chinsky and P. Y. Turpin, *Photochem. Photobiol.*, **36**, 503 (1982).

[16]   Y. Mukai-Kuroda, R. Fujii, N. Ko-chi, T. Sashima and Y. Koyama, *J. Phys. Chem. A.*, **106**, 3566 (2002).

[17]   N. Ohashi, N. KoChi, M. Kuki, T. Shimamura, R. J. Cogdell and Y. Koyama, *Biospectroscopy*, **2**, 59 (1996).

[18]   M. Chen, A. Telfer, S. Lin, A. Pascal, A. W. D. Larkum, J. Barber and R. E. Blankenship, *Photochem. Photobiol. Sci.*, **4**, 1060 (2005).

[19]   B. Loll, J. Kern, W. Saenger, A. Zouni and J. Biesiadka, *Nature*, **438**, 1040 (2005).

[20]   A. Telfer, *Photochem. Photobiol. Sci.*, **4**, 950 (2005).

[21]   K. N. Ferreira, T. M. Iverson, K. Maghlaoui, J. Barber and S. Iwata, *Science*, **303**, 1831 (2004).

[22]   P. Jordan, P. Fromme, H. T. Witt, O. Klukas, W. Saenger and N. Krauss, *Nature*, **411**, 909 (2001).

[23]   A. Ben-Shem, F. Frolow and N. Nelson, *Nature*, **426**, 630 (2003).

[24]   A. Amunts, O. Drory and N. Nelson, *Nature*, **447**, 58 (2007).

[25]   N. Pfennig, in *The Photosynthetic Bacteria* (ed. R. K. Clayton and W. R. Sistrom), p. 3, Plenum
        Publishing Corporation, New York (1978).

[26]   H. Zuber and R. J. Cogdell, in *Anoxygenic Photosynthetic Bacteria* (ed. R. E. Blankenship, M. T.
        Madigan and C. E. Bauer), p. 315, Kluwer Academic Publishers, Dordrecht (1995).

[27]   A. Freer, S. Prince, K. Sauer, M. Papiz, A. Hawthornthwaite-Lawless, G. McDermott, R. Cogdell and
        N. W. Isaacs, *Structure*, **4**, 449 (1996).

[28]   J. Koepke, X. Hu, C. Muenke, K. Schulten and H. Michel, *Structure*, **4**, 581 (1996).

[29]   G. McDermott, S. M. Prince, A. A. Freer, A. M. Hawthornthwaite-Lawless, M. Z. Papiz, R. J. Cogdell
        and N. W. Isaacs, *Nature*, **374**, 517 (1995).

[30]   A. W. Roszak, T. D. Howard, J. Southall, A. T. Gardiner, C. J. Law, N. W. Isaacs and R. J. Cogdell,
        Science, **302**, 1969 (2003).

[31]   R. J. Cogdell, H. Zuber, J. P. Thornber, G. Drews, G. Gingras, R. A. Niederman, W. W. Parson and G.
        Feher, *Biochim. Biophys. Acta*, **806**, 185 (1985).

[32]   R. J. Cogdell, A. Gall and J. Köhler, *Q. Rev. Biophys.*, **39**, 227 (2006).

[33]   S. Karrasch, P. A. Bullough and R. Ghosh, *EMBO J.*, **14**, 631 (1995).

[34]   A. Gall, S. Henry, S. Takaichi, B. Robert and R. Cogdell, *Photosynth. Res.*, **86**, 25 (2005).

[35]   M. Z. Papiz, S. M. Prince, T. Howard, R. J. Cogdell and N. W. Isaacs, *J. Mol. Biol.*, **326**, 1523 (2003).

[36]   J. B. Arellano, B. B. Raju, K. R. Naqvi and T. Gillbro, *Photochem. Photobiol.*, **68**, 84 (1998).

[37]   V. Cherezov, J. Clogston, M. Z. Papiz and M. Caffrey, *J. Mol. Biol.*, **357**, 1605 (2006).

[38]   A. Gall, A. T. Gardiner, R. J. Cogdell and B. Robert, *FEBS Lett.*, **580**, 3841 (2006).

[39]   G. Cohen-Bazire and R. Y. Stanier, *Nature*, **181**, 250 (1958).

[40]   R. C. Fuller and I. C. Anderson, *Nature*, **181**, 252 (1958).

[41]   C. N. Hunter, B. S. Hundle, J. E. Hearst, H. P. Lang, A. T. Gardiner, S. Takaichi and R. J. Cogdell, *J.
        Bacteriol.*, **176**, 3692 (1994).

[42]   H. P. Lang, R. J. Cogdell, A. T. Gardiner and C. N. Hunter, *J. Bacteriol.*, **176**, 3859 (1994).

[43]   H. P. Lang and C. N. Hunter, *Biochem. J.*, **298**, 197 (1994).

[44]   K. R. Miller, *Proc. Natl. Acad. Sci. USA*, **76**, 6415 (1979).

[45]   Z. Liu, H. Yan, K. Wang, T. Kuang, J. Zhang, L. Gui, X. An and W. Chang, *Nature*, **428**, 287 (2004).

[46]   J. Standfuss and W. Kühlbrandt, *J. Biol. Chem.*, **279**, 36884 (2004).

[47]   F. G. Plumley and G. W. Schmidt, *Proc. Natl. Acad. Sci. USA*, **84**, 146 (1987).

[48]   R. Croce, S. Weiss and R. Bassi, *J. Biol. Chem.*, **274**, 29613 (1999).

[49]   A. V. Ruban, P. J. Lee, M. Wentworth, A. J. Young and P. Horton, *J. Biol. Chem.*, **274**, 10458 (1999).

[50]   F. Ros, R. Bassi and H. Paulsen, *Eur. J. Biochem.*, **253**, 653 (1998).

[51]   J. Nield and J. Barber, *Biochim. Biophys. Acta*, **1757**, 353 (2006).

[52]   A. E. Yakushevska, W. Keegstra, E. J. Boekema, J. P. Dekker, J. Andersson, S. Jansson, A. V. Ruban
        and P. Horton, *Biochemistry*, **42**, 608 (2003).

[53]   R. Croce, M. Mozzo, T. Morosinotto, A. Romeo, R. Hienerwadel and R. Bassi, *Biochemistry*, **46**, 3846
        (2007).

[54]   V. K. Schmid, K. V. Cammarata, B. U. Bruns and G. W. Schmidt, *Proc. Natl. Acad. Sci. USA*, **94**, 7667
        (1997).

[55]   D. G. Durnford, J. A. Deane, S. Tan, G. I. McFadden, E. Gantt and B. R. Green, *J. Mol. Evol.*, **48**, 59
        (1999).

[56]   G. Guglielmi, J. Lavaud, B. Rousseau, A.-L. Etienne, J. Houmard and A. V. Ruban, *FEBS J.*, **272**, 4339
        (2005).

[57]   E. Papagiannakis, I. H. M. van Stokkum, H. Fey, C. Büchel and R. van Grondelle, *Photosynth. Res.*, **85**,
        241 (2005).

[58]   M. Chen and T. S. Bibby, *Photosynth. Res.*, **86**, 165 (2005).

[59]   M. Chen, R. G. Quinnell and A. W. D. Larkum, *FEBS Lett.*, **514**, 149 (2002).

[60]   T. S. Bibby, J. Nield, M. Chen, A. W. D. Larkum and J. Barber, *Proc. Natl. Acad. Sci. USA*, **100**, 9050 (2003).

[61]   M. Chen, T. S. Bibby, J. Nield, A. Larkum and J. Barber, *Biochim. Biophys. Acta*, **1708**, 367 (2005).

[62]   M. Chen, T. S. Bibby, J. Nield, A. W. D. Larkum and J. Barber, *FEBS Lett.*, **579**, 1306 (2005).

[63]   N. U. Frigaard and D. Bryant, *Arch. Microbiol.*, **182**, 265 (2004).

[64]   N.-U. Frigaard, H. Li, P. Martinsson, S. Das, H. Frank, T. Aartsma and D. Bryant, *Photosynth. Res.*, **86**, 101 (2005).

[65]   T. Nozawa, K. Ohtomo, M. Suzuki, H. Nakagawa, Y. Shikama, H. Konami and Z. Y. Wang, *Photosynth. Res.*, **41**, 211 (1994).

[66]   J. Psencik, J. B. Arellano, T. P. Ikonen, C. M. Borrego, P. A. Laurinmaki, S. J. Butcher, R. E. Serimaa and R. Tuma, *Biophys. J.*, **91**, 1433 (2006).

[67]   J. Psencik, T. P. Ikonen, P. Laurinmaki, M. C. Merckel, S. J. Butcher, R. E. Serimaa and R. Tuma, *Biophys. J.*, **87**, 1165 (2004).

[68]   E. Hofmann, P. M. Wrench, F. P. Sharples, R. G. Hiller, W. Welte and K. Diederichs, *Science*, **272**, 1788 (1996).

[69]   A. Oren, *Saline Sys.*, **1**, 2 (2005).

[70]   T. Tomo, T. Okubo, S. Akimoto, M. Yokono, H. Miyashita, T. Tsuchiya, T. Noguchi and M. Mimuro, *Proc. Natl. Acad. Sci. USA*, **104**, 7283 (2007).

[71]   Q. Hu, H. Miyashita, I. Iwasaki, N. Kurano, S. Miyachi, M. Iwaki and S. Itoh, *Proc. Natl. Acad. Sci. USA*, **95**, 13319 (1998).

[72]   N. Adir, H. Zer, S. Shochat and I. Ohad, *Photosynth. Res.*, **76**, 343 (2003).

[73]   M. Havaux, F. Eymery, S. Porfirova, P. Rey and P. Dörmann, *Plant Cell*, **17**, 3451 (2005).

[74]   P. J. Nixon, M. Barker, M. Boehm, R. de Vries and J. Komenda, *J. Exp. Bot.*, **56**, 357 (2005).

[75]   M. Havaux and K. K. Niyogi, *Proc. Natl. Acad. Sci. USA*, **96**, 8762 (1999).

[76]   M. P. Johnson, M. Havaux, C. Triantaphylides, B. Ksas, A. A. Pascal, B. Robert, P. A. Davison, A. V. Ruban and P. Horton, *J. Biol. Chem.*, **282**, 22605 (2007).

[77]   H. T. Witt, *Biochim. Biophys. Acta*, **505**, 355 (1979).

[78]   C. H. Foyer and J. Harbinson, in *The Photochemistry of Carotenoids* (ed. H. A. Frank, A. J. Young, G. Britton and R. J. Cogdell), p. 305, Kluwer Academic Publisher, Dordrecht (1999).

[79]   E. J. G. Peterman, F. M. Dukker, R. Van Grondelle and H. van Amerongen, *Biophys. J.*, **69**, 2670 (1995).

[80]   S. L. S. Kwa, F. G. Groeneveld, J. P. Dekker, R. van Grondelle, H. van Amerongen, S. Lin and W. S. Struve, *Biochim. Biophys. Acta*, **1101**, 143 (1992).

[81]   J. Barber and M. D. Archer, *J. Photochem. Photobiol. A: Chem.*, **142**, 97 (2001).

[82]   A. Telfer, T. C. Oldham, D. Phillips and J. Barber, *J. Photochem. Photobiol. B*, **48**, 89 (1999).

[83]   V. Barzda, E. J. G. Peterman, R. van Grondelle and H. van Amerongen, *Biochemistry*, **37**, 546 (1998).

[84]   N. Mochizuki, J. A. Brusslan, R. Larkin, A. Nagatani and J. Chory, *Proc. Natl. Acad. Sci. USA*, **98**, 2053 (2001).

[85]   A. Krieger-Liszkay, *J. Exp. Bot.*, **56**, 337 (2005).

[86]   A. Telfer, *Phil. Trans. Roy. Soc. Lond. B.*, **357**, 1431 (2002).

[87]   N. Keren, A. Berg, P. J. M. van Kan, H. Levanon and I. Ohad, *Proc. Natl. Acad. Sci. USA*, **94**, 1579 (1997).

[88]   H. J. van Gorkom and J. P. M. Schelvis, *Photosynth. Res.*, **38**, 297 (1993).

[89]   A. Telfer, S. Dhami, S. M. Bishop, D. Phillips and J. Barber, *Biochemistry*, **33**, 14469 (1994).

[90]   R. Edge and T. G. Truscott, in *The Photochemistry of Carotenoids* (ed. H. A. Frank, A. J. Young, G. Britton and R. J. Cogdell), p. 223, Kluwer Academic Publisher, Dordrecht (1999).

[91]   K. Maxwell and G. N. Johnson, *J. Exp. Bot.*, **51**, 659 (2000).

[92]   B. Demmig-Adams and W. W. Adams, *Trends Plant Sci.*, **1**, 21 (1996).

[93]   X. P. Li, O. Björkman, C. Shih, A. R. Grossman, M. Rosenquist, S. Jansson and K. K. Niyogi, *Nature*, **403**, 391 (2000).

[94]   P. Horton, A. V. Ruban, D. Rees, G. Noctor, A. A. Pascal and A. J. Young, *FEBS Lett.*, **292**, 1 (1991).

[95]   A. A. Pascal, Z. Liu, K. Broess, B. van Oort, H. van Amerongen, C. Wang, P. Horton, B. Robert, W. Chang and A. V. Ruban, *Nature*, **436**, 134 (2005).

[96]   A. V. Ruban, R. Berera, C. Ilioaia, I. H. M. van Stokkum, J. T. M. Kennis, A. A. Pascal, H. van Amerongen, B. Robert, P. Horton and R. van Grondelle, *Nature*, **450**, 575 (2007).

[97]   M. Aspinall-O'Dea, M. Wentworth, A. Pascal, B. Robert, A. Ruban and P. Horton, *Proc. Natl. Acad. Sci. USA*, **99**, 16331 (2002).

[98]   N. E. Holt, D. Zigmantas, L. Valkunas, X. P. Li, K. K. Niyogi and G. R. Fleming, *Science*, **307**, 433 (2005).

[99]   A. V. Ruban, J. Lavaud, B. Rousseau, G. Guglielmi, P. Horton and A.-L. Etienne, *Photosynth. Res.*, **82**, 165 (2004).

[100] G. Kurisu, H. Zhang, J. L. Smith and W. A. Cramer, *Science*, **302**, 1009 (2003).

[101] D. Stroebel, Y. Choquet, J. L. Popot and D. Picot, *Nature*, **426**, 413 (2003).

[102] H. M. Zhang, D. R. Huang and W. A. Cramer, *J. Biol. Chem.*, **274**, 1581 (1999).

[103] E. J. G. Peterman, S. O. Wenk, T. Pullerits, L. O. Palsson, R. van Grondelle, J. P. Dekker, M. Rögner and H. van Amerongen, *Biophys. J.*, **75**, 389 (1998).

[104] R. Edge, E. J. Land, D. J. McGarvey, M. Burke and T. G. Truscott, *FEBS Lett.*, **471**, 125 (2000).

[105] A. Telfer, D. Frolov, J. Barber, B. Robert and A. Pascal, *Biochemistry*, **42**, 1008 (2003).

[106] C. A. Tracewell and G. W. Brudvig, *Biochemistry*, **42**, 9127 (2003).

[107] C. C. Moser and P. L. Dutton, *Biochim. Biophys. Acta*, **1101**, 171 (1992).

[108] P. A. Loach and P. S. Parkes-Loach, in *Anoxygenic Photosynthetic Bacteria* (ed. R. E. Blankenship, M. T. Madigan and C. E. Bauer), p. 437, Kluwer Academic Publishers, Dordrecht (1995).

[109] G. E. Bartley and P. A. Scolnik, *J. Biol. Chem.*, **264**, 13109 (1989).

[110] S. Takaichi, in *The Photochemistry of Carotenoids.* (ed. H. A. Frank, A. J. Young, G. Britton and R. J. Cogdell), p. 39, Kluwer Academic Publisher, Dordrecht (1999).

[111] P. M. Bramley, in *Carotenoids in Photosynthesis* (ed. A. J. Young and G. Britton), p. 127, Chapman and Hall, London (1993).

[112] A. Gall, R. J. Cogdell and B. Robert, *Biochemistry*, **42**, 7252 (2003).

[113] J. D. Olsen, B. Robert, C. A. Siebert, P. A. Bullough and C. N. Hunter, *Biochemistry*, **42**, 15114 (2003).

[114] C. M. Davis, P. L. Bustamante and P. A. Loach, *J. Biol. Chem.*, **270**, 5793 (1995).

[115] L. Fiedor and H. Scheer, *J. Biol. Chem.*, **280**, 20921 (2005).

[116] S. Hobe, H. Niemeier, A. Bender and H. Paulsen, *Eur. J. Biochem.*, **267**, 616 (2000).

[117] L. N. M. Duysens, *Science*, **120**, 353 (1954).

[118] B. Chance and L. Smith, *Nature*, **175**, 803 (1955).

[119] H. T. Witt, in *Bioenergetics of Photosynthesis* (ed. Govindjee), p. 493, Academic Press, New York (1975).

[120] J. B. Jackson and A. R. Crofts, *FEBS Lett.*, **4**, 185 (1969).

[121] A. R. Crofts, V. P. Shinkarev, D. R. J. Kolling and S. J. Hong, *J. Biol. Chem.*, **278**, 36191 (2003).

[122] M. A. Palacios, S. Caffarri, R. Bassi, R. van Grondelle and H. van Amerongen, *Biochim. Biophys. Acta*, **1656**, 177 (2004).

Chapter 15

Functions of Carotenoid Metabolites and Breakdown Products

George Britton

A. Introduction

It is not only intact carotenoids but also fragments of carotenoid molecules that have important natural functions and actions. The electron-rich polyene chain of the carotenoids is very susceptible to oxidative breakdown, which may be enzymic or non-enzymic. Central cleavage gives C_{20} compounds, retinoids, as described in *Chapter 16*. Cleavage at other positions gives smaller fragments, notably C_{10}, C_{13} and C_{15} compounds that retain the carotenoid end group. The formation of these is described in *Chapter 17* and in *Volume 3, Chapter 4*. Oxidative breakdown can also take place during storage, processing and curing of plant material, and the products contribute to the desired aroma/flavour properties of, for example, tea, wine and tobacco. The importance of vitamin A (C_{20}) in animals is well known. Vitamin A deficiency is still a major concern in many parts of the world. It can lead to blindness and serious ill-health or death, especially in young children. Volatile smaller carotenoid fragments ('norisoprenoids') are widespread scent/ flavour compounds in plants.

A book about carotenoids would therefore not be complete without at least an outline of the functions and actions of these carotenoid-derived products. The treatment in this Chapter will concentrate on the main features and principles and will not be comprehensive. Detailed accounts of particular topics are presented in specialized books and reviews cited in the appropriate Sections below.

B. C$_{20}$ Compounds: Retinoids

1. Vitamin A

The main dietary source of vitamin A for most people is the provitamin β-carotene (**3**) and other carotenoids with one unsubstituted β ring. In principle, central cleavage of a C$_{40}$ molecule can give rise to two C$_{20}$ fragments (Scheme 1). The cleavage of β-carotene in this way, and by excentric cleavage, to give vitamin A aldehyde, retinal (*1*), in animals is described in detail in *Chapter 16*.

The reversible enzymic interconversion of retinal (*1*) and retinol (*2*), and the irreversible further oxidation of retinal to retinoic acid (*3*), also shown in Scheme 1, have been demonstrated in various tissues. All these three forms of vitamin A are of fundamental importance, but in totally different ways.

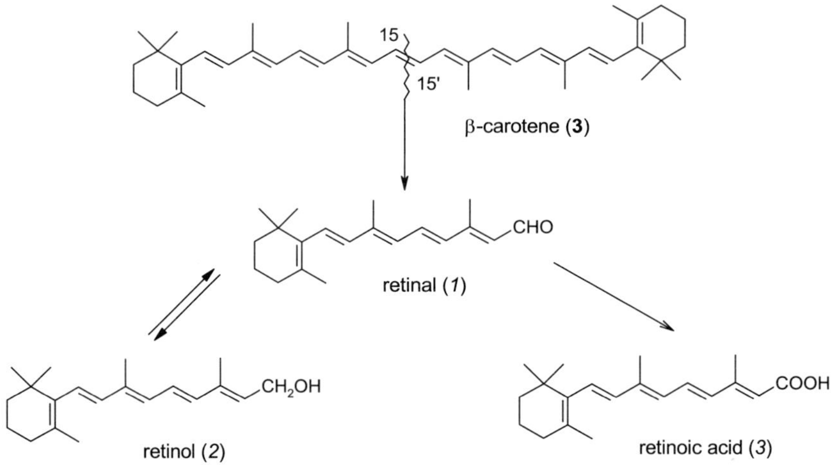

Scheme 1

The literature on vitamin A and retinoids is vast, and includes some comprehensive monographs [1-3]. The importance of vitamin A in human health and nutrition is discussed in *Volume 5, Chapters 8* and *9*. Here, the discussion is focused on the mechanistic principles of how the various C$_{20}$ compounds function, not only in animals but also in some microorganisms where they play major roles.

2. Retinol

Although it is well established that its derivatives retinal (*1*) and retinoic acid (*3*) play essential roles in human health, until recently no direct role was recognized for retinol (*2*) itself. Now, however, it is becoming clear that retinol does have a key cytoplasmic role in regulating some serine/threonine kinase enzymes, especially protein kinase C and Raf proteins that are involved in crucial molecular signalling processes [4]. These enzymes have a 'zinc finger' domain which retinol can enter to activate the enzyme. Retinol delivered to the cells by plasma retinol-binding protein (pRBP) can also be converted into other derivatives, notably 14-hydroxy-*retro*-retinol (*4*), 13,14-dihydroxyretinol (*5*) and anhydroretinol (*6*), which are found widely in vertebrate and invertebrate animals and are reported to function in regulating cell survival and apoptosis [4].

This topic is discussed in *Volume 5, Chapter 9*.

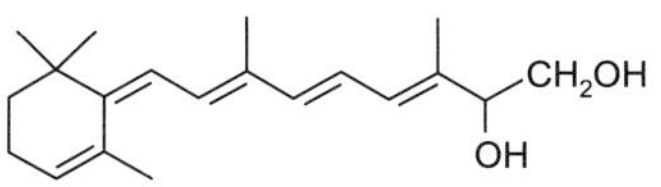

14-hydroxy-*retro*-retinol (*4*)

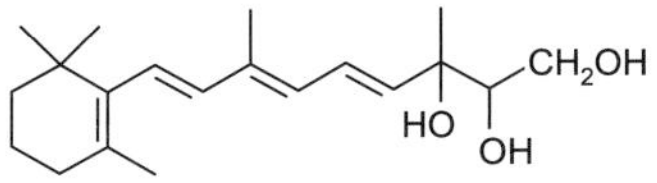

13,14-dihydroxyretinol (*5*)

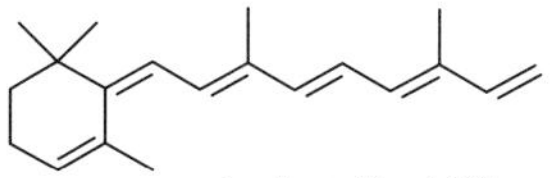

anhydroretinol (*6*)

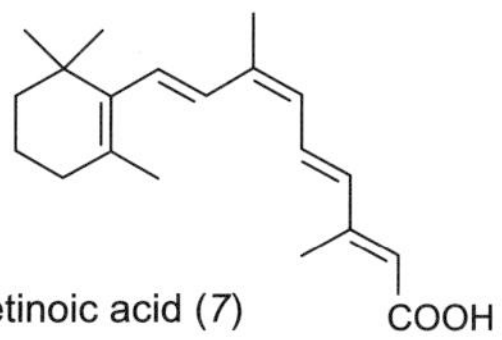

(9-*cis*)-retinoic acid (*7*)

3. Retinoic acid

Since the 1980s it has been recognized that retinoic acid, formed by the irreversible oxidation of retinal, is required by mammals to maintain reproduction, embryogenesis and cell differentiation. It is also an essential factor in mediating responses to various hormones and regulatory factors, such as vitamin D, thyroid hormone, and oestrogens, *etc.* Its role is in regulating gene activation, *via* specific receptors [5-7]. Three 'retinoic acid receptors' RARα, RARβ, and RARγ specifically bind (all-*trans*)-retinoic acid (*3*), and three 'retinoid receptors', RXRα, RXRβ, and RXRγ, mainly bind (9-*cis*)-retinoic acid (*7*). The receptors, activated by retinoid binding, form dimers. The active complex may be an RXR homodimer (RXR-RXR) or a heterodimer of RXR with another ligand-activated receptor, *e.g.* RAR, vitamin D receptor (VDR), thyroid hormone receptor (TR), various steroid hormone receptors or peroxysome proliferator activated receptor (PPAR). The dimer pair recognizes, binds to and activates its specific response element within a gene, leading to the required response. One of the dimer partners must be an RXR, so vitamin A, through retinoic acid and RXR, serves as the master regulator of a variety of hormone responses. Further discussion of this in relation to human health is presented in *Volume 5*.

4. Retinal and photoreceptors

Evolution has made good use of combinations of retinal with proteins (opsins) to provide highly efficient and versatile photoreceptors, not only as visual pigments in animals but also for other purposes in microorganisms and unicellular algae [8]. The retinal is bound covalently to the primary amino group of the side chain of a lysine residue by a Schiff base linkage (*8*). Interconversion of the protonated (*9*) and unprotonated (*8*) Schiff base (Scheme 2) is an essential feature of the functioning. In all cases, the primary photochemical event is the extremely rapid *cis* → *trans* or *trans* → *cis* isomerization of the covalently bound retinylidene chromophore, and this photoisomerization provides the energy to drive conformational changes in the protein that lead to the response. The main features are summarized below. More details can be obtained from various reviews, Symposium reports and books [8-14]. Broader accounts of vision, including the anatomy of the eye, the structure of rod and cone cells and the location of the visual pigments, can be found in text and specialist books [15-17].

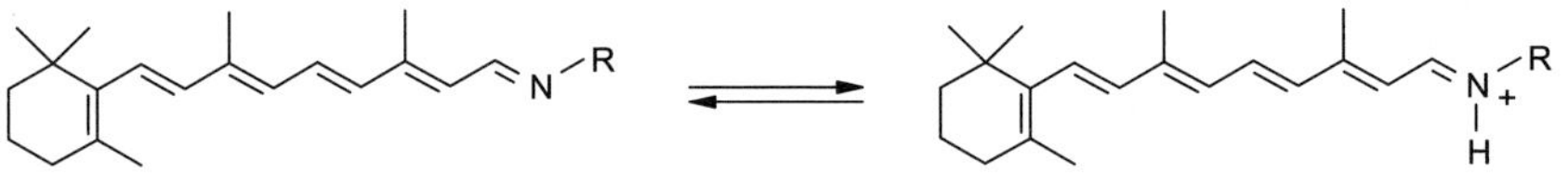

retinylidene Schiff base (*8*)   protonated retinylidene Schiff base (*9*)

Scheme 2

a) Visual pigments

In all animals, the retina of the eye contains a large number of photoreceptor cells which are responsible for the processes of vision. The photoreceptor pigments are proteins containing a covalently bound (11-*cis*)-retinal (*10*) chromophore which contains a sterically hindered *cis* double bond [18]. Although in mammals and most other animals, the chromophore is retinal itself, some animals employ other retinal derivatives, *e.g.* 3,4-didehydroretinal (*11*) in porphyropsins in some fish [19], 3-hydroxyretinal (*12*) in xanthopsin in some insects [20] or 4-hydroxyretinal (*13*) in squid [21].

In humans, the 90 million rod cells are responsible for scotopic vision ('night vision'), *i.e.* vision in dim light. This extremely sensitive, essentially monochromatic process uses a single photoreceptor, rhodopsin. The 4.5 million cone cells contain related photoreceptors (photopsins) and are responsible for photopic (colour) vision (Section 4.c). Mammalian rhodopsin is the most extensively studied example, but the principles of structure, and the structural changes that take place during the 'visual cycle' are similar in all cases.

(11-cis)-retinal *(10)*

3,4-didehydroretinal *(11)*

3-hydroxyretinal *(12)*

4-hydroxyretinal *(13)*

b) Rhodopsin

i) Structure. The rhodopsin molecules are located in extensive internal membranes in the outer segment of the rod cells. Opsin is a 40kDa protein with seven transmembrane helix domains and, on the outer side of the membrane, a C-terminal loop that can interact with several other proteins, notably transducin, as part of the signal transduction mechanism [8,22]. The chromophore, retinal, is bonded covalently to the ε-amino group of lysine residue 296, by a Schiff base linkage, and sits more or less horizontally in the hydrophobic core of the membrane [8,22-24] (Fig. 1).

Fig. 1a.

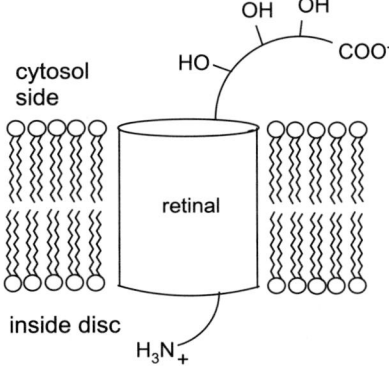

Fig. 1b.

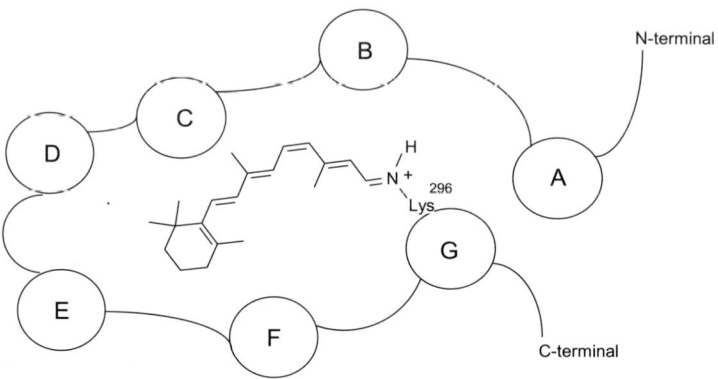

Fig. 1. (a) Simplified diagram of the membrane-bound dark-adapted rhodopsin. The cylinder represents the seven transmembrane helices of the protein, with the retinal chromophore located in the middle, perpendicular to the helices. The C-terminal part of the protein, on the cytosol side of the membrane, contains several hydroxy groups, phosphorylation of which is an essential regulatory feature. (b) View from the surface of the membrane, showing the location of the chromophore relative to the seven transmembrane helices A-G, and the Schiff base linkage to lysine 296 in helix G.

In the relaxed, dark-adapted form, the retinylidene Schiff base is protonated and held in a distorted 11-*cis*, 15-*anti* configuration. The ring-chain conformation is 6-s-*cis*, but the molecule is distorted substantially out of planarity by twisting about the C(6,7) and C(12,13) single bonds. 'External point charge' negatively charged glutamic acid side chains serve to delocalize the positive charge, which is formally on the nitrogen atom of the protonated Schiff base, along the conjugated double bond system. This gives rhodopsin a λ_{max} of 498-500 nm, in contrast to that of free retinal (375 nm) and a simple protonated retinylidene Schiff base (*9*, R = *n*Bu) (440 nm) [8]. Rhodopsin ('visual purple') therefore has greatest sensitivity to yellow light.

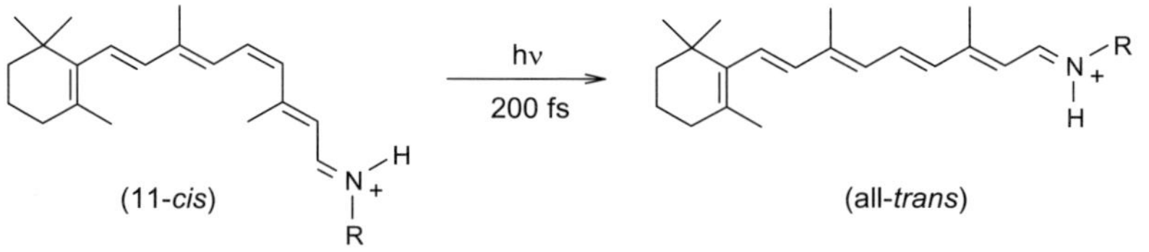

(11-*cis*) $\xrightarrow[200\ \text{fs}]{h\nu}$ (all-*trans*)

Scheme 3

ii) Visual cycle. The functioning of rhodopsin involves a complex 'visual cycle' [18] (Fig. 2). In the primary event, a photon of light is absorbed by rhodopsin and causes isomerization of the (11-*cis*)-retinylidene chromophore to all-*trans* (Scheme 3). This is the fastest photobiological reaction known (200 fs) [25] and occurs too rapidly to allow any significant overall change in the shape of the retinal. The quantum yield is very high (65-70%) [18] and the process is extremely efficient; 80% of the energy of the photon is stored in the primary product [26]. In the product, bathorhodopsin, the retinylidene chromophore is thus in a highly strained conformation. It has been suggested that bathorhodopsin may not be the immediate product of photoisomerization but is formed by the rapid relaxation of this product, 'photorhodopsin' [27]. A succession of conformational changes then occurs, in which the highly strained chromophore gradually relaxes to a stable form, and the release of strain energy drives conformational changes in the protein. Intermediates can be trapped at low temperature and identified by UV/Vis, resonance Raman and other spectroscopic techniques. The key step is the formation of metarhodopsin II, in which the Schiff base is no longer protonated. In metarhodopsin II the C-terminal loop has moved away from the surface of the membrane to allow access to the signal transduction protein transducin [13], and this activates the signal transduction cascade leading to the closing of ion channels and stimulating the nerve impulse that reports the detection of light to the brain [28].

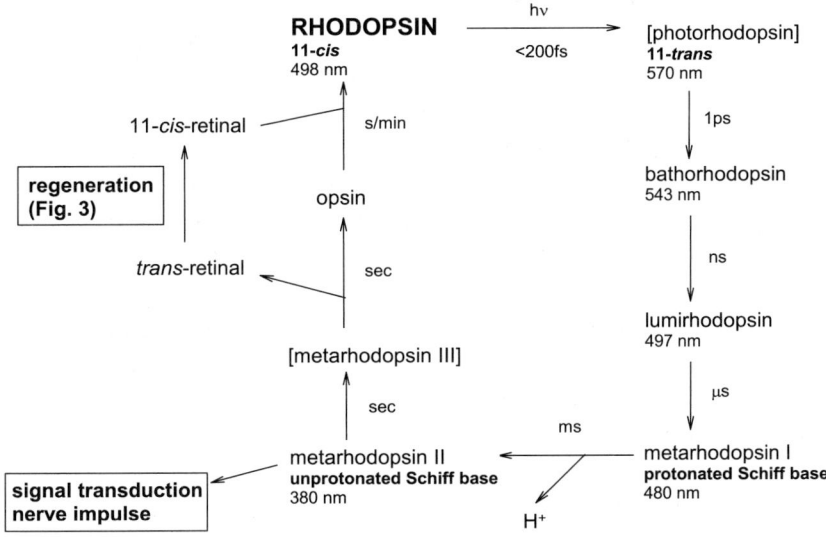

Fig. 2. The rhodopsin visual cycle, showing the sequence of events and intermadiates following the primary photoreaction. The λ_{max} values of the intermediates that can be trapped at low temperature are shown, and the timescale of each step is indicated.

Phosphorylation of a series of hydroxy groups in amino acid side chains in the C-terminal loop is effective in allowing release of transducin and deactivating the system [13]. Once the stage of metarhodopsin III is reached, the (*trans*)-retinal–opsin complex is unstable and dissociates irreversibly. The (all-*trans*)-retinal must be converted back to 11-*cis* before it can combine with opsin to regenerate the visual pigment. This is not a simple process but requires several enzyme-catalysed steps [29,30] (Fig. 3).

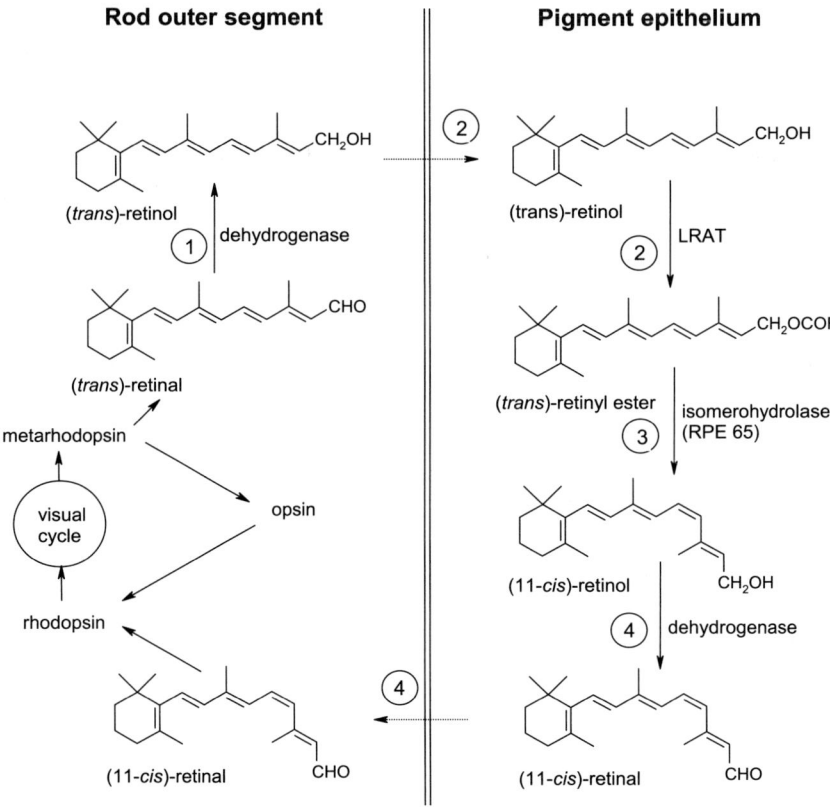

Fig. 3. The sequence of isomerization and other reactions in the regeneration of rhodopsin. The numbers in circles refer to the steps described below.

1. (*trans*)-Retinal is reduced to (*trans*)-retinol in the rod outer segment by a dehydrogenase.

2. The (*trans*)-retinol is transported from the rod cell into the pigment epithelium and ester-ified with fatty acid from phospholipid by lecithin:retinol acyl transferase (LRAT).

3. In the pigment epithelium the retinyl ester is hydrolysed by an isomerohydrolase (RPE 65). Cleavage of the ester provides the energy for the non-photochemical isomerization to (11-*cis*)-retinol.

4. The (11-*cis*)-retinol is oxidized by a dehydrogenase in the pigment epithelium to (11-*cis*)-retinal which is transported back into the rod cell, where it combines with opsin to regenerate dark-adapted rhodopsin.

This multi-step process allows any losses of retinal/retinol to be made up from vitamin A stores. In cases of vitamin A deficiency, this is not possible, so losses are not replaced, leading to failure of the visual process.

An alternative photochemical isomerization of (all-*trans*)-retinal to (11-*cis*)-retinal, in which the retinal is bound to a protein RGR in the pigment epithelium, has recently been reported [31,32].

In the squid *Ommastrephes sloanii pacificus*, the regeneration of (11-*cis*)-retinal requires a second retinal-opsin protein, retinochrome. (all-*trans*)-Retinal is transferred to retinochrome, isomerized photochemically, and then released, as (11-*cis*)-retinal, to the rhodopsin protein [33].

c) Colour vision

The photopsin pigments in cone cells are structurally similar to rhodopsin and take part in similar photocycles. There are three different related photoreceptors, which are most sensitive to yellow (570 nm), green (540 nm) and violet (420 nm) light, respectively [34]. The pigments differ from rhodopsin and from each other in only a few amino acid residues, but this is sufficient to bring about the small changes to the interactions with the retinal chromophore that lead to the different absorption maxima. Although less sensitive than scotopic vision, this provides a means of wavelength discrimination and is the basis of a trichromatic colour vision. There are variations on this theme throughout the animal kingdom. Some insects, notably bees, have four different photoreceptors, one of them being sensitive to short wavelength light, into the UV region, thus enabling them to 'see' UV light. Other animals may have only one photoreceptor and are colour blind.

d) Bacteriorhodopsin

The same structural and mechanistic principles, though with totally different outcomes, form the basis of the functioning of bacteriorhodopsin and other photoreceptors in some members of the Archaea, notably *Halobacterium* species. The extreme halophile *Halobacterium salinarum* is a characteristic dominant species in environments with a very high concentration of salt (NaCl). Although it can live by normal aerobic fermentation metabolism, it can adapt to severe stress (high light intensity, high salinity, low oxygen) by harvesting light energy and changing to phototrophic metabolism. Normally *Halobacterium* is pink-red in colour, due to the presence of C_{50} carotenoids in its membranes. Under stress conditions, however, it develops a 'purple patch' (purple membrane) which contains the purple (λ_{max} 570 nm) pigment bacteriorhodopsin, so named because of its considerable similarity to rhodopsin. Bacteriorhodopsin also consists of an opsin protein (26 kDa), which is anchored in the purple membrane by seven transmembrane helices and covalently bonded to retinal by (protonated) Schiff base linkage *via* a lysine residue (216) [15].

In a similar manner to rhodopsin, the retinal sits in the middle of the protein, parallel to the faces of the membrane. In bacteriorhodopsin, however, the dark-adapted form has the retinal in the all-*trans* configuration with 6-s-*trans* conformation.

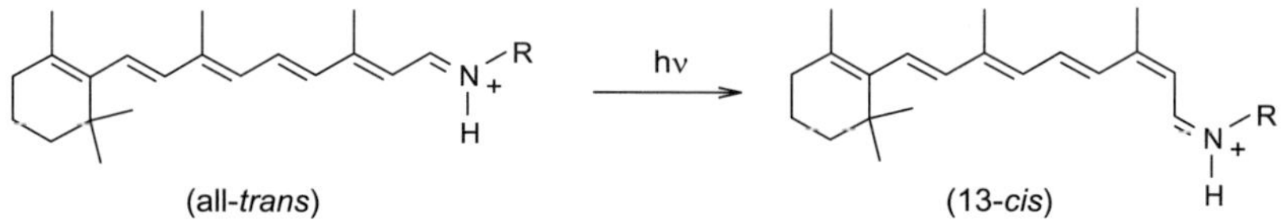

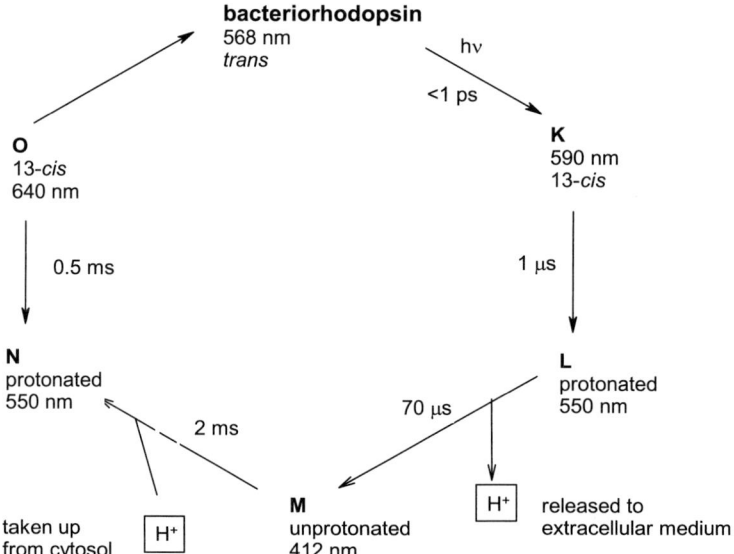

Scheme 4

Fig. 4. The photocycle of bacteriorhodopsin, showing the sequence of events and intermediates following the primary photoreaction. The λ_{max} values of intermediates that can be trapped at low temperature and the timescale of each step are indicated.

Like rhodopsin, bacteriorhodopsin participates in a photocycle (Fig. 4). The primary event is again a very rapid photoisomerization, this time from all-*trans* to a distorted 13-*cis* form (Scheme 4), and this is followed by a series of conformation changes driven by the relaxation of the strain energy *via* a series of intermediates designated K, L, M, N and O. The conformational changes in the protein are accompanied by the loss and gain of protons. During the cycle, the protonated Schiff base of intermediate M loses its proton, which is effectively passed on to an aspartate and released to the outside of the membrane *via* an arginine residue. A proton is then taken up from inside the cell and passed on *via* an aspartate residue to reprotonate the retinal Schiff base (intermediate N). The light energy absorbed is thus converted into a proton gradient across the membrane and the energy of the proton gradient is used to drive ATP synthesis by a coupled ATPase enzyme. The conversion of the (13-*cis*)-retinal chromophore back to all-*trans* appears to be a simple photoisomerization process. Details of the structure and functioning of bacteriorhodopsin are given in a number of reviews [*e.g.* 8,9,35-37], and this topic is well covered in textbooks [*e.g.* 15].

Related photoreceptor pigments serve other functions in these organisms, *e.g.* the light-driven chloride ion pump halorhodopsin. The dark-adapted form is again all-*trans* and the primary step is a photoisomerization to 13-*cis*. The principles of the structure and functioning are similar [8]. In another example, xanthorhodopsin, in the Eubacterium *Salinibacter ruber*, the participation of an intact carotenoid as a light-harvesting antenna that passes its absorbed light energy on to the retinal-protein has been reported [38] (see *Chapter 10*).

e) Chlamydomonas

In *Chapter 10*, mention was made of the phototactic response of the unicellular green alga *Chlamydomonas reinhardtii*, in which carotenoids play a role as screening pigments. The photoreceptor is now known to be a retinal-opsin protein, operating *via* a simple photocycle, and involving as the primary photochemical step a *trans* to 13-*cis* isomerization, as with bacteriorhodopsin [8,39].

C. $C_{10} - C_{15}$ Norisoprenoids

1. Formation and diversity

Oxygenase enzymes that cleave carotenoids at other positions, especially C(7,8), C(9,10) and C(11,12) to give a diversity of C_{15}, C_{13} and C_{10} norisoprenoid fragments, respectively, are widespread in plants and microorganisms (Fig. 5).

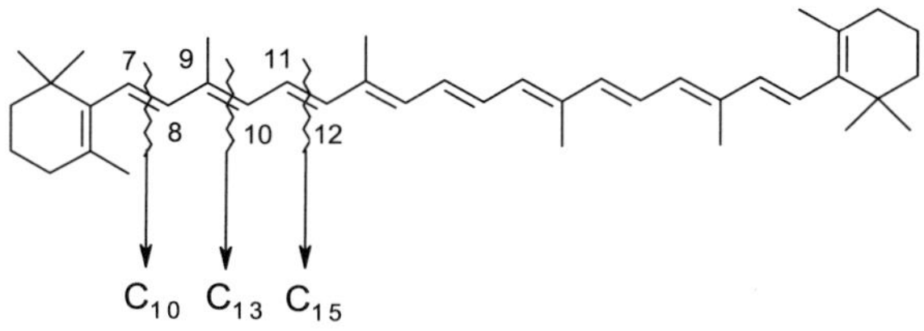

Fig. 5. Cleavage of a carotenoid molecule to give C_{10}, C_{13} and C_{15} fragments containing the carotenoid end group.

These enzymes and the cleavage processes are described in *Chapter 17*. Oxidizing conditions can also lead to unspecific, non-enzymic cleavage of the carotenoid molecule to form similar fragments. Among the large collection of carotenoid breakdown products that have been identified, many are volatile and make an important contribution to perfume and aroma; some have other specialized functions.

2. Perfume/flavour/aroma compounds

Sensory stimulants variously described as flavour, aroma, perfume or fragrance are very important in our modern way of life. Generally, many compounds contribute to the complex mixture that characterizes a particular perfume or aroma, and volatile carotenoid breakdown products can make a significant and sometimes decisive contribution. A monograph on carotenoid-derived aroma compounds gives a comprehensive treatment of the topic [40]. These compounds are extremely powerful; the fruity signature of β-ionone (*14*) is recognizable even at concentrations as low as 0.007 ppm, and β-damascenone (*15*) at even lower concentration (2ng/litre), making it the most potent of all flavour-active organic molecules. Among the most extensively studied examples are flowers of rose (*Rosa* species) and sweet olive (*Osmanthus fragrans*), which contain ionones and damascenones and related compounds, and fruit such as quince (*Cydonia oblonga*), passion fruit (*Passiflora edulis*), raspberry (*Ribes idaeus*) and grapes (*Vitis vinifera*), which contain a variety of C_{13} norisoprenoids, including ionones, ionols (*16*), theaspiranes (*17*) and edulanes (*18*). Low levels of some of these norisoprenoids are also produced from endogenous carotenoids during wine-making [*e.g.* vitispiranes (*19*), 'Riesling acetal' (*20*)] and curing of tea (theaspiranes, *17*) and tobacco (megastigmenes, *21*), and are important to the quality of the product.

Many of these compounds are important constituents of commercial perfumes and flavours.

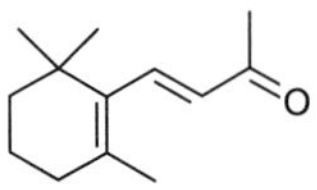

β-ionone (*14*) 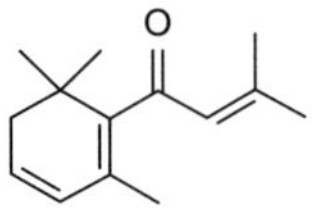   β-damascenone (*15*) 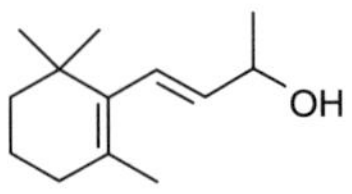   β-ionol (*16*) 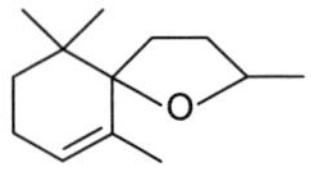   theaspirane (*17*)

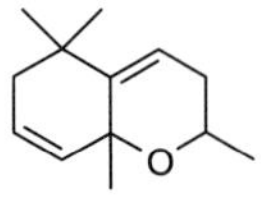

edulane (*18*) 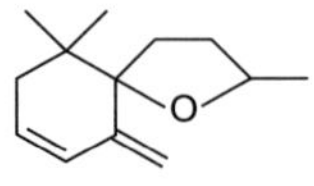   vitispirane (*19*) 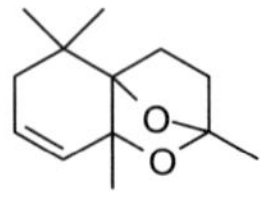   Riesling acetal (*20*) 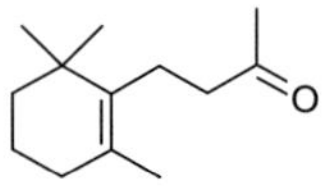   megastigmene (*21*)

More information about the structural diversity of these norisoprenoids, their formation by enzymic and non-enzymic processes and their flavour/fragrance properties is given in *Chapter 17* and in *Volume 3, Chapter 4.*

Long before man came along and learned to appreciate the aesthetic pleasures of aroma and perfume, these compounds had important ecological roles as sensory signals. The compounds that give rise to the scent of flowers can be detected at low concentration by insects, which are attracted to the flower and thus ensure pollination. Flavours and aromas that we enjoy in fruit are also appreciated by birds and other animals, thus ensuring efficient seed dispersal.

Some volatile small molecules structurally related to the end groups and other fragments of the carotenoid molecule may have pheromone (sexual signalling) properties. The relevant insect species can detect them at extremely low concentration over long distances and follow the trail to find a mate. β-Ionone (14), for example, is a strong sensory attractant for the leafminer *Liriomyza sativae* and other insects [41].

3. Abscisic acid

It is now recognized that abscisic acid (ABA, *22*) is one of the main hormone-like growth-regulating substances in plants. First identified as the substance that promotes the senescence and abscission of leaves and induces dormancy in buds and seeds, ABA is now inplicated in the regulation of a number of other processes [42,43]. Particular effects that have been reported are stimulating the closure of stomata, inducing seeds to synthesize storage proteins, and an apparent role in pathogenic defence by inducing gene transcription for protease inhibitors in response to wounding. The effects of ABA generally oppose those of the main growth-promoting hormones auxins, gibberellins and cytokinins.

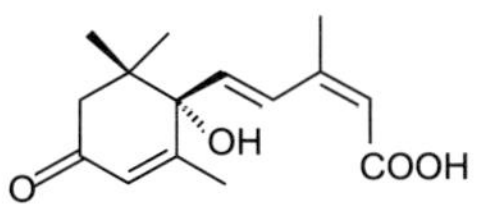

abscisic acid (22) 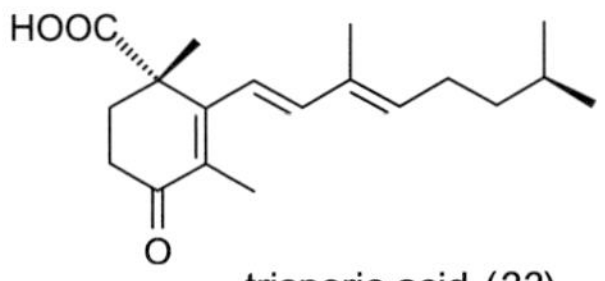   trisporic acid (23) 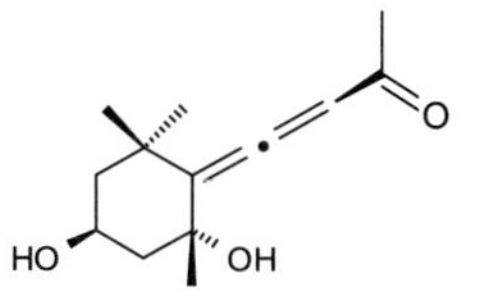   grasshopper ketone (24)

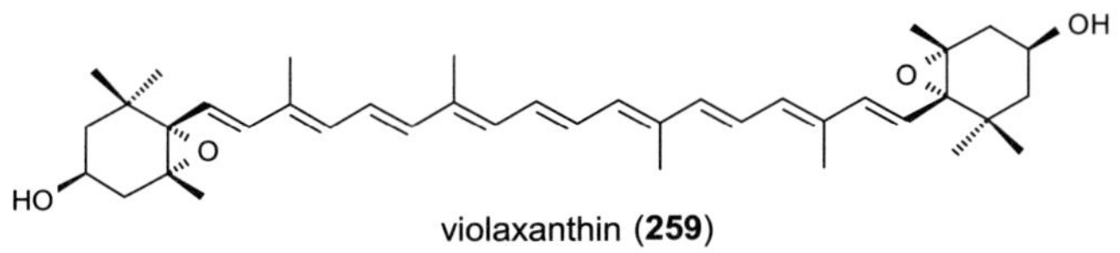

violaxanthin (**259**)

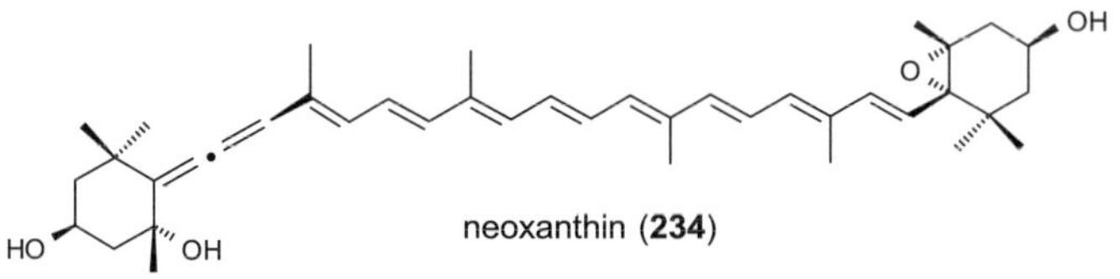

neoxanthin (**234**)

The active form of ABA is the 9-*cis* isomer, and its formation by specific enzymic cleavage of the epoxyxanthophylls (9-*cis*)-violaxanthin [(9Z)-**259**] and (9'-*cis*)-neoxanthin [(9'Z)-**234**] is described in *Chapter 17*.

4. Trisporic acid

Among compounds which function as regulators of sexual reproduction in fungi, the carotenoid metabolite trisporic acid (*23*) controls gametogenesis in the mould *Blakeslea trispora* and in other mucoraceous genera such as *Mucor* and *Phycomyces* [44]. Its main physiological role is in eliciting morphological changes that occur during the sexual process, especially zygophore formation. Mating of the (+) and (-) sexual forms of these heterothallic fungi frequently results in an increased synthesis of β-carotene (**3**). This is the basis of the large-scale commercial production of β-carotene by *B. trispora* (see *Volume 5, Chapter 5*). The biosynthesis of trisporic acid from β-carotene *via* retinol, involves enzymes from both the (+) and (-) strains, and is outlined in *Volume 3, Chapter 2*.

5. Grasshopper ketone

The 'grasshopper ketone' was identified in the frothy liquid exuded from a species of grasshopper (*Romalea microptera*) when the insect is disturbed [45]. It was reported to have a repellent effect on ants and other predators. The structure of grasshopper ketone (*24*) is effectively that of the C$_{13}$ allenic end group of neoxanthin (**234**), and its formation from neoxanthin in plants has been demonstrated (see *Chapter 17*).

D. Polymers: Sporopollenin

Oxidizing conditions that lead to peroxidation and cleavage of the carotenoid polyene chain can also lead to cross-linking and polymerization of the carotenoid molecules, or to co-polymerization with other compounds. The natural material thus produced, *e.g.* the outer layer (exine) of pollen grains and algal and fungal spores, is an inert polymer known as sporopollenin [46-48]. This material is very difficult to work with and details of the structure have not been elucidated but evidence from studies with *Lilium henryii* plants supports the idea that the material of pollen exine is in large part a polymer formed by oxidative copolymerization of carotenoids and carotenoid acyl esters with polyphenolic compounds [47]. The deposition of the exine during pollen formation was preceded and accompanied by a dramatic increase in the biosynthesis and concentration of carotenoids and their esters in the anthers. Considerable similarities have been noted between sporopollenin and products of chemical polymerization of carotenoids.

Sporopollenin has been described as the strongest known biological 'plastic' and renders pollen, spores *etc.* highly resistant to microbial and chemical destruction.

E. Other Carotenoid Breakdown Products and Human Health

Oxidative degradation of carotenoids, enzymic and non-enzymic, gives rise to many products, depending on conditions. Some such breakdown can occur in fruit and vegetables, especially during processing and cooking. Oxidizing free radicals can cause similar breakdown of carotenoids in the body. Even though they are present in very small amounts, there is discussion about whether some of the products may have biological activity and effects on human health. There are reports that some apocarotenals and smaller fragments can strongly affect molecular processes associated with serious diseases and conditions, and could be the compounds actually responsible for some of the biological actions reported for carotenoids. Aspects of this are discussed in various *Chapters* in *Volume 5*.

References

[1]   M. B. Sporn, A. B. Roberts and D. S. Goodman, (eds.), *The Retinoids: Biology, Chemistry and Medicine, 2nd Edn.*, Raven Press, New York (1994).

[2]   R. Blomhoff (ed.), *Vitamin A in Health and Disease*, Dekker, New York (1994).

[3]   L. Packer, U. Obermüller-Jevic, K. Kraemer and H. Sies (eds.), *Carotenoids and Retinoids: Molecular Aspects and Health Issues*, AOCS Press, Champaign, IL (2004).

[4]   B. Hoyos and U. Hammerling, in *Carotenoids and Retinoids: Molecular Aspects and Health Issues* (ed. L. Packer, U. Obermüller-Jevic, K. Kraemer and H. Sies), p. 42, AOCS Press, Champaign, IL (2004).

[5]   P. Kastner, P. Chambon and M. Leid, in *Vitamin A in Health and Disease* (ed. R. Blomhoff), p. 189, Dekker, New York (1994).

[6]   D. J. Mangelsdorf, K. Umesono and R. M. Evans, in *The Retinoids: Biology, Chemistry and Medicine, 2nd Edn.*, (ed. M. B. Sporn, A. B. Roberts and D. S. Goodman), p. 319, Raven Press, New York (1994).

[7]   P. Chambon, *FASEB J.*, **10**, 940 (1996).

[8]   K. Nakanishi, *Pure Appl. Chem.*, **63**, 161 (1991).

[9]   W. J. de Grip and A. Watts (eds.), *Biophys. Chem., Special Issue*, **56**, 1 (1995).

[10] M. G. Holmes (ed.), *Photoreceptor Evolution and Function*, Academic Press, London (1991).

[11] D. J. Cosens and D. Vince-Price (eds.), *The Biology of Photoreception*, Cambridge University Press, Cambridge (1983)

[12] K. Nakanishi, *Pure Appl. Chem.*, **57**, 769 (1985).

[13] K. Nakanishi, A.-H. Cheng, F. Derguini, P. Franklin, S. Hu and J. Wang, *Pure Appl. Chem.*, **66**, 981 (1994).

[14] W. J. de Grip, F. de Lange, P. Bovee, P. J. E. Verdegem and J. Lugtenburg, *Pure Appl. Chem.*, **69**, 2091 (1997).

[15] L. Stryer, *Biochemistry, 4th Edn.*, p. 318, 332, Freeman, New York (1995).

[16] N. A. Campbell and J. B. Reece, *Biology, 6th Edn.*, p. 1063, Benjamin Cummings, San Francisco (2002).

[17] C. W. Oyster, *The Human Eye: Structure and Function*, Sinauer, Sunderland (1999).

[18] G. Wald, *Nature*, **219**, 800 (1968).

[19] C. D. B. Bridges, in *Handbook of Sensory Physiology, Vol. II/I: The Photochemistry of Vision* (ed. H. J. A. Dartnall), p. 417, Springer, Berlin (1972).

[20] K. Vogt and K. Kirschfeld, *Naturwiss.*, **71**, 211 (1984).

[21] S. Matsui, M. Seidou, I. Uchiyama, N. Sekiya, K. Hiraki, K. Yoshihara and Y. Kito, *Biochim. Biophys. Acta*, **966**, 370 (1988).

[22] P. A. Hargrave, H. E. Hamm and K. P. Hoffmann, *Bioessays*, **15**, 43 (1993).

[23] K. Nakanishi, H. Zhang, K. A. Lerro, S.-I. Takekuma, T. Yamamoto, T. H. Lien, L. Sastry, D.-J. Baek, C. Moquin-Pattey, M. F. Boehm, F. Derguini and M. A. Gawinowicz, *Biophys. Chem.*, **56**, 13 (1994).

[24] S. T. Menon, M. Han and T. P. Sakmar, *Physiol. Rev.*, **81**, 1659 (2001).

[25] R. W. Schoenlein, L. A. Peteanue, R. A. Mathies and C. V. Shank, *Science*, **254**, 412 (1991).

[26] R. R. Birge, *Biochim. Biophys. Acta*, **1016**, 293 (1990).

[27] Y. Shichida, S. Matuoka and T. Yoshizawa, *Photobiochem. Photobiophys.*, **7**, 221 (1984).

[28] P. A. Hargrave and J. H. McDowell, *Int. Rev. Cytol.*, **137B**, 49 (1992).

[29] R. R. Rando, *Biochemistry*, **30**, 595 (1991).

[30] R. R. Rando, *Pure Appl. Chem.*, **66**, 989 (1994).

[31] P. Chen, W. Hao, L. Rife, X. P. Wang, D. Shen, J. Chen, T. Ogden, G. B. van Boemel, L. Wu, M. Yang and H. F. W. Fong, *Nature Genet.*, **28**, 256 (2001).

[32] D. R. Pepperberg and R. K. Crouch, *Lancet*, **358**, 2098 (2001).

[33] T. Seki, *J. Gen. Physiol.*, **84**, 49 (1984).

[34] J. K. Bowmaker and H. J. Dartnall, *J. Physiol.*, **298**, 501 (1980).

[35] R. A. Mathies, S. W. Lin, J. B. Ames and W. T. Pollard, *Ann. Rev. Biophys. Chem.*, **20**, 491 (1991).

[36] H. G. Khorana. *J. Biol. Chem.*, 263, 7439 (1988).

[37] D. Oesterhelt, C. Bräuchle and N. Hampp, *Quart. Rev. Biophys.*, **24**, 425 (1991).

[38] S. P. Balashov, E. S. Imasheva, V. A. Boichenko, J. Anton, J. M. Wang and J. K. Lanyi, *Science*, **309**, 2061 (2005).

[39] R. D. Smyth, J. Saranak and K. W. Foster, *Progr. Phycol. Res.*, **6**, 255 (1988).

[40] P. Winterhalter and R. L. Rouseff (eds.), *Carotenoid-derived Aroma Compounds*, American Chemical Society, Washington (2002).

[41] M. Wei, X. Deng and J. Du, *Ying Yong Sheng Tai Xue Bao*, **16**, 907 (2005).

[42] F. T. Addicott (ed.), *Abscisic Acid*, Praeger, New York (1983).

[43] D. C. Walton and Y. Li, in *Plant Hormones* (ed. P. J. Davies), p. 140, Kluwer, Dordrecht (1995).

[44] J. D. Bu'Lock, in *Biosynthesis of Isoprenoid Compounds, Vol. 2* (ed. J. W. Porter and S. L. Spurgeon), p. 437, Wiley, New York (1983).

[45] J. Meinwald, K. Erickson, M. Hartshorn, Y. C. Meinwald and T Eisner, *Tetrahedron Lett.*, 2959 (1968).

[46] J. Brooks and G. Shaw, *Nature*, **220**, 678 (1968).

[47] G. W. Gooday, *Ann. Rev. Biochem.*, **43**, 35 (1974).

[48] W. J. Guilford, D. M. Schneider, J. Labovitz and S. J. Opella, *Plant Physiol.*, **86**, 134 (1988).

Carotenoids
Volume 4: Natural Functions
© 2008 Birkhäuser Verlag Basel

Chapter 16

Cleavage of β-Carotene to Retinal

Adrian Wyss and Johannes von Lintig

A. Introduction

Elucidating the physiological roles played by vitamins has always been a major goal of nutritionists and biochemists. In humans, vitamin A deficiency disorder (VADD) in milder forms leads to night blindness, whilst more severe progression can lead to corneal malformations, *e.g.* xerophthalmia (See *Volume 5, Chapters 8* and *9*). This deficiency also affects the immune system, leads to infertility and causes malformations during embryogenesis. The molecular basis for these diverse effects lies in the dual role of vitamin A (retinol, *1*) derivatives. In all visual systems, retinal (*2*), or a closely related compound such as 3-hydroxyretinal (*3*), is the chromophore of the visual pigments (*e.g.* rhodopsin) [1,2]. In vertebrates, the derivative retinoic acid (RA, *4*) is a major signalling molecule that controls a wide range of processes. Retinoic acid is the ligand of the nuclear retinoic acid receptors (RARs) and retinoid X receptors (RXRs) [3-6] (see *Chapter 15*).

Vitamin A deficiency disorder is still a major problem, particularly in developing countries, leading as it does to blindness and childhood mortality [7]. The requirement for vitamin A can be satisfied by natural foods of either animal or plant origin. Certain animal fats were shown to contain a factor that was essential for the growth of rats and also cured eye disorders [8,9]. Later work revealed that certain yellow plant pigments had the same activity. Carotenes, but not xanthophylls, induced growth in deficient rats [10]. This phenomenon was explained in 1930 by Moore [11] who described a conversion of β-carotene (*3*) into vitamin A in the small intestine of the rat, thus providing the first evidence that a plant-derived carotenoid is the direct precursor of vitamin A in animals. Karrer [12] elucidated the structure of β-carotene and proposed a central cleavage of the C(15,15') double bond of β-carotene to form two molecules of vitamin A. In 1954, Glover proposed that an excentric cleavage process also exists, followed by a stepwise β-oxidation-like process, which leads ultimately only to one molecule of vitamin A per molecule of carotene consumed [13].

The primary source of vitamin A is the plant-derived provitamin A carotenoids, especially β-carotene (3), α-carotene (7) and β-cryptoxanthin (55). The central and excentric cleavage pathways and the subsequent formation of retinol and retinoic acid are illustrated in Fig. 1.

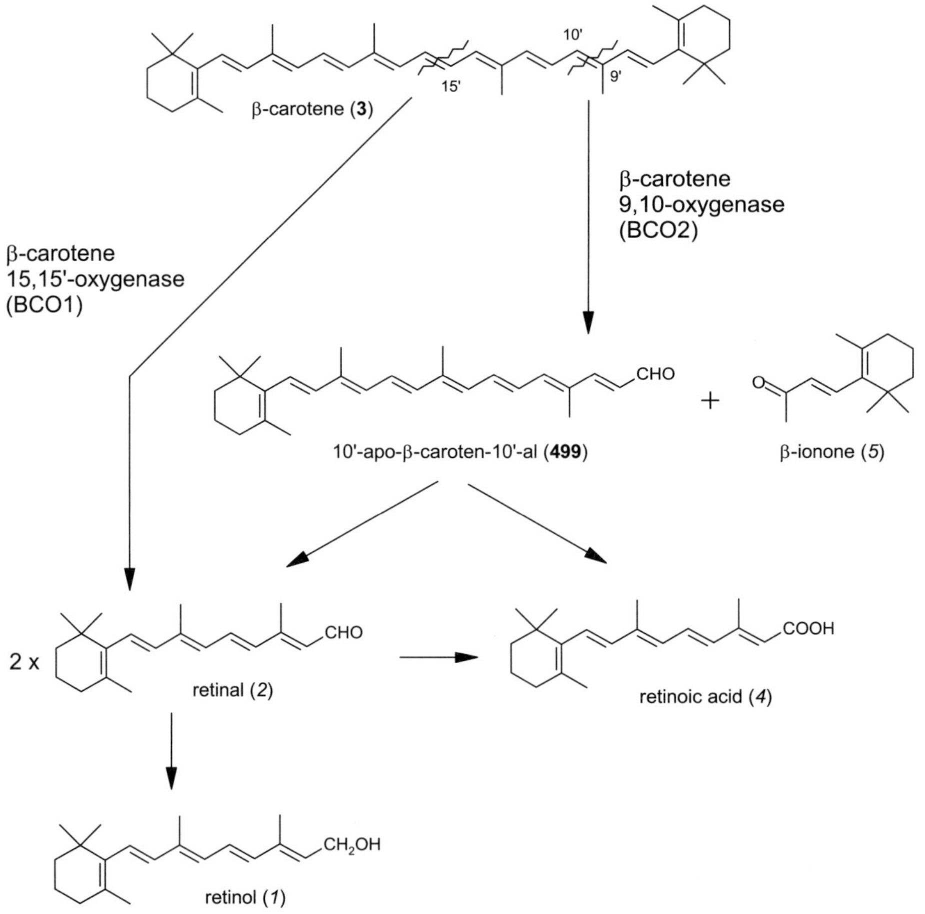

Fig. 1. Pathways for central cleavage of β-carotene (3) leading to two molecules of retinal (2), and excentric cleavage to 10'-apo-β-caroten-10'-al (499) and β-ionone (5). Similar cleavage of other double bonds gives apocarotenals of different chain length. Retinal is either reduced to retinol (1) which is stored in the liver as retinyl esters, or oxidized to retinoic acid (4). Apo-β-carotenals can be shortened to retinal or to retinoic acid in a process similar to the β-oxidation of fatty acids.

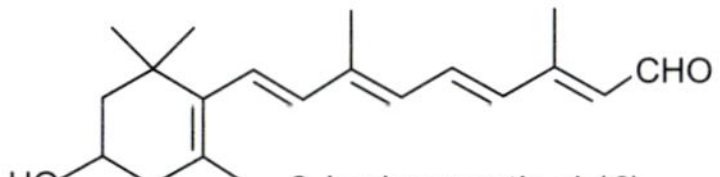

3-hydroxyretinal (*3*)

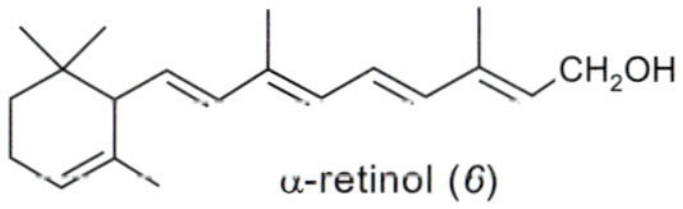

α-retinol (*6*)

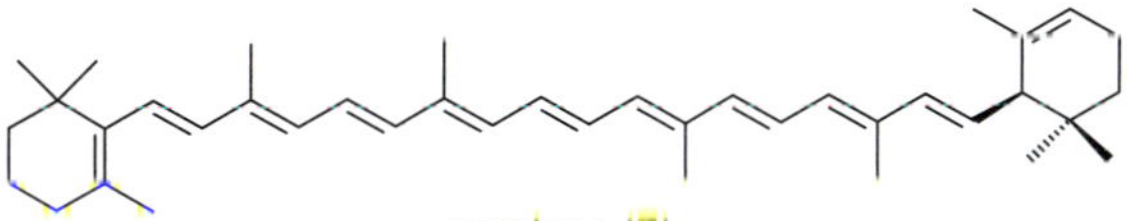

α-carotene (*7*)

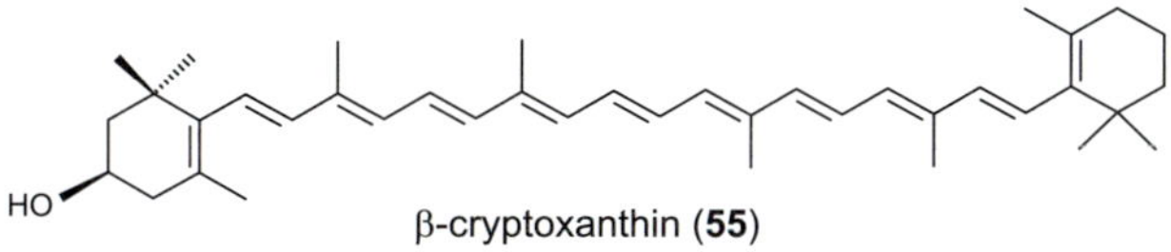

β-cryptoxanthin (**55**)

The long and controversial debate about the relative importance of the two cleavage pathways was resolved fifty years later by the molecular identification of enzymes for both the central and the excentric cleavage of β-carotene. Thanks to new methods in molecular biology, biochemistry and analysis, cleavage enzymes from various species and with different substrate specificity could be identified. These enzymes constitute a new enzyme family with common biochemical and structural features.

This Chapter covers the enzymic cleavage of β-carotene to retinal, mainly in animals. Plant enzymes that cleave carotenoids at other positions in the chain are described in *Chapter 17*.

B. Pioneering Work

The presence of a β-carotene 15,15'-oxygenase in cell-free systems was first demonstrated independently in two studies in 1965. In one, the enzymic cleavage of β-carotene by a soluble enzyme preparation from rat liver and intestine was demonstrated [14]; the other [15] described the same enzyme from cell-free homogenates of rat intestinal mucosa. Both reports claimed a central cleavage with a yield of approximately two molecules of retinal per molecule of β-carotene. Retinal was the only product found in these experiments. Sensitive HPLC analysis, which might have enabled detection of other minor products, was not available at that time. Due to the requirement for molecular oxygen, the enzyme was termed β-carotene 15,15'-dioxygenase. It was almost 40 years before strong evidence was obtained for a mono-oxygenase mechanism [16]. Throughout this Chapter, therefore, this and related enzymes are referred to simply as oxygenases, *e.g.* β-carotene 15,15'-oxygenase (BCO1).

The β-carotene 15,15'-oxygenase was shown to be a soluble, cytosolic enzyme [14,15,17,18], though another report [19] suggests that the enzyme activity is associated with a high-molecular weight lipid-protein aggregate fraction. The enzyme required molecular oxygen and was inhibited by ferrous ion, chelating agents and sulphydryl-binding reagents. With β-carotene as substrate, retinal was the only product that could be identified. The K_m value of the enzyme was estimated [20] to be in the range of 2-10 μM and the pH optimum around 8.0.

In several animal species studied, the highest β-carotene 15,15'-oxygenase activity was found in the intestinal mucosa, with a gradient of decreasing activity from the duodenum to the colon [21,22]. Activity was also detected in liver, lung, kidney and brain [23]. Guinea-pigs [24], rabbits [25], and rats [26] showed the highest levels, but activity was also detected in intestinal mucosa of pigs [27], monkeys, chicken, fish and turtles [28]. In humans, only low levels of the enzyme were described [29], but in the cat, a carnivorous species, no enzymic activity could be detected [28]. Consistently, it has been reported that cats develop VADD when they are fed a diet lacking vitamin A, but supplemented with β-carotene [30]. Cats, therefore, strictly depend on pre-formed dietary retinoids to meet their demand for vitamin A.

Many attempts were made to purify or enrich the enzyme. In the earlier work [14,15,17,18], purification by a factor of 20-30 was achieved. Refinement of the purification protocol, to include ammonium sulphate fractionation, heat treatment and acetone precipitation, gave a 238-fold improvement of specific activity [28]. Treatment with protamine sulphate, before ammonium sulphate fractionation and acetone precipitation, led to a purification factor of 38 and a quite high specific activity of 3300 pmol retinal/h/mg protein [31]. Other work, however, reported substantial losses in almost every purification step [24], and inclusion of PMSF (phenylmethanesulphonyl fluoride) and mercaptoethanol in the buffers did not overcome these problems; a purification factor of only 2 was reached.

C. Cloning the Genes Coding for the Cleavage Enzymes

The first molecular identification of a carotenoid cleavage enzyme, the plant enzyme VP14, was achieved in 1997 [32] by analysis of the abscisic acid (ABA)-deficient maize mutant *viviparous 14* (*vp14*) (see *Chapter 17*). The gene product VP14 catalyses the oxidative cleavage of 9-*cis* epoxycarotenoids to form xanthoxin, the direct precursor of ABA, an important plant hormone and growth factor (*Chapter 15*). It was proposed that there may be similarities between this enzyme and those involved in carotenoid metabolism in animals.

In 2000, two research groups independently cloned the genes coding for the key enzyme in vitamin A formation [33-35]. One approach [34] relied on detecting areas of sequence conservation between the animal enzyme BCO1 and the plant carotenoid-cleaving enzyme

VP14. An expression strategy in an *Escherichia coli* strain, genetically engineered to contain all the enzymes needed to synthesize β-carotene *de novo*, allowed a β-carotene 15,15'-oxygenase from the fruit fly *Drosophila melanogaster* to be identified. The purified recombinant (cloned) carotene oxygenase catalysed exclusively the central cleavage of β-carotene (C_{40}) to yield retinal (C_{20}) and depended on ferrous ion as cofactor [34]. Direct genetic evidence that this enzyme catalyses the key step in vitamin A formation was provided by mutant analysis. Among the various available *Drosophila* mutants affected in visual performance, the *ninaB* mutant lacks the visual pigment chromophore, when raised on standard media with carotenoids as the sole source of vitamin A. The *ninaB* mutation has been cytologically mapped in the *Drosophila* genome on chromosome 3 at position 87E-F [36], coinciding with the physical location of the gene encoding the β-carotene 15,15'-oxygenase. Analysis of the molecular basis of the blindness in *ninaB* mutants showed that this phenotype is due to mutations in the *bco1/ninaB* gene, thus unequivocally demonstrating that the gene product BCO1/NinaB does catalyse vitamin A formation *in vivo* [37].

Confirmation followed that the first step in vitamin A metabolism in metazoans is generally catalysed by this type of enzyme [33,38]. The first β-carotene 15,15'-oxygenase from higher animals to be cloned was that from chicken [33]. Enriched enzyme fractions were obtained from chicken intestinal mucosa extracts so that partial amino acid sequences could be determined by MALDI-TOF (Matrix-Assisted Laser Desorption Ionization-Time of Flight) mass spectrometry. Two peptide sequences were subsequently used to obtain cDNA sequences for the chicken β-carotene 15,15'-oxygenase gene by reverse genetics. Following one round each of PCR (polymerase chain reaction) and RT-PCR (reverse transcriptase polymerase chain reaction), a cDNA band of 597 bp was obtained and used to screen a cDNA library from chicken duodenum. A full-length cDNA of 3.1 kb, encoding a 60.4 kDa protein, was successfully cloned and expressed in *E. coli*. The recombinant protein was tested for cleavage activity by incubation with β-carotene *in vitro*. Retinal was the only reaction product in the oxygenase activity assay; no other apo-β-carotenals were detected.

Later, cloning and characterization of the homologue from mouse was achieved [38-40] and the first cloning of the human β-carotene 15,15'-oxygenase gene from an RPE (retinal pigment epithelium) cDNA library was reported [41]. The allele was characterized on chromosome 16 and all the exon-intron boundaries were sequenced. The correct identification of the enzyme was demonstrated in an oxygenase activity assay. A thorough biochemical analysis of the recombinant human β-carotene 15,15'-oxygenase was published in 2002 [42]. Human BCO1 was expressed as hexa-HIS-tagged protein in the baculovirus system, the assay for this enzyme *in vitro* was optimized by testing different detergents and sulphydryl reducing agents, and kinetic and inhibition studies *in vitro* were reported (Section **F**).

D. Identification of an Excentric Cleavage Enzyme

In *Drosophila*, only one gene for a member of the polyene chain oxygenase family is found in the entire genome, namely the *ninaB* gene which encodes BCO1. In vertebrates, however, besides the β-carotene 15,15'-oxygenase, another protein with significant sequence identity has been described, namely RPE65, which is exclusively expressed in the retinal pigment epithelium (RPE) [43,44]. The physiological role of RPE65 in the visual cycle (*Chapter 15*) was elucidated recently; several groups reported an isomerohydrolase activity of this protein [45-47]. A search of mammalian EST (expressed sequence tags) databases revealed an EST-fragment, from the mouse, with significant peptide sequence similarity to both RPE65 and BCO1. This was not identical with the mouse RPE65 or BCO1, however, and thus represented a candidate cDNA for an enzyme catalysing the excentric oxidative cleavage of carotenoids [48]. To obtain its full-length cDNA, up-stream primers deduced from the EST-fragment were designed. Then, RACE (Rapid Amplification of cDNA Ends)-PCR was performed on a total RNA preparation and the resulting fragment was cloned. Sequence analysis revealed that the cDNA encoded a protein of 532 amino acids, and the deduced amino acid sequence shared *ca.* 40% identity with the mouse β-carotene 15,15'-oxygenase. From a commercial cDNA library, the human homologue was identified and cloned. Thus, in mammals, a third type of polyene chain oxygenase [BCO2 (β-carotene-9,10-oxygenase)] exists besides BCO1 and RPE65.

For functional characterization, BCO2 of the mouse was expressed as a recombinant protein in *E. coli* and tested *in vitro* for enzymic activity. HPLC analysis revealed that no retinoids were formed from β-carotene. However, a compound was detected with a UV/Vis spectrum consistent with an apo-β-carotenal structure. To obtain larger amounts of this material for further analysis, the *E. coli* engineered to produce β-carotene was used, with BCO1 from mouse as a control. Whilst the carotenogenic *E. coli* strain expressing the BCO1 from mouse became white because of cleavage of endogenous β-carotene to retinal, in the *E. coli* strain expressing the BCO2 no such pronounced colour change occurred. The β-carotene content of the *E. coli* strain expressing the BCO2 was significantly reduced, however, compared to a control strain. The carotenoids were extracted and subjected to HPLC analysis. Besides β-carotene, 10'-apo-β-caroten-10'-al (**499**, Fig. 1) and significant amounts of 10'-apo-β-caroten-10'-ol (**495.1**) were detected and identified by their UV/Vis spectra and by LC-MS. Thus, 10'-apo-β-caroten-10'-al is formed from β-carotene. Oxidative cleavage of β-carotene at the C(9,10) double bond should also produce β-ionone (*5*), but this was not detectable by HPLC. Analysis of the bacterial growth medium by GC-MS after solid phase extraction of lipophilic compounds allowed a significant amount of β-ionone to be detected. The BCO2, therefore, catalyses the excentric cleavage of β-carotene at the C(9,10) double bond, resulting in the formation of 10'-apo-β-caroten-10'-al (**499**) and β-ionone. Note that, according to the rules of carotenoid nomenclature, this enzyme (BCO2) should be termed β-carotene-9,10-oxygenase, though the reaction product is 10'-apo-β-caroten-10'-al (**499**).

In a test with a lycopene-accumulating strain of *E. coli*, the recombinant enzyme also catalysed the oxidative cleavage of lycopene (**31**). The cleavage of lycopene and the formation of apolycopenals are indicative of a putative role in vertebrate physiology. In vertebrates, several nuclear receptors with unknown ligands (orphan receptors) exist [49]. Besides being precursors for RA formation in the case of β-carotene cleavage, it may be speculated that the compounds formed by the excentric cleavage reaction of β-carotene and/or lycopene could also represent putative ligands for some of these orphan receptors.

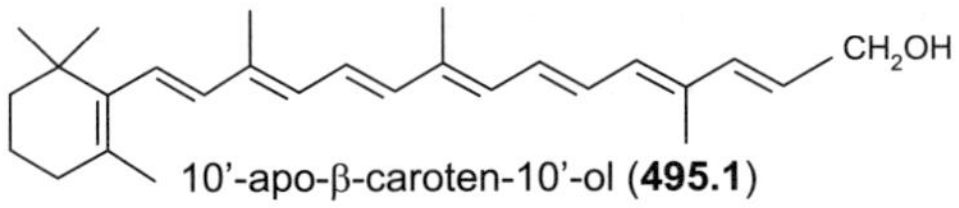

10'-apo-β-caroten-10'-ol (**495.1**)

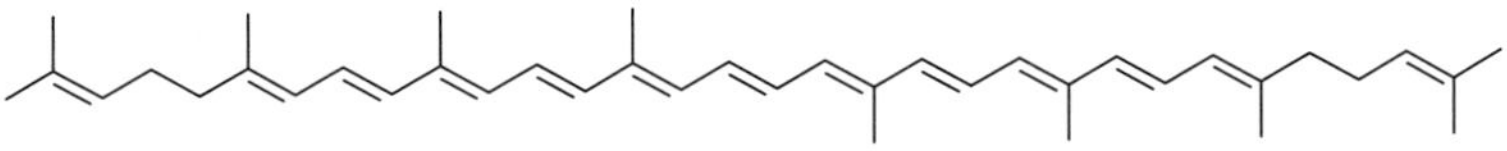

lycopene (**31**)

E. A Novel Family of Sequence-related and Structure-related Non-haem Iron Oxygenases

After the first production of the plant carotenoid cleavage enzyme, VP14, by cloning, it became clear that it showed similarity, on the level of the deduced amino acid sequence, with other proteins in the databases, including RPE65 [44,45,50]. Mutations in the gene*RPE65* are associated with specific forms of blindness in humans, such as Leber's congenital disease, autosomal recessive retinitis pigmentosa and rod-cone dystrophy [51,52]. A direct involvement of RPE65 in the visual cycle of the eyes (see *Chapter 15*) was demonstrated by the analysis of *Rpe65*-null mice, which lack rhodopsin despite the presence of the opsin apoprotein in the rod outer segments. In the eyes of these mice, (11-*cis*)-retinal is not detectable; instead, there is an accumulation of (all-*trans*)-retinyl esters [50], intermediates in the cycle that regenerates (11-*cis*)-retinal in the RPE. (all-*trans*)-Retinyl esters would be the substrate for the proposed isomerohydrolase which was believed to catalyse the key step in the regeneration of the visual pigments in a combined isomerization and ester-hydrolase reaction in the RPE [53]. In 2005, it was shown that RPE65 is identical with this long-sought after isomerohydrolase [46,47,54].

When the symmetrical and excentric β-carotene cleavage enzymes from animals were obtained by cloning [48], it became evident that they are members of a novel but ubiquitous protcin family of non-haem iron oxygenases. Comparison of the deduced amino acid sequences revealed that BCO1, BCO2 and RPE65 share approximately 40% sequence identity within a species. Furthermore, these animal enzymes share structural similarities with a large subset of plant and bacterial enzymes that modify carotenoids and apocarotenoids (Fig. 2).

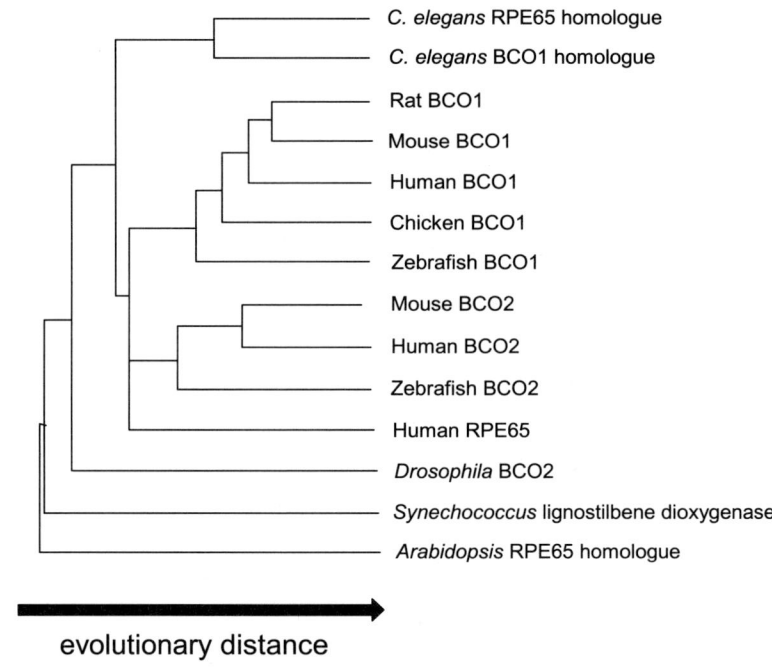

evolutionary distance

Fig. 2. Phylogenetic tree with important members of the new family of double-bond cleavage enzymes. The dendrogram was created by use of the PileUp program, included in the GCG software package (University of Madison, WI). The distance along the horizontal axis is proportional to the differences between the sequences of the various species. The shorter the distance between two species, the higher the homology between the two carotenoid cleavage enzymes. The BCO1 homologues build a sub-group, as do the excentric cleavage enzymes.

Various amino acids, *e.g.* histidine residues, are conserved in position throughout the sequences, suggesting conserved binding motifs for carotenoids or cofactors, as well as a conserved catalytic centre which has not changed significantly during evolution. In plants, besides VP14, other members of this oxygenase family, that cleave various carotenoids at C(5,6), C(7,8) or C(9,10), have been identified. These are described in *Chapter 17*.

Another family member to be cloned and characterized is one from the cyanobacterium *Synechocystis*; it converts apo-β-carotenoids into retinal [55]. The structure of this enzyme was recently resolved by X-ray diffraction at a resolution of 2.4 Å [56]. The active centre is confirmed as an octahedral arrangement of four histidines holding one Fe^{2+} ion. This centre lies at the axis of a 'seven-bladed β-propeller' structure (Fig. 3). With two exceptions, each blade consists of four antiparallel strands. The protein consists of a hydrophobic tunnel which is perpendicular to the propeller axis and which holds the substrate in place. This three-dimensional structure is likely to be representative of the whole family of double-bond cleavage enzymes. The role of conserved histidine residues for the binding of the cofactor, ferrous ion, was recently confirmed for BCO1 and RPE65 by a mutagenesis approach [54,57].

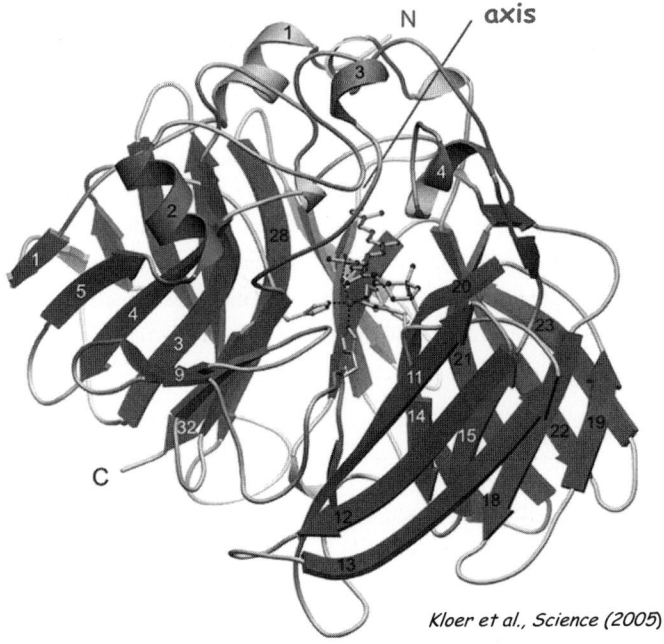

Kloer et al., Science (2005)

Fig. 3. Ribbon plot of the ACO chain fold consisting of 490 residues with a mass of 54,286 daltons. The N-terminal eleven residues are disordered. The seven-bladed β-propeller has the usual topology. The four histidines holding the Fe^{2+} ion in an octahedral arrangement are shown together with two water molecules and the substrate as ball-and-stick models. The view is along the active centre tunnel, the rear exit of which is marked by a straight line. Strands β1 and β2 (light grey) are additions to the common propeller. (From [56], with permission).

F. Biochemical Properties of β-Carotene 15,15'-Oxygenase 1 (BCO1)

Although early work with the mammalian β-carotene oxygenases used only cell-free homogenates, some of the characteristics and predictions described for the enzyme are still valid, *e.g.* that it is a soluble cytosolic enzyme with a metal-containing active centre. Later, after the genes had been cloned and expressed in heterologous systems, extensive biochemical characterization was undertaken of the recombinant human BCO1, expressed in a baculovirus system [42]. Of several detergents tested in the incubations, 1-*S*-octyl-β-D-thiogluco-pyranoside (1%) gave the highest enzyme activity. Among many sulphydryl reducing agents, tris(2-carboxyethyl)phosphine hydrochloride (0.5 mM) was the most effective. The pH optimum was 7.8, in good agreement with all previously reported data. Metal-chelating agents and sulphydryl alkylating agents inhibited the enzyme in a dose-dependent manner. Classical gel filtration chromatography and use of the same buffer conditions as for the activity measurement indicated that the active enzyme is a tetramer in solution. Intracellular localization studies confirmed that the β-carotene 15,15'-oxygenase is a cytosolic enzyme; immunostaining gave no hint of a localization in the nucleus or in other organelles.

Finally, a kinetic study with β-carotene as substrate gave a K_m of 7.1 ± 1.8 μM and V_{max} 10.4 ± 3.3 nmol retinal/mg/min, and a conversion rate (yield of retinal after cleavage) of 100%. The enzyme also used β-cryptoxanthin (**55**) as a substrate, with K_m 30 ± 3.8 μM, V_{max} 0.9 ± 0.2 nmol retinal/mg/min, and conversion rate 40%. Neither zeaxanthin (**119**) nor lycopene (**31**) was accepted as substrate under these conditions. This supports the long-held view that, to be a substrate, the carotenoid must have at least one unsubstituted β ring.

G. Mechanisms of Cleavage

1. Central *versus* excentric cleavage

There has been long debate about the relative importance of the two possible cleavage pathways leading to vitamin A [58]. Despite the evidence for an excentric cleavage pathway [59,60], the central cleavage mechanism leading to the formation of two molecules of retinal was the more widely accepted pathway for retinoid formation, until the production of apocarotenals and retinoic acid as major products of β-carotene cleavage *in vivo* and *in vitro* was demonstrated [29,61,62]. It was proposed that the apocarotenals formed by the excentric cleavage are subsequently shortened through a β-oxidation-like mechanism, resulting in the formation of one molecule of retinoic acid (*4*) per molecule of β-carotene.

With the molecular characterization of β-carotene 9,10-oxygenase, both pathways have been demonstrated in vertebrates, so the debate about an excentric *versus* a central β-carotene cleavage mechanism has been settled. Depending on the specific tissue, however, there may be a preference for one pathway over the other or even the interaction of both enzymes.

2. Dioxygenase or monooxygenase?

Early experimental work showed that (i) molecular oxygen was required for enzyme activity, (ii) the two hydrogen substituents at the central C(15,15') double bond were retained in the product, retinal and (iii) the molar ratio of retinal formed from β-carotene was approximately 2:1. This led to the conclusion that the enzyme is a dioxygenase, and it was termed β-carotene 15,15'-dioxygenase (EC 1.13.11.21) [14,63]. After the enzyme became available in an enriched form from chicken duodenum, it was possible to reveal that it is a monooxygenase rather than a dioxygenase [16]. In a double-labelling experiment it was shown that both ^{17}O (derived from $^{17}O_2$) and ^{18}O (derived from $H_2^{18}O$) were incorporated into retinal, The crucial step in this experiment, however, was the choice of an asymmetric carotenoid, namely α-carotene (β,ε-carotene, 7) as an enzyme substrate. This allowed two different products to be distinguished. The two products, retinol (1) and α-retinol (6), each contained 48-52% ^{17}O and 41-44% ^{18}O, a result which is in good agreement with a monooxygenase reaction mechanism. The first step is an epoxidation of the central double bond, followed by an unselective ring opening with water and a final diol cleavage to give two aldehyde molecules. A dioxygenase mechanism would require the incorporation of one complete oxygen molecule into the two retinals and the absence of any ^{18}O label (from ^{18}O water).

In contrast, a dioxygenase mechanism has been demonstrated recently for *At*CCD1 (a carotenoid cleavage enzyme from the plant *Arabidopsis thaliana*) [64].

H. Regulation

Before the molecular characterization of the enzymes, most publications on regulation were limited to the enzyme activity. Later, a few papers dealt with the regulation at the transcriptional and translational level. The influence of nutritional factors on the conversion of β-carotene into vitamin A is discussed in *Volume 5, Chapter 8*. Vitamin A status regulates the enzyme activity, which is increased by vitamin A deficiency but down-regulated by supplementation with vitamin A [65]. Similar findings were reported later [66], when the effect of vitamin A on absorption and cleavage of β-carotene was published.

Several research groups investigated the effect of other carotenoids on enzyme activity [21,67,68]. Lutein (133), lycopene (31), 15,15'-dihydro-β-carotene (7) and, to a lesser extent, astaxanthin (406) inhibited the carotene cleavage activity of BCO1. Whereas the inhibition by lycopene and 15,15'-dihydro-β-carotene is competitive, the inhibition by lutein and astaxanthin includes a non-competitive component. The interaction with other dietary ingredients or micronutrients was also studied.

The nutritional status, especially the vitamin A level, is a key factor that influences the conversion of β-carotene into retinal (*Volume 5, Chapter 8*). Rats with low vitamin A plasma concentrations showed higher BCO1 activity (14.9 ± 2.43 pmol/h/mg protein *versus* 6.3 ± 1.37 pmol/h/mg protein for normal vitamin A status) [69]. Restricted protein in the diet led to a decrease of BCO1 activity (3.6 ± 1.30 pmol/h/mg protein *versus* 13.8 ± 1.60 pmol/h/mg protein for a protein-sufficient diet).

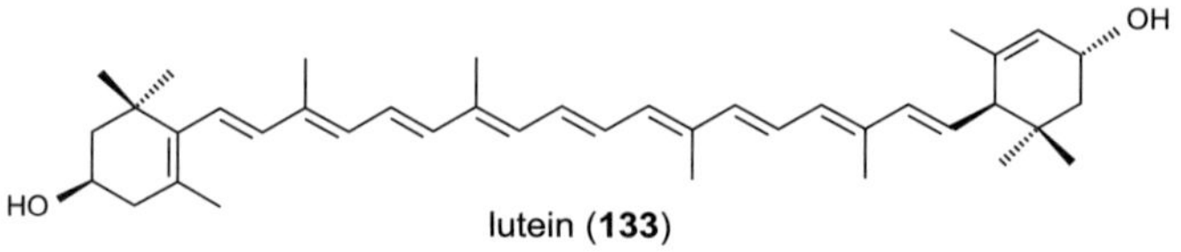

lutein (**133**)

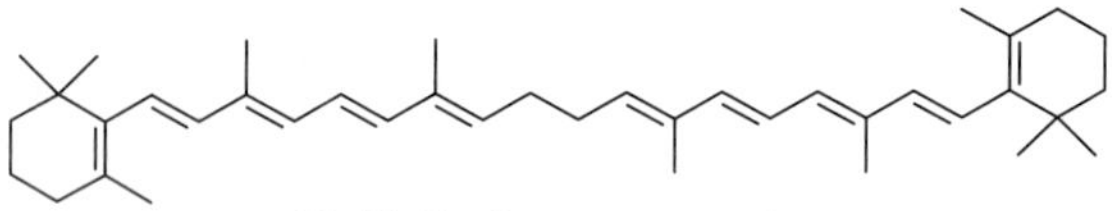

15,15'-dihydro-β-carotene (**7**)

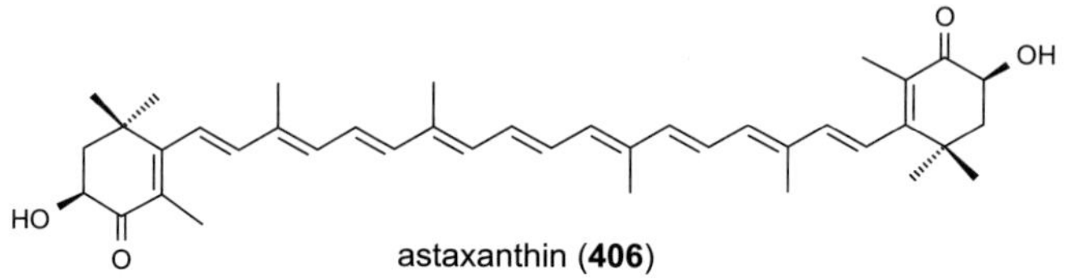

astaxanthin (**406**)

There was speculation about a possible transcriptional regulation of the *Bco1* gene. The first detailed analysis of the *Bco1* promoter region for the mouse gene [70] showed that, besides the core promoter elements (TATA and CACA boxes), an AP2 (activator protein 2), a bHLH (basic helix-loop-helix) and a PPAR (peroxysome proliferator activated receptor) response element (PPRE) were identified. Among those promoter sites, the PPRE is potentially the most interesting. It was shown that PPARγ and RXRα bind as a heterodimer to this response element and were also able to activate a luciferase reporter gene under the control of the *Bco1* promoter in cell lines that normally express BCO1 (namely TC7, PF11 and monkey retinal pigment epithelium cells). The functionality and specificity of this PPRE were also proven by EMSA (electromobility shift assay) and by use of the PPARγ-specific agonist ciglitazone and the RXRα-specific agonist (9-*cis*)-retinoic acid.

The cellular retinol-binding protein 2 gene (*Crbp2*) is the only other gene in vitamin A metabolism so far known to contain a PPRE. Like *Bco1*, this gene is up-regulated by ligands of PPAR and also by (9-*cis*)-retinoic acid [71]. In adult mammals, CRBP2 is expressed in large amounts in the small intestine, the major site of vitamin A formation. CRBP2 may act downstream of BCO1 in binding retinal, the primary cleavage product of provitamin A conversion. Thus common mechanisms in the regulation of the genes involved in the formation of vitamin A may contribute to vitamin A homeostasis, and the involvement of PPARs may interlink vitamin A formation to the regulation of lipid metabolism overall.

I. The Role of Carotenoid Cleavage Enzymes during Development

Retinoids are essential for vertebrate development. The present model in developmental biology assumes that the retinoids needed for development are derived mainly from maternal, preformed vitamin A. Vertebrate reproduction demands an elevated vitamin A supply for the egg or embryo. Under laboratory conditions, test animals are normally kept on a diet that is fortified with vitamin A but low in carotenoids. An interesting issue, therefore, is the embryonic supply of vitamin A under conditions when the mothers are fed a diet that provides mainly the provitamin and not preformed vitamin A. In most vertebrates, including man, these are natural diets. In addition, before molecular data were available, it could not be elucidated whether the vertebrate embryo itself is able, at least partially, to convert the non-toxic provitamin into the retinoids needed during development. With the identification of the carotene oxygenases, these issues can now be addressed on the molecular level.

The zebrafish (*Danio rerio*) was first used as a vertebrate model system to investigate the expression patterns of *Bco1* during development. By whole mount *in situ* hybridization with anti-sense RNA probes, the spatial and temporal expression patterns were established. In this experiment, whole embryos were fixed and hybridized with labelled anti-sense probes. Expression of the zebrafish *Bco1* homologue was seen at the beginning of segmentation stages and could be detected in several embryonic structures including the developing eye and neural crest. By use of the anti-sense morpholino-oligonucleotide technique, the consequence of a loss of function of BCO1 during zebrafish development could be analysed. This treatment led to malformations in the architecture of the branchial arch skeleton and the eye, thus indicating a crucial developmental role of this enzyme in zebrafish. Even though only small amounts of β-carotene, together with large amounts of retinoids, are found in the zebrafish yolk, these studies reveal that some tissues may rely on the local formation of vitamin A from the provitamin during development [72].

In the mouse embryo, Northern blot analysis detected high expression levels of *Bco1* mRNA at E7 whilst, at later stages of development (E11-E15), mRNA levels decreased but were still detectable [39]. In contrast, by *in situ* hybridization experiments, *in utero*, *Bco1* expression could be found mainly in the maternal tissues surrounding the embryo, but not in detectable levels in embryonic tissues (after 7.5 and 8.5 days) [40]. The expression of *Bco1* in the maternal tissues surrounding the mammal embryo indicates that this BCO1 may contribute directly to satisfy the elevated vitamin A demand of the embryo. These results are contradictory, however, so the putative role of an endogenous embryonic vitamin A formation should be addressed in more detail, in early and late stages of mouse embryogenesis.

J. Carotenoid Cleavage Enzymes in Different Tissues

In adult mammals, most of the vitamin A is formed in epithelial cells of the intestinal mucosa by the conversion of provitamin A carotenoids, and is then transported to the liver for storage. The tissue-specific expression patterns of *Bco1* have been analysed in several vertebrate species [38-42]. In chicken, the tissue-specific expression patterns of *Bco1* were analysed by a combination of Northern blot and *in situ* hybridization experiments. Its mRNA was mainly localized in liver, in duodenal microvilli, and in tubular structures of the lung and the kidney [38]. In the mouse, *Bco1* mRNA was detectable mainly in small intestine and liver but also in kidney, testes, uterine tissues, skin, and skeletal muscle [39,40,48]. Analyses of *Bco1* mRNA expression in humans revealed a comparable picture [42]. In addition, it was reported [41] that *Bco1* is highly expressed in the retinal pigment epithelium of the human eye. This RPE-specific expression of *Bco1* in mammals was confirmed in monkeys [73] and in a human RPE cell line [74].

The surprising result of these recent investigations is that the steady-state *Bco1* mRNA levels are quite high in peripheral tissues not involved in digestion. Testes, for example, require retinoids for spermatogenesis, and vitamin A is needed for retinoid signalling in almost all tissues. Thus, *Bco1* expression in peripheral tissues indicates that, besides an external vitamin A supply *via* the circulation, provitamin A carotenoids and conversion *in situ* may be important in retinoid metabolism and functions in various cell types and tissues.

K. Outlook

The molecular identification of the different metazoan members of this novel gene family established the existence of an ancient family of non-haem iron oxygenases in animals. Through these enzymes, animals have access to and can modulate their retinoids as needed for biological processes such as vision, cell differentiation and development. With the increasing number of sequences established (more than 100 are now available in the data base) together with the elucidation of the molecular structure of a family member from *Synechocystis*, common structural features can be predicted and functional domains and active site residues identified. It is thus possible to predict the consequences of single base pair polymorphism in these genes, as found in the human population. A question for the future is whether excentric cleavage products like β-ionone or apo-β-carotenals have any physiological role in vertebrates. It will be of particular interest to elucidate functions or physiological interactions of both central cleavage and excentric cleavage enzymes. For this, suitable animal models are necessary. The*Bco1* and*Bco2* knock-out mice that have been generated by targeted homologous recombination will be an excellent tool to elucidate the regulatory factors of the bioavailability of carotenoids, their subsequent conversion into biologically active derivatives, and the regulatory factors that interlink these processes with lipid metabolism as a whole. This interest inherently arises from the fact that carotenoids in staple foods are the major source to satisfy the world population's vitamin A demand.

References

[1]   K. Vogt, *Z. Naturforsch.*, **38**, 329 (1983).
[2]   G. Wald, *Nature*, **219**, 800 (1968).
[3]   V. Giguere, E. S. Ong, P. Segui and R. M. Evans, *Nature*, **330**, 624 (1987).
[4]   M. Petkovich, N. J. Brand, A. Krust and P. Chambon, *Nature*, **330**, 444 (1987).
[5]   D. J. Mangelsdorf and R. M. Evans, *Cell*, **83**, 841 (1995).
[6]   P. Chambon, *FASEB J.*, **10**, 940 (1996).
[7]   B. A. Underwood, *J. Nutr.*, **134**, 231S (2004).
[8]   T. B. Osborne and L. B. Mendel, *J. Biol. Chem.*, **15**, 311 (1913).
[9]   E. V. McCollum and M. Davis, *J. Biol. Chem.*, **15**, 167 (1913).
[10] H. Steenbock, *Science*, **50**, 352 (1919).
[11] T. Moore, *Biochem. J.*, **24**, 692 (1930).
[12] P. Karrer, A. Helfenstein, H. Wehrli and A. Wettstein, *Helv. Chim. Acta.*, **13**, 1084 (1930).
[13] J. Glover and E. R. Redfearn, *Process Biochem*, **58**, xv (1954).
[14] J. A. Olson and O. Hayaishi, *Proc. Natl. Acad. Sci. USA*, **54**, 1364 (1965).
[15] D. S. Goodman and H. S. Huang, *Science*, **149**, 879 (1965).
[16] M. G. Leuenberger, C. Engeloch-Jarret and W. D. Woggon, *Angew. Chem. Int. Ed.*, **40**, 2613 (2001).
[17] D. S. Goodman, R. Blomstrand, B. Werner, H. S. Huang and T. Shiratori, *J. Clin. Invest.*, **45**, 1615 (1966).
[18] N. H. Fidge, F. R. Smith and D. S. Goodman, *Biochem. J.*, **114**, 689 (1969).
[19] D. Sklan, *Br. J. Nutr.*, **50**, 417 (1983).
[20] J. A. Olson, *J. Nutr.*, **119**, 105 (1989).
[21] T. van Vliet, F. van Schaik, W. H. Schreurs and H. van den Berg, *Int. J. Vitam. Nutr. Res.*, **66**, 77 (1996).
[22] C. Duszka, P. Grolier, E. M. Azim, M. C. Alexandre-Gouabau, P. Borel and V. Azais-Braesco, *J. Nutr.*, **126**, 2550 (1996).
[23] A. During, A. Nagao, C. Hoshino and J. Terao, *Anal. Biochem.*, **241**, 199 (1996).
[24] J. Devery and B. V. Milborrow, *Br. J. Nutr.*, **72**, 397 (1994).
[25] A. A. Dmitrovskii, V. Iu. Ershov and V. Bykhovskii, *Prikl Biokhim Mikrobiol*, **28**, 199 (1992).
[26] R. Lakshman, I. Mychkovsky and M. Attlesey, *Proc. Natl. Acad. Sci. USA*, **86**, 9124 (1989).
[27] A. Nagao, A. During, C. Hoshino, J. Terao and J. A. Olson, *Arch. Biochem. Biophys.*, **328**, 57 (1996).
[28] M. R. Lakshmanan, H. Chansang and J. A. Olson, *J. Lipid Res.*, **13**, 477 (1972).
[29] X. D. Wang, G. W. Tang, J. G. Fox, N. I. Krinsky and R. M. Russell, *Arch. Biochem. Biophys.*, **285**, 8 (1991).
[30] S. N. Gershoff, S. B. Andrus, D. M. Hegsted and E. A. Lentini, *Lab. Invest.*, **6**, 227 (1957).
[31] H. Singh and H. R. Cama, *Biochim. Biophys. Acta*, **370**, 49 (1974).
[32] S. H. Schwartz, B. C. Tan, D. A. Gage, J. A. Zeevaart and D. R. McCarty, *Science*, **276**, 1872 (1997).
[33] A. Wyss, G. Wirtz, W. Woggon, R. Brugger, M. Wyss, A. Friedlein, H. Bachmann and W. Hunziker, *Biochem. Biophys. Res. Commun.*, **271**, 334 (2000).
[34] J. von Lintig and K. Vogt, *J. Biol. Chem.*, **275**, 11915 (2000).
[35] J. von Lintig and A. Wyss, *Arch. Biochem. Biophys.*, **385**, 47 (2001).
[36] R. S. Stephenson, J. O'Tousa, N. J. Scavarda, L. L. Randall and W. L. Pak, in *The Biology of Photoreception* (ed. D. J. Cosens and D. Vince-Price), p. 477, Cambridge University Press, Cambridge (1983).
[37] J. von Lintig, A. Dreher, C. Kiefer, M. F. Wernet and K. Vogt, *Proc. Natl. Acad. Sci. USA*, **98**, 1130

(2001).
[38] A. Wyss, G. M. Wirtz, W. D. Woggon, R. Brugger, M. Wyss, A. Friedlein, G. Riss, H. Bachmann and W. Hunziker, *Biochem. J.*, **354**, 521 (2001).
[39] T. M. Redmond, S. Gentleman, T. Duncan, S. Yu, B. Wiggert, E. Gantt and F. X. Cunningham, Jr., *J. Biol. Chem.*, **276**, 6560 (2001).
[40] J. Paik, A. During, E. H. Harrison, C. L. Mendelsohn, K. Lai and W. S. Blaner, *J. Biol. Chem.*, **276**, 32160 (2001).
[41] W. Yan, G. F. Jang, F. Haeseleer, N. Esumi, J. Chang, M. Kerrigan, M. Campochiaro, P. Campochiaro, K. Palczewski and D. J. Zack, *Genomics*, **72**, 193 (2001).
[42] A. Lindqvist and S. Andersson, *J. Biol. Chem.*, **277**, 23942 (2002).
[43] C. O. Bavik, F. Levy, U. Hellman, C. Wernstedt and U. Eriksson, *J. Biol. Chem.*, **268**, 20540 (1993).
[44] C. P. Hamel, E. Tsilou, B. A. Pfeffer, J. J. Hooks, B. Detrick and T. M. Redmond, *J. Biol. Chem.*, **268**, 15751 (1993).
[45] M. Jin, S. Li, W. N. Moghrabi, H. Sun and G. H. Travis, *Cell*, **122**, 449 (2005).
[46] G. Moiseyev, Y. Chen, Y. Takahashi, B. X. Wu and J. X. Ma, *Proc. Natl. Acad. Sci. USA*, (2005).
[47] T. M. Redmond, E. Poliakov, S. Yu, J. Y. Tsai, Z. Lu and S. Gentleman, *Proc. Natl. Acad. Sci. USA*, **102**, 13658 (2005).
[48] C. Kiefer, S. Hessel, J. M. Lampert, K. Vogt, M. O. Lederer, D. E. Breithaupt and J. von Lintig, *J. Biol. Chem.*, **276**, 14110 (2001).
[49] A. Chawla, J. J. Repa, R. M. Evans and D. J. Mangelsdorf, *Science*, **294**, 1866 (2001).
[50] T. M. Redmond, S. Yu, E. Lee, D. Bok, D. Hamasaki, N. Chen, P. Goletz, J. X. Ma, R. K. Crouch and K. Pfeifer, *Nature Genet.*, **20**, 344 (1998).
[51] S. M. Gu, D. A. Thompson, C. R. Srikumari, B. Lorenz, U. Finckh, A. Nicoletti, K. R. Murthy, M. Rathmann, G. Kumaramanickavel, M. J. Denton and A. Gal, *Nature Genet.*, **17**, 194 (1997).
[52] F. Marlhens, C. Bareil, J. M. Griffoin, E. Zrenner, P. Amalric, C. Eliaou, S. Y. Liu, E. Harris, T. M. Redmond, B. Arnaud, M. Claustres and C. P. Hamel, *Nature Genet.*, **17**, 139 (1997).
[53] P. S. Bernstein, W. C. Law and R. R. Rando, *J. Biol. Chem.*, **262**, 16848 (1987).
[54] Y. Takahashi, G. Moiseyev, Y. Chen and J. X. Ma, *FEBS Lett.*, **579**, 5414 (2005).
[55] S. Ruch, P. Beyer, H. Ernst and S. Al-Babili, *Mol Microbiol*, **55**, 1015 (2005).
[56] D. P. Kloer, S. Ruch, S. Al-Babili, P. Beyer and G. E. Schulz, *Science*, **308**, 267 (2005).
[57] E. Poliakov, S. Gentleman, F. X. Cunningham, Jr., N. J. Miller-Ihli and T. M. Redmond, *J. Biol. Chem.*, **280**, 29217 (2005).
[58] G. Wolf, *Nutr. Rev.*, **53**, 134 (1995).
[59] R. V. Sharma, S. N. Mathur and J. Ganguly, *Biochem. J.*, **158**, 377 (1976).
[60] R. V. Sharma, S. N. Mathur, A. A. Dmitrovskii, R. C. Das and J. Ganguly, *Biochim. Biophys. Acta*, **486**, 183 (1976).
[61] X. D. Wang, N. I. Krinsky, G. W. Tang and R. M. Russell, *Arch. Biochem. Biophys.*, **293**, 298 (1992).
[62] X. D. Wang, R. M. Russell, R. P. Marini, G. Tang, G. G. Dolnikowski, J. G. Fox and N. I. Krinsky, *Biochim. Biophys. Acta*, **1167**, 159 (1993).
[63] D. S. Goodman, H. S. Huang and T. Shiratori, *J. Biol. Chem.*, **241**, 1929 (1966).
[64] H. Schmidt, R. Kurzer, W. Eisenreich and W. Schwab, *J. Biol. Chem.*, **281**, 9845 (2006).
[65] L. Villard and C. J. Bates, *Br. J. Nutr.*, **56**, 115 (1986).
[66] T. van Vliet, M. F. van Vlissingen, F. van Schaik and H. van den Berg, *J. Nutr.*, **126**, 499 (1996).
[67] V. Iu. Ershov, A. A. Dmitrovskii and V. Bykhovskii, *Biokhimiia*, **58**, 733 (1993).
[68] P. Grolier, C. Duszka, P. Borel, M. C. Alexandre-Gouabau and V. Azais-Braesco, *Arch. Biochem. Biophys.*, **348**, 233 (1997).
[69] S. G. Parvin and B. Sivakumar, *J. Nutr.*, **130**, 573 (2000).
[70] A. Boulanger, P. McLemore, N. G. Copeland, D. J. Gilbert, N. A. Jenkins, S. S. Yu, S. Gentleman and T. M. Redmond, *FASEB J.*, **17**, 1304 (2003).
[71] K. Suruga, K. Mochizuki, M. Kitagawa, T. Goda, N. Horie, K. Takeishi and S. Takase, *Arch. Biochem. Biophys.*, **362**, 159 (1999).
[72] J. M. Lampert, J. Holzschuh, S. Hessel, W. Driever, K. Vogt and J. von Lintig, *Development*, **130**, 2173 (2003).
[73] R. A. Bhatti, S. Yu, A. Boulanger, R. N. Fariss, Y. Guo, S. L. Bernstein, S. Gentleman and T. M. Redmond, *Invest. Ophthalmol. Vis. Sci*, **44**, 44 (2003).
[74] G. R. Chichili, D. Nohr, M. Schaffer, J. von Lintig and H. K. Biesalski, *Invest. Ophthalmol. Vis. Sci.*, **46**, 3562 (2005).

Carotenoids
Volume 4: Natural Functions
© 2008 Birkhäuser Verlag Basel

Chapter 17

Enzymic Pathways for Formation of Carotenoid Cleavage Products

Peter Fleischmann and Holger Zorn

A. Introduction

Degraded carotenoids (apocarotenoids, norisoprenoids) have been a subject of intensive research for several decades. From the perspective of human physiology and nutrition, the retinoids, acting as vitamins, signalling molecules, and visual pigments, attracted the greatest attention (*Chapters 15* and *16*). Plant scientists, however, detected a wealth of different apocarotenoids, presumably derived by the excentric cleavage of carotenoids in various species, the plant hormone abscisic acid (*1,* Scheme 6) being the best-investigated example. With the onset of fruit ripening, flower opening or senescence of green tissues, carotenoids are degraded oxidatively to smaller, volatile compounds. The natural biological functions of the reaction products are outlined in *Chapter 15*. As many of these apocarotenoids act as potent flavour compounds, food chemists and flavourists worldwide have investigated meticulously their structural and sensory properties. Many aspects of carotenoid metabolites and breakdown products as aroma compounds are presented in a comprehensive book [1].

Until the end of the 20th century, research was focused mainly on the discovery and description of new members of this so-called norisoprenoid family. Though several lines of evidence suggested that they may be formed *via* cleavage of tetraterpenoid precursor molecules, final proof of this was missing. With the identification of the maize enzyme VP14 which cleaves (9Z)-epoxycarotenoids specifically at the C(11,12) double bond, the situation changed fundamentally. Since then, several carotenoid-degrading enzymes from various plant and animal species have been isolated and characterized on a molecular level. By heterologous expression of the genes, the pathways proposed previously could be substantiated in experiments *in vitro*. Pathways for the formation of carotenoid degradation products are described in *Volume 3, Chapter 4* [2]. The focus of this Chapter, therefore, is on the enzymes that catalyse the cleavage reactions.

   The current nomenclature for these oxygenases can be misleading. The carotenoid-cleaving enzymes have traditionally been referred to as 'dioxygenases', even though, with only one exception, *At*CCD1 from *Arabidopsis thaliana* [3], a dioxygenase mechanism has not been demonstrated experimentally. Indeed, as described in *Chapter 16*, the much-studied central cleavage enzyme, important for the conversion of β-carotene (3) into vitamin A in animals, and traditionally known as 'β-carotene 15,15′-dioxygenase', has recently been shown to be a monooxygenase [4]. The general 'dioxygenase' designations used in the literature will be used in this Chapter. This should not be taken as an indication of the reaction mechanism.

   The formation of flavour-active carotenoid breakdown molecules is complicated. In most cases, enzymic processes co-exist with non-enzymic chemical reactions that form the same compounds. Furthermore, volatiles can be derived from the central part of the conjugated carotenoid chain as well as from the end groups. This gives a great structural diversity of products.

B. Structural Diversity of Carotenoid-derived Flavour Compounds

Among the many norisoprenoids identified as, or suspected to be, breakdown products of carotenoids are some of the most powerful flavour compounds. β-Ionone (2), a molecule first found in the Bulgarian rose, has an odour threshold of 0.007 ppm [5] and β-damascenone (3), also very important for rose scents, is one of the most potent of all flavour-active organic molecules [6-9]. Besides these two examples of C_{13} norisoprenoids, C_9, C_{10} and C_{11} compounds can also be obtained by oxidative degradation of precursor carotenoids. Safranal (4), a C_{10} compound, is of key importance for the flavour of saffron and can be formed by non-enzymic processes [10] as well as enzymically. The structures of some of these molecules, which are all derived from the terminal parts of a precursor carotenoid, are shown in Fig. 1. Only some of them, however, can be formed in a one-step reaction by excentric oxidative cleavage of the carotenoid. Even the formation of β-damascenone (3) needs several additional steps, following the initial cleavage of its precursor carotenoid neoxanthin (234) [11] (Scheme 1).

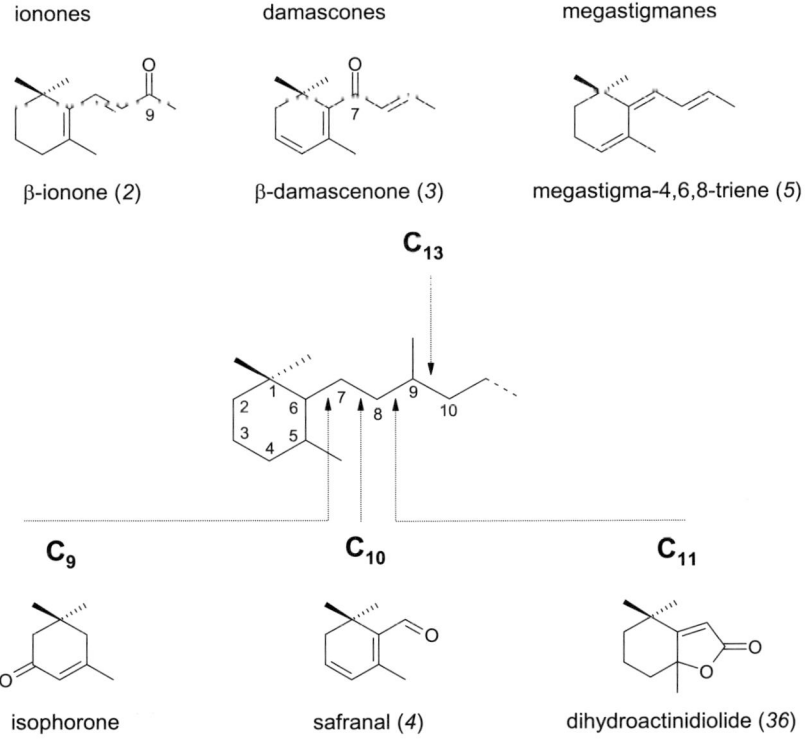

Fig. 1. Examples of C_9, C_{10}, C_{11}, and C_{13} norisoprenoids which are derived directly from carotenoids, either by non-enzymic reactions or by the action of carotenoid cleavage enzymes. Most norisoprenoids, however, require further steps for their formation from the initially formed precursor.

Other examples (Fig. 2) of carotenoid breakdown compounds which result from modifications after the initial cleavage of the precursor carotenoids in order to generate the flavour-active compounds are megastigmanes (*5*) found in tobacco (*Nicotiana tabacum*) [12,13], various theaspiranes (*6*) and their derivatives found in tea (*Camellia sinensis*) [14,15], and cyclic ethers (*7*) and oxygenated theaspiranes (*8*) in flowers of the sweet olive *Osmanthus fragrans*, an East-Asian shrub that also contains large amounts of α-ionone (*9*) and β-ionone (*2*) in its scent [16]. All these compounds have been well known for several years, but little is known about the biological pathways that lead from the primary carotenoid cleavage product to the final flavour molecule. In many cases, however, their formation by co-oxidation (either enzymic or non-enzymic, *e.g.* by free radicals or light-driven isomerization) [17] or by thermal carotenoid degradation [18] has been demonstrated.

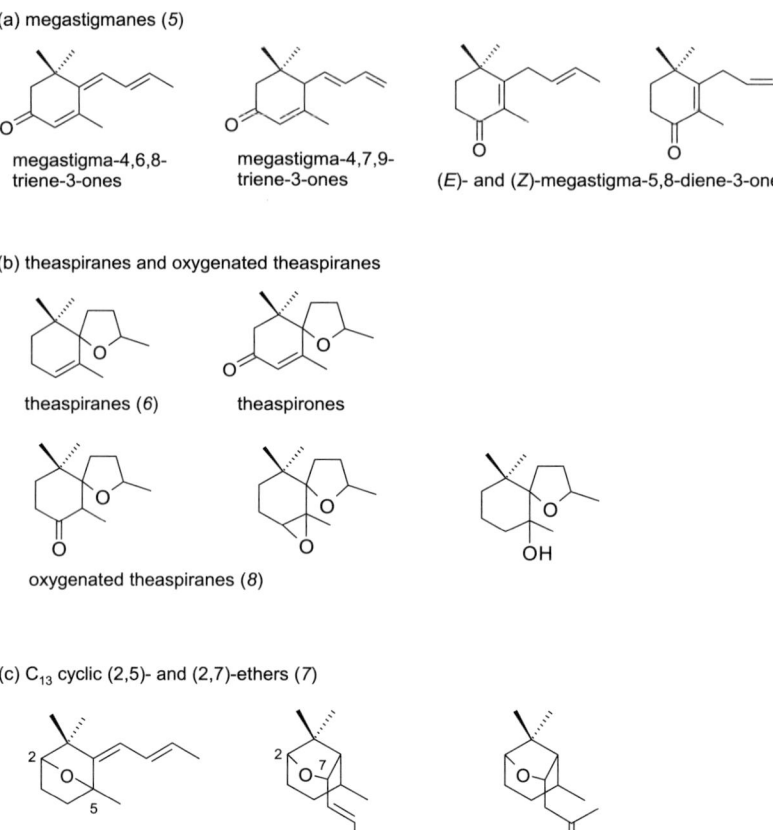

(a) megastigmanes (5)

megastigma-4,6,8-
triene-3-ones

megastigma-4,7,9-
triene-3-ones

(E)- and (Z)-megastigma-5,8-diene-3-one

(b) theaspiranes and oxygenated theaspiranes

theaspiranes (6)   theaspirones

oxygenated theaspiranes (8)

(c) C_{13} cyclic (2,5)- and (2,7)-ethers (7)

Fig. 2. Flavour-active carotenoid breakdown compounds: (a) megastigmanes (5) from tobacco (*Nicotiana tabacum*) [12,13]; (b) theaspiranes (6) and their oxygenated derivatives (8) from tea (*Camellia sinensis*) [14,15] and flowers of *Osmanthus fragrans* [16]; (c) C_{13} cyclic (2,5)- and (2,7)-ethers (7) from *O. fragrans* [16]. These products are are formed by secondary modification of the primary oxidative carotenoid cleavage products.

A second class of flavour-active carotenoid breakdown products can be formed from the middle part of the carotenoid chain. Examples are marmelo oxides (*10*) [19], marmelo lactones (*11*) [20] and quince oxepine (*12*) [21], all found in quince fruit (*Cydonia oblonga*) (Scheme 2). These norisoprenoids are formed from C_{10} or C_{12} breakdown compounds from the central isoprenoid chain of precursor carotenoids, but the biological processes responsible for their formation have not been described. It is known, however, that the direct precursors of the free volatiles are glucosides of the primary cleavage products, that are converted into free norisoprenoids by acid catalysis [22-25]. It is likely that, in the plant, this hydrolysis is carried out by glycosidases, even if this has not been demonstrated experimentally.

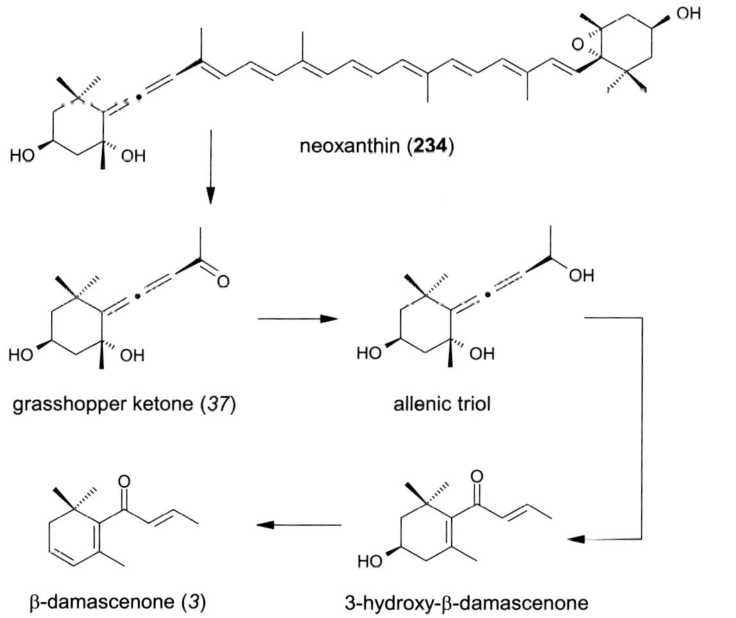

neoxanthin (**234**)

grasshopper ketone (*37*)

allenic triol

β-damascenone (*3*)

3-hydroxy-β-damascenone

Scheme 1

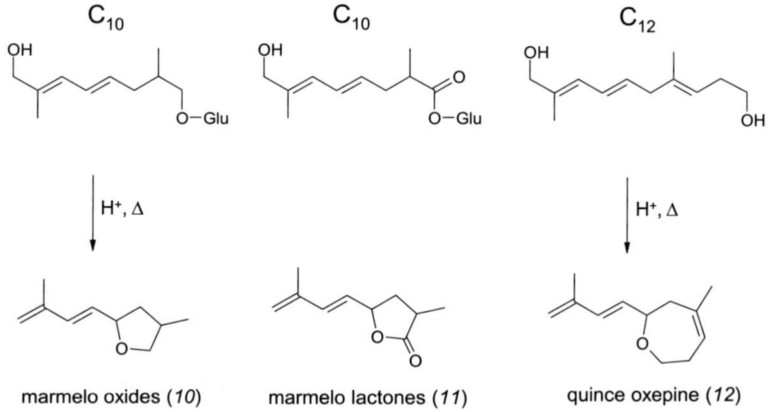

C$_{10}$

C$_{10}$

C$_{12}$

H$^+$, Δ

H$^+$, Δ

marmelo oxides (*10*)

marmelo lactones (*11*)

quince oxepine (*12*)

Scheme 2

C. General Model for Carotenoid-related Flavour Formation and Release

Among all flavour compounds that may be derived from carotenoids, only some can be formed in a one-step reaction from their precursor carotenoids. Whereas in many cases it is still not elucidated whether the subsequent steps necessary for flavour formation are biological or not, it is well known that enzymes are involved in storage and release of norisoprenoids. Ionones and damascenones, the most abundant of the carotenoid-derived volatiles, are present in many flavours of fruits, flowers and green leaves [22,26,27]. In most cases, however, their formation is not directly correlated with flavour release [28,29].

Figure 3 shows a general scheme for the formation and release of carotenoid-based flavour. In all cases, oxidative breakdown reactions, often but not necessarily enzymic, are the first step. These degradation reactions are described in detail below.

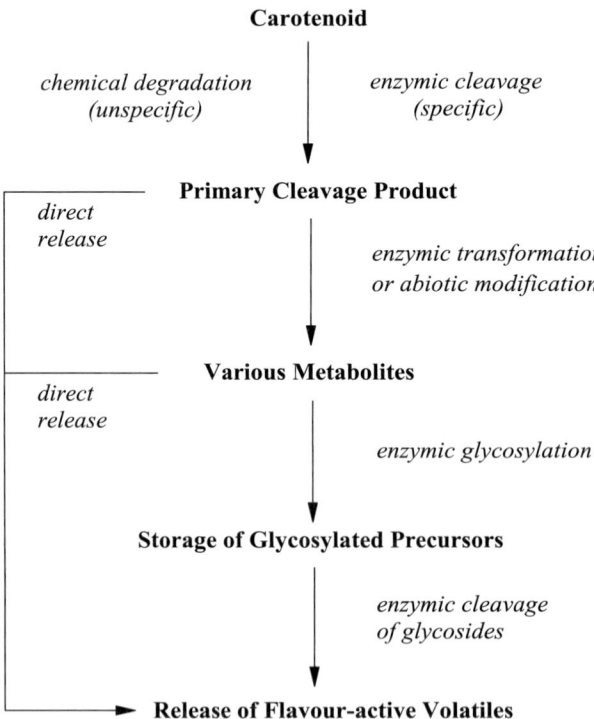

Fig. 3. General scheme of oxidative carotenoid degradation leading to norisoprenoid flavour compounds in plant tissues.

In many cases, the primary reaction products are already identical with the final flavour compounds, but they are not always released directly they are formed. Norisoprenoid glycosides are often formed and stored until flavour release is initiated by maturation or ripening [29-31]. Flavour release is then uncoupled from carotenoid degradation and initiated by enzymic cleavage of the norisoprenoid glycosides, not by direct oxidative cleavage of precursor carotenoids [29-32]. Whereas the formation of these glycosides is still under investigation, the release of free norisoprenoids by the action of specific glycosidase enzymes is already well established [33].

For other norisoprenoid flavours (Figs. 1 and 2, Schemes 1 and 2), the primary products undergo a more complex cascade of reactions until the final flavour compound is released. For β-damascenone (3) and quince oxepine (12), for example, the primary cleavage products may undergo multiple enzymic or non-enzymic modifications until a stable, non-volatile storage precursor, most commonly a glycoside, is formed [34]. In other cases, though, the metabolites formed by modification of the initial carotenoid breakdown compounds are released without further storage.

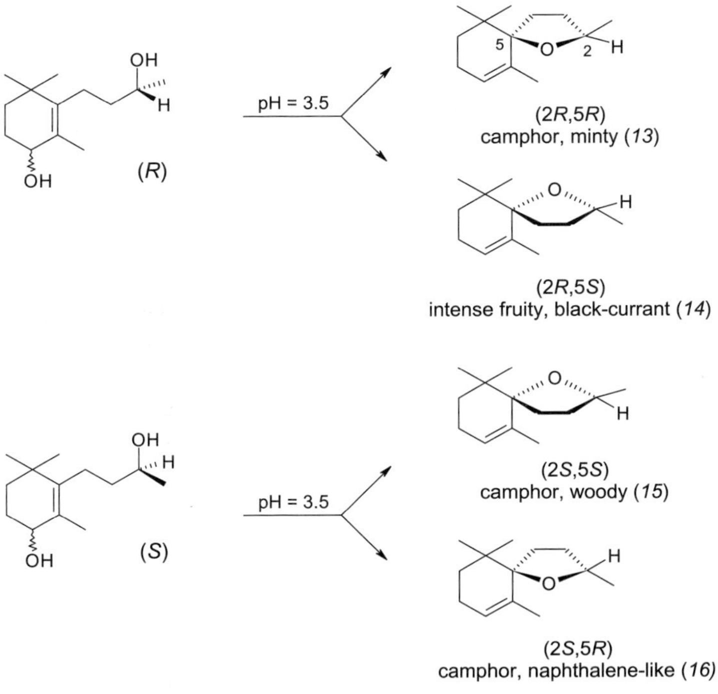

Scheme 3

Almost nothing is known about the biology of these secondary modification reactions of norisoprenoids, though chemical pathways may be known. Some living systems, however, *e.g.* raspberries (*Rubus idaeus*) [35] and quince (*Cydonia oblonga*) [36] are highly specific, and are able to synthesize stereospecifically flavour compounds such as the four theaspirane isomers *13-16* (Scheme 3) [35], each with its own perfume/flavour characteristics. This strongly suggests that specific enzyme-catalysed reactions may at least contribute to the modification of carotenoid breakdown compounds.

In principle, the reactions that degrade carotenoids may occur with or without the influence of biological processes. Fragrant norisoprenoids are frequently formed in plant tissues, without any enzyme activity. These non-enzymic degradation reactions are unspecific and result in a random mixture of products, depending on the precursor carotenoid and the reaction conditions. In a similar way, co-oxidation of fatty acids and carotenoids leads to the unspecific formation of carotenoid breakdown products. Co-oxidations, however, often are processes accelerated by specific enzymes, *i.e.* peroxygenases (lipoxygenases) [37,38] (see Section **H**). Finally, a family of enzymes exists which specifically degrades carotenoids to C_9 or C_{13} norisoprenoids by cleaving the C(7,8) or the C(9,10) double bond of the conjugated carotenoid chain. These enzymes ('carotenases') are all related to the same highly homologous 'carotenoid-cleaving dioxygenase' gene family (*CCD1-7*). Schemes 4 and 5 show general examples of such cleavage reactions involving a C(9,10)- or a C(7,8)-carotenoid dioxygenase, respectively. Such enzymes, which are the main focus of this Chapter, are found in a large number of plant families and species and are able to metabolize most common carotenoids [39-41]. Note that, as pointed out in *Chapter 16*, strict application of the rules of carotenoid nomenclature means that a cleavage enzyme must be designated as, for example, a C(9,10)-dioxygenase though the product is an apo-10'-carotenal.

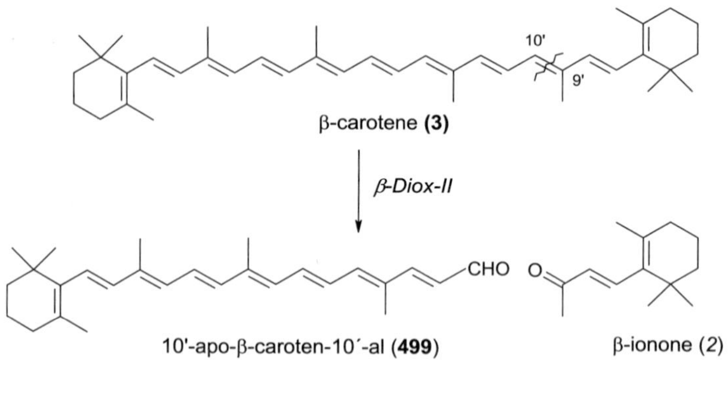

β-carotene (**3**)

β-Diox-II

10'-apo-β-caroten-10′-al (**499**)   β-ionone (*2*)

Scheme 4

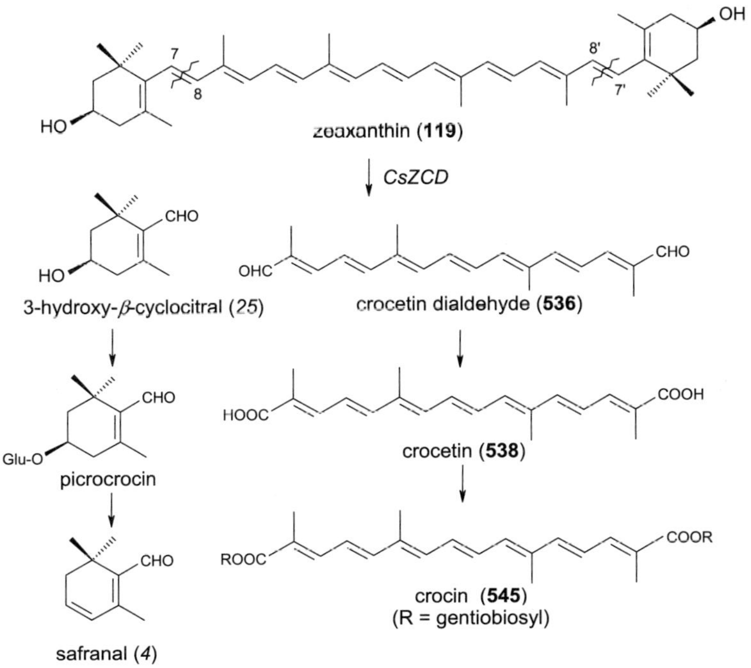

Scheme 5

Additionally, many findings related to carotenoid-mediated flavour formation and release in living plant tissues point to a more complex pattern of physiologically and chemically regulated processes. The formation of primary carotenoid cleavage products (C_{13} norisoprenoids) in *Vitis vinifera* has been described [28] as being closely related to the expression of *CCD1* genes. There is, however, a long delay of more than one week in flavour release following gene expression. Work [42] on circadian enzymic carotenoid cleavage in petunia (*Petunia hybrida*) shows that, from gene expression to flavour release, a delay of no more than six hours is needed, enabling petunia flowers to change petal colour and release flavour in a circadian rhythm which clearly depends on external lighting conditions of the cultivated plants. This proves that the physiology-based system of gene expression, *i.e.* the formation of primary carotenoid cleavage products through active enzymes and flavour release, is fast enough to follow diurnal changes of environmental conditions. Furthermore, it is evident that the delay between gene activation and flavour release in *Vitis vinifera* cannot be explained by the action of a single enzymic carotenoid cleavage reaction alone. In such cases, it is still unclear if the norisoprenoid flavours are transformed to storage products (perhaps glycosides) or if the gene products themselves have to pass an additional activation step before carotenoid cleavage and flavour release start at the same time. For *Petunia* flowers, the regulatory connection between carotenoid cleavage enzyme formation (and action) and the light sensory systems is not elucidated.

A close connection was shown between the colouring of flower petals and the presence of *CCD* genes in the flower tissues of *Chrysanthemum morifolium* [43]. All white-flowering *Chrysanthemum* varieties showed high *CCD4* expression levels whereas yellow-flowering varieties did not. No differences in the flavour of these flowers were reported. In contrast to *Petunia hybrida*, this may be caused by secondary transformation of the enzyme products to more stable and transportable storage compounds rather than the initiation of instant release of norisoprenoids immediately following their formation. A similar model of carotenoid-based flavour formation and release is known from quince fruit (*Cydonia oblonga*). Free norisoprenoids could be found in parallel with glycosylated storage compounds, even if only in smaller amounts [36]. On the other hand, enzymic carotenoid cleavage activity could only be traced in fully ripened fruit after a cold induction period of at least one week [30]. The flower petals of *Chrysanthemum* most likely do not show changes in flavour, and carotenoid cleavage (and flower colour) is determined by the presence of expressed CCD4-type enzymes. In quince fruit, direct flavour release was observed following the activation of carotenoid cleavage enzymes through cold induction, but stored norisoprenoid precursors could also be found. As with *Petunia*, the mechanism of enzyme induction may include close connections to the temperature- and/or light-sensory systems in these plants, but is not yet elucidated.

Finally, there is evidence that, in the same plant, there may be seasonal changes in the structure and kinetic properties of carotenoid cleavage enzymes. The carotenases isolated from Japanese tea plants (*Camellia sinensis*) show changes in activity, isoelectric point (pK$_i$) and even in molecular weight when isolated from the same plant in different seasons [44].

Even if their regulatory and molecular background is not completely understood, these changes show that the physiology of carotenoid catabolism in *Camellia sinensis*, and other examples mentioned above, is far more complex than might at first seem apparent from the existence of a single family of genes coding for all carotenoid dioxygenases. However, the data available for cloned carotenoid cleavage genes and their related enzymes currently still give the best insight into the field. The following sections describe, at the molecular level, the primary enzymic cleavage of carotenoids at various positions in the polyene chain.

D. Cleavage of the C(11,12) Double Bond: (9*Z*)-Epoxycarotenoid Dioxygenases (NCED)

Pioneering work [45] on the biosynthesis of abscisic acid (ABA, *1*) permitted the first insights into the family of carotenoid-degrading enzymes on a molecular level. The characterization of a viviparous, ABA-deficient maize mutant (*vp14*) resulted in the identification of the enzyme VP14 (synonym *Zm*NCED). This enzyme catalyses the oxidative cleavage of (9*Z*)-violaxanthin (**259**) and (9'*Z*)-neoxanthin (**234**) (C$_{40}$) to the sesquiterpenoid (C$_{15}$) compound xanthoxin (*17*) and the corresponding C_{25} apocarotenals. Subsequently, xanthoxin is readily converted into ABA aldehyde (*18*) and thence ABA [46] (Scheme 6). The stoichiometric conversion of the precursor carotenoids into the two products indicates a specific cleavage between C(11) and C(12) of the polyene chain. The enzyme has a strict requirement for a (9Z)-double bond adjacent to the site of cleavage; the (all-*E*)-isomers of violaxanthin and neoxanthin are not cleaved. As recombinant purified VP14 also cleaved (9Z)-zeaxanthin (**119**) at the C(11,12) double bond, an epoxide group does not seem to be essential for the acceptance of the substrate [45] For its catalytic activity, VP14 requires molecular oxygen, ferrous (non-haem) iron and, *in vitro*, additionally a detergent.

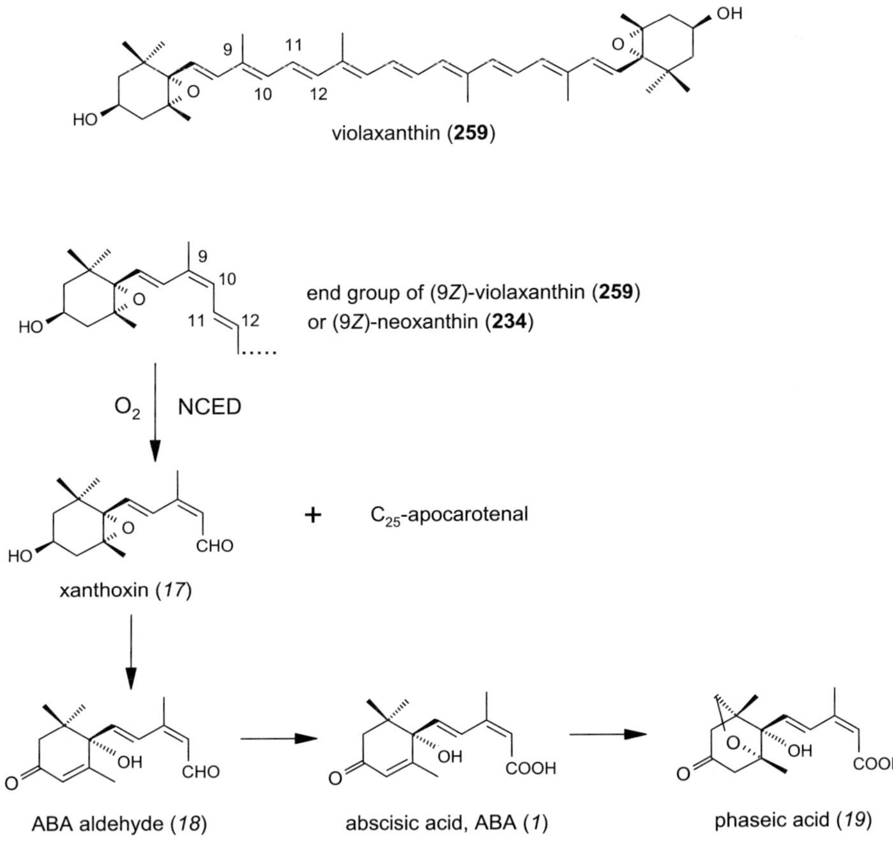

violaxanthin (**259**)

end group of (9Z)-violaxanthin (**259**)
or (9Z)-neoxanthin (**234**)

O_2 | NCED

+ C_{25}-apocarotenal

xanthoxin (*17*)

ABA aldehyde (*18*)

abscisic acid, ABA (*1*)

phaseic acid (*19*)

Scheme 6

VP14 is a nuclear-encoded chloroplast-localized protein, which is strongly induced by water stress [47]. The localization within the chloroplast is consistent with its function in carotenoid degradation, as plant carotenoids are synthesized within plastids and are associated with the thylakoid and envelope membranes. Inside the chloroplasts, the enzyme was detected predominantly in the stroma as a soluble protein, but approximately 35% of the imported VP14 was found associated with thylakoids. Thus, mature VP14 exists in two sub-organellar forms, one of which could reflect an enzymically active form and the other an inactive form [48,49], though experimental proof for this hypothesis is lacking.

In recent years, orthologous genes of VP14 have been identified throughout the plant kingdom. The enzymes share sequence homologies not only among each other but also with bacterial lignostilbene dioxygenases (LSDs), which catalyse a double-bond cleavage reaction similar to that of ABA formation [50]. Based on their substrate specificity observed *in vitro* and presumed *in vivo*, the name 'nine <u>c</u>is epoxycarotenoid <u>d</u>ioxygenase' (NCED) has become generally adopted for this group of carotenoid degrading enzymes, though the dioxygenase mechanism has not been established. NCEDs are typically encoded by multigene families, and they play a key role in the modulation of ABA levels.

The recombinant *Arabidopsis thaliana* proteins *At*NCED2, *At*NCED3, *At*NCED6, and *At*NCED9 have been shown to cleave (9Z)-epoxycarotenoids *in vitro*. Considering the high sequence homology, *At*NCED5 is probably also able to catalyse the cleavage reaction. All five *At*NCEDs are plastid targeted, but their sub-organellar localization is different. Whilst *At*NCED2, *At*NCED3 and *At*NCED6 are found in both stroma and thylakoid membrane-bound compartments, *At*NCED4 and *At*NCED9 remain soluble in the stroma and *At*NCED5 is exclusively bound to thylakoids [51]. Under drought stress, only the expression of the *AtNCED1* and *AtNCED3* genes was strongly induced [52-54].

From avocado (*Persea americana* cv. Lula) three potential *PaNCED* genes were cloned. The recombinant proteins *Pa*NCED1 and *Pa*NCED3 were capable of cleaving (9Z)-epoxycarotenoids into xanthoxin (*17*) and C_{25}-apocarotenoids *in vitro*, whereas *Pa*NCED2 was not. Whilst *PaNCED2* was expressed constitutively in fruit during ripening as well as in leaves, *PaNCED1* and *PaNCED3* were strongly induced as the fruit ripened. The enzyme structure deduced from the *PaNCED2* gene sequence lacks a predicted chloroplast transit peptide region, so *Pa*NCED is unlikely to be involved in ABA biosynthesis [55].

Further *NCED*s have been described from, for example, tomato (*Lycopersicon esculentum*, *LeNCED1*) [56-58], cowpea (*Vigna unguiculata*, *VuNCED1*) [59], bean (*Phaseolus vulgaris*, *PvNCED1* and *PvNCED2*) [60] and *Citrus* species [61]. When *PvNCED1* was over-expressed in tobacco (*Nicotiana plumbaginifolia*), increasing concentrations of ABA and phaseic acid (*19*), an ABA catabolite, were observed [62]. The enzyme *Pv*NCED1 is imported into chloroplasts, where it is strictly associated with the thylakoids. In detached bean leaves under water stress, ABA accumulation was preceded by a large increase in *PvNCED1* mRNA and protein levels. Altogether, the EMBL (European Molecular Biology Laboratory) nucleotide databank currently lists 41 *NCED* sequences from 16 plant species (http://srs.ebi.ac.uk/).

Sequence alignments of *NCEDs* from various species show that several highly conserved histidine and acidic residues may be necessary for coordinating iron in the active site [50]. Characteristics of the NCEDs discussed above are summarized in Table 1.

Table 1. Characteristics of some (9Z)-epoxycarotenoid dioxygenases (NCED).

Enzyme	Accession no.	Source	Amino acids	Size [kDa]	Refs.
VP14	O24592	Maize	604	65.6	45,47
*Pv*NCED1	Q9M6E8	Bean	615	68.1	60
*Le*NCED1	O24023	Tomato	605	67.3	58
*Pa*NCED3	Q9AXZ4	Avocado	625	69.7	55
*Vu*NCED1	Q9FS24	Cowpea	612	67.1	59
*At*NCED3	Q93ZU5	*Arabidopsis*	599	65.8	53,54

E. Cleavage of the C(9,10) or C(9′,10′) Double Bond: Carotenoid Cleavage Dioxygenases (CCD)

A large number of C_{13} norisoprenoid aroma compounds, released by the oxidative cleavage of carotenoid precursor molecules at the C(9,10) double bond, are produced by flowers, fruit and vegetables [22]. However, the first enzyme that was shown to cleave carotenoids asymmetrically at the C(9,10) position (β-carotene 9,10-oxygenase), was obtained from mouse, human, zebrafish and rat, as described in *Chapter 16*.

A comparable product profile was observed with the plant protein *At*CCD7, an *Arabidopsis* enzyme capable of β-carotene degradation. When *At*CCD7 was expressed in a β-carotene-accumulating *E. coli* strain, the concentration of β-carotene was significantly reduced compared to that in a reference strain carrying an empty plasmid. In assays *in vitro*, the affinity-purified recombinant protein catalysed a specific C(9,10)-cleavage of β-carotene (**3**) to produce β-ionone (**2**) and 10′-apo-β-caroten-10′-al (**499**) (Scheme 4). No products were detected when carotenoids other than β-carotene were used as substrates. *At*CCD7 contains a probable chloroplast targeting sequence of 31 amino acids, which is consistent with a role in metabolism of carotenoids that are located in chloroplasts [63,64].

With *At*CCD1, another putative carotenoid cleavage enzyme of *Arabidopsis thaliana*, a different product profile was obtained. Expressed in *E. coli*, the recombinant enzyme was able to cleave a variety of carotenoids, namely β-carotene (**3**), lutein (**133**), zeaxanthin (**119**), and violaxanthin (**259**), symmetrically at both the C(9,10) and C(9′,10′) positions of the polyene backbone to produce a C_{14} dialdehyde and two C_{13} norisoprenoid products [65] (Scheme 7).

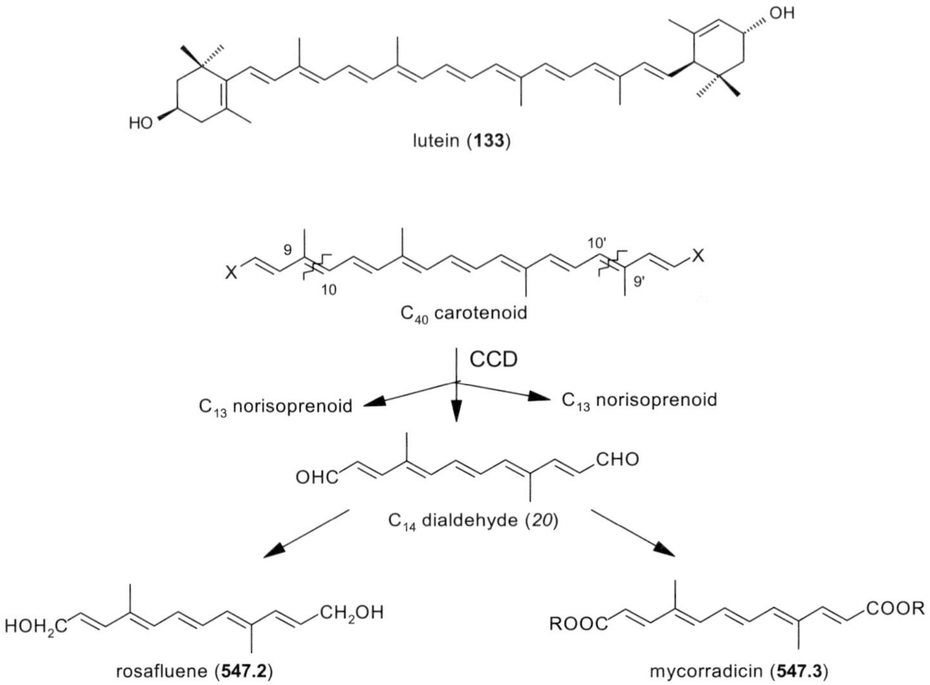

Scheme 7

The C_{14} dialdehyde (20) is the presumed precursor of rosafluene (547.2), a compound found in some varieties of roses [2], and of mycorradicin (547.3), a C_{14} yellow pigment that accumulates in the roots of wheat and barley at the onset of a mycorrhizal relationship. The complementary C_{13} cyclohexenone compound blumenin (21) possesses antifungal properties [66,67]. When the 9Z isomers of neoxanthin and violaxanthin served as substrates for AtCCD1, the C_{27} apocarotenals were the main reaction products; cleavage adjacent to a Z double bond or the allenic bond of neoxanthin appears to be impaired [65]. Recent mechanistic studies [3] have shown that AtCCD1 is indeed a dioxygenase.

Various CCDs and the corresponding enzymes with properties similar to those of AtCCD1 have been characterized e.g. from tomato (Lycopersicon esculentum, LeCCD1A and LeCCD1B) [68], crocus (Crocus sativus, CsCCD) [69,70], Suaeda salsa (SsCCD1) [71], petunia (Petunia hybrida, PhCCD1) [42], rice (Oryza sativa, OsCCD1), maize (Zea mais, ZmCCD1) [72], and grape (Vitis vinifera, VvCCD1) [55]. The gene corresponding to the tomato enzyme LeCCD1B is highly expressed in ripening fruit and the enzyme cleaves various carotenoid substrates excentrically at the C(9,10) and C(9',10') positions. Transgenic tomato plants with reduced expression of LeCCD1A and LeCCD1B were produced to

determine whether the corresponding enzymes are responsible for production of norisoprenoid flavour compounds *in vivo*. A single constitutively expressed antisense construct
was used to reduce the expression of both genes. The transgenic plants showed no obvious
morphological alterations, but the production of the carotenoid-derived flavour compounds β-
ionone (*2*) and geranylacetone (*22*) was significantly reduced [68]. Although no data were
available for ψ-ionone (*23*), the presumed primary cleavage product of lycopene (*31*), these
findings clearly demonstrate that *LeCCD1* genes are involved in the formation of flavour
compounds *in vivo*. This is also supported by the detection of changes in tomato flavour
composition, related to the carotenoid content of the respective varieties [73,74].

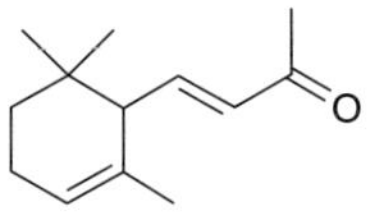

α-ionone (*9*) 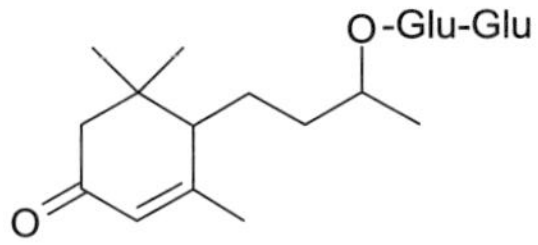   blumenin (*21*) 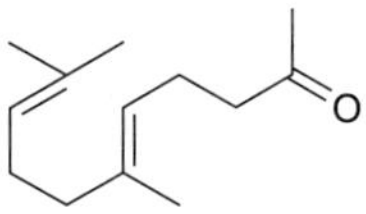   geranylacetone (*22*)

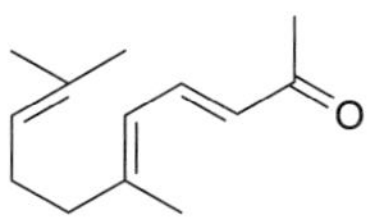

ψ-ionone (*23*) 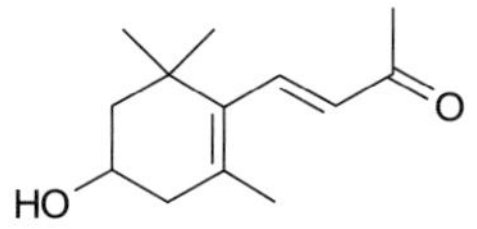   3-hydroxy-β-ionone (*24*) 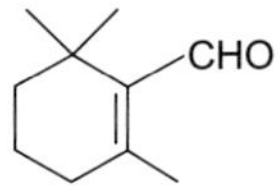   β-cyclocitral (*26*)

Likewise, reduction of *PhCCD1* levels in transgenic petunia plants led to a 58-76% decrease
in β-ionone formation [42]. The amino acid sequences of *LeCCD1A* and *LeCCD1B* do not
indicate the presence of plastid transit peptides, and assays of import into pea chloroplasts
showed that the *LeCCD1* proteins are not localized in the plastid. A cytosolic location has
also been reported for crocus *CsCCD* [70]. Recombinant *Vitis vinifera VvCCD1* cleaves
zeaxanthin (**119**) symmetrically, yielding the C_{13} 3-hydroxy-β-ionone (*24*) and the C_{14}
dialdehyde *20*. The expression of the gene was studied by real-time PCR at different
developmental stages of Muscat and Shiraz grapes. A significant induction of the gene
expression was observed as the grapes approached maturity, and the level of C_{13}
norisoprenoids increased in parallel, though the increase was delayed by more than one week
[28]. In the halophyte *Suaeda salsa,* the expression of *SsCCD1* was significantly up-regulated
by NaCl and drought stress in roots and aerial part tissues.

Table 2. Characteristics of some C(9,10) and C(9',10') double-bond cleaving carotenoid dioxygenases (CCDs).

Enzyme	Accession no.	Source	Amino acids	Size [kDa]	Refs.
Bco2	Q3KNZ2	mouse	532	60.1	*Chap. 16*
*At*CCD7	Q7XJM2	*Arabidopsis*	618	69.5	63,64
*At*CCD1	Q8GRI2	*Arabidopsis*	501	56.9	65
*Le*CCD1B	Q6E4P4	tomato	545	61.0	68
*Ph*CCD1	Q6E4P3	petunia	546	61.3	42
*Vv*CCD1	Q3T4H1	grape	542	61.1	28
*Cs*CCD	Q84KG5	crocus	546	61.7	70
*Ss*CCD1	Q4ZJB4	*Suaeda salsa*	556	62.6	71

In addition to the *CCDs* that have been cloned from animals and plants, biochemical evidence for enzymes with polyene C(9,10) cleavage specificity has been obtained for several plant materials in recent years. Protein fractions with enzyme activity were purified from skins of star fruit (*Averrhoa carambola*) [75] and nectarines (*Prunus persica*) [76]. A partial characterization revealed acidic isoelectric points for both enzymes. A cytosolic carotenoid cleavage enzyme was isolated from quince (*Cydonia oblonga*) [30].

F. Cleavage of C(7,8) and C(5,6) Double Bonds

An oxidative cleavage of both the C(7,8) and C(7',8') double bonds of carotenoid precursor molecules should result in the release of two C_{10} aldehydes and one C_{20} dialdehyde. This predicted pattern of products was formed by a C(7,8) CCD (*CsZCD*, zeaxanthin cleavage dioxygenase) obtained by cloning the *CsZZD gene* from crocus (*Crocus sativus*). *CsZCD* specifically catalyses the formation of crocetin dialdehyde (**536**) (C_{20}) and two molecules of 3-hydroxy-β-cyclocitral (*25*) (C_{10}) from zeaxanthin (**119**). By an aldehyde oxido-reductase, crocetin dialdehyde is subsequently oxidized to crocetin (**538**), which is finally esterified with monosaccharides or disaccharides to yield crocin (**545**) (Scheme 5). The sequence identity between *CsZCD* and *CsCCD*, which catalyses the C(9,10) and C(9',10') bond breakage, is 27%. For localization of *CsZCD in situ*, antibodies developed against synthetic peptide sequences were used, and tissue sections of crocus style branches were analysed by immunocytochemical staining and confocal laser-scanning microscopy. The labelling pattern observed was consistent with a plastid localization of *CsZCD*. With polyclonal antibodies prepared against purified peptide sequences, anti-*CsZCD* revealed a single 40 kDa protein in the style branches only. RNA gel blot analysis of various crocus tissues additionally confirmed that *CsZCD* was expressed specifically in the style branches [70].

A further enzyme that cleaves β-carotene and zeaxanthin specifically at C(7,8) and C(7',8') has been purified, but the gene not yet cloned, from the cyanobacterium *Microcystis*. β-Cyclocitral (*26*) and crocetin dialdehyde (**536**) resulted from the degradation of β-carotene (**3**), whilst zeaxanthin (**119**) was cleaved to 3-hydroxy-β-cyclocitral (*25*) and crocetin dialdehyde (**536**). The enzyme was shown to be membrane associated, and iron seemed to be essential for its activity. No additional cofactors were required [77].

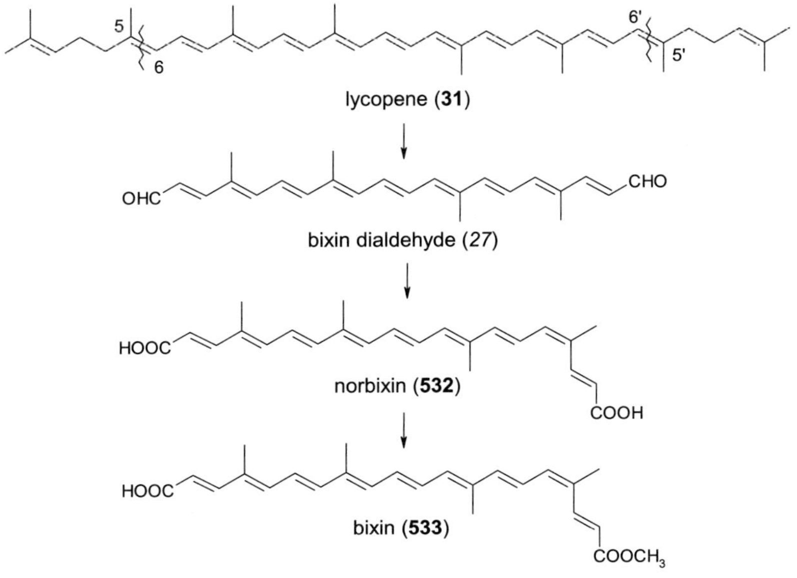

Scheme 8

From the fruit of the annatto tree (*Bixa orellana*), the gene controlling a chromoplast-localized putative lycopene C(5,6) cleavage dioxygenase, was isolated. The translated protein sequence showed a 97% identity to *CsZCD* from *Crocus sativus* [78]. Overexpressed in *E. coli,* the recombinant enzyme cleaved the acyclic carotenoid lycopene (**31**) to (all-*E*)-or (9*Z*)-6,6'-diapocarotene-6,6'-dial (bixin aldehyde, *27*), which represents the initial step of norbixin (**532**) and bixin (**533**) synthesis (Scheme 8). The characteristics of these enzymes are summarized in Table 3.

Table 3. Characteristics of some lycopene (**31**)-cleaving enzymes

Enzyme	Accession no.	Source	Amino acids	Size [kDa]	Refs.
CsZCD	Q84K96	crocus	369	40.8	70
BoLCD	Q70YP8	annatto	369	40.8	78

G. Cleavage of Apocarotenals

Recently, an additional class of enzymes has been discovered which further degrade apo-carotenals that are formed as primary carotenoid cleavage products. When the *Arabidopsis* gene coding for the enzyme *At*CCD8 was expressed in carotenoid-accumulating *E. coli* strains, no apocarotenoids were detected. The *At*CCD8 protein catalyses a secondary cleavage of 10'-apo-β-caroten-10'-al (**499**) at the C(13,14) position of the polyene chain (Scheme 9) to produce the C_{18} ketone 13-apo-β-caroten-13-one (*28*) [63]. Orthologous genes to *At*CCD8 have since been cloned from petunia (*Petunia hybrida, PhCCD8*) [79] and pea (*Pisum sativum, RMS1*) [80], where these genes are involved in the regulation of shoot branching [81].

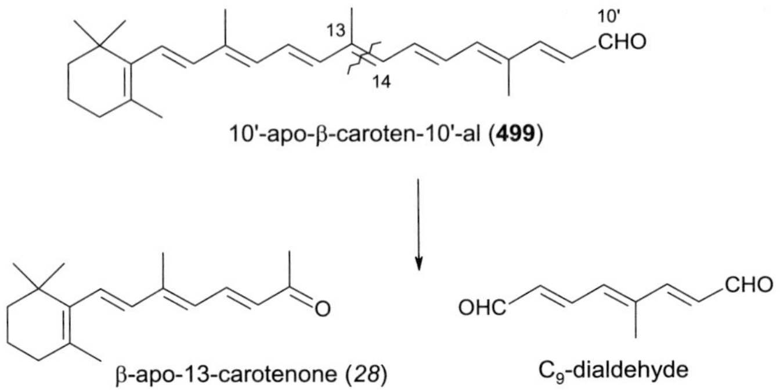

10'-apo-β-caroten-10'-al (**499**)

β-apo-13-carotenone (*28*) C_9-dialdehyde

Scheme 9

A gene (*Diox1*) coding for the enzyme that cleaves apocarotenals specifically has been cloned from the cyanobacterium *Synechocystis* sp. PCC6803. The enzyme shows significant homology to animal and plant β-carotene-cleaving enzymes. All sequenced cyanobacterial genomes reveal at least one gene with striking similarity to *Diox1*, indicating a vital importance of this enzyme class for cyanobacteria. Based on the sequence homologies to bacterial LSDs, *Diox1* and homo-logues thereof have been annotated as LSD or RPE65 (retinal pigment epithelium 65) in the public databases. *Diox1* was expressed as a GST fusion protein and assays with the purified enzyme *in vitro* showed that (all-*E*)-8'-apo-β-caroten-8'-al (**482**) (C_{30}) was readily degraded to retinal (*29*) and the dialdehyde, 2,6-dimethylocta-2,4,6-trienedial (*30*) (Scheme 10).

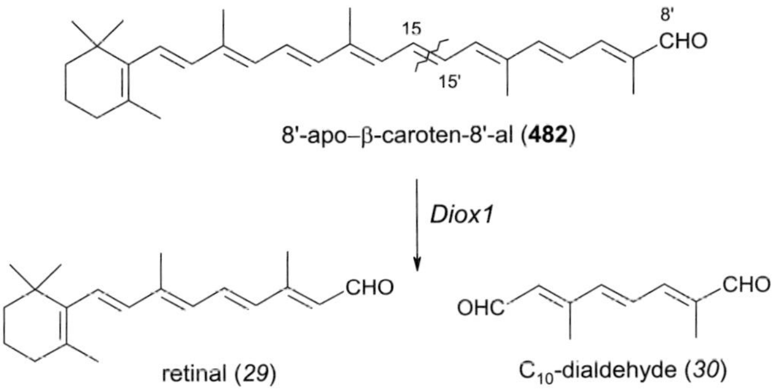

8'-apo–β-caroten-8'-al (**482**)

Diox1

retinal (**29**)

C_{10}-dialdehyde (**30**)

Scheme 10

The respective *Z* isomers were not converted by *Diox1*, indicating specificity for the all-*E* form of apocarotenals. Several further substrates [8'-apo-β-caroten-8'-ol (*31*), apo-8'-lycopenal (**491**), apo-8'-lycopenol (*32*), (3*R*)-3-hydroxy-8'-apo-β-caroten-8'-al (β-citraurin, **483**), (3*R*)-3-hydroxy-8'-apo-β-caroten-8'-ol (**481**), 4-oxo-8'-apo-β-caroten-8'-al (*33*), 4-oxo-8'-apo-β-caroten-8'-ol (*34*)], formally derived from different carotenes and xanthophylls, were subjected to the enzymic cleavage reaction. In all cases, the substrates were cleaved at the C(15,15') position of the polyene chain to yield the corresponding C_{20} compounds.

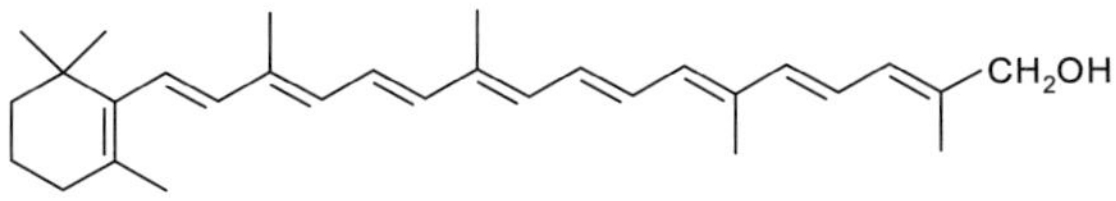

8'-apo-β-caroten-8'-ol (*31*)

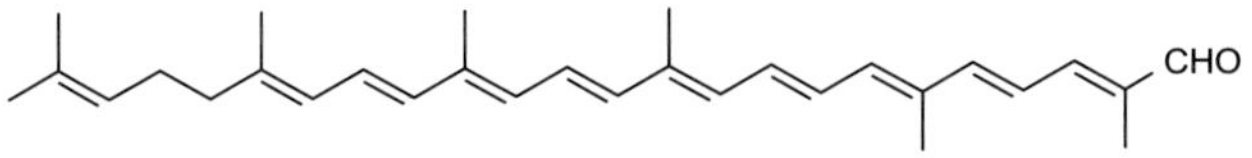

8'-apo-ψ-caroten-8'-al (**491**)

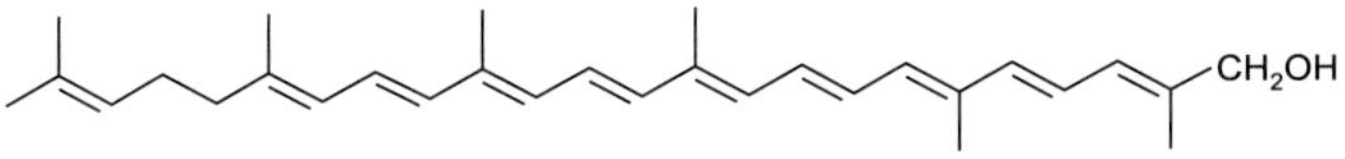

8'-apo-ψ-caroten-8'-ol (*32*)

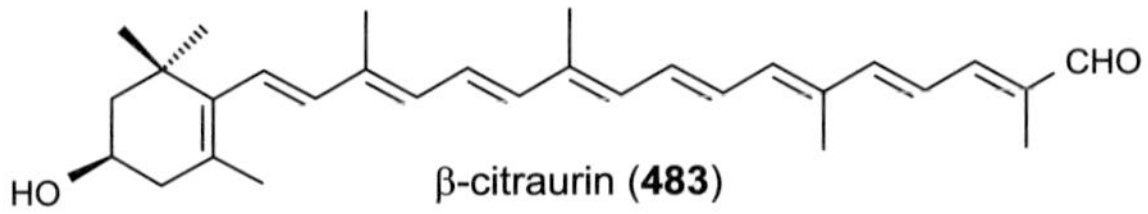

β-citraurin (**483**)

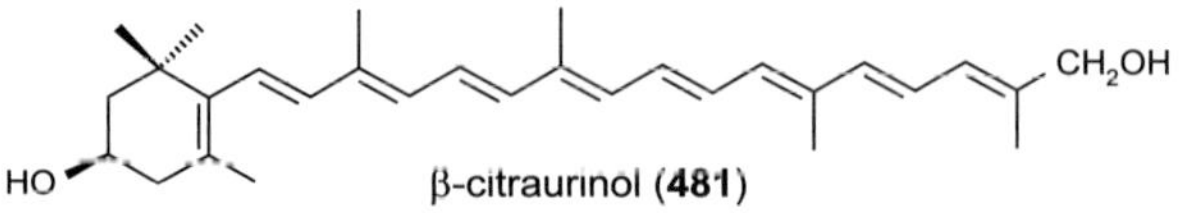

β-citraurinol (**481**)

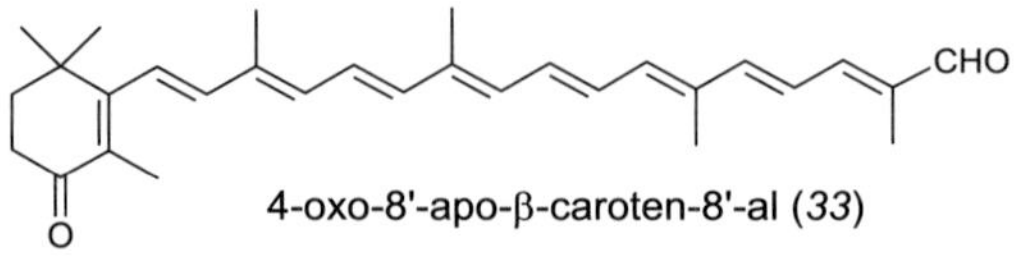

4-oxo-8'-apo-β-caroten-8'-al (*33*)

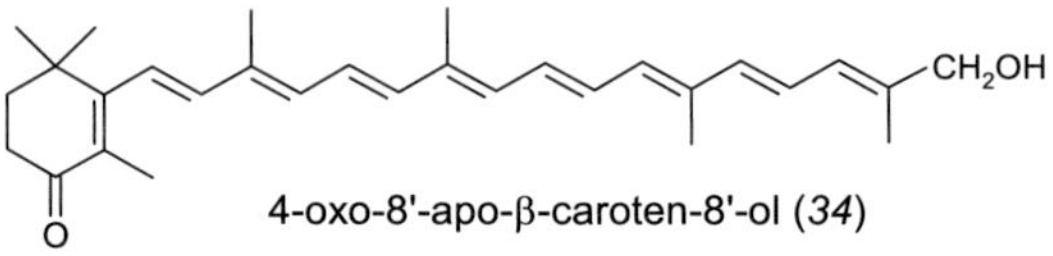

4-oxo-8'-apo-β-caroten-8'-ol (*34*)

Furthermore, apocarotenals of varying chain length [4'-apo-β-caroten-4'-al (*35*), 8'-apo-β-caroten-8'-al (**482**), 10'-apo-β-caroten-10'-al (**499**), 12'-apo-β-caroten-12'-al (**507**) and their corresponding alcohols) were cleaved by Diox1. The preferred chain length of Diox1 is probably in the range of C_{27} or C_{30}, as a further reduction in chain length increased the apparent K_m values [82]. Characteristics of the enzymes are given in Table 4.

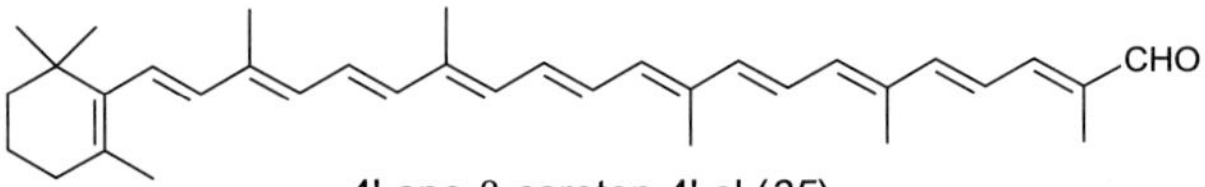

4'-apo-β-caroten-4'-al (*35*)

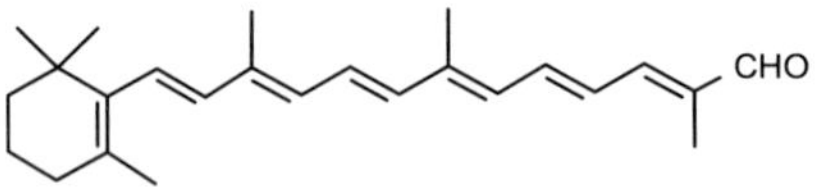

12'-apo-β-caroten-12'-al (**507**)

Table 4. Characteristics of some apocarotenoid-cleaving enzymes.

Enzyme	Accession no.	Source	Amino acids	Size [kDa]	Refs.
AtCCD8	Q8VY26	*Arabidopsis*	570	64.0	63
Diox1	P74334	*Synechocystis*	490	54.3	82
PhCCD8	Q5C8T4	petunia	556	62.3	79
RMS1	Q6Q623	pea	561	62.4	80,81

To summarize the information on specific enzymic carotenoid cleavage reactions presented in Sections **D** to **G**, Fig. 4 shows an unrooted phylogenetic tree of deduced protein sequences of carotenoid-cleaving enzymes. The tree diagram gives an overview of sequence similarities and differences within the highly homogeneous group of plant enzymes (*e.g.* CCDx, NCEDx, VP14, RMS1 and Diox1) involved in specific excentric carotenoid cleavage.

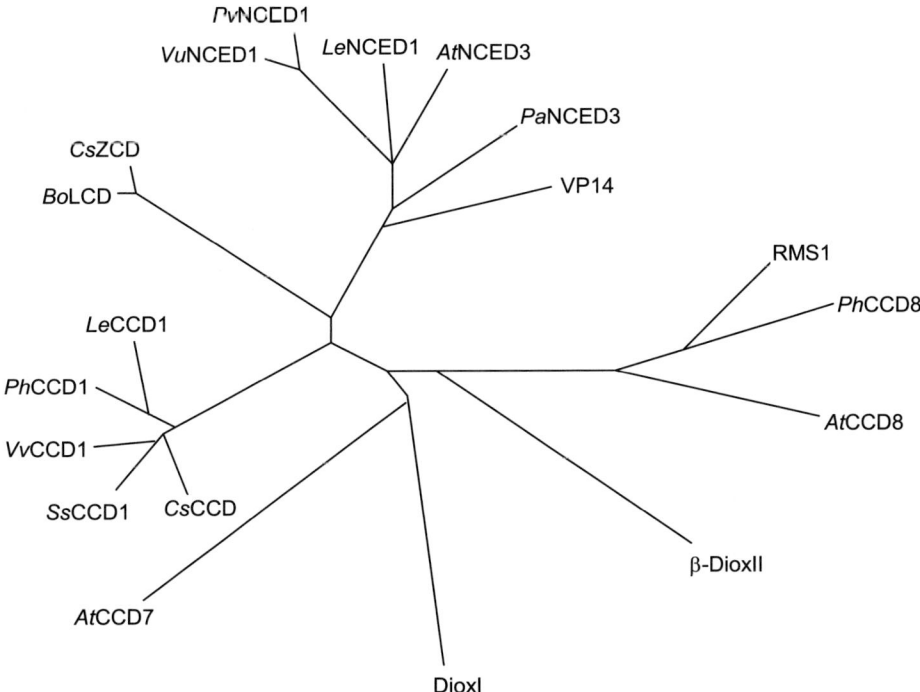

Fig. 4. Unrooted phylogenetic tree of deduced protein sequences of carotenoid cleaving enzymes. Alignments were performed with ClustelW [85] and visualized with PHYLIP [86]. Amino acid sequences and Accession Nos.: *At*CCD1 Q8GRI2; *At*CCD7 Q7XJM2; *At*CCD8 Q8VY26; *At*NCED3 Q93ZU5; *Bo*LCD Q70YP8; β-diox-II Q3KNZ2; *Cs*CCD Q84KG5; *Cs*ZCD Q84K96; Diox1 P74334; *Le*CCD1B Q6E4P4; *Le*NCED1 O24023; *Pa*NCED3 Q9AXZ4; *Ph*CCD1 Q6E4P3; *Ph*CCD8 Q5C8T4; *Pv*NCED1 Q9M6E8; RMS1 Q6Q623; *Ss*CCD1 Q4ZJB4; VP14 O24592; *Vu*NCED1 Q9FS24; *Vv*CCD1 Q3T4H1

H. Unspecific Enzyme-based Cleavage

1. Co-oxidation

Evidence for a carotenoid-bleaching activity of the enzyme lipoxygenase was obtained as early as 1934 [85]. The so-called co-oxidation of various carotenoids, mainly with lipoxygenase (LOX) or alternatively xanthine oxidase systems, has been studied intensively since then. The carotenoids are oxidized by free-radical species generated from another substrate by enzymic reactions [86,87]. A soybean extract containing LOX isoforms is used commercially in the bakery industry to improve flour properties and to bleach the carotenoids in the flour, especially lutein (133) and its monoacyl and diacyl esters.

The mechanism of co-oxidative carotenoid degradation is thought to involve LOX-generated peroxy radicals from which a carotenoid radical and then a peroxycarotenoid radical are produced.

$$\beta\text{-carotene} + \text{ROO}^\bullet \longrightarrow \beta\text{-carotene}^\bullet + \text{ROOH}$$

$$\beta\text{-carotene}^\bullet + O_2 \longrightarrow \beta\text{-carotene}\!-\!OO^\bullet$$

In co-oxidation systems of β-carotene with either soybean or recombinant pea lipoxygenases, more than 50 products absorbing at >350 nm were detected by HPLC. Most probably, the LOX-initiated oxidation of β-carotene occurs randomly both at the ring double bonds and at different positions along the conjugated polyene chain, releasing a number of 5,6-epoxy and 5,8-epoxy products as well as apocarotenals and apocarotenones from the hydrocarbon skeleton [87]. The geometrical (E/Z) configuration affected the product profile in a xanthine oxidase/acetaldehyde co-oxidation system. Higher proportions of (9Z)-β-carotene in the reaction mixture increased the concentration of the flavour compound dihydroactinidiolide (36), while the yield of β-ionone (2) was not affected [86]. Apart from β-carotene, neoxanthin (234) was readily degraded in this system, yielding inter alia β-damascenone (3) and its proposed precursor grasshopper ketone (37) [17].

In general, the product yields obtained from the co-oxidative degradation of carotenoids are too low for industrial flavour production. One reason may be that the primary products undergo further degradation reactions. The fact that retinol (38), β-ionone (2), and 4-hydroxy-β-ionone (39) were oxidized to racemic 5,6-epoxy and 5,8-epoxy derivatives by several soybean lipoxygenase isoenzymes supports this hypothesis [88].

retinol (38)

OH

4–hydroxy-β-ionone (39)

2-hydroxy-2,6,6-tri-
methylcyclohexanone (40)

2. Plant and Fungal Peroxidases

Though some 100 million tons of carotenoids are biosynthesized and subsequently degraded naturally every year, knowledge about natural carotenoid degradation in the environment is still fragmentary, and surprisingly few data have been obtained on the biotic carotenoid degradation by microorganisms.

Mixed cultures of *Trichosporon asahii* and *Paenibacillus amylolyticus* degraded lutein (**133**) derived from marigold flowers (*Tagetes erecta*). The presence of both microbial species was indispensable for the release of norisoprenoid flavour compounds, but the enzymes involved in the degradation pathways have not yet been characterized [89,90].

An extra-cellular versatile peroxidase from the edible fungus *Pleurotus eryngii* (previously erroneously identified as *Lepista irina*) was found to degrade efficiently β-carotene (**3**) to β-ionone (**2**), β-cyclocitral (**26**), dihydroactinidiolide (**36**), and 2-hydroxy-2,6,6-trimethylcyclo-hexanone (**40**). 10'-Apo-β-caroten-10'-al (**499**) was identified as a non-volatile degradation product [37]. Fungal versatile peroxidases are key enzymes of natural lignin degradation, and they have been reported to share catalytic properties of lignin peroxidase and manganese peroxidase. A wide variety of phenols and non-phenolic aromatic compounds is oxidized either directly, as with horseradish peroxidase, or indirectly through Mn^{3+} formed from Mn^{2+} as with manganese peroxidases [91]. From the basidiomycete *Marasmius scorodonius*, a new type of peroxidase capable of carotenoid degradation was isolated [92]. The *M. scorodonius* enzymes share about 50% sequence identity with a dye-decolourizing haem peroxidase (DyP) of the yeast *Geotrichum candidum* [93].

Data on the oxidative degradation of carotenoids by plant peroxidases are rather scarce. A carotenoid-bleaching enzyme has been identified in aqueous paprika (*Capsicum annuum*) extracts [94]. After partial purification, it exhibited typical characteristics of plant peroxidases. Commercial soybean and horseradish peroxidases, lipoxygenases, and catalases were tested for their potential to bleach flour. Partial degradation of β-carotene was observed with horseradish and soybean peroxidase as well as with lactoperoxidase. Addition of H_2O_2 enhanced the activity of lactoperoxidase and horseradish peroxidase, but was not essential for the carotenoid bleaching [95].

References

[1]   P. Winterhalter and R. L. Rouseff (eds.), *Carotenoid-derived Aroma Compounds*, Americal Chemical Society, Washington (2002).

[2]   I. Wahlberg and A.-M. Eklund, in *Carotenoids, Vol.3: Biosynthesis and Metabolism* (ed. G. Britton, S. Liaaen-Jensen and H. Pfander), p. 195, Birkhäuser, Basel (1998).

[3]   H. Schmidt, R. Kurzer, W. Eisenreich and W. Schwab, *J. Biol. Chem.*, **281**, 9845 (2006).

[4]   M. G. Leuenberger, C. Engeloch-Jarret and W.-D. Woggon, *Angew. Chem. Int. Ed.*, **40**, 2613 (2001).

[5]   P. Werkhoff, W. Bretschneider, M. Güntert, R. Hopp and H. Surburg, *Z. Lebensm. Unters. Forsch.*, **192**, 111 (1991).

[6]   R. G. Buttery, R. Teranishi and L.C. Ling, *Chem. Ind.*, **7**, 238 (1988).

[7]   B. M. Lawrence, *Perfum. Flavor.*, **3**, 11 (1978).

[8]   K. Kumazawa and H. Masuda, *J. Agric. Food Chem.*, **47**, 5169 (1999).

[9]   F. Mayer, M. Czerny and W. Grosch, *Eur. Food Res. Technol.*, **211**, 272 (2000).

[10]   B. L. Raina, S. G. Agarwal, A. K. Bhatia and G. S. Gaur, *J. Sci. Food Agric.*, **71**, 27 (1996).

[11]   G. Skouroumounis, in *Carotenoid-derived Aroma Compounds* (ed. P. Winterhalter and R. L. Rouseff), p. 214, American Chemical Society, Washington (2002).

[12]   E. Demole, P. Enggist, M. Winter, A. Furrer, K.-H. Schulte-Elte, B. Egger and G. Ohloff, *Helv. Chim. Acta*, **62**, 67 (1979).

[13]   A. J. Aasen, B. Kimland, S. Almquist and C. R. Enzell, *Acta Chem. Scand.*, **26B**, 2573 (1972).

[14]   W. Renold, R. Müller, U. Keller, B. Willhalm and G. Ohloff, *Helv. Chim. Acta*, **57**, 1301 (1974).

[15]   K. Ina, T. Takano, Y. Imai and Y. Sakato, *Agric. Biol. Chem.*, **36**, 1033 (1972).

[16]   R. Kaiser and D. Lamparsky, *Helv. Chim. Acta*, **61**, 373 (1978).

[17]   Y. Wache, A. Bosser-DeRatuld and J.-M. Belin, in *Carotenoid-derived Aroma Compounds* (ed. P. Winterhalter and R. L. Rouseff), p. 102, American Chemical Society, Washington (2002).

[18]   J. Crouzet, in *Carotenoid-derived Aroma Compounds* (ed. P. Winterhalter and R. L. Rouseff), p. 115, American Chemical Society, Washington (2002).

[19]   Y. Nishida, H. Ohrui and H. Meguro, *Agric. Biol. Chem.*, **47**, 2969 (1983).

[20]   T. Tsuneya, M. Ishihara, H. Shiota and M. Shiga, *Agric. Biol. Chem.*, **47**, 2495 (1983).

[21]   S. Escher and Y. Niclass, *Helv. Chim. Acta*, **74**, 179 (1991).

[22]   P. Winterhalter and R. Rouseff, in *Carotenoid-derived Aroma Compounds* (ed. P. Winterhalter and R. L. Rouseff), p. 1, American Chemical Society, Washington (2002).

[23]   P. Winterhalter, A. Lutz and P. Schreier, *Tetrahedron Lett.*, **32**, 3669 (1991).

[24]   R. Näf and A. Velluz, *Tetrahedron Lett.*, **32**, 4487 (1991).

[25]   A. Lutz, P. Winterhalter and P. Schreier, *Tetrahedron Lett.*, **32**, 5943 (1991).

[26]   R. Kaiser and P. Kraft, *Chem. Zeit*, 35, 9 (2001).

[27]   R. Kaiser, in *Carotenoid-derived Aroma Compounds* (ed. P. Winterhalter and R. L. Rouseff), p. 160, American Chemical Society, Washington (2002).

[28]   S. Mathieu, N. Terrier, J. Procureur, F. Bigey and Z. Günata, *J. Exp. Bot.*, **56**, 2721 (2005).

[29]   H. MacTavish, N. W. Davies and R. C. Menary, in *Carotenoid-derived Aroma Compounds* (ed. P. Winterhalter and R. L. Rouseff), p. 183, American Chemical Society, Washington (2002).

[30]   P. Fleischmann, K. Studer and P. Winterhalter, *J. Agric. Food Chem.*, **50**, 1677 (2002).

[31]   M. Suzuki, S. Matsumoto, P. Fleischmann, H. Shimada, Y. Yamano, M. Ito and N. Watanabe, in *Carotenoid-derived Aroma Compounds* (ed. P. Winterhalter and R. L. Rouseff), p. 89, American Chemical Society, Washington (2002).

[32]   N. Oka, H. Ohishi, T. Hatano, M. Hornberger, K. Sakata and N. Watanabe, *Z. Naturforsch. C*, **54**, 889 (1999).

[33]   P. Winterhalter and P. Schreier, *Food Rev. Int.*, **11**, 237 (1995).

[34]   V. M. Dembitzky, *Lipids*, **40**, 535 (2005).

[35]   G. Schmidt, G. Full, P. Winterhalter and P. Schreier, *J. Agric. Food Chem.*, **40**, 1188 (1992).

[36]   P. Winterhalter and P. Schreier, *J. Agric. Food Chem.*, **36**, 560 (1988).

[37]   H. Zorn, S. Langhoff, M. Scheibner, M. Nimtz and R. G. Berger, *Biol. Chem.*, **384**, 1049 (2003).

[38]   P. Luning, A. Carey, J. Roozen and H. Wichers, *J. Agric. Food Chem.*, **43**, 1493 (1995).

[39]   D. P. Kloer and G. E. Schulz, *Cell. Mol. Life Sci.*, **63**, 2291 (2006).

[40]   F. Bouvier, J.-C. Isner, O. Dogbo and B. Camara, *Trends Plant Sci.*, **10**, 187 (2005).

[41]   M. E. Auldridge, D. R. McCarty and H. E. Klee, *Curr. Opinion Plant Biol.*, **9**, 315 (2006).

[42]   A. J. Simkin, B. A. Underwood, M. Auldridge, H. M. Loucas, K. Shibuya, E. Schmelz, D. G. Clark and H. J. Klee, *Plant Physiol.*, **136**, 3504 (2004).

[43]   A. Ohmiya, S. Kishimoto, R. Aida, S. Yoshioka and K. Sumitomo, *Plant Physiol.*, **142**, 1193 (2006).

[44]   P. Fleischmann, S. Baldermann, M. Yamamoto, N. Watanabe and P. Winterhalter, in *State-of-the-Art in Flavour Chemistry and Biology* (ed. T. Hofmann, M. Rothe and P. Schieberle), p. 234, Deutsche Forschungsanstalt für Lebensmittelchemie, Garching, (2005).

[45]   S. H. Schwartz, B.-C. Tan, D. A. Gage, J. A. D. Zeevaart and D. R. McCarty, *Science*, **276**, 1872 (1997).

[46]   T. Oritani and H. Kiyota, *Nat. Prod. Rep.*, **20**, 414 (2003).

[47]   B.-C. Tan, S. H. Schwartz, J. A. D. Zeevaart and D. R. McCarty, *Proc. Natl. Acad. Sci. USA*, **94**, 12235 (1997).

[48]   B.-C. Tan, K. Cline and D. R. McCarty, *Plant J.*, **27**, 373 (2001).

[49]   S. Liotenberg, H. North and A. Marion-Poll, *Plant Physiol. Biochem.*, **37**, 341 (1999).

[50]   S. H. Schwartz, X. Qin and J. A. D. Zeevaart, *Plant Physiol.*, **131**, 1591 (2003).

[51]   H. Naested, A. Holm, T. Jenkins, H. B. Nielsen, C. A. Harris, M. H. Beale, M. Andersen, A. Mant, H. Scheller, B. Camara, O. Mattsson and J. Mundy, *J. Cell Sci.*, **117**, 4807 (2004).

[52]   S. J. Neill, E. C. Burnett, R. Desikan and J. T. Hancock, *J. Exp. Bot.*, **49**, 1893 (1998).

[53]   B.-C. Tan, L. M. Joseph, W.-T. Deng, L. Liu, Q.-B. Li, K. Cline and D. R. McCarty, *Plant J.*, **35**, 44 (2003).

[54]   S. Iuchi, M. Kobayashi, T. Teruaki, M. Naramoto, M. Seki, T. Kato, S. Tabata, Y. Kakubari, K. Yamaguchi-Shinozaki and K. Shinozaki, *Plant J.*, **27**, 325 (2001).

[55]   J. T. Chernys and J. A. D. Zeevaart, *Plant Physiol.*, **124**, 343 (2000).

[56]   A. Burbidge, T. Grieve, A. Jackson, A. Thompson and I. Taylor, *J. Exp. Bot.*, **48**, 2111 (1997).

[57]   A. J. Thompson, A. C. Jackson, R. A. Parker, D. R. Morpeth, A. Burbidge and I. B. Taylor, *Plant Mol. Biol.*, **42**, 833 (2000).

[58]   A. Burbidge, T. M. Grieve, A. Jackson, A. Thompson, D. R. McCarty and I. B. Taylor, *Plant J.*, **17**, 427 (1999).

[59]   S. Iuchi, M. Kobayashi, K. Yamaguchi-Shinozaki and K. Shinozaki, *Plant Physiol.*, **123**, 553 (2000).

[60]   X. Qin and J. A. D. Zeevaart, *Proc. Natl. Acad. Sci. USA*, **96**, 15354 (1999).

[61]   M. Kalala, A. K. Cowan, P. Molnár and G. Tóth, *S. Afr. J. Bot.*, **67**, 376 (2001).

[62]   X. Qin and J. A. D. Zeevaart, *Plant Physiol.*, **128**, 544 (2002).

[63]   S. H. Schwartz, X. Qin and M. C. Loewen, *J. Biol. Chem.*, **279**, 46940 (2004).

[64]   J. Booker, M. Auldridge, S. Wills, D. R. McCarty, H. Klee and O. Leyser, *Curr. Biol.*, **14**, 1232 (2004).

[65]   S. H. Schwartz, X. Qin and J. A. D. Zeevaart, *J. Biol. Chem.*, **276**, 25208 (2001).

[66]   G. Giuliano, S. Al-Babili and J. von Lintig, *Trends Plant Sci.*, **8**, 145 (2003).

[67]   T. Fester, W. Maier and D. Strack, *Mycorrhiza*, **8**, 241 (1999).

[68]   A. J. Simkin, S. H. Schwartz, M. Auldridge, M. G. Taylor and H. J. Klee, *Plant J.*, **40**, 882 (2004).

[69]   B. Camara and F. Bouvier, *Arch. Biochem. Biophys.*, **430**, 16 (2004).

[70]   F. Bouvier, C. Suire, J. Mutterer and B. Camara, *Plant Cell*, **15**, 47 (2003).

[71]   Y. Cao, X.-L. Guo, Q. Zhang, Z.-Y. Cao, Y.-X. Zhao and H. Zhang, *Plant Growth Regul.*, **46**, 61 (2005).

[72]   R. McCarty and B.-C. Tan, *PCT Int. Appl.* WO 2003-US38669 20031205 (2004).

[73]   E. Lewinsohn, Y. Sitrit, E. Bar, Y. Azulay, M. Ibdah, A. Meir, E. Yosef, D. Zamir and Y. Tadmor, *Trends Food Sci. Technol.*, **16**, 407 (2005).

[74]   E. Lewinsohn, Y. Sitrit, E. Bar, Y. Azulay, A. Meir, D. Zamir and Y. Tadmor, *J. Agric. Food Chem.*, **53**, 3142 (2005).

[75]   P. Fleischmann, N. Watanabe and P. Winterhalter, *Phytochemistry*, **63**, 131 (2003).

[76]   S. Baldermann, M. Naim and P. Fleischmann, *Food Res. Int.*, **38**, 833 (2005).

[77]   F. Jüttner and B. Höflacher, *Arch. Microbiol.*, **141**, 337 (1985).

[78]   F. Bouvier, O. Dogbo and B. Camara, *Science*, **300**, 2089 (2003).

[79]   K. C. Snowden, A. J. Simkin, B. J. Janssen, K. R. Templeton, H. M. Loucas, J. L. Simons, S. Karunairetnam, A. P. Gleave, D. G. Clark and H. J. Klee, *Plant Cell*, **17**, 746 (2005).

[80]   E. Foo, E. Bullier, M. Goussot, F. Foucher, C. Rameau and C. A. Beveridge, *Plant Cell*, **17**, 464 (2005).

[81]   K. Sorefan, J. Booker, K. Haurogne, M. Goussot, K. Bainbridge, E. Foo, S. Chatfield, S. Ward, C. Beveridge, C. Rameau and O. Leyser, *Genes Dev.*, **17**, 1469 (2003).

[82]   S. Ruch, P. Beyer, H. Ernst and S. Al-Babili, *Mol. Microbiol.*, **55**, 1015 (2005).

[83]   J. D. Thompson, D. G. Higgins and T. J. Gibson, *Nucleic Acids Res.*, **22**, 4673 (1994).

[84]   J. Felsenstein, *Cladistics*, **5**, 164 (1989).

[85]   L. W. Haas and R. M. Bohn, *U.S. Patent* 1957-333-1957-337 (1934).

[86]   Y. Wache, A. Bosser-DeRatuld, J.-C. Lhuguenot and J.-M. Belin, *J. Agric. Food Chem.*, **51**, 1984 (2003).

[87]   S. Aziz, Z. Wu and D. S. Robinson, *Food Chem.*, **64**, 227 (1999).

[88]   D. Waldmann and P. Schreier, *J. Agric. Food Chem.*, **43**, 626 (1995).

[89]   A. Sanchez-Contreras, M. Jiminez and S. Sanchez, *Appl. Microbiol. Biotechnol.*, **54**, 528 (2000).

[90]   E. Rodriguez-Bustamante, G. Maldonado-Robledo, M. A. Ortiz, C. Diaz-Avalos and S. Sanchez, *Appl. Microbiol. Biotechnol.*, **68**, 174 (2005).

[91]   F. J. Ruiz-Duenas, M. J. Martinez and A. T. Martinez, *Mol. Microbiol.*, **31**, 223 (1999).

[92]   H. Zorn, S. Langhoff, M. Scheibner and R. G. Berger, *Appl. Microbiol. Biotechnol.*, **62**, 331 (2003).

[93]   Y. Sugano, K. Sasaki and M. Shoda, *J. Biosci. Bioeng.*, **87**, 411 (1999).

[94]   J. Kanner and H. Mendel, *J. Food Sci.*, **42**, 1549 (1977).

[95]   P. Gelinas, E. Poitras, M. McKinnon and A. Morin, *Cereal Chem.*, **75**, 810 (1998).

Index